The Encyclopedia of
LAND INVERTEBRATE BEHAVIOUR

The Encyclopedia of LAND INVERTEBRATE BEHAVIOUR

Rod and Ken Preston-Mafham

The MIT Press
Cambridge, Massachusetts

Acknowledgements
A book of this size and content would not have been possible without the assistance and co-operation of a number of individuals and institutions. Most of the literature search was carried out in Birmingham University Library; the Hope Department Library, the Radcliffe Science Library and the Zoology Department Library at Oxford; and the Natural History Museum Library in London. At each visit we were treated with great kindness and consideration by the library staff and without their co-operation our task would have been that much more onerous. We must also thank Alan Rollason for his painstaking work in producing the superb line drawings for this book.

First MIT Press edition, 1993

First published in the UK by
Blandford, a Cassell imprint, London.

This book was printed and bound in Hong Kong
by Dah Hua Printing Co.

Library of Congress Cataloging-in-Publication Data
Preston-Mafham, Rod.
The encyclopedia of land invertebrate behaviour/Rod A. Preston-Mafham and Ken G. Preston-Mafham.—1st MIT Press ed.
p. cm.
Includes bibliographical references and index.
ISBN 0-262-16137-0
1. Invertebrates—Behavior. I. Preston-Mafham, Ken. II. Title.
QL364.2.P74 1993 92-47092
595′.20451—dc20 CIP

Line drawings by Alan Rollason

Typeset by MS Filmsetting Limited, Frome, Somerset
Printed in Hong Kong by Dah Hua Printing Co.

Contents

Introduction

The aim of this book is to present, in a readable form, some of the most complex and bizarre behaviours in the animal world. It does not attempt to be an A–Z of invertebrate behaviour; nor is it a comprehensive review which skims the surface of the subject but fails to delve deeply into any one aspect. Neither is its purpose to make a minute analysis of the evolution of the multifarious behaviours of the invertebrates, or of the advantages of one form of behaviour over another. Rather, it is a series of essays describing in considerable detail as many types of behaviour as feasible. Where research has made it possible to weigh the consequences of one mode of behaviour against its alternatives, or to speculate on the evolution of an aspect of behaviour, details have been included where appropriate. It is hoped that the reader will thereby gain an insight into the reasons why invertebrates do some of the peculiar things for which they are noted.

As this is a book on behaviour rather than biology, the extent to which each individual group is covered relates to the amount of research which has been carried out. Inevitably those groups which have a direct impact upon our lives have been investigated more than those which do not. Again, and for obvious reasons, much more research has been carried out on the behaviour of the arthropods than of the other invertebrates, simply because arthropods occur in the greatest abundance. Even within the insects, certain orders have been left out altogether. For example, the reader wishing to learn more of the behaviour of fleas or lice can easily look this up in a conventional encyclopedia. On the other hand, insects such as flies (Diptera) and bees (Hymenoptera) exhibit a range of extremely complex and fascinating behaviours, earning them a substantial place in the overall treatment.

We have deliberately used straightforward language, keeping scientific terms to a minimum. Since we expect this volume to be read and used by a wide spectrum of readers, from the interested amateur to the dedicated professional biologist, we have kept the biological language as simple as possible. This has meant on occasion using some terms which are not strictly speaking precise biologically, such as abdomen when discussing wasps, when gaster would be more appropriate, foot instead of tarsus, face instead of frons etc. When discussing insects, the concept of the front, middle and hind legs is quite obvious but, in arachnids, the four pairs of legs are less easy to describe. Consequently they are referred to as the first (front), second, third and fourth (hind) pairs of legs.

It is important to be clear about the vocabulary used to describe how or why certain behaviours are carried out. Thus we frequently talk about insects 'adopting' certain 'strategies', of damselflies which 'switch' to 'sneak' tactics, of flies which 'assess'. This is not meant to imply that the organism concerned is capable even in any remotely human way of weighing up the pros and cons and then making a conscious decision about the best way to do something. It is the limitations of language which oblige us to use terms which may seem rather anthropomorphic, but this is unavoidable. Invertebrates do not 'make' decisions, but this is the only easily understandable way of dealing with the stimulus-response, 'instinctive' series of actions which govern the lives of invertebrates.

We have restricted ourselves to the 'terrestrial' invertebrates in order to avoid the enormous increase in bulk which would be needed to cover the behaviour of marine invertebrates. 'Terrestrial' in the sense used here implies organisms which spend their adult lives on land, or which live in water but are capable of leaving it to walk or fly to fresh habitats. This enables us to include such interesting insects as water skaters and giant water bugs which it would not be appropriate to exclude from such a work.

Wherever possible we have drawn our information directly from the original research papers, as will be obvious from the extensive bibliography. It is to these many tireless workers that we owe our knowledge of the workings of the natural world, and it is to all of them that we dedicate this work. We hope that this book will communicate their deserving and original work to a wide audience, most of whom would never have access to the treasurehouse of information buried in the scientific literature. Some of the information is based exclusively or mainly on Ken's own personal observations, abbreviated in the text to 'pers. obs.'

Within each section the information is generally arranged in a 'biological' sequence. However, within a single order this usually switches to a behavioural sequence, which enables more easy comparison of like behaviours. However, for the section on defensive behaviour an arrangement based on strategies was thought more appropriate and this has been adopted, although within each strategy the subjects are treated in the usual biological sequence.

Clearly a good picture often says more than a thousand words and to this end as many forms of behaviour as possible have been illustrated, either by photographs, all of which have been taken in the wild or with line drawings.

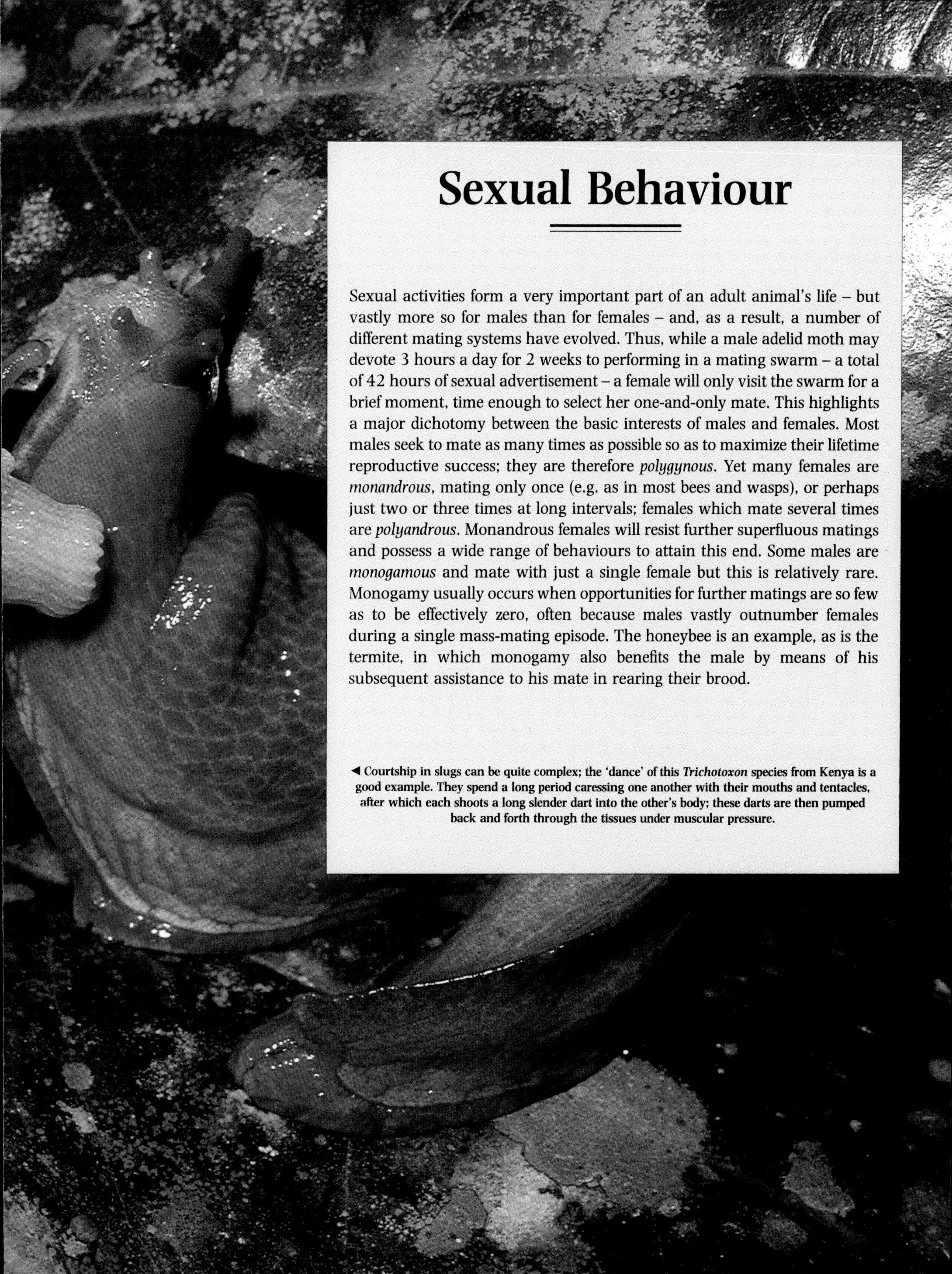

Sexual Behaviour

Sexual activities form a very important part of an adult animal's life – but vastly more so for males than for females – and, as a result, a number of different mating systems have evolved. Thus, while a male adelid moth may devote 3 hours a day for 2 weeks to performing in a mating swarm – a total of 42 hours of sexual advertisement – a female will only visit the swarm for a brief moment, time enough to select her one-and-only mate. This highlights a major dichotomy between the basic interests of males and females. Most males seek to mate as many times as possible so as to maximize their lifetime reproductive success; they are therefore *polygynous*. Yet many females are *monandrous*, mating only once (e.g. as in most bees and wasps), or perhaps just two or three times at long intervals; females which mate several times are *polyandrous*. Monandrous females will resist further superfluous matings and possess a wide range of behaviours to attain this end. Some males are *monogamous* and mate with just a single female but this is relatively rare. Monogamy usually occurs when opportunities for further matings are so few as to be effectively zero, often because males vastly outnumber females during a single mass-mating episode. The honeybee is an example, as is the termite, in which monogamy also benefits the male by means of his subsequent assistance to his mate in rearing their brood.

◀ Courtship in slugs can be quite complex; the 'dance' of this *Trichotoxon* species from Kenya is a good example. They spend a long period caressing one another with their mouths and tentacles, after which each shoots a long slender dart into the other's body; these darts are then pumped back and forth through the tissues under muscular pressure.

However, the act of insemination does not necessarily guarantee fatherhood, due to a special factor which dictates the behaviour of most male insects (and probably also spiders) – the phenomenon of *sperm precedence*. In the vast majority of insects the female can store sperm for later use. She can then regulate the amount of sperm which can fertilize the eggs as they are laid. As a general rule, it is the sperm of the last male with whom she mated which fertilizes all or most of the eggs on a last-in-first-out basis. This simple fact has exerted an enormous effect on the mating systems of male insects. It demands not only that they should mate as often as possible, but also that they find some way of ensuring that their females do not mate with other males until their eggs have been laid. The need to solve this problem has produced a variety of mating strategies in male insects:

(1) The consort may stay with his mate and guard her until she has safely oviposited;

(2) He may attach to her a plug which will prevent further matings, at least until some eggs have been laid;

(3) He may render her chemically unattractive to other males;

(4) He may donate a nuptial gift which induces a refractory period during which she will not re-mate.

Less is known about sperm precedence in other invertebrates, although it is probably much as in insects. There are exceptions, however, such as the bowl-and-doily spider, *Frontinella pyramitela*, in which first-male precedence operates.

MATING SYSTEMS

Males and females do not generally meet by accident, but come together within the context of a mating system. The parameters of the system depend on the nature of the organism and the ecological constraints under which it lives. This has made it possible to define mating systems on an ecological basis (Emlen & Oring, 1977) which also takes into account whether the animal mates once or many times. Monogamy in males will be likely to occur when two parameters are satisfied:

(1) The operational sex ratio, i.e. the relative number of males versus females at any given period, is heavily biased towards the males.

(2) The occurrence of sexually receptive females in space and time is unpredictable.

If the operational sex ratio is so consistently biased towards males that competition effectively eliminates any chance of more than one mating, then monogamy is probably the best bet. If females are not predictably discoverable because they do not cluster in one place, or at any one time, then a male who does manage to locate one will be wise to hang on to her.

Within invertebrates, two kinds of male monogamy can be postulated (Thornhill & Alcock, 1983):

(1) *Female-guarding monogamy*, in which the male writes off any further chances of mating in order to make certain that his one and only mate is not stolen by a rival male.

(2) *Male-assistance monogamy*, which is similar but the male actually stays with his mate in order to increase the survival of his offspring – burying beetles (Silphidae) are an example.

In most insects, the chances of obtaining more than one mating are high enough for polygyny to be the norm. Even so, there is usually a severe skew towards relatively few males (often the biggest) obtaining most of the matings. Several strategies are available. In species whose virgin females emerge predictably from a nest which can be viably defended against rival males, then *female-defence polygyny* could be the answer, as practised by certain bees and wasps.

It is much more common for a male to be in a position to defend, and therefore monopolize, not the females themselves, but certain resources which they need, e.g. for oviposition. This system, called *resource-defence polygyny* is only practicable when the resources are neither so ubiquitous that their monopolization is physically unrealizable, nor so thinly scattered that the chances of encountering a female are unacceptably low. To be *economically* defendable, the resource should be distributed in discrete patches. Resource defence is a high-cost (in terms of the energy expended in fighting off rivals), high-yield strategy. Defending oviposition resources has the advantage that the female can also be guarded until she lays her eggs. However, resource-defence systems must break down under increasing population pressure, when the costs incurred in defence come to outweigh any possible benefits; they may then be replaced by a free-for-all. As the only way to gain or maintain ownership of a resource is to fight for it, it is hardly surprising that weapons such as horns and enlarged mandibles have arisen for the purpose in a number of insects.

It is often impractical to monopolize either the females or any resource used by them. In this case, the males race one another to the females, either through regular searching or by waiting on a non-defendable resource for females to arrive. This kind of free-for-all non-territorial behaviour is called *scramble-competition polygyny* (Thornhill & Alcock, 1983). Most insect swarms, such as those of gnats and mosquitoes, are typical examples of this strategy.

One of the strangest mating systems of all is *lek polygyny*. Males gain access to receptive females by competing for the 'best' position in special mating territories. The important point about this system is the purely symbolic nature of the territories, which serve solely for arranging a rendezvous between male and female. Territories are usually based upon some visual landmark – such as a tree or bush – which can be recognized by both sexes. They contain no resources of value to the females, who 'know' how to visit the lek when they need a mate. There is no mad scramble for a female when she arrives – instead the males let her do the choosing. An important factor in the evolution of lek behaviour is that the females should not all visit the lek simultaneously. Instead, individual females visit the lek at long intervals, staying just long enough to select the 'fittest' mate from those on show. The males may remain in the lek for many weeks and leks are often established in the same place year after year, although successive generations can have no prior knowledge of earlier lekking sites. This means that certain physical characteristics of the site can be instinctively recognized by both males and females.

SEXUAL SELECTION

Sexual selection operates within the confines of the overall strategies outlined above. The concept of sexual selection was first postulated by Darwin as long ago as 1871. He pointed out that two kinds of selective forces come into play during reproduction:

(1) *Intrasexual selection* derived from competition between one sex (almost always the males) for access to the other.

(2) *Intersexual selection* produced by preferences shown by one sex (usually the females) for certain individuals of the other.

Intrasexual competition becomes particularly intense when the opposite sex – usually the female – is in short supply, which accounts for much of the need to try to monopolize females either indirectly (e.g. through resources) or directly.

One reason for the intense nature of sexual competition lies in the differences in reproductive investment made by the sexes. Males generally invest little with each mating, sperm being cheap to produce in terms of energy, while relatively small amounts are lost during each copulation. A male therefore has little to lose and everything to gain from trying to mate with as many females as possible. Females on the other hand invest far more in their offspring – eggs are much larger than sperm and a female will have to commit substantial reserves to the production of her lifetime reproductive output. As a single mating will often provide her with sufficient sperm to fertilize all the eggs which she is ever likely to lay, it may not be in her interests to mate more than once. This puts intense pressure on a male to be her first and only consort, and explains why it is the males who generally carry out the energy- and time-consuming task of searching for females. As searching may also involve additional risks for the male compared with the female, who most frequently merely waits for a male to arrive, it is logical for the males to do it – they have most to gain and least to lose. It also places a premium on males to be 'good' at courtship once a female has been located.

From this, it will be clear that males can be expected to be promiscuous and unfussy, while females, even when mating more than once, should be choosy about their partners and have ways of assessing their qualities as potential fathers. Just how, or indeed if, females do really regularly 'choose' between males has been much discussed and disputed. The matter is complicated by the way in which the forces of intrasexual and intersexual selection operate, i.e. not always alone, but often in combination. A female may therefore 'choose' a male not directly, by picking him out from a number of rivals, or by kicking all the others off her back until she is mounted by the one she 'likes', but indirectly by mating with a male who has already 'proved' himself fit during intrasexual conflict. Thus, a female hover fly seeking out a male at a 'landmark' site will unquestioningly mate with the 'owner'. She might therefore appear not to be choosing between partners. However, in order to maintain ownership of such a highly coveted site, her mate will have invested vast amounts of time and effort in defending his position against rival males. These have in effect done the female's 'choosing' for her. Similarly, females who mate immediately with a resource-owner are carrying out the same form of 'proxy' selection. In fact, if expenditure of effort in order to provide females (and therefore their offspring) with 'fit' genes is regarded as a form of male investment, then the disparity between the degree of investment made by each sex narrows considerably, particularly where males are territorial.

Even in scramble-competition situations, intrasexual competition may still 'choose' a female's mate for her. Although she herself may mate with the first male to seize her, he may still be deposed by a 'fitter' male. However, some experiments have established that direct intersexual selection can indeed apply, as females often mate preferentially with certain males, usually the larger ones. There may also be other subtle forms of female 'choosiness' which at present cannot be detected. However, not all males make a small investment consisting of just a modicum of sperm. Some make a donation to the female in the form of a 'nuptial gift' which may be sufficiently large and nutritious to make a contribution towards egg development. Occasionally, these male investments are so large, or his protected resource so valuable and in such short supply, that they exceed those made by the females. When this happens, the 'attitude' of the sexes changes and it is now the males who become choosy and the females who fight over access to them. This *sex-role reversal* is known from relatively few insects, but is well studied in certain katydids, belostomatid giant water bugs and in the gift-bearing empidid fly, *Empis borealis*.

One effect of competition is the tendency for relatively few males to achieve most of the copulations; in some species, for example, small males are at a distinct disadvantage. This gives rise to the opportunity to resort to alternative tactics, something which is especially common in territorial systems. If a male is consistently unable to establish a territory, his chances of gaining access to females are severely limited. In this case he could use an alternative strategy, such as *sneaking*, within an owner's territory. This also has the advantage of avoiding the energetic and physical costs

▼ In some invertebrates (particularly insects), males make a substantial reproductive investment beyond the modicum of sperm passed to the female in any mating. Cases in which such a 'nuptial gift' consists of the male himself are rare. However, in the European beetle *Cantharis rustica* (Cantharidae), females quite often seem to twist around so as to be able to graze on their mate's nutritious thorax while he copulates. The fact that he does not uncouple and flee as soon as the female starts to nibble may be due to the shortage of receptive females and his poor chances of ever mating a second time.

associated with the conflict needed to be an 'owner'. 'Sneaking' is an attempt by a disadvantaged male to maximize his own success in the only practicable way open to him. However, only rarely (if ever) are the success rates of the two strategies equal – usually the alternative is less rewarding and very much a case of making the best of a bad job. 'Sneaking' is not, however, confined to born 'losers', for whom it may be the only viable strategy – even formerly successful territory-owners may resort to 'sneaking' tactics under certain conditions, while a former 'sneak' may eventually seize a territory. Sexual behaviour tends to be very flexible; using so-called 'conditional' tactics, which take account of variations in environmental conditions at the time, is very common.

FIGHTING

Another result of intrasexual selection is the frequent need to adopt physical violence as a necessary means to an end. Wherever possible, really damaging conflict is avoided, as the costs are likely to outweigh the benefits. Contests may be purely ritualized or there may be some kind of precombat display which allows the prospective combatants to assess their chances of winning. Such 'risk assessment' takes a number of forms, usually involving the display of body parts, which gives the contestants an idea of each other's relative overall size. Males are likely to fight harder and longer for high-value resources – such as a prime spot in a lek or a large, heavily gravid female – than those of lower value. This however assumes that males are, in fact, able to assess relative values. In many cases only the current 'owner' has enough 'inside information' about the resource to be in a position to tailor his response to its value; the challenger will be at a disadvantage through not having this information, which may help to explain why 'owners' tend to win fights more often than challengers.

Fighting behaviour can be classified into two types. In 'wars of attrition' the costs associated with conflict – such as potential injury – escalate linearly with the duration of conflict. Success is therefore mainly decided by persistence, rather than pugilistic excellence or superior weaponry. The outcome may be determined by the nutritional state of the contestants. In the damselfly *Calopteryx maculata*, lengthy contests over territorial ownership are of the 'war-of-attrition' type, and the result may be decided not by superior fighting ability but by whichever contestant runs low on fat reserves and is forced to retire (Marden & Waage, 1990). In 'hawk–dove' contests, the chances of injury jump considerably with each turn of the escalatory screw. Such contests are usually found in species having weapons, such as the 'stabbing' front legs of certain thrips (Crespi, 1986), which are capable of inflicting serious injury, or even death, if fighting escalates to a high level of intensity. 'Fights to the death' are relatively rare, and usually occur when the males have a negligible chance of gaining access to a female in any other way. This could be because females are extremely scarce, or because males are unable to disperse and locate unguarded females (as happens in many parasitic wasps, which breed inside the egg batches of other invertebrates, or in fig wasps).

GASTROPODA
Slugs, snails, limpets, etc.

PULMONATA Slugs, snails

All pulmonate molluscs are hermaphrodite, thus ensuring that when any two members of the same species meet they will be physically able to mate, a useful feature in such slow-moving animals. Mating is often preceded by a courtship which may be remarkably complex. In some slugs, the two partners merely 'chase' one another in ever-decreasing circles until they end up snugly curled together, with their reproductive openings opposite each other.

The bizarre mid-air copulation of the great grey slug, *Limax maximus*, first impressed observers at the end of the last century. The initial stages of courtship usually take place on a branch, or perhaps on a bracket fungus – a favourite food. Delicate interplay with the tentacles is followed by 30–150 minutes of 'follow-my-leader', during which one slug pursues the other 'mouth-to-tail' around and around until the circle begins to shrink. At this point the slugs, with an obvious sense of increasing arousal, pitch themselves earthwards for a distance of some 38–46 cm before being brought up short on a broad rope of mucus which they have belayed behind them. As they swing to a halt, each slug unleashes its penis to a length of some 10 cm and their bodies closely entwine. Becoming quiescent, they exchange sperm and split up, either returning back whence they came, by crawling up the mucus cable, or else dropping to the ground.

In snails, the procedure is less acrobatic but nevertheless of considerable interest. The garden snail, *Helix aspersa*, evinces a procedure whose main elements are common to many snails. After touching tentacles (Adamo & Chase, 1987), the two snails rise upwards slightly on their creeping soles until their mouths are in contact. They 'kiss' for a minute or two and then lean over slightly to mouth one another's genitals, which by now will be semi-everted. The next major step, about 30 minutes into the ritual, is mutual stimulation of the genital areas. This causes one of the snails to propel a 'love dart' into its partner's body.

Darts are intrinsic to the mating rituals of 11 of the 65 families of pulmonates, and are composed mainly of a form of calcium carbonate in a protein matrix. Their exact function is disputed: they may serve to arouse the other animal, as dart-shooting usually precedes copulation, or they may function as an 'anti-aphrodisiac', inhibiting further matings for a period (Jeppesen, 1976). The dart may even possibly fulfil both roles. After being 'shot', the recipient normally retracts briefly, then re-joins the action and shoots its own dart, usually around half an hour later. If the first snail to strike is a bad shot and misses its partner, this slows down the whole process, as the return fire is considerably delayed. Once both darts have been fired, copulation follows quickly, regardless of the accuracy of the second dart. It thus appears that the first snail to shoot is the one who decides the timing of subsequent events. The actual coupling of the genitalia may require numerous attempts, for both sets must be engaged simultaneously. Once in position each snail transfers a spermatophore to the other. This usually takes several hours, during which the head and tentacles shrink down and the snails sink into a quiescent state.

The rather more complex courtship repertoire of the European Roman snail, *Helix pomatia*, includes a fascinating 'mating dance'. Exploratory contact consists of mutual titillation with the tentacles and lips, soon escalating into virtual whole-body contact as each snail rears up sole to sole. The long tentacles extend to their maximum length and curve over in mutual fondling, accompanied by mouth-to-mouth nuzzling. Quite unexpectedly, the participants quickly wind down the action and slump to half their former height so that the top part of their soles and their 'faces' are no longer in contact. However, the former intensity is soon resumed until one snail communicates an intent to shoot its dart by adopting a down-curved posture to one side of its partner, followed by expulsion of the dart into the opposing sole, causing a sudden withdrawal. Copulation and a long quiescent phase then follow.

► The 'dance' of the Roman snail, *Helix pomatia*, includes a great deal of mutual fondling with the tentacles and mouth before mutual dart-shooting takes place.

▼ These Roman snails, *Helix pomatia*, have just withdrawn partly into their shells after engaging in mutual dart-shooting; the tip of a dart is visible protruding from the animal on the right.

DIPLOPODA Millipedes

▲ **In the Kenyan giant millipede, *Epibolus pulchripes*, courtship is purely tactile; the male nibbles at the female's antennae and face while mounted on her back. His ability to stay in place without slipping off is enhanced by the female's matt, non-slip finish, unlike the glossy exterior of the male. Here a second male is climbing on board, a frequent occurrence.**

With one or two exceptions, male millipedes make direct contact with the female and transfer sperm via a pair of modified front legs (gonopods) which have been previously charged with semen from the genitalia situated towards the rear of the body. Sexual communication is often quite complex (Haacker, 1974). In several genera, the male signals his purpose by beating against the female with his antennae. This requires bodily contact, something which is not necessary in the much more sophisticated acoustic communication employed by *Chordeuma silvestre*. This is an unusually fleet-of-foot species, whose general aversion to any kind of company usually leads to instant flight when two individuals chance to meet. When a male encounters a female, he therefore needs to inhibit any flight reaction quickly and at a 'non-flight' distance. He achieves this by drumming his head against the substrate. The female shows her interest by approaching rather than fleeing, and begins to lick a glandular area around the male's sixteenth tergite. Absorbed in her licking, and sexually 'aroused' by pheromones released by the gland, she assumes a posture which facilitates mating.

Some species produce acoustic signals using specialized stridulatory structures. The male *Loboglomeris pyrenaica* stridulates while grasping his mate, transmitting the resulting vibrations directly through her antennae or vulva. The signal seems to be species-specific, for although *L. pyrenaica* males will court females of the sympatric *L. rugifera*, the latter cut short the proceedings by coiling up as soon as they detect the 'incorrect' courtship stridulations. The stridulatory efforts of some of the large South African *Sphaerotherium* are loud enough to be detected by the human ear (at up to 5 m in *S. punctulatum*). The male's noisy output serves to persuade a coiled female to open up and probably also functions to prevent hybrid matings where several species occur together.

Although pheromonal communication probably has a part to play in all millipede courtship, it is particularly important in certain species. *Polyxenus lagurus* is unique in being the only millipede not to employ some kind of direct sperm transfer. Instead, the male constructs a zigzag web above a small hollow in the bark of a tree and exudes onto it a droplet of sperm. As he makes his departure, he draws along behind him two parallel threads about 2 cm long. These are heavily embellished with a chemical 'dew' secreted by glands on his seventh segment. The female homes in on the correct spot via the tell-tale threads, which act as a chemical signpost leading to the sperm droplet (Schömann, 1954; Schömann & Schaller, 1964). *Julus* males advertise themselves chemically by raising their front ends and exposing pheromone-producing legs.

In *Glomeris marginata*, the male adopts a kind of pre-mating position, gripping one antenna and one vulva of the female. He then douses the area immediately beneath the female's mouthparts with a glandular secretion (from his post-gonopodial glands). When she starts to feed on the secretion, he takes this as his cue to grip the other vulva, thereby achieving the actual mating posture. The secretion appears to act as an arrestant, retaining the female in one spot and in the correct position long enough for the male to do his work. Glandular feeding with a similar purpose is also seen in some arachnids and certain insects, notably crickets and cockroaches. It is also found in some other millipedes, such as *Trachysphaera pyrenaica*, *T. noduligera*, *Haploglomeris multistriata* and *Sphaerotherium ancillare*. *Julus scandinavius* females are induced to adopt and maintain a suitable posture by feeding on glandular secretions from spoon-shaped structures on the male's second pair of legs (Haacker, 1969). He rears up and presents these to the female, if necessary maintaining this stance for several hours if she at first fails to take the bait. A receptive female responds by lifting her body to face the male, then presses her mouthparts against the glands. She becomes instantly 'hooked', lapping up the secretions while oblivious to the male inserting his gonopods into her vulvae.

Actual mating techniques are of great interest and variety. The male of the African pill millipede *Sphaerotherium dorsale* lacks gonopods, so his normal walking legs serve to transfer his sperm (Haaker, 1968 & 1968a). His method for conveying the sperm from the genitalia at his rear end to the vulvae at the female's front end is highly unusual to say the least. As they lie belly to belly, with their bodies curved into a C-shape, the genital openings of the two partners are almost 2 cm apart. The male grips the female's vulvae with his pincer-like front legs while his genitalia produce a droplet of sperm. This is passed from foot to foot, from his rear end up into the female's vulvae. The method used by the common European pill millipede *Glomeris marginata* and some other Glomeridae is perhaps even more bizarre. The male picks up some small nearby object with his front legs and then rolls it on the ground and gnaws at its surface until it assumes a generally globular shape. The chosen object is usually a particle of their own frass, which is highly amenable to such modification. Once the 'tool' is ready, the male presses it against his genital pores where it picks up a droplet of sperm. With remarkable dexterity, he then 'walks' this back along his body until he can introduce the sperm droplet into the female.

Unreceptive females often have to put up with persistent suitors for long periods. However, in *Leptoiulus*, the female can persuade an overzealous male to give up by venting a dose of defensive quinones. *Alloporus* females employ similar tactics, which are particularly effective when aimed directly at the male's front end, which is doused in a gush of repellent defensive secretion.

CHILOPODA Centipedes

Male centipedes lack an intromittent organ and rely on depositing a spermatophore for the female to pick up. In the slender, soil-living geophilomorph centipedes, the two sexes meet head to tail and tap one another with their antennae and anal legs. The male now moves away and constructs a crude web, zigzagging down into the mouth of his gallery. He expels a droplet of sperm on to the silk, leaving the female to locate it, which may not be for several hours. She eventually ferrets out the sperm web with her antennae and crawls over it, wiping her rear end back and forth to sweep up the sperm droplet.

The stockier Scolopendromorpha employ a more complex courtship. In *Scutigera coleopatra*, the male and female stroke one another with trembling antennae and may position their bodies to form a ring (Klingel, 1960). The male now shoves himself beneath the female's front and gently rocks her. After several rocking sessions, he extrudes a lemon-shaped spermatophore, turning round and guiding the female over it with his front legs. In *Thereuopoda decipiens cavernicola*, the male takes an even more active role. After a courtship involving vibrations of the antennae and front legs, the male stimulates the female with several jerky 'body lifts' (Klingel, 1962). After depositing a spermatophore, he picks it up in his mouthparts and presses it directly into the female's genital aperture. After an hour or so, the female turns and consumes it, presumably after its contents have been absorbed. Treating the empty spermatophore as a post-nuptial meal is common practice in many katydids and crickets.

ARACHNIDA Scorpions, spiders, mites, etc.

SCORPIONES Scorpions

Scorpions are noted for the massive development of the pedipalps with their pincer-like chelae. These serve an important function in courtship. Sperm transfer takes place via a spermatophore which the male deposits on the substrate, from where it is picked up by the female. Courtship in *Paroroctonus mesaensis* (Vaejovidae) from California is unusual in that the initial contact is made by the female, upon detecting the movements of a passing male (Polis & Farley, 1979). This contrasts with observations of male-initiated courtship in other scorpions – possibly an artefact of captivity. The female *P. mesaensis* briefly grabs the male with her chelae, then clubs him with her retracted sting and possibly even tries to sting him. This initial 'female attack behaviour' causes the male to 'strut defensively' on stilted legs, his sting raised vertically while his pectines sweep the ground. This behaviour does not seem to be sexually motivated, but is normal after any unexpected assault. Female attack and male strutting are now repeated alternately a couple of times before the male 'judders', pitching his body back and forth. Finally they grasp one another's chelae face to face and launch into a *promenade à deux*, the 'dance of the scorpions'. The object of the dance is to locate a suitable attachment point for the spermatophore. Then the male leads the female around by the chelae, his pectines sweeping the ground until they happen to pass over a favourable spot. He overcomes occasional female resistance by massaging her with his chelicerae, or 'kissing'. The dance may take them some distance – 8 m or so – before the male stops and extrudes his spermatophore, usually (in this species) cementing it to a stick lying on the ground. The male must now position the female accurately above the spermatophore, so a certain amount of jiggling goes on before the female does a 'head-stand' enabling valves in her genitalia to absorb the contents of the spermatophore. The male now withdraws, probing the female briefly with his sting while letting go of her chelae. He makes off briskly, for cannibalism is common and mature males are often killed by females during the mating season. However, it is thought that little or no cannibalism takes place in a sexual context.

So-called 'sexual stinging' takes place in some chactids, bothriurids, scorpionids and vaejovids, and presumably acts as a stimulant. In the chactid *Megacormus gertschii* from Mexico, the 'aphrodisiac sting' seems to be vital, at least under captive conditions (Francke, 1979). After grabbing the female's chelae, the male stings her on a pedipalp. She seems totally unmoved, even though the sting may be sunk home for as long as 9 minutes. After several such lengthy stinging episodes, the male finally begins to tug the female backwards, his 'tail' now stowed away and pointing away from the female. After depositing the spermatophore, he pulls her over it by her pedipalps. Suddenly they break into a 'stabbing match', lashing out wildly with their stings until the male smartly slaps his mate's sides with his pedipalps and she picks up the spermatophore.

PSEUDOSCORPIONES Pseudoscorpions

In the Cheliferidae and Chernetidae, there is a *promenade à deux*; in other families there is no male–female contact. *Serianus carolinensis* from the USA is one of several species whose males deposit a spermatophore when stimulated by the presence of a nearby female. The male *S. carolinensis* uses his unique silk glands (in the rectal region) to spin two lines linking the spermatophore to the ground below and to some object such as a twig above (Weygoldt, 1966). It is likely that the silk is saturated with male pheromones, enabling the female to locate the spermatophore more easily.

SOLIFUGAE Sun spiders or solifuges

Sexual routines are broadly similar, but differ in some important points, not only between species, but also between the various Old World families and the purely New World Eremobatidae. *Metasolpuga picta* (Solpugidae) from the Namib Desert is strictly diurnal which is unusual for a solifuge. Males run across the hot sand looking for burrows containing females (Wharton, 1986), who betray their presence by either sound or pheromones (or both). The male immediately starts to dig, which quickly lures the occupant to the surface. He instantly and rather roughly grabs her rear end with his long needle-like chelicerae, but also delicately strokes her prothorax and chelicerae with his palps. This seems to have a dramatic effect, instantly inducing a trance-like state as the female draws in her appendages close to her body. A receptive female will remain like this throughout the rest of the procedure; if unreceptive, she will struggle during the next step. The suitor chews his way down the length of her abdomen in a massage in which he ends up kneading her chelicerae. He is now standing directly over her, perfectly positioned to clutch her legs and palps and eject on to her back a spermatophore. He picks this up in both chelicerae, hoists the female's abdomen up vertically and places the tips of his chelicerae, along with the spermatophore, into the female's genital opening. He now spends 2–3 minutes delving around inside her, presumably to ensure the thorough liberation of the sperm. At this point the female usually awakes from her dream-like state and struggles free.

Other Old World solifuges have similar habits, but the nocturnal species place the spermatophore on the ground. This would be safe enough in the cool of the night, but the sperm would be rapidly cooked on the hot desert sand in daytime, hence the *M. picta* male's habit of using the female's relatively cool back to receive the spermatophore. Several species of American *Eremobates* (Eremobatidae) show some important departures from this 'norm' (Muma, 1966). The female does not collapse into a state of apparent catalepsy at the male's first physical advances. There is a period of initial defensive posturing, after which the female may or may not assume a submissive posture and its accompanying trance-like state. Even if she remains fully alert, she may still accept a male. The male gently kneads a path down her abdomen with his chelicerae, ending up at the genital segment. This relaxes the female, and then the male can tip her over on to her back or side and plunge his chelicerae up to the hilt in her genital aperture. Locked in place, he turns her so that she is in a normal walking posture, but with her abdomen cocked right over so that its tip sits above her head. He then chews away energetically at her genital region, sometimes even picking her up and carting her around. He smooths away any objections by caressing her with his legs and palps. Finally he withdraws his chelicerae and climbs onto her bowed abdomen, where he makes a direct transfer of a blob of semen. Backing off, he uses his chelicerae to stir the semen inside her body, securing its thorough percolation into the complex cavities within her.

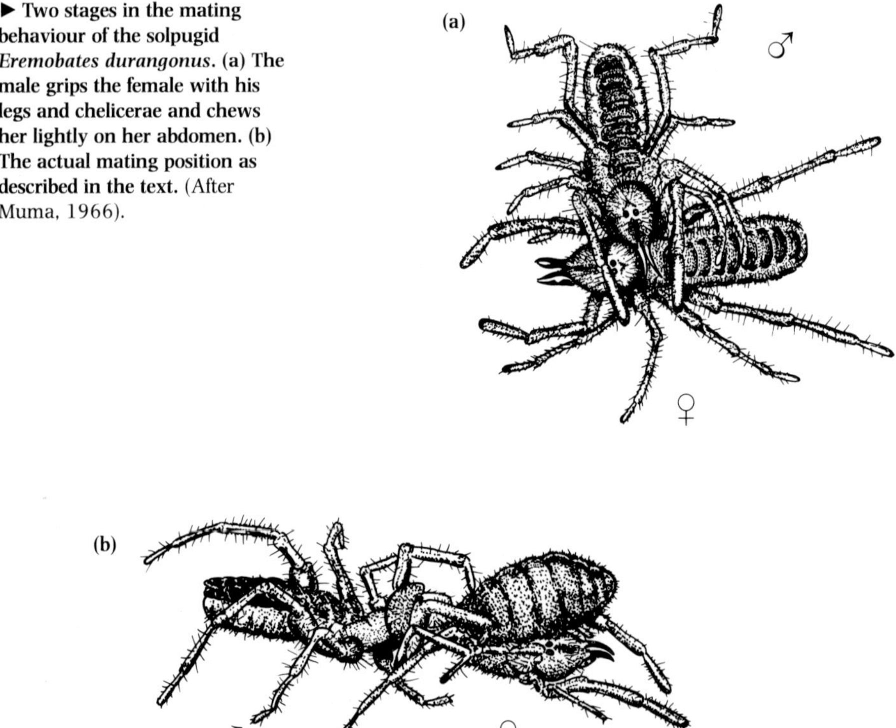

▶ **Two stages in the mating behaviour of the solpugid *Eremobates durangonus*. (a) The male grips the female with his legs and chelicerae and chews her lightly on her abdomen. (b) The actual mating position as described in the text.** (After Muma, 1966).

UROPYGI Whip scorpions

Whip scorpions use a spermatophore deposited on the substrate, as do scorpions. In some genera, this is accomplished very simply; in others it is more complex. In the American giant whip scorpion, *Mastigoproctus giganteus*, the male launches himself at a female and struggles to seize her antenniform front legs (these are very long and not used for walking) with the pincers on his palps (Weygoldt, 1970). By retreating backwards, the female slides her antenniform legs through his grip, enabling him to grasp their tips, usually in a crossed position. He now leads her in a courtship 'dance' during which the female's resistance is gradually worn down. After some 2–3 hours of this *promenade à deux*, the female takes the lead and folds her pedipalps around her partner's abdomen in a firm hug. This is the male's cue to stroke his genitalia on the ground and extrude just the stalk of the spermatophore. Only after a further 2 hours or so is the intricately shaped spermatophore fully developed and ready for extrusion. The male now inches forward, carefully guiding the female over the spermatophore. Her genital aperture opens fully, gripping the hooked tips of the sperm-carriers and drawing them free of the spermatophore. The male now climbs on to his mate's front end, reaches down on either side of her abdomen with his pedipalps and, with the mobile fingers on his chelae (an exclusively male device), meticulously presses home the sperm-carriers. He keeps this up for 2 hours or more, presumably to ensure a thorough dispersal of the sperm. Similar procedures are known in *Thelyphonellus* and *Typopeltis*.

The courtship and spermatophore collection in *Thelyphonus linganus* from Malaysia are very similar (Weygoldt, 1988), but the male does not push home the sperm-carriers. His spermatophore is a very sophisticated 'self-loading' device operated solely by the female. As she simultaneously depresses a hooked projection on either side and a cushion-like 'plunger' in the centre, the two horn-like sperm packages on either side are automatically pushed upwards into her. Once inside her body, the sperm-carriers eventually fragment and release the semen.

RICINULEI Ricinuleids

These heavily armoured robust arachnids have been little studied, but the general mating habits are known. In the west African *Ricinoides hanseni*, the male first loads the emboli (sperm-carriers) on his specially adapted third legs with semen from his penis (Legg, 1977). Crawling on to the back of a co-operative female, he inserts a tarsus into the female's genital aperture. This has a number of pockets which fit snugly around a corresponding series of lobes on the male's tarsus, holding him in place during copulation.

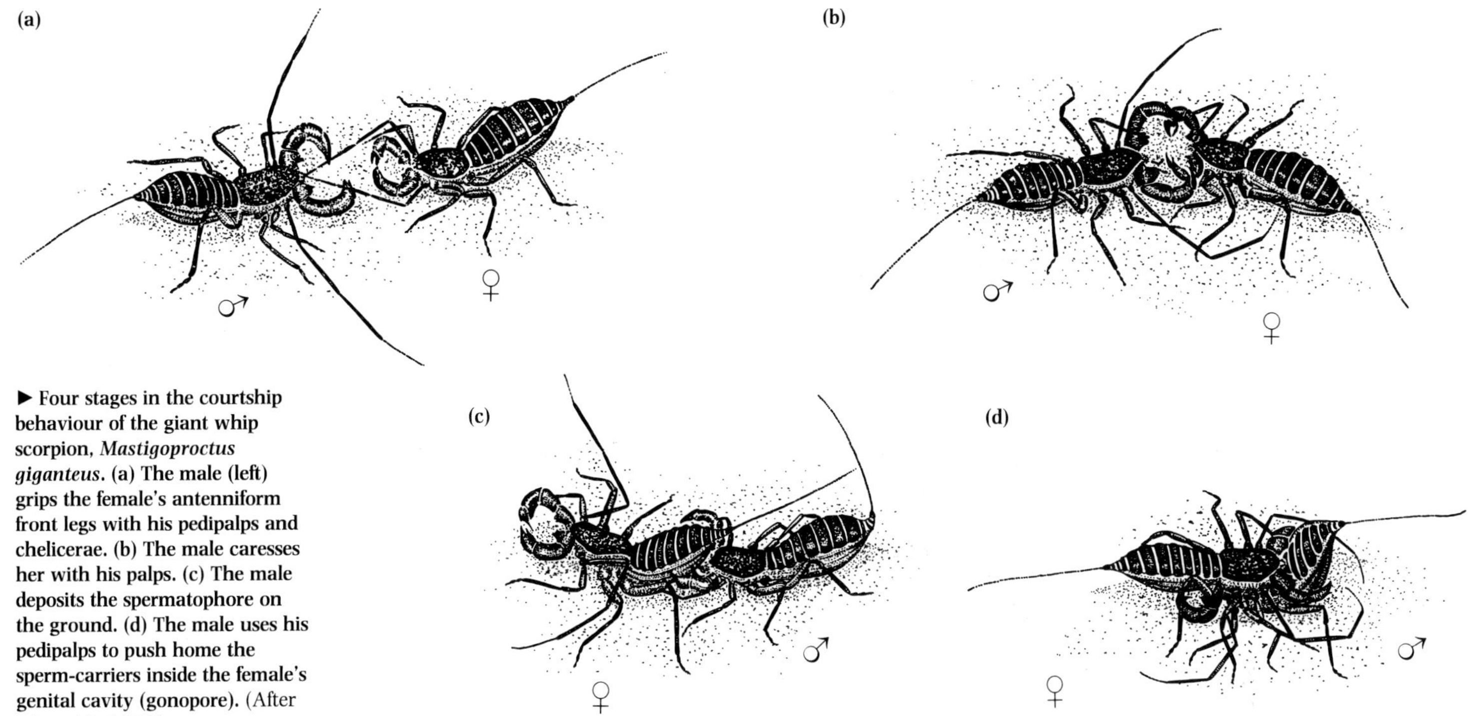

► **Four stages in the courtship behaviour of the giant whip scorpion, *Mastigoproctus giganteus*. (a) The male (left) grips the female's antenniform front legs with his pedipalps and chelicerae. (b) The male caresses her with his palps. (c) The male deposits the spermatophore on the ground. (d) The male uses his pedipalps to push home the sperm-carriers inside the female's genital cavity (gonopore).** (After Weygoldt, 1970).

OPILIONES Harvestmen

Relatively little is known about how the sexes come together in the wild. In the North American *Leiobunum longipes*, males and females cluster on 'courting grounds' – damp raised hummocks of moss or rotting logs and stumps – where mating and oviposition take place (Edgar, 1971). When females are present, the males often fight within these areas, clamping their chelicerae on to one another's legs and attempting to pull them off. Large males usually emerge as victors and assume top position in a kind of pecking order which decides which male will eventually mate with a nearby female. The successful male will try and remain with her and copulate repeatedly over a few hours. His temporary absence while fending off challengers may allow the smallest males to sneak in and secure a copulation but this is unlikely to be productive as the 'owning', or 'alpha', male will immediately carry out an 'insurance' mating upon his return. In most harvestmen courtship is absent, and the sexes merely meet face to face so that the male's penis (the male has a penis instead of an accessory organ) can enter the female's genital aperture. In *Leiobunum vittatum*, the female has been seen using her palpi and chelicerae to guide the male's penis into the correct place. Soon after mating, the male *L. longipes* shields the female with his second or third pair of legs and ushers her towards a suitable spot for oviposition. He may even persuade her to do her duty by stroking her gently with his legs. This behaviour is not found in *L. vittatum*, *L. calcar* or *L. politum*.

The European *Ischyropsalis hellwigi* is unusual in having a simple courtship. The basal segments of the males' chelicerae are furnished with a dense array of bristles which are bathed in a glandular secretion. When the sexes meet they tap one another with their front legs and pedipalps; this seems to be mostly a confirmation of identity. The male then rears up and presents his chelicerae to the female, who steps forwards so that her mouthparts can press against the male's bristles and lap up the secretion. Only by enticing the female to raise the front part of her body in this way can the male's penis seat home correctly. Sexual feeding on such 'excitators' is also seen in other invertebrates, such as millipedes, crickets, cockroaches and malachiid beetles. The procedure in *I. luteipes* is broadly similar, but the male launches himself at the female and tries to force his chelicerae against her mouthparts. She usually struggles furiously at first, but only until her mouthparts touch the secretion. This instantly exerts such a calming effect that she immediately adopts the correct posture for copulation (Martens, 1969).

The Neotropical *Zygopachylus albomarginis* has a highly unusual male-controlled mating system based on nests which is described in detail on page 162.

ACARI Ticks, mites

Few species have been studied in detail. Most female ixodid ticks mate only once, usually while on the animal host, produce a large clutch of eggs and then die. In the Australian reptile tick *Aponomma hydrosauri*, the males respond to a female-released pheromone by ceasing to feed and heading for the source (Andrews & Bull, 1980 & 1981). Unfed males will not respond until they have had a meal from the host. During courtship, the male pats the female's back with his front legs, then mounts and embraces her with his legs, facing forwards. Then he turns around and clasps the rear of her body. A receptive female tilts up at an angle of 70°, making room for the male to creep beneath her, belly to belly, and pull himself forwards until his mouthparts can be anchored in her genital pore. Now he can start to manufacture the spermatophore, a process which takes from $1\frac{1}{2}$ to 2 hours. In ixodid and argasid ticks, the male normally reaches forward to pick up the emerging spermatophore in his mouthparts which then transfer it to the female. In argasids, the spermatophore is held between the chelicerae and palps, while in ixodids it is glued to a drop of hardening saliva exuded copiously from the mouthparts during copulation (Feldman-Muhsam & Borut, 1971). Most males leave the female shortly after copulation, but in *Ixodes holocyclus* and *I. pilosus* the males are often parasitic upon their former mates, who provide them with a blood meal. Up to three males may batten upon a single female, and *I. holocyclus* males probably never feed from any other source (Norval, 1974).

The two-spotted spider mite, *Tetranychus urticae*, of the family Tetranychidae is a destructive plant-pest. The males are the first to become sexually mature and head for a pheromone emitted by the immature females in their final-instar quiescent stage (Potter *et al.*, 1976). As sperm precedence overwhelmingly favours the first male to mate, it is in every male's interest to be on hand when the female moults and becomes sexually receptive. A prospective suitor stations himself on or by the immobile female and guards her until she emerges, possibly even assisting her in this task by peeling away her old skin with his front legs. During his sojourn on guard he will probably have to defend his property against rivals. His dedication as a guardian depends on whether the female is close to her final vital moult. He is able to assess her state of development, and therefore 'knows' how much effort he should expend to defend her. If the final emergence is still some time away, he might even allow a rival to join him as a co-guarder. However, when her emergence is close, he puts up a fierce resistance. Savage fights develop during which the antagonists vie for position, circling one another with flailing legs and charging in with extruded stylets held like lances. One trick is to trip up and snag an opponent by smearing him with silk from the palps, obliging him to disengage for a clean-up. Size tends to influence the outcome of these battles, so that the biggest males win the most fights and gain access to the largest number of females (Potter, 1981). Injury is rare, but the occasional fatal stab with the stylets does occur. Even victory may not always be rewarding, as the winner occasionally seems disoriented by the hectic nature of the battle and may wander off, deserting the female whom he fought so hard to defend.

Such fights may turn out to be quite common in mites. *Sancassania anomalus* is known to stage contests among stored granary products, while old mature males of *Macrocheles parapisentii* thin out the numbers of any future competition by slaughtering freshly emerged males in their rearing cells.

ARANEAE Spiders

Sperm induction

In common with all other arachnids except harvestmen, male spiders have no penis for introducing sperm directly into the female's body. Spiders employ an indirect method using their palps, which are situated below the head. Before mating, these have to be charged with sperm, although never by direct contact, even in spiders whose palps are long enough to contact the genitalia at the tip of the abdomen. (Contrast this with dragonflies, in which the males' accessory genitalia *are* charged directly from a penis at the tip of the abdomen). Spiders perform this task in a roundabout way known as *sperm induction*, using a silken 'sperm web' as the intermediary between genitalia and palps.

The sperm web is at its simplest in spiders such as the 'primitive' *Pholcus* daddy-long-legs (Pholcidae), in which the third pair of legs are used to draw a single silk thread across the genital opening until it picks up a drop of semen. This is then picked up in the chelicerae and it is from here that the palpi are dipped alternately into the globule until it is all absorbed. A number of other primitive spiders do something similar, although the minute web is often held between the front legs or palpi until the droplet is absorbed. Most spiders attach a small rectangular or triangular sperm web to some convenient object.

In the European *Araneus quadratus*, the male spins his small rectangular web between a few blades of grass. He deposits a small droplet of semen on its upper surface, then hangs to one side of the web and dips one palp at a time into the semen, which is exhausted within about 10–15 minutes (an 'average' time for spiders). In other species, the stance is reversed, and the male reaches down beneath the web to absorb the sperm through its lower surface. Normally the male chews his palps both before and during the procedure, apparently to prepare these complex organs for their duty.

In some mygalomorph spiders, such as the tarantulas (Theraphosidae), the process is much more complex, lasting some 3 hours or more. The male constructs an extensive and very dense sheet of silk large enough to cover about two-thirds of his body (Minch, 1979), then exudes a drop of semen on to the underside. He moves around to the upper side and reaches beneath with alternate palps. Most male spiders will spin several sperm webs during their active sexual life. Sperm induction usually first takes place soon after the male spider matures sexually, but before he goes off in search of a female. The trigger is thus generally an internal one, rather than the presence of a female, except in the Linyphiidae, in which contact with a female or her web stimulates the act of induction.

Courtship

Who is afraid of whom?

The function of spider courtship with its often elaborate nature has been much debated. The favourite explanation has always proposed that males are seeking to avoid being mistaken for a meal and eaten. This has probably arisen because male spiders are often smaller – sometimes much smaller – than the females. However, in many families, the males are as big as the females and well able to take care of themselves due to their longer reach and greater agility. Even very recently (Suter, 1990), it was suggested that the female's 'normal' first reaction on detecting a male is to display aggression. Any thorough examination of the voluminous literature on spider mating habits will reveal few grounds for this view. Far from being aggressive, a female's first response may be to collapse into a state bordering on catalepsy. Even when this does not happen, she may be cautious or even nervous rather than aggressive. In fact, blatant hostility is probably quite rare *in nature*, although when it does occur it is probably mainly when it is not a *virgin* female who is being courted.

Under wild conditions virgins are in short supply, quickly being mated as soon as they mature. Males therefore lavish huge amounts of time and energy in often futile courtship of fertilized females who may be close to laying eggs. Not surprisingly, such heavily gravid mothers-to-be do not grant late-coming suitors a friendly reception – although there is still a chance that a really pushy male will yet achieve his aims (see *Agelena*, page 27). However, little is known about sperm precedence in spiders, so it is generally impossible to know whether such last-minute matings succeed in fertilizing many eggs. Unlike the situation in many

▶ The dilemma faced by a virgin female spider when a courting male trespasses in her web is well illustrated in the European araneid *Nuctenea cornuta*. The female usually responds quickly and positively to the male's vibratory overtures, rushing down from her lair to meet him with an alacrity which seems to reflect her innate need to mate. However, she does not normally (top) immediately adopt her mating posture, but instead engages in one or more bouts of vigorous sparring with the male. Despite his much smaller size, he may come off best in such encounters to the extent that it is the *female* who may retreat by dropping out of her *own* web on her safety line. These sparring matches seem to be an essential part of the courtship procedure in this species and may serve to confirm the male's identity with an added degree of certainty. The extremely vulnerable posture finally adopted by the female to allow the male free access to her epigyne is obvious from the lowermost illustration. The female remains in this position for several minutes while the male makes repeated, rather brief, stabbing insertions of his palps. When mating has finished the female trots back to her lair and normally makes no attempt to harry the male from her web.

insects, which do not bother with courtship, such lack of finesse is simply not practical in spiders, as the full co-operation of the female is usually essential right from the start. In order to insert his complex palps fully and correctly in the female's epigyne the male generally requires the female to:

(1) Remain still long enough for him to get into position.

(2) Adopt a special posture which allows access to the epigyne, which is on her vulnerable *underside*.

If the female fails to respond with the correct posture, the male cannot mate. This, however, does mean that, once a male has been allowed 'on board', he is in a commanding position and could theoretically administer a fatal bite.

It is common for the female's co-operation to take the form of a 'collapse' into a state akin to a trance. Such a vulnerable dream-like condition puts her at the mercy of allcomers, so it behoves the female to be absolutely 'sure' about the correct identity of an intruder. The male's complex species-specific courtship gives her that security, which she – almost uniquely among animals – must have. Spider courtship is thus probably dedicated to the gradual wearing down of the female's protective barriers, re-assuring her that it is a male with 'honourable' intentions who confronts her, and that she can afford to yield in safety – it is for *her* protection. A significant pointer to the correctness of this viewpoint is that, in most known cases of sexual cannibalism, males eat females almost as often as the reverse. When males *are* eaten it is often by non-virgins, or under experimental circumstances, which may not completely reflect what happens in the wild (see comments on *Araneus diadematus*, page 28). Females most positively do not take advantage of males, as they probably could. Instead of rushing out to attack an unwelcome suitor, an unreceptive female warns him off, usually by vigorously jerking the web or lunging at him from a distance. Males who repeatedly ignore the message and refuse to take 'no' for an answer are the ones who might occasionally end up as meals.

The premise that courtship serves to 're-assure' females is backed up by the following points:

(1) Where the males and females are of approximately equal sizes – e.g. Salticidae, Lycosidae, Amaurobiidae – the female would have most to fear and the male least. In line with this, these families tend to have the most complex and species-specific courtships, sometimes in conjunction with the use of chemical (pheromonal) recognition agents.

(2) Females who have nothing to fear from their mates because of their tiny size – e.g. *Nephila, Argiope, Micrathena, Gasteracantha* – seem to have a simple courtship more concerned with keeping the female immobile or preparing her physically for insemination than with lulling her into a state of catalepsy. It is significant that, in both *Nephila* and *Argiope*, the female keeps her wits about her during copulation to such an extent that she can rush off and catch an insect which enters the web, leaving the unfortunate male to be brushed off against the silk or jump for his life. *Argiope* males may be allowed to court females of a different species *without being driven away or eaten*, despite *giving all the wrong courtship signals*. This powerfully suggests that the females have so little to fear from their midget suitors that correct species-specific identification is not vital. *Argiope* males which have lost several legs, rendering them incapable of signalling correctly, are also able to complete copulation successfully. *Gasteracantha* males are among the

smallest of all in relation to the females, yet sexual cannibalism seems to be virtually absent (Robinson & Robinson, 1980). This is also true of thomisids such as *Misumena vatia* and *Epicadus heterogaster* – the diminutive males are allowed to walk around with impunity upon the huge females. Courtship is notable by its absence, yet, according to traditional theory, such tiny males should be fearful of their lives. But courtship is redundant because it is the *females* who have nothing to fear.

Types of courtship

Courtship procedures can broadly be divided into three categories, each linked to the anatomy and life-styles of the spider. These are:

(1) Little or no courtship; typical of the *short-sighted hunters* whose mediocre vision precludes the extravagant visual displays seen in the next group. Included here are mygalomorph families, such as the Theraphosidae and Atypidae, and araneomorphs, such as the Clubionidae, Thomisidae and Sparassidae.

(2) Visual displays employing movements of the palps, legs or abdomen – or all three in combination. These are the *longsighted hunters* whose excellent eyesight permits such highly visual methods. Included here are families such as the Lycosidae, Pisauridae and Salticidae.

(3) Web-tweaking and tapping, as employed by the varied families of *web-builders*, including the Agelenidae, Theridiidae, Araneidae and Linyphiidae. The females live in a tactile world, communication being effected via the silken lines of their webs. Not surprisingly, the males utilize the web itself as the primary transmitter of identity and intentions.

Strategies

A female spider is not an immutable entity for she changes as she grows and matures, being more likely to submit to a male at some times than at others. A young virgin may be a willing mate, yet a few weeks later, when nearly ready to lay her eggs, her 'mood' will have changed and she will be reluctant to accept a last-minute suitor. In view of this gradual decline in female availability, there are three possible strategies open to the males, each being tailored to the particular circumstances of the encounter – and this in itself is heavily dependent on the time of year.

(1) The male moves in with an immature female just before her final moult to maturity. He may have to repel other males, making this a form of *female-defence polygyny*. This may turn out to be the most common method by which females lose their virginity. It certainly occurs widely in web-builders of several families and also in the Salticidae and probably in many Lycosidae, whose females live under stones etc. It obviously cannot occur where females are free-living. This strategy has the advantage of guaranteeing the male a virgin – which in itself may remove any need for time-consuming courtship.

(2) The male approaches an adult female and conducts his full courtship routine. The problem here is that, *at least in the wild*, a good percentage of such females will no longer be virgins, having already been deflowered by males adopting strategy 1. Courtship is therefore liable to be both protracted and of uncertain outcome. In some spiders this may therefore constitute more of an alternative strategy, although in others (e.g. *Pardosa*) it will be the only strategy on option.

(3) The male approaches a female who has already mated one or more times and is close to laying her eggs. Late in the season all females will fall into this category. Every female approached by a suitor now greets him with outright hostility, and it is by this that he 'knows' that females are a fast-dwindling resource, making this possibly his last opportunity to secure any more copulations. He therefore has little to lose by throwing caution to the winds and adopting a so-called 'kamikaze tactic' in which he throws himself at the female, regardless of the consequences. Such a strategy has only been observed in a few spiders (e.g. *Agelena labyrinthica*), but it will probably turn out to be widespread.

▲ The tiny male of the common crab spider, *Misumena vatia* (Thomisidae), can stroll around over the female's body with impunity. If she has already mated and is therefore unreceptive, she will warn him off before he comes too close.

THE HUNTERS

Courtship in the short-sighted hunters

In these mostly drab-bodied spiders, courtship tends to be minimal or even absent. For tarantulas, the sequence of events seems to be remarkably similar in all species. The male ventures forth at night in search of females' burrows. Upon contacting the silk around the burrow entrance with his front legs, he announces his arrival by rapping on it. The ability to recognize the silk of a conspecific female merely by touch has now been established experimentally in many families. It is facilitated by a finely-developed chemotactic sense situated in fine hairs on the front legs. This enables most male spiders to detect the recent presence of a female merely through touching the silk which she constantly trails behind her. Indeed, it is quite common to see male wolf spiders courting an empty space formerly occupied by a recently departed female.

The female tarantula responds to the male's tapping at her door by coming out to meet him and rearing up on her hind legs in what seems to be an aggressive posture. However, this rampant stance is essential because, if the female fails to assume it by herself, the male will prompt her by tapping vigorously with his front legs. As she rears back she exposes her fangs, not as a threat, but to enable the male to catch them deftly and secure them behind the spur on each front leg (provided for just this purpose). This fang-snaring operation has probably more to do with physical practicalities than with supposed male safety, for now he can gently lever her body backwards to lay open her genital region. He pushes in beneath her and inserts his palps, one at a time, the right one into the opening on the right, the left one into that on the left. The actual mating in tarantulas is relatively brief, normally lasting only a minute or so.

Once finished, the male disengages rapidly and quickly distances himself from the female. The need for such haste seems far from obvious, female aggression at any stage being rare. The front-to-front posture and alternate insertion of the palps has been noted in all mygalomorph spiders studied. However, it is not always the female's fangs which the male clasps in his front legs. In the Antrodiaetidae and Ctenizidae, it is the chelicerae above the fangs which are gripped, while in the Dipluridae it is the bases of the female's pedipalps or her second pair of legs.

The problems for the *Atypus* male (Atypidae) in making his initial approach to a female would seem particularly perplexing. He must somehow contact a mate who spends her entire life inside a sealed silken tube. He must be careful not to blunder unknowingly on to this tube, for he may suffer the same fate as an insect which does likewise – i.e. he will be instantly transfixed by the occupant's long fangs, which are plunged through the silk and into the victim blindly, yet with uncanny accuracy. A male *Atypus* avoids such a grisly end by being able to recognize a female's tube at the lightest touch through his chemotactic sense. His instant reponse is to freeze for a moment before announcing his identity in a manner which will keep him from harm – he strikes up a rapid burst of tapping with his palps and legs against the silk. It is the *lack* of any response which gives the green light to his next action. He tears a hole in the silk, having first eased his path by softening the tough material with a blob of saliva, and enters. By now the occupant will have been reduced to a state of passive receptivity by his initial tappings. It merely remains for him to press her against the silken wall, using his front legs and partially gaping fangs. So what would happen should the tube be occupied by an unreceptive female gravid with eggs? Would she take advantage of her invisibility by instantly spearing him as an easy meal? Apparently not, for instead she warns him off by tugging at the tube, sending a life-saving message to which he responds with a swift withdrawal.

The *Scytodes thoracica* male (Scytodoidea) is permitted to clamber unceremoniously over the female immediately upon their first meeting, sliding beneath her from the rear in an inverted position. She makes this easy for him by standing up high on her legs while he tips her slightly backwards, stroking the underside of her abdomen with his chelicerae in a quest for two small chitinized pockets just to the rear of her genital aperture. These are destined to receive his fangs and, once they are safely sunk home and he is securely locked in place beneath her, he inserts both palps at the same time.

The beautiful green, yellow and scarlet male of the European sparassid *Micrommata virescens* is considerably smaller than his bulky green mate, yet he jumps straight at her on their first meeting and grasps a leg or her abdomen in his chelicerae. Far from reacting aggressively, she permits him to stroke her and climb on to her back, from where he lifts her abdomen and inserts a single palp. Mating is prolonged, with each palp being inserted for 3 hours or more before the male leaps off and departs. Similarly hurried departures from females who have remained placidly immobile for anything up to 6 or 7 hours (depending on the species) are common in most male spiders. It may reflect the imminent ending of the female's dream-like state which has kept her quiescent. Even so, there is little evidence that females suddenly develop cannibalistic inclinations at this stage, so why the males make themselves scarce in such haste is a mystery.

Sexual dimorphism has developed to an extreme stage in many Thomisidae, the crab spiders. The males are always smaller – sometimes astonishingly so – as in *Epicadus heterogaster* from Brazil, in which the male is positively lilliputian, perhaps only one-hundredth the weight of the female. Sexual relations in *Misumena vatia*, a common crab spider in Europe and North

America, usually go smoothly. The female deals instant death to such formidable stinging insects as bees, yet the tiny male can proceed with scant caution. The female apparently recognizes him immediately, probably via chemotactic senses, which in the male are so acute that he becomes visibly excited upon encountering a flower draped with a female's silk. Even a completely unreceptive female heavy with eggs will give fair notice to an unwelcome suitor by raising her front legs in a menacing stance. If he still fails to take the hint, she violently twitches her body, sending a tactile message through the flower. Note that she does *not* cash in on the males' powerful sexual drive to use them as a succession of easy meals, adopting the theoretically feasible role of *femme fatale* to lure them up close before pouncing. A receptive female will allow her tiny suitor to spend an hour or more scrambling around over her, without showing any acknowledgement of his presence. In *Thomisus, Misumena, Misumenopsis, Diaea* and *Epicadus*, the females are so bulky that the male has to hang underneath to insert his palps. In *Xysticus*, the male is at least half the size of the female. Prior to mating he ties her down with skeins of silk, the 'bridal veil'. Since its discovery in *X. cristatus* the use of a similar bridal veil has been discovered in two other families of spiders (see page 24).

Sound is produced in at least 27 families of spiders. The buzzing spider, *Anyphaena accentuata* (Anyphaenidae), is one of the few whose output is audible to the human ear. During a dynamic courtship, the male taps the tip of his abdomen and the tarsi of his front legs against a leaf, producing a distinctly audible buzzing.

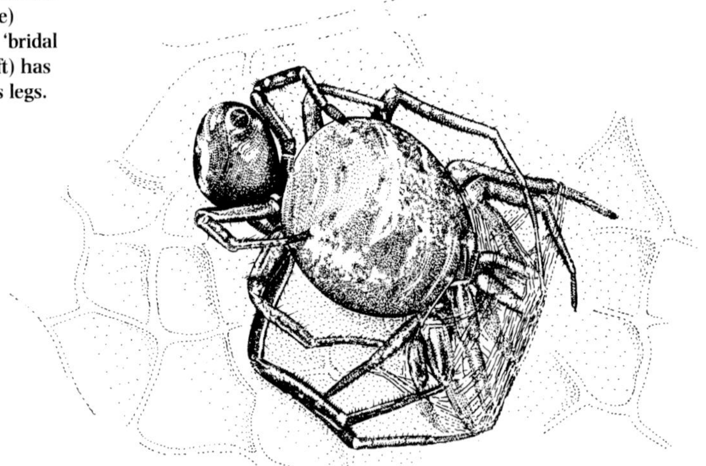

▶ A pair of *Xysticus cristatus* crab spiders (Thomisidae) mating. Note the silken 'bridal veil' which the male (left) has spun across the female's legs.

Courtship in the long-sighted hunters

In these spiders courtship is mainly visual, although some, such as the buzzing spider, use percussive sound. In the wolf spider *Lycosa rabida* from the USA, the male takes up a characteristic courtship pose with body lowered and forelegs flexed outwards (Rovner, 1967). Then, one at a time, he raises his palps and swings them in a rotary action, which causes a file on the stem of one palp to rub against a scraper on the opposing stem. The stridulation thus produced is amplified through contact with a dead leaf on the forest floor. After giving several 'rattles', the male stretches forward and taps away with one of his front legs, quivering his abdomen and rapping vigorously with his palps against the leaf. Each set of movements constitutes a sequence which is interrupted by brief pauses. The female signals compliance by giving a 'leg-waving' display and then takes the initiative by walking towards the male to make the first contact, a reversal of the more normal situation in spiders in which it is the male who steps forward. It seems that the 'rattling' is not essential for success during daylight, when the visual display suffices. But at night, when this species is active, the acoustic component alone is adequate to lure in any nearby females.

In the wolf spiders *Schizocosa rovneri* and *S. ocreata* from eastern North America, similarly produced sounds seem to be vital in keeping these two closely related sibling species reproductively separate (Stratton & Uetz, 1983). The males will happily court females of either species, indicating an apparent inability to distinguish their own females. The latter, by contrast, are more discerning and only become receptive when given the 'correct' series of courtship signals. The males are superficially identical, save for a conspicuous tuft of black bristles on the tibiae of the adult *S. ocreata*. These are displayed during courtship, which incorporates a strong visual element. As he parades around the female, the suitor taps his forelegs against the substrate and jerks his body to and fro. The acoustic ingredient consists of simultaneous palpal movements which produce stridulation. Both the visual and acoustic elements are important in provoking female receptivity, but it is the visual component, as characterized by the tufted legs, which seems to play the leading role. Bereft of such conspicuous adornments as leg tufts, which can be flaunted on parade, the male *S. rovneri* stays put and shoves his abdomen forwards between his legs, often banging it against the substrate as he does so. Simultaneously, he windmills his palps to produce a stridulation which is clearly audible to the human ear and consists of a series of sound pulses called 'bounces'. The 'bounces' are produced at intervals of about 4.5 seconds. Although the males of both species produce their stridulations using the same mechanism, the result is entirely different. *S. ocreata* produces much longer bursts of continuous sound than *S. rovneri* in what amount to species-specific courtship 'songs' on a par with those of grasshoppers, birds or frogs. Sound alone is sufficient to induce a receptive posture in the female *S. rovneri*, but not in *S. ocreata*; she needs the additional stimulus of the male's leg movements. The differences in the ingredients of courtship thereby keep these two species apart due to the 'fussiness' of the females, who must receive the correct stimulus.

In the nocturnal *S. mccooki* from the western USA, sound seems to have taken over completely and there is no specific visual display. The male drums a rapid tattoo against a dead leaf with his palps, and it is this percussion which exerts the main seductive effect. Although a stridulatory mechanism is present on the palps, its use has not so far been recorded (Stratton & Lowrie, 1984).

The strongly diurnal *Pardosa* wolf spiders, often abundant on both sides of the Atlantic, specialize in highly visual methods of courtship. As is typical of the family, the male locates the female through his well-developed chemotactic sense, often carrying out his full courtship displays to a 'chemical shadow' when a female has just run off. Several species often occur together and it is divergence in courtship signals and male ornamentation which are assumed to prevent hybridization. The black palps are large and play an important role in courtship signalling. In the European *Aulonia albimana*, it is an eye-catching white ring contrasting with the basic black of the palp which is prominently flaunted before the female, each palp being alternately angled upwards to best effect. In *Trochosa*, the palps are unadorned, but the front legs are often conspicuously black, being waved in various species-specific routines.

Nuptial gifts in pisaurids

The Pisauridae are also popularly called wolf spiders. Agility and vision are both highly developed, so much so that the author has seen *Pisaura mirabilis* juveniles launch themselves vertically upwards to snatch passing flies out of the air. This very common European species is generally known as the nursery-web spider. It enjoys the major distinction of being the only spider in the world (so far known) in which the male presents a nuptial gift to the female during courtship, a habit it shares with insects such as certain bittacid hangingflies (Mecoptera) and empidid dance flies (Diptera). According to most literature, the primary function of the nuptial gift in *P. mirabilis* is to placate the predatory female and prevent her from eating the male. Anyone who has observed the nuptial behaviour of this species will recognize the absurdity of such a notion. In fact, in most of the very few known cases of sexual cannibalism, it was a dispute over *ownership* of the gift which caused the problem, with either the male *or* the female doing the killing; the gift may therefore *cause* fatalities, not *prevent* them (Austad & Thornhill, 1986). During May, the nursery-web spider can reach densities of up to 14 spiders per square metre. The males are therefore in regular chemotactic contact with a great deal of female silk in the days preceding courtship. A mature male is more or less equal in stature to a virgin female.

After sperm induction, his first act is to catch an insect. The victim is usually a fly, often the dance fly *Empis tessellata*. (Ironically enough, often a male is caught while busy searching for a nuptial gift to present to *his* own females; the male flies are more susceptible to predation at this time.) The *P. mirabilis* male treats his capture in a special way which deviates markedly from his normal feeding methods. Prey is usually consumed naked and 'as-caught', but the corpse destined to become a gift is trussed up in a dense wrapping of white silk, forming roughly the shape of a ball. Why the difference? Well, in *P. mirabilis* there is no complex visual or tactile courtship to establish unambiguously the male's identity. Therefore it seems likely that the nuptial gift may serve as a token, a kind of 'badge of identity', its chemical message written in the silken wrapping, which is almost certainly impregnated with a species-specific male pheromone. There is certainly another good reason for the nuptial gift – its potential food value to the donor's mate. Well-fed female spiders grow larger, lay more eggs and produce their egg batches more quickly after mating than females on a poor diet (Kessler, 1971). In *P. mirabilis*, large females survive better and eventually go on to produce more spiderlings, so it is surely in the male's interests to fatten up his mate by providing her with a meal (Austad & Thornhill, 1986). The male wraps the gift by standing over it on tiptoes, pointing his abdomen downwards as he spins a weft of threads. It is common to see several males just sitting around, holding these little silken bundles in their chelicerae, all primed and ready to trot off and commence courtship whenever circumstances (probably weather conditions and time of day) are suitable.

▲ The male *Pisaura mirabilis* (Pisauridae) from Europe is unique among spiders in presenting the female with a nuptial gift, an insect wrapped in silk. An unreceptive female may try to fend off an unwelcome suitor with her front legs; more often, she will simply retreat from him.

While searching for a female, the would-be suitor walks with a strange jerky gait. Stopping in front of his quarry, he assumes a rather bizarre posture. His abdomen is pointed vertically downwards, with its tip resting on the leaf, while his long front legs are directed upwards, although the tibiae and tarsi are bent out to either side in a rather 'limp-wristed' pose. Finally, his large, long-stalked palps are held upwards and outwards, well out of the way of the gift. In such a position the prey, eye-catchingly gift-wrapped in white silk, is prominently presented before the perceptive gaze of the female. It is interesting that the strange arrangement of the male's legs bears an uncanny resemblance to the 'legs-in' submissive posture adopted by many female spiders; it may thus function as an 'assurance' of non-aggressive intentions.

A receptive female creeps towards the male, who tilts his body backwards as she grasps the fly in her chelicerae. Once it is securely in her grip (although it is still linked by a 'security line' to the male's spinnerets), the male ducks beneath her and inserts a palp from one side. Copulation may continue for an hour or more, the female quietly chewing on the fly while the male inserts each palp once or twice. During palpal change-overs he may take a quick chew on the gift, an intrusion which generally passes unrebuked by the female.

That is how things go under 'ideal' conditions, as when a virgin female is lounging on a broad leaf large enough to accommodate both spiders with room to spare. Under natural conditions, the females often sit on small leaves, making it impossible for the male to approach from the front in the classic manner. He is therefore obliged to advance from beneath up over the edge of the leaf and, at first, only his face and the tips of his front legs may be visible. In such a tricky situation, he shows his flexibility by jerking the leaf with his front legs while vigorously waving his palps, in an effort to gain the female's attention.

When confronted by a male, the average *P. mirabilis* female almost always shows signs of extreme nervousness, usually manifested by a reluctance to approach the male and take the gift, or a strong tendency to flee at the slightest hint of threat. In fact, the usual reaction to the arrival of a courting male is to flee (so much for the 'placating-the-predatory female' theory). The male's main problem therefore is not to avoid being eaten, but to prevent the female from bolting before she notices the gift. Even when she does show a flicker of interest, the male may have to jerk the fly enticingly for several minutes before she finally overcomes her 'apprehension' and takes it. In marked contrast to many other spiders, even plump gravid females usually do not lunge at unwelcome suitors but run from them, often with the male in hot pursuit, still brandishing his gift. However, females may mate several times, even consenting to copulation when close to laying eggs, so male persistence is understandable.

Surface waves in fishing spiders

In the *Dolomedes* swamp or fishing spiders, found in marshy situations in both Europe and the USA, the male's tactics are more conventional, mainly comprising waggling movements of the front legs. This probably conveys information to the female via vibrations of the water or plant. The male *D. triton* performs a jerking action, which generates bursts of concentric waves on the water's surface (Bleckmann & Bender, 1987). In duration and frequency, these signals differ markedly from those caused by a variety of other sources, such as potential prey struggling on the water, a predator approaching or a falling leaf. They thus appear to be suitably coded to convey information to the female about the male's identity and intentions, preventing her from fleeing or attacking.

Pisaurids with bridal veils

The simple courtship of *Ancylometes bogotensis* from Colombia is triggered by contact with the female's silk and merely consists of vibrations with the legs and palps as the male approaches the static female (Merrett, 1988). If receptive, she responds by drawing in all her legs, much like other female spiders, from *Xysticus* (Thomisidae) to *Agelena* (Agelenidae). With his front legs quivering, the male now crawls over the female and begins to lay down a series of threads across the tips of her front tibiae, then across her second and third pairs of legs, until her legs are sufficiently enswathed for him to tip her on to one side. After mating, which lasts 10–15 minutes, the male rapidly departs, leaving the swaddled female to break out of her bonds and disencumber her limbs from the clinging silk, a process which may take 2–3 hours.

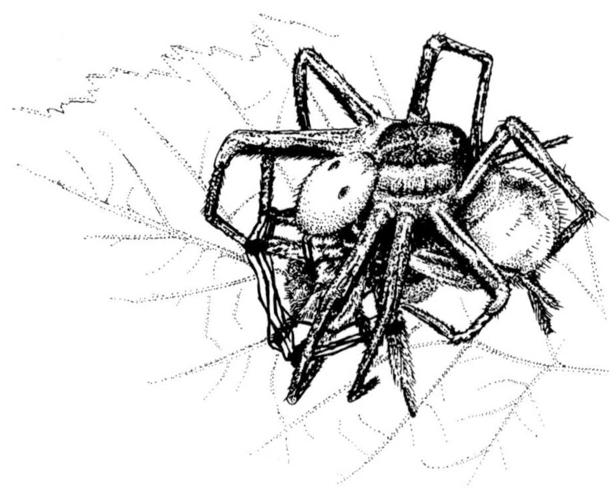

▶ **The South American pisaurid *Ancylometes bogotensis* male (top) ties the female legs with a silken 'bridal veil'.** (After Merrett, 1988).

Sexual interest in the American *Pisaurina mira* is also triggered by contact with a female's silken dragline. This is followed to its source using regular contact with the palpi, much as in members of the Lycosidae (Bruce & Carico, 1988). A receptive female stays put until the male's front legs actually touch her, when she may react with a little harmless sparring. However, her active connivance in what is to follow is exceptionally important. She must be sufficiently 'aroused' to drop from the leaf on a silken dragline, for *P. mira* is one of only a few cursorial spiders known to adopt a mid-air mating position. Once the female is suspended beneath the leaf, the male drops down on a separate thread, following the female's line with one outstretched front leg until he is facing her; they are both now head-downwards. By this juncture the female is in a state of apparent catalepsy, with her front two pairs of legs drawn inwards, suitably positioned to accept the silken bonds which the male now weaves around them. As he spins, he twirls her around several times. Once she is securely trussed, he adds his own dragline to his handiwork, so that the female is now suspended by two threads. Such strengthening is essential, for he now adds his own weight by crawling on to her body in order to mate. After several palpal insertions the female's trance-like state gives signs of evaporating as she begins to move slightly. The male quickly adds some extra silk to her bonds and departs up her dragline, leaving his mate trussed and dangling to unfetter herself alone, a process which takes but a few moments.

In both these pisaurids, the males are nearly as large as the females, have longer legs, are more agile and induce a state of catalepsy before embarking on the close contact necessary during mating. The function of the bridal veil is therefore uncertain, although it may be useful when the female is no longer a virgin and is not 'wholehearted' about her continued co-operation.

Air-borne mating in the green lynx spider

Mid-air mating is also practised by certain Oxyopidae, such as the North American green lynx spider, *Peucetia viridans*, another species with large males well able to look after themselves (Whitcomb *et al.*, 1966). With much leg-raising and palpal drumming the suitor edges closer to his intended mate until he can touch her forelegs. If compliant, she briefly strokes his front legs, then darts away and dives head-first off the edge of the leaf on her dragline. From his position back up on the leaf the male twirls her dragline with his front legs until her underside is facing him, then drops on his own line to hang facing her. He pulls up just short of the end of her abdomen, as this permits him to tap it vigorously with his palps and forelegs. This causes the female to bend into a U-shaped curve, upon which the male lunges bodily at her and, quick as a flash, inserts first one palp and then the other. The complete mating act takes less than $\frac{1}{10}$ second, but it may be repeated several times before the pair climb back up their respective threads to the leaf. Even this may not be the end of the affair, for the females seem amenable to submitting to this whole ritual four or five times in rapid succession, at least until the male is finally 'satisfied' that he has transferred sufficient sperm and wanders off to seek further conquests. By contrast, the female will never take another mate. Nor could she, as her first and only mate has stymied any future rivals by plugging her epigyne with a hard blackish substance, possibly consisting of congealed semen.

▼ This male jumping spider *Evarcha falcata* (Salticidae) from Europe has finally managed to mate after a long and complex visual courtship. Many jumping spiders can gear their courtship styles to whether the female is in the open, as here, or in her silken nest.

Courtship in jumping spiders

The prime exponents of visual courtship are the jumping spiders (Salticidae), which include the sharpest-eyed of all arachnids. Many species, especially in the tropics, are brilliantly coloured, although this is confined mainly to the males, who may be positively jewel-like. The females can also be quite pretty, but their colours may be so different from the male's that they could be mistaken for separate species – and indeed, in the past, they often have been. The males are often decorated with special ornamentations such as tufts of brightly coloured hairs and scales on the legs, palps, face or abdomen; the courtship displays are often specifically tailored to expose these to best effect. These displays are many and varied. Among the North American species, *Habrocestum pulex* conducts a mad whirling dance around the stationary female; *Euophrys monadnock*, like many salticids, shows off certain colourful areas on his legs in a kind of 'dance sequence' whose steps, movements and costume are all species-specific. *Tutelina elegans*, in similar vein, manages to bring several appendages into play. He rings up the curtain by standing a few centimetres in front of the female and windmilling with his front legs, which are adorned with plumes of white hairs. If he manages to get closer, usually after being driven away several times, he stretches out his front legs and palpi before him. Yet another rebuttal seems to stimulate the second part of his display as he launches into a pirouette, breaking off briefly to approach the female and vibrate his iridescent abdomen as he stands on tiptoe before her. If she raises her abdomen, he rushes up to stroke her, persisting in the face of numerous rebuffs until her resistance is broken and he can crawl on to her back and insert a palp. The European *Aelurillus v-insignitus* contrives to draw attention to several prominently coloured parts of his anatomy. He may take up his first stance when as much as 15 cm from the female, which demonstrates the sharp vision in these spiders. Posed in an arched position, stilted up high on his second pair of legs, and with his abdomen bent downwards to touch the ground, he points his front legs skywards, flourishing the yellow femurs. At the same time, he renders conspicuous a light-coloured V against the dark background of his head, along with the pale underside of his drooping abdomen – all characters not present in the drab grey female. Holding this pose for a while, he sets up a vigorous shivering in his front legs and palps, stands up ever higher, then breaks into a jerky advance with both front legs sticking out lance-like before him.

A growing number of jumping spiders are now known to possess a courtship repertoire of such versatility that they can adjust their actions to circumstances, depending on whether the females are out in the open or in the nests where both juvenile and adult females spend part of their time. The American *Phidippus johnsonii* deploys three different display sets, each with several individual signals which, when added to displays given to other males, plus displays of various kinds used by females, gives the remarkable total of around 24 distinct signals which can be executed by this species (Jackson, 1977 & 1978).

Such utilization of either visual or tactile courtship methods, depending on circumstances, is probably widespread in the Salticidae. It has recently been observed in detail in a *Menemerus* species and two species of *Pseudicius* from Kenya; *Mopsus mormon*,

Myrmarachne lupata (Jackson, 1982a) and *Cosmophasis micarioides* from tropical Queensland; and two highly unusual salticids, *Simaetha paetula* and *S. thoracica* (Jackson, 1985), from the same area. These two spiders are aberrant because they build large webs which are used to catch prey, rather than employing the purely visual 'cat-and-mouse' techniques normal in the family. Both inside and outside the females' nest webs, the males employ visual movements which include to-and-fro and zigzag dancing, coupled with clawing and waving actions of the legs. On the nest, this is mostly carried out near the door, being accompanied by clawing of the silk at this point. An unreceptive female responds by pulling the silk inwards so that the male's legs cannot touch her through it, but such delaying tactics often become weaker and eventually the male enters the nest. Inside the tube, he adopts the normal salticid mating posture, on top of the female and facing towards her rear. Outside, on a leaf, he takes up an unusual stance, sitting beside the female and often maintaining contact only with a single palp. This posture (so-called position 5 of spiders) was originally described in the ant-mimicking Australian *Myrmarachne lupata* in which the huge projecting chelicerae of the male are so large that he is more or less forced to sit beside his mate. The practice of cohabiting with juvenile females is also common in this species, as well as in *Simaetha* and several other *Myrmarachne*.

As in many other families of spiders, the salticids utilize sound during sexual communication. The European *Euophrys frontalis* taps a leaf with his front legs, while the American *Phidippus mystaceus* stridulates rather in the manner of a lycosid by drawing plectra on the tibia of the palps across files on the tarsi. This produces a clearly audible trilling which is repeated in an apparently systematic scheme (Edwards, 1981). For sheer volume of sound, these cannot compare with the buzz produced by the Australian *Saitis michaelseni*. This can be heard at a distance of several metres and is so penetrating that it was originally mistaken for the calling of a grasshopper (Gwynne & Dadour, 1985). The courtship arena always seems to be a fallen leaf, and the male first tries to grab a female's attention by brandishing his front legs, which are conspicuously tipped with white. During this introductory phase, both male and female are usually beneath the leaf, but now the male trots around to the upper side, stations himself accurately above the female, stilts high on his legs and strikes up his stridulation. His 'musical instrument' is situated at the junction between the cephalothorax and the abdomen, each of which bears a file on its opposing edge (as found in various Clubionidae, Gnaphosidae, Theridiidae and Hahniidae). He rapidly oscillates his abdomen up and down, producing a loud buzzing sound. This is amplified by the leaf below which, usually being dry and crisp, makes the perfect sounding-board. During the first phase the performer produces short volleys of sound every 7.5 seconds, remaining quiescent during the intervals. Then he shifts up a gear and starts to generate much longer bursts at approximately half-minute intervals, during which he embarks upon brisk bouts of body-waggling. About 5 minutes into this second act, he scuttles back to join the female beneath the leaf, and she is then bombarded with a combination of visual and percussive displays, with the male now also rapping his front legs loudly against the leaf.

Male jumping spiders are noted for the unusual frequency of their contacts with other males and the highly ritualized nature of the 'contests' which take place as a result. Most male spiders try to avoid one another, perhaps after a brief initial scuffle, but salticids often engage in strange face-to-face stand-offs in which one 'contestant' may withdraw without any actual physical contact having been made. Despite the threatening appearance of bared fangs during these contests, injuries are rare or absent and they are ritualistic to an extraordinary degree. Such engagements, which do not necessarily take place on or near a female's nest, may serve to establish a dominance hierarchy among the males. It is particularly common in species which habitually cohabit with immature females in their nests, implying a degree of competition for available females.

Myrmarachne lupata males seem to employ a 'gauge-strength-by-length' kind of assessment contest. The *M. lupata* male is notable for the bizarre length and stoutness of his chelicerae and his amazingly long lance-like fangs. After some initial posturing with the front legs, the two contestants spread their chelicerae, unhinging their huge fangs in an intimidating but harmless gesture. It is the length of the opponent's gaping chelicerae which is to be assessed, as they press against one another face to face. One of the two may also 'bite' by folding his fangs behind his opponent's chelicerae; or they may grip one another's chelicerae (with fangs folded safely away) in a kind of 'cheliceral handshake'. From this, it can be seen that at no time are the fangs deployed in an offensive mode, and one spider usually gives way and retreats after one or more such ritualistic sessions. The losers tend to be the smaller, from which it seems that it is dominance status which is being decided. However, real

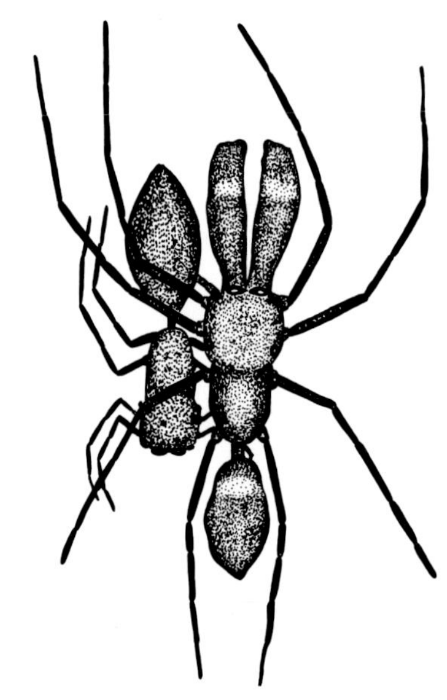

▲ The unusual side-by-side mating position of the Australian salticid *Myrmarachne lupata*; note the male's enormous chelicerae.

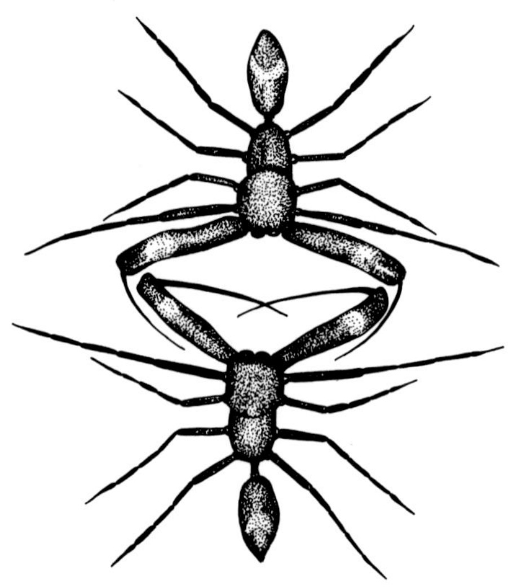

▲ A ritualistic contest between two male *Myrmarachne lupata* salticids from Australia. One male uses his long fangs to 'hold' his opponent's chelicerae. (Both after Jackson, 1982).

fighting in salticids can erupt. In one observed instance of combat in *Myrmarachne foenisex* from Ghana, a large male threw his weight around to evict a smaller rival from his nest (Edmunds, 1978). Real fighting, in which injury can result, seems to be more common between females, as in the Australian *Portia fimbriata* in which fights can lead to loss of a leg or two (Jackson, 1982). In other respects female–female encounters are similar to those of males, but may occur more rarely under natural conditions.

In *Pardosa* males (Lycosidae), the normal tendency to avoid one another (except during brief bouts of mistaken-identity homosexual courtship, which are quickly rebuffed) changes to outright rivalry when a female is present. Then the males assume characteristic threat postures and may even engage in a scuffle. The first to press home the attack usually emerges as victor, and he duly emphasises his newly won status by strutting ostentatiously around with jerky steps. This flaunts a dominance which seems to endure, as the winner in one match will always be victorious in any future encounters with the same opponent.

THE WEB-BUILDERS

The web itself conveys the male's message as he plucks and twangs the silk with his legs and body in a species-specific code. This may be minimal. In the primitive *Pholcus phalangoides*, the male just approaches the female in her web with the faintest of vibrations of his abdomen, creeping up so that eventually he embraces her. Copulation lasts as long as 3 hours, and the female remains fully co-operative and static, despite the meagreness of the male's introduction. As in most spiders, the males are footloose, engaging in brief assignations with as many partners as possible. By contrast, in at least four other pholcids, three species of *Modismus* and one *Blechroscelis* from Colombia, the males cohabit with a female for a considerable period (Eberhard & Briceño, 1983). They compete actively for this privilege, often losing a leg or two in the process, and it is the larger males who usually emerge as victors. A male lodging in a female's web virtually takes it over, becoming dominant in all respects save feeding. In fact, an instinct for chivalrous self-denial seems often to govern the males' actions when a fly blunders in. Although both occupants may deal with this co-operatively, if it is the male who has bitten the fly first and started to feed, he will probably yield it to the female if she performs a distinctive vibration of the abdomen and front pairs of legs, in a style closely resembling the male's courtship. This appears to function as a form of 'begging' (as seen in many female birds), which is most effective in persuading the male to relinquish his meal to the supplicant. Such 'unselfish' behaviour is probably quite common, as cohabiting males are known to feed less than solitary males. This would seem to indicate the regular occurrence of 'chivalrous' behaviour.

Courtship in sheet-web spiders

Many spiders build a more or less horizontal sheet web, often tapering into a funnel in which the occupant resides, dashing out to bite any insects which show up. The need to capture the insect before it contrives to scramble free of the web's maze of trip-lines places a premium on speed of response. When a male (who may be as big as the female) trespasses upon a female's web, he must therefore be sure to announce his identity immediately to avoid 'mistakes'. He does this by tapping or tweaking the silk in a series of coded signals, which vary between species. Thus the male's output in several Amaurobiidae from Europe differs in several ways (Krafft, 1978). *Amaurobius similis* commences courtship when merely in the presence of a conspecific female's web, but shows no interest in webs of the related *A. fenestralis* and *A. ferox*, indicating a pheromone-based detection system capable of taking its cues from the female's silk alone. During courtship, the male drums vigorously against the web with the palps. In *A. similis*, this transmits a signal with a frequency of around 4.1 Hz at around 30-second intervals. In *A. fenestralis*, the male also responds to the silk of his own female, which stimulates the production of two different signals. These consist firstly of rapid vibrations of the palps and legs, emitting a rather random signal, and secondly of vibrations of the abdomen having a mean frequency of 150 Hz. By contrast, *A. ferox* is left unmoved by a female's empty web and requires the presence of the pheromone-producing female herself before he responds with courtship. This is more complex than in the other two species, and involves four kinds of vibratory signals. Palpal drumming produces a mean frequency of 4.6 Hz, instantly followed by abdominal vibrations which produce waves with a mean frequency of 68 Hz, beginning high and ending low. The intervals between these associated signals are filled with single or twin taps of the pedipalps and low-amplitude abdominal quivering, with a mean frequency of 47 Hz. As vibrations of the web caused by an incoming insect are completely random, these stereotyped signals presumably act simultaneously to induce receptivity. Although at least two of the above species are often found living close together in the same habitat, the prevention of accidental hybridization by means of the females' ability to distinguish between 'right' and 'wrong' courtship codes may not be so important as in some other spiders. This is because the males' species-specific reaction to the females' silk ensures that courtship behaviour is only released in the correct context.

In *Agelena labyrinthica*, the male plucks the web with his front legs, only venturing closer across the open web when the occupant signals her compliance by drawing her legs closely up to her sides. He seldom mates at once, being fussy about the exact spot where this should take place. He often grasps one of her legs in his chelicerae and hauls her roughly across her web to a suitable position, usually just inside the mouth of her funnel. He can afford to treat her in this careless manner, for his courtship pluckings have induced a state of catalepsy which is particularly profound, and she is unconscious of his slapdash treatment. She remains in her trance during the 30 minutes or so of palpal insertions, not finally coming round until after he has departed. This is one species in which the author has observed late-season 'kamikaze' tactics by males entering the webs of females only a day or two away from laying their eggs. No longer does the suitor wait cautiously at the edge of the web until his pluckings induce the female to fold her legs. This is highly unlikely to happen at this late date. Instead he advances boldly across the silken mat and rears up face to face with the resident in an aggressive-stanced stand-off. Being little inferior in stature, he may even intimidate her sufficiently to force her back inside her own funnel and will even follow her inside and seek to 'seduce' her within its narrow confines. Such devil-may-care tactics are probably worthwhile, as the males' reckless persistence does indeed sometimes lead to a copulation, even with a female who is within a day or two of laying her eggs. 'Kamikaze' tactics make sense when only a really pushy approach can have any chance of securing a last-minute mating and the males are anyway soon approaching a natural death through old age.

Courtship in the black widows and other theridiids

Most theridiids build rather scrappy-looking scaffold webs which have a mesh of criss-crossing lines. A male communicates his intentions with a series of vibrations created by movements of the legs, palps or abdomen; in the latter case, he also sends stridulations down the silk by rubbing tiny teeth at the abdomen's base against files at the corresponding apex of the cephalothorax. In some species, the male first constructs a special mating thread within the female's web, and courtship is initially designed to entice her on to this thread and adopt an inverted mating position. This is similar to the procedure found in many araneids. Mating may also be interrupted while the male moves to one side to recharge his palps, perhaps up to seven times (e.g. in the very common European *Theridion sisyphium*) before he finally leaves the female. The *Latrodectus* 'widows' derive their soubriquet from their supposed habit of treating the male as a meal shortly after mating. Whether this happens regularly under natural conditions is uncertain, but there are good reasons why it should. The male is said to part with the tip of the palpal embolus during mating; this would render him impotent. If he is incapable of inseminating further females, it would be sensible for him to donate his body to nourish the eggs which he has just fertilized. In the western black widow, *Latrodectus hesperus*, from the USA, the male does not rush away from the female immediately after mating, but strolls off rather casually and then hangs around in the close vicinity 'inviting' the female to finish him off (Ross & Smith, 1979).

Within a few days, his empty husk will be his only remains. This 'negligent' post-mating behaviour is in marked contrast to his first approach, when he vibrates his abdomen, taps the silk rhythmically with his palps and 'throws' fine silk on to the web. Females are usually unreceptive at first and tend to rush at the males, who usually decamp by dropping on a silken line. However, persistence eventually pays off, and the male prosecutes his cause by severing some of her lines at strategic points, cutting off a number of potential routes by which she could evade him. A particularly 'ardent' female (presumably a virgin 'overdue' for mating) will hurry matters along. She does this by encouraging the male with jerky movements of her whole body and vigorous wiggles of her abdomen.

Courtship in the orb-web spiders. The araneids exhibit a broad spectrum of sexual strategies. The males are often smaller – sometimes very much smaller – than their mates. Against this, the males are often longer-legged (in proportion to body size) and less portly, making them more mobile in their quest for female webs and more agile when conducting courtship manouevres. Once he has attained maturity, the male devotes the rest of his life to searching for mates. His only source of food during this drifting existence is prey caught in females' webs, although he may also rob other smaller spiders of prey.

Much has been written about the terrible risks run by these small males when venturing on to a female's web. Under *experimental* conditions nearly half the males of the garden spider, *Araneus diadematus*, were killed by the female either before or after mating. The risk was slightly greater during the courtship phase, when 62 per cent of deaths took place (Elgar & Nash, 1988). However, these results were obtained by placing males in contact with the webs of *adult* virgin females. There is good reason to believe that this situation is so artificial that events do not reflect what usually happens under natural conditions. From the middle of July onwards, males of *A. diadematus* can often be found close to the lair of a subadult juvenile. The consort may remain in close proximity for up to a week, making regular visits to the female and even entering her lair (at this stage he is rather bigger than she is) to check on her reproductive status (pers. obs.) He is therefore on hand just as she moults, when he can copulate with little need for extensive preliminaries.

Under natural conditions, most *virgin* females are probably mated in this way. This means that the drawn-out and hazardous courtship seen later is almost always directed towards a mature gravid female, who has no need of a male and is therefore liable to act aggressively towards a suitor. Males who 'push their luck' in such a negative situation are therefore liable to end up being eaten but continue to take their chances because even gravid females may still yield to a persistent male just before egg-laying.

A. quadratus males adopt a similar strategy but actually move right in with their virginal host. This usually happens after the male has tried his normal courtship routine without success (as in salticids). A sub-adult female is normally slow to react to his courtship pluckings but eventually trots down and attempts to drive away the interloper. After a bout of sparring she retreats to one side, while he heads straight up to her lair and sits chewing his palps. When she eventually returns, she has little choice but to tolerate his presence and they live peacefully together until her final moult, when they mate.

▶ In the European *Araneus quadratus*, the immature female often has little choice but to share her residence with a mature male (right). He is powerful enough to move in with her regardless of her objections and will be on hand to mate with her as soon as she moults to adulthood.

▼ Later in the season most courtship in the garden spider, *Araneus diadematus*, is directed at gravid females, who have little interest in accepting a further mating. Not surprisingly, they give any further suitors a hostile reception and it is only through sheer persistence that a male might possibly win through. It could, however, cost him his life if he consistently refuses to take 'no' for an answer.

The cephalothorax of *Argyrodes* males is often ornamented with two large protruding knobs whose purpose has only recently been established with any certainty. In *A. antipodiana* from New Zealand, the two sexes meet head to head, grasp one another's front legs and push them outwards. The female meanwhile uses her palps to detect a special groove situated between the male's knobs (Whitehouse, 1986). During copulation, her chelicerae (and possibly also her mouthparts) rest snugly in this groove, from which a liquid (presumed to be a male pheromone) is secreted. Even during breaks in copulation, the female often hangs with her palps reposing on the groove. The exact function of this male pheromone is not known, but it probably serves to stimulate the female sexually and keep her in place. As in many other spiders, theridiid males take advantage of newly moulted females who are newly adult. The author has seen a male *Enoplognatha ovata* consorting with a female still hanging, soft and pliant, beneath her recently shed skin. The presence of no fewer than four more males waiting 'in the wings' around the periphery of the web indicated their ability to detect when a juvenile female is about to make her final moult. Such a rush to be in at the start hints that first-male sperm precedence probably operates in this species.

Courtship in the orb-web spiders

Courtship methods within the Araneidae can basically be split into three categories (Robinson & Robinson, 1978a):

Type a. Courtship which takes place on the hub of the female's web. There is a great deal of preliminary contact between the sexes, and this provokes a fairly simple response in the female.

Type b. Courtship which also starts off with a bout of male–female contact at the hub of the web. However, the male subsequently severs the silk in several places, leaving a gap in which he inserts a special mating thread. He now switches to vibratory courtship on this thread, to which he must lure the female and induce her to adopt an acceptance posture which permits him to mate.

Type c. Courtship in which the male does not trespass on the female's web at all but instead constructs a special mating thread at its periphery. He now performs a complex vibratory routine designed to coax her across her web and on to the mating thread, where she adopts a receptive pose.

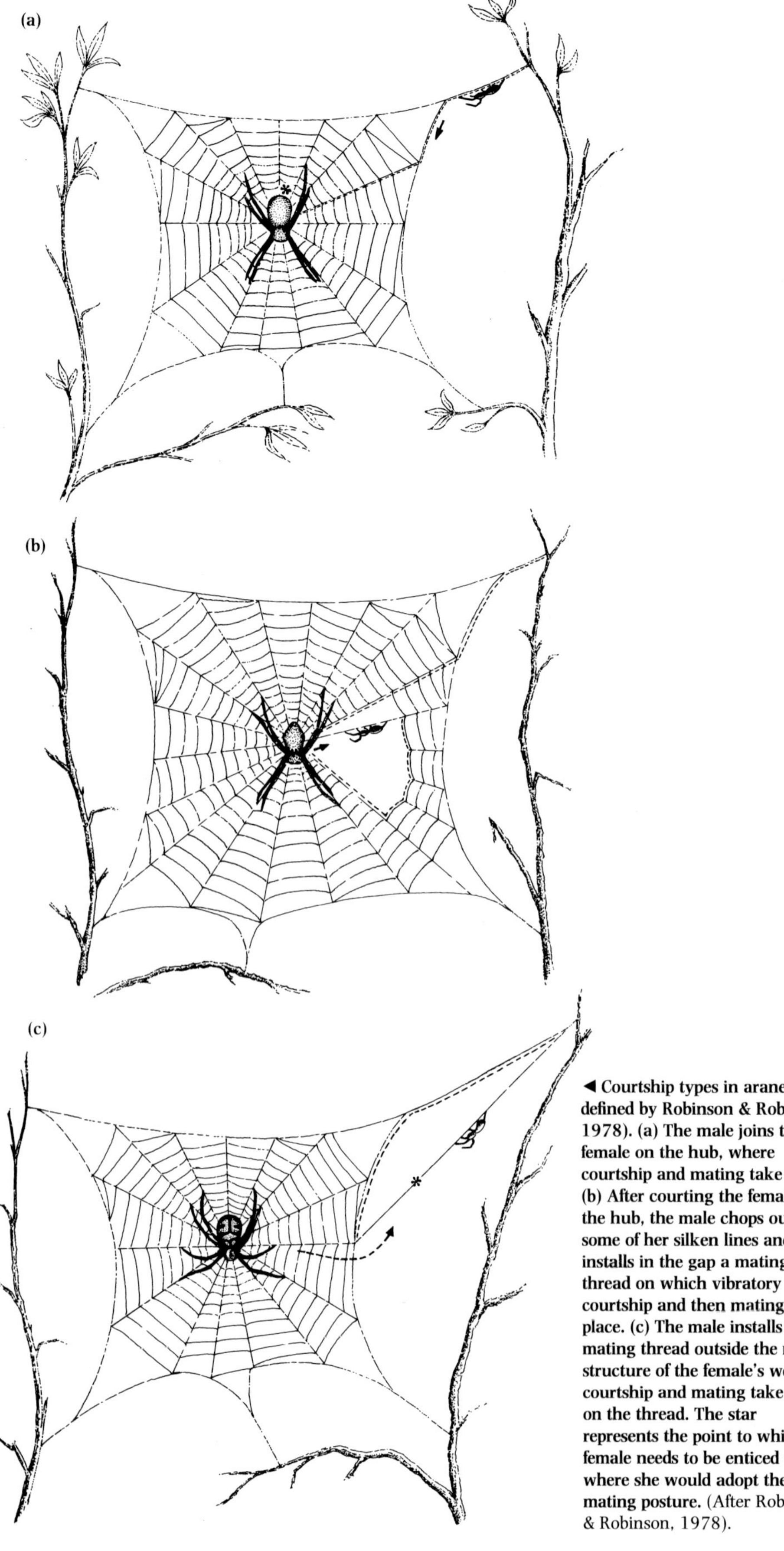

◀ **Courtship types in araneids (as defined by Robinson & Robinson, 1978). (a) The male joins the female on the hub, where courtship and mating take place. (b) After courting the female at the hub, the male chops out some of her silken lines and installs in the gap a mating thread on which vibratory courtship and then mating take place. (c) The male installs a mating thread outside the main structure of the female's web; courtship and mating take place on the thread. The star represents the point to which the female needs to be enticed and where she would adopt the mating posture.** (After Robinson & Robinson, 1978).

The problem for the male garden spider, and many *Gasteracantha, Isoxya* and *Aetrocantha*, is twofold. Firstly, he has to tempt the female to move on to the mating thread, and, secondly, he has to leave her to recharge his palp after the first insertion before he can make a second. In order to perform the latter, he has to repeat the whole courtship process. Often he just does not bother, as about one-third of *A. diadematus* males retire after just one insertion This is not surprising in view of the agonizingly protracted courtship when trying to tempt an unreceptive partner. Dozens of times he will edge forwards, plucking the web, only to be charged by the female, whereupon he swings down to safety on the dragline, which he always attaches to some adjacent vegetation before commencing his approach.

An initially unreceptive female may require an hour or more of this repetitive cat-and-mouse game before she 'changes her mind' and trots on to the mating thread. There she hangs down on her two hindmost pairs of legs, exposing her underside, and hunches her front legs close to her body. The male usually darts 'quick as a flash' towards her, as if inserting a palp, in several so-called 'pseudo-copulations'; these are more often seen in *A. quadratus*. Once the male has finally inserted a single palp, the female returns to her lair while her suitor either makes off or recharges a palp. It is because of his diminutive stature that the *Araneus pallidus* male can only ever insert but a single palp. He has to embark upon copulation by leaping at the female's underside and stabbing one palp into her epigyne. Because he is so small, and has nothing to cling on to while the female is in her head-down receptive pose, he immediately falls backwards, bringing his abdomen directly beneath the female's face. He now has no problem in staying put long enough to transfer a full supply of sperm, for his mate's fangs quickly sink into his body, anchoring him securely in the correct position. She commences to suck him dry, but his task of fertilization has been accomplished. His self-sacrifice seems to be essential if the mating is to be successful. Without the female's fangs as an anchor, he keeps on sliding off her slippery underside and is incapable of keeping his palp in place.

A similar sacrifice is demanded of the male *Cyrtophora cicatrosa* from India, and for the identical reason (Blanke, 1975). The male is a midget and simply cannot bridge the yawning gap between himself (on the

▲ **In *Argiope* (Araneidae), the tiny male often spends several days in the female's web without being harmed; this is the African *A. lobata*.**

web) and the female, as her mating posture angles her well out from the silk. All he can do is to spring at her in the hope of planting home a palp, which will lock him in position. If he misses, he falls down to the bottom of the web and has to clamber back up and have another try. The male is not very good at this, usually needing several (up to 20) attempts, during which the female 'helpfully' remains in the necessary position; she does *not* take advantage of his incompetence and try to bite him prematurely. Yet once he manages to strike home with his palp, she immediately renders essential assistance by holding him in place with her fangs, wrapping him in silk and sucking his juices while he mates. Successful transfer of sperm is therefore only possible with the deliberate but fatal support of the female. Virgin females actively attract males from a distance of a metre or more by emitting a pheromone, thus guaranteeing a positive reception to the initial suitor. A few days after her first mating, the female ceases to attract males and responds aggressively to any tardy arrivals.

Courtship in Argiope

In relation to their males, *Argiope* females are particularly large and bulky, yet show great tolerance toward them, allowing them to lodge within the web and even steal food. Immature females host most of these lodgers, so that the males are on hand for the final moult, but even in a mature female's web the males are largely ignored. Most *Argiope* males differ from *Araneus* in not making a special mating thread. In the European *A. bruennichii*, the male approaches the female at her hub, his legs quivering (type a courtship). Unlike in some *Araneus*, female aggression at this stage is absent and a receptive female signals her acceptance by angling her body away from the web, making room for the male to slide in beneath her. Absorbed in mating, he can do nothing about the silken threads with which the female now shrouds him. Although the alternate insertion of each palp only takes 30 seconds or so, his bonds may still be dense enough to seal his fate, although he may manage to break free and escape.

Such a risky termination of mating is also seen in *Argiope aemula* from New Guinea. There is a lengthy pre-copulatory period as the male walks back and forth over the female, tying her down with a copious cat's-cradle of silk. This appears to stimulate the female (probably because it is impregnated with pheromones) and eventually causes her to arch her body away from the web, allowing the male to creep in (Robinson & Robinson, 1980). During copulation, the female clasps the male lightly, but then pushes him down towards her chelicerae so that he can be packed in silk and eaten. An agile male can spring out of the trap and drop to safety on his silken dragline, usually at the meagre cost of one or two legs. However, as he will normally return to try and insert the second palp, his chances of getting off so lightly next time are much lower. In *A. argentata* from the Americas, the male spider's difficulties are compounded by his ineptitude at aiming correctly when trying to insert his palps, since he frequently misses the epigyne on successive attempts.

Eunuchs as guardians. The minuscule males of *Nephilengys* and *Herennia* also cohabit with their large handsome females for a considerable period. As in many spiders with tiny males, courtship is very simple. In fact, it seems designed initially to persuade the female not to flick her suitor away as an irritation and then to adopt a receptive stance with her body stilted away from the silken cup (often hard up against a trunk) on which she tightly sits. During copulation the male's palps become twisted and useless, so they are bitten off one at a time and hung in the female's web. This self-trimming occurs after each of the sole two copulations (one palp inserted each time) which a male can ever achieve. As he cannot be promiscuous, the eunuch should make every effort to protect his single priceless investment and take up the role of bodyguard in his mate's web. Given the chance, she will accept another mate, so it is in the resident eunuch's interests to stay and intercept any prospective suitors before they can reach the female. With so much at stake, a guard will frequently engage in fierce combat with intruders, although they also often resort to ritualistic web-shaking duels. Another tactic, which benefits from entailing less chance of combat, is to interrupt a mating by jostling the rival just as he finally climbs on to the female, so that he wastes the maximum amount of time and effort.

▲ **During copulation with his first mate, the tiny male *Herennia ornatissima* (Araneidae) from Asia so damages his palps that he has to remove them himself. As a eunuch, he then remains in the female's web (top right) and guards her against competing males.**

▲ It is probably because of his small size that the male of *Nephila clavipes* (Araneidae) is able to walk around over the female and mate without any prior courtship. Males compete fiercely for the 'hub' spot near the female; dominant 'hub males' enjoy the majority of copulations. Note the 'peripheral' male at the top of the picture; he will enjoy few or no copulations unless he can displace the 'hub' male.

Courtship in Nephila

Differences in size between male and female reach a peak in *Nephila* in which a large heavily gravid female can weigh several hundred times as much as an adult male. Taking body length as a measure, females of *N. maculata* average eight times the length of the males. It is this very smallness which may allow the males to spend much of their adult lives in the females' webs. However, the level of toleration shown by the female varies from species to species, as does the degree of courtship. In *Nephila clavipes*, a wide-ranging New World species, courtship is virtually absent. The male is able to walk around on the female and mate almost at will, generally being ignored, regardless of whether or not the female is occupied by feeding. There is considerable male competition for the prime position close to the female at the hub of her web (Christenson & Goist, 1979). The aim is to occupy a spot about 5 cm above the female but on the reverse side of the web. Dominant or 'hub' males will defend this area by fighting one or more low-status 'peripheral' males, who occupy the less favourable outer strands. The peripheral males regularly seek to improve their lot by trying to depose the hub male. The latter reacts by jinking from side to side while violently tweaking the web in a warning to withdraw. If the challenger still presses on, the hub male advances with outstretched front legs and tries to grab his opponent in his chelicerae. The outcome is reached in a mere 1–2 seconds and the loser, usually the smaller of the antagonists, withdraws. The peripheral males do not fight among themselves and an evicted hub male suffers a permanent demotion and cannot win back his former status. Peripheral males often give up and wander off to try their luck elsewhere, perhaps remaining only a single day in any one web, although others might remain for several weeks.

The establishment of such a dominance hierarchy is important for two reasons. Firstly, hub males enjoy an overwhelming advantage in the number of copulations, for mating attempts by peripheral males are either absent or extremely rare. When they do take place, they tend to be quick 'sneak' affairs, too brief to be reproductively rewarding. The hub male also enjoys access to the female's meals, eating rather better than his subordinates by being in a position to poach some of the prey, which the female suspends around the hub. The risks are minimal, for despite their constant close proximity to the females, hub males are killed no more often than peripheral males and all such cases are, in any case, very rare and tend to be in a non-sexual context (Cohn & Christenson, 1987)

In both *Nephila clavipes* and the African *N. senegalensis*, the males' general immunity from attack is probably due to their small size. In *N. senegalensis*, the minimum size of prey which the female bothers to attack is considerably larger than any male (Clausen, 1987). During courtship, the male vibrates his legs and palps before he mounts her abdomen and taps it several times with his palps. This is apparently stimulatory and widespread in the genus. The female usually sits quietly for most of the 90 seconds needed for copulation but she often becomes touchy towards the end. This may galvanize the male into baling out on his silken line or he may escape on to the web above the female, after scaling the heights of her giant abdomen. Mating may be frequent – as many as 11 times in 2 days – indicating a broad measure of tolerance and receptivity on behalf of the female.

The dwarf male of the giant wood spider, *Nephila maculata*, from the Far East, matches his behaviour to circumstances (Robinson & Robinson, 1973 & 1976). As a newcomer, he proceeds very cautiously down towards the hub and its alert resident, instantly dropping on his dragline if she plucks the web in warning. By contrast, a well-established male 'familiar' to the occupant treats her with scant respect, being ignored most of the time. Males are seldom attacked and the lack of any cannibalistic intent is indicated by the way in which the female may flick the male's corpse summarily from her web – he has been killed as an annoyance, to be disposed of, not as an easy meal. Much of the courtship process is devoted to the male laying down a tangle of silken threads around the female's body and between the bases of her legs. He takes a break from his weaving sessions by sitting motionless on the female's abdomen and it is often several days before everything is ready and mating can finally commence. As in other spiders, the purpose of the female's 'bonds' is uncertain but as the silk is probably soaked with male pheromones, it probably induces a state of receptivity, once the quantity is sufficient. This would explain why the tiny male takes several days to lay down a large enough amount, during which time he seems virtually immune from attack.

The trigger for the commencement of courtship in the Australian *N. edulis* is the arrival of an insect in the female's web (Austin & Anderson, 1978). The males are relatively larger than those of *N. clavipes* and *N. senegalensis*, so are potentially at greater risk of being mistaken for prey. It therefore makes sense to mate while the female is already preoccupied with a meal. If instead of feeding immediately the female stashes the insect away in her food store, the advancing suitor aborts his approach and retreats. If she starts feeding at once he keeps on coming, but with extreme caution, often plucking at the web with his legs, as if testing the strength of her interest in her meal. She may respond to this by turning towards him threateningly, or she may continue to be engrossed in her meal. If the latter, he crawls on to her and inserts a palp. The male's wariness is well advised, for the females often abandon their prey and turn on the advancing suitor. His alertness and agility usually save him but, if he is caught, he will be treated as prey and eaten. It appears that, in this species, the

The tetragnathid fang-lock. Spiders of the genus *Tetragnatha* are long and slim, which earns them the name of 'grass spiders'. On warm windless evenings in late May or early June, conditions are right for the male *T. extensa* to set out in search of females' webs. He always arrives unannounced, rushing straight at the occupant with no hesitation whatever. He can afford to be so precipitate as he can rely on the female to let him know her 'mood'. If unreceptive, she instantly warns him off by tugging violently at the silk with her two front legs (pers. obs.). Only if he fails to take heed and still approaches will she lunge at him open-jawed, more in further warning than in genuine attack. A receptive female responds with instant co-operation. Despite the apparent absence of courtship, she seems able to recognize him and his purpose immediately. Turning to meet him, she opens her chelicerae wide, not in threat but in invitation, for the male hurries confidently forwards and grips the proffered chelicerae with two spurs on the leading edge of his own. Sliding his fangs neatly around the bases of the female's, he locks them in position behind two prongs situated on the rear of his chelicerae. These 'fang-clamps' are specifically designed for this purpose and are only found in the males. As they connect – and it takes several seconds for everything to slip correctly into place – the spiders strain against one another, their front legs pointing straight out to either side. They then descend on draglines until they are suspended vertically, before the male inserts a palp. When the male has finished, he instantly snaps free and drops to the ground as if in danger, although, far from ever showing any signs of hostility, the female has 'eagerly' co-operated right from the start.

◀ In the grass spider *Tetragnatha extensa* (Tetragnathidae), the female 'willingly' presents her gaping fangs to the male to be grasped in a special 'fang-lock' during copulation.

▼ The male *Meta segmentata* (Araneidae) from Europe is far from being a 'hen-pecked husband'. After taking up residence in a female's web he is quite capable of 'bossing' the rightful owner around. This male (bottom right) was able to deal capably with two females simultaneously. He courted the web's owner (bottom left) across the corpse of a crane fly (*Tipula* sp.), while also taking time out to chase away the intruding female (top right), who was trying to creep in and snatch a quick bite from the fly. He was completely dominant over *both* females.

males have lost the capacity to 'advise' the females of their identity and intention, either through courtship or via pheromones; hence the reliance on a dead insect to keep the female occupied.

Courtship in Meta

The arrival of a fly in the female's web is also the trigger for courtship in *Meta segmentata* (often placed in the Metidae), a very common European species. The male, with his long spiny front legs and approximately equal size, is more than a match for the web's owner and he may even become dominant over her in most matters during his lengthy uninvited sojourn. However, he will often remain immobile as she rushes in and bites an insect but, as she starts to wrap her prey in silk, he edges near, softly plucking at the silken lines and twitching his abdomen. Finally, he is hanging face to face with the female but separated from her by the corpse of the insect, around which he reaches with his long front legs to touch her. Although she will often fend him off at first, she soon gives ground and yields her meal rather than respond to his overtures. The male must now re-kindle her interest, so he jerks the fly with his chelicerae as if it is struggling to escape. This usually works well, enticing the female back, only to end quickly with more sparring. This may be repeated many times before the female finally allows her suitor to creep past the trussed-up insect. With trembling legs, he walks over her, paying out behind him a thread which he attaches to a web-radius a little further on. Now he crosses to and fro, adding more silk to produce a mating thread, on to which he must tempt her with the usual vibrations. He may even start to twine a few strands around the female, possibly as a pheromone-carrier, before he finally mates. There is a regular change-over of would-be suitors, although a resident cohabiting with a virgin will savagely defend his position against invading rivals; the loser may pay with his life and be eaten by the victor.

Courtship in the Linyphiidae

Linyphiids are mostly very small spiders. One of the larger European species, *Linyphia triangularis*, often occurs in vast numbers in late summer, when a male can be seen to have taken up residence in the web of nearly every female, usually a day or two before her final moult. He quickly assumes the dominant role in her web, claiming the best of any food. Sometimes two or three males may take up 'guarding' positions simultaneously. This usually sparks off fights which establish one male as dominant; the losers are relegated to the web periphery. Courtship is generally typical for the family, with a complex mixture of web vibrations produced by varied movements of the legs. This usually takes place immediately after the female's final moult, when she is still pale and soft; hence the struggle between the males to be chief consort. Stridulation also probably plays an important part in courtship and is generated by rubbing pegs on the insides of the palpal base against files on the opposed outer surfaces of the chelicerae. Linyphiid males are unusual in not performing sperm induction until they have ascertained the female's state of receptivity by performing a bout of 'pre-courtship' and 'pseudo-copulation' in the female's web.

In *L. tenuipalpis*, a species which may occur with *L. triangularis*, the start of the 'serious' stage of courtship (during which the male starts to construct his sperm web) often acts as the 'call to arms' for a subordinate male whom the dominant guarder has previously relegated to the edge of the web. This is the moment for the subordinate to seize his last chance to ruin his oppressor's sexual prospects by rushing in and interrupting the smooth running of the mating sequence (Toft, S., 1989). Such 'final-call' spoiling tactics may postpone the dominant male's mating act by several hours. Even better, it may provoke him into deserting the female after a lengthy and now futile courtship, leaving her to be claimed by the victorious subordinate. 'Spoiling' is obviously a potentially rewarding alternative mating tactic for this species and, to a lesser extent, for *L. triangularis*. Spoiling behaviour appears to be finely tuned, judging by the waiting subordinate's acutely perceptive response. He instantly detects (and then reacts appropriately to) his rival's switch from pseudo-copulatory exploratory tactics to the real thing. Thus, although large males always enjoy a distinct mating advantage, this can be reduced by the successful use of spoiling tactics by their smaller rivals. This seems to be a potentially more profitable strategy than the one employed by subordinate *Frontinella* males (see next example).

The bowl-and-doily spider

The preliminary courtship in the North American bowl-and-doily spider, *Frontinella pyramitela*, is probably to ascertain whether or not the female is a virgin. Only if the male receives a positive response from the female does he proceed to charge his palps by building a tiny sperm web within the female's web structure; this is followed by the 'proper' courtship (Suter, 1990). As in *L. triangularis*, a male will hang around in a female's web for considerable periods, during which time any invading rivals are enjoined in battle, which may lead to the injury or death of one combatant. A losing male who has already managed to mate will quickly quit the scene. If, however, he has yet to couple with the female, he will loiter on the edge of her web until the winner has finished mating and then copulate with her himself. Given the order of sperm precedence in linyphiid spiders (first male's sperm takes priority), being second in the queue is less gainful than being first in line and vastly fewer eggs are fertilized; so being a winner in a fight yields a direct reproductive pay-off (Austad, 1983). A male *F. pyramitela* will begin courtship on mere contact with the female's silk, regardless of whether or not she is present. This behaviour is species-specific, indicating the presence of a characteristic female pheromone on the silk (Suter & Renkes, 1982).

In a concentrated form, this may serve to attract males from a long distance for mating, as seems to be the case in the Sierra dome spider, *Linyphia litigiosa*, from North America. A virgin female who is long overdue for mating seems to advertise her spinsterhood and summon a mate by saturating her web in an attractant. The first male to reach her reacts by folding up the whole web into a compact lump, which has greatly reduced evaporative qualities. By so doing, he virtually snuffs out the source of the attractant pheromone, reducing the chances of any more males showing up to compete for the highly compliant virgin (Watson, 1986). That the male sets to work bundling up her web before he mates with the occupant indicates the importance of cutting out the competition. For the female, such male-inspired vandalism is less positive, for she loses her web and will have to commit considerable resources to rebuilding it. Unlike other spider families, linyphiids cannot recycle any of the nutrients in a web by eating it. However, this situation probably seldom arises in the dense concentrations typical of this species. Only abnormally isolated females in unfavourable low-density sites would be driven to advertising themselves so destructively.

Linyphiid heads

The heads of many male linyphiids are remarkable for their bizarre ornamentations, often rounded knobs above the eyes associated with grooves on either side. As with many secondary sexual characters, these seem to serve a special function during sexual encounters. In the European *Hypomma bituberculatum*, the female greets the male's courtship overtures with bared fangs as she launches herself at him with apparently fatal intent. However, instead of dropping from the web to safety, the male reacts by clasping his legs to his sides. Her chelicerae then slide neatly around his head-lobes, allowing her fangs to fit snugly in the associated lateral grooves. Several bouts of courtship and sperm induction may follow, and each time the female responds in the correct manner by grabbing the male's protrusions. In some species, the area just beyond the groove is furnished with a carpet of very fine hairs covered with a secretion. The female's fangs probably contact this secretion during mating and she may even imbibe some, thus assuring the maintenance of her acquiescence and immobility during mating (Meijer, 1977). In some linyphiids, the female secretes saliva on to the male's head and then re-absorbs the liquid (presumably now laced with dissolved male pheromones) as mating proceeds.

The European *Baryphyma pratense* male perhaps goes one step further by actually providing a blood-meal which may be akin to the nuptial contributions made by some male insects (e.g. certain tettigoniids, gryllids, cockroaches). Such a gift of body fluids may represent a form of paternal investment by helping to nourish the female's developing eggs (Blest, 1987). As the female clasps these cephalic grooves, they ooze blood which she immediately consumes. The males seem to be unharmed by the loss of blood and prove that they can easily cope with this level of sacrifice of their body fluids by mating again within 24 hours. Body parts as nuptial gifts are common in insects, but such an occurrence in *B. pratense*, if conclusively proven, would be a 'first' among spiders.

EPHEMEROPTERA Mayflies

Mating takes place in swarms in which each male soars upwards and then slowly flutters down, looking for a female above him with the large ommatidia in the upper region of the eyes, which can render a particularly sharp image. This specifically modified region may even form a separate eye. As a female flies into a swarm, the quickest male instantly grabs her from below. He embraces her thorax with his very elongated front legs, whose tarsal claws make a close fit in made-to-measure recesses on the female. They consummate their union as they drift earthwards, the male's twin penes operating simultaneously inside the female's paired genital apertures in a process which takes a mere few seconds.

▼ Whilst guarding his territory, this male *Trithemis kirbyi* darter dragonfly from India has adopted the 'obelisk' pose. By pointing his abdomen directly towards the sun he is able to minimize heat absorption.

ODONATA Damselflies, dragonflies

The rendezvous and territoriality

These relatively large insects are ideal subjects for behavioural studies. Generally, it is the mature males who arrive at the water both earlier in the season and earlier in the day than the females, thereby being on hand to intercept them when they arrive. Males are often highly territorial, defending a specific resource, usually an oviposition site favoured by the females (resource-defence polygyny). Other species act in a 'casually' territorial way, partitioning up sections of a river or pond on a day-to-day or even hour-by-hour basis. Individual territory size becomes compressed when large numbers of males crowd around the water. Most territorial damselflies and many of the smaller dragonflies monitor their territories from a strategically placed perch on a leaf, stick or rock, the 'monitoring post'. The 'owner' sallies forth to intercept and repel intruding males of his own species, while often ignoring 'foreign' males. Most large hawkers (Aeshnidae) and many Corduliidae mount a continuous aerial patrol over their territories. Most territorial species adopt a dual strategy and males who have (often temporarily) failed to establish territories act as 'wanderers' or 'satellites'. These are making the best of a bad job by 'picking up the sexual crumbs' from the territory-owners, either by 'sneaking' matings within their territories or through trying to latch on to females moving between territories. Many odonates are non-territorial and the males compete for females more or less equally in a sexual free-for-all (scramble-competition polygyny). Under such competitive conditions, dual strategies may help to maximize reproductive success. Even so, because males always greatly outnumber females beside the water, the cut-throat nature of the contest usually ensures that only a proportion (perhaps 50 per cent or less) of the males will ever get to mate; by contrast, all the females will be inseminated, often several times.

Copulation – releasers and mechanics

The releasers for male sexual behaviour vary between species and involve factors such as distinctive flight modes by the female and characteristic colours and markings on her abdomen and wings, plus their shape and size. Despite this suite of clear-cut visual clues, the males often tenaciously chase, and even try to mate with, females of the wrong species. This happens most often when female colour in co-occurring species is similar but even quite divergently patterned females may prove to be remarkably irresistible to a determined male. In such cases, mating seldom proceeds to its climax, because species-specific structures on the male's abdominal claspers can be detected by the female once she has been caught. If she picks up the the wrong tactile stimulus from the claspers, the female usually fails to lift her abdomen to connect with the male (as described on page 36).

Unlike all other insects (but like spiders and millipedes), the dragonfly male cannot transfer sperm directly to the female from the genitalia situated near the tip of his abdomen (on the ninth segment). Before mating, he must first transfer a supply of semen into a set of accessory genitalia situated at the base of his abdomen (on the second segment). The point at which this 'sperm-charging' takes place varies from species to species.

The full sequence of events in most damselflies (Zygoptera) is basically as follows. After landing on a female's thorax, the male bends his abdomen forwards and upwards so as to grasp the apex of the thorax in his paired anal claspers, then lets go with his legs so that he is now projecting ahead of the female, connected only by his claspers. They are now in the *tandem position*.

The procedure in dragonflies (Anisoptera) is similar, save that the anal claspers grip the female behind the head, rather than on the front of the thorax. Next the male bends his abdomen tip forwards, with the female attached, and briefly charges his accessory genitalia with sperm. In some species, sperm-charging takes place before the male first grasps a female (e.g. in *Erythromma viridulum, Hetaerina americana* and many large dragonflies) or when *perched* in the tandem position (e.g. probably in most Calopterygidae). The process takes 1–27 seconds, depending on species. The female must make the next move, this being driven to a large extent by the nature of her 'capture' by the male's claspers. Bending her abdomen forwards and upwards, she quests around with its tip until it connects with the male's accessory genitalia. They are now in the *wheel position*, during which the male's abdomen can be seen flexing regularly up and down in a *pumping* action. This is not, as had long been thought, devoted to the transfer of sperm. Rather, it is designed to remove as much sperm as possible (derived from previous matings) from the female's genital receptacle. The penis in many (perhaps most) odonates bears structures capable of scooping out a competitor's semen. This operation, which is 88–100 per cent effective in *Calopteryx maculata* (Waage, 1979) – and probably in many other damselflies – may occupy more than 90 per cent of the total time spent *in copula*. Only the short remaining period is devoted to the actual transfer of fresh sperm. Some male dragonflies are known to have inflatable 'bulbs' on the penis which push a rival's sperm to the far reaches of the female's spermathecae where it will have little chance of being used, allowing the copulating male's sperm to be 'first out' near the exit.

Once in the 'wheel' position, a male dragonfly can exert a considerable amount of control over his mate. His claspers are powerful and may even bear sharp tips, sometimes capable of inflicting wounds in the back of the head or on the eyes of a female who struggles too much; such wounds are noticeably common in the Gomphidae (Dunkle, 1991). The male also appears to have some sanction over the point at which a female breaks the wheel position. In any species in which the bulk of mating time is devoted to sperm removal, any premature uncoupling will mean that no new sperm will have been inserted. Females who show signs of restlessness are therefore given instant 'punishment', being slapped smartly around the sides of the head with the male's hind wings. He is able to admonish her thus by bending his hind wings backwards through 90°. Several such slaps in quick succession are often sufficient to cut short a female's struggles. Lone females may be able to ward off the unwelcome approaches of persistent males by delivering a clout with their first pair of legs; these can be directed upwards to kick out strongly above the top of the thorax. Such defensive tactics are also used by ovipositing females and by copulating females who are assaulted by gatecrashers.

Post-copulatory male strategies

Once a fresh supply of sperm has been transferred and the genitalia uncoupled, the male can adopt one or more of three different strategies. He can

(1) Desert the female and leave her to oviposit alone. This runs the risk that the egg-laying female will be spotted by

▲ The male *Calopteryx splendens* damselfly from Europe rapidly flaps his banded wings as he lifts the female into the 'wheel' position; this may possibly help stimulate her into responding positively.

◀ As a preliminary to assuming the tandem position, the male damselfly lands on the female's thorax and grips its front edge with his anal claspers. This *Coenagrion mercuriale* male from Europe is mistakenly (but only briefly) grabbing another male of his own species. Such 'homosexual' mating attempts are rather rare in odonates, due to the different colours of male and female, but common in some other insects.

▲ If a female dragonfly tries to uncouple prematurely she can be 'disciplined' by her mate; he slaps her smartly around the sides of her head by bending his hind wings backwards through 90°. This is the ruddy darter, *Sympetrum sanguineum*, from Europe.

another male and re-mated, with subsequent loss of the first male's sperm via scooping. 'Deserting' as a strategy is probably only sensible in low-density species, in which the possibility of a second male finding the ovipositing female is low, or in species whose females are very cryptic while egg-laying (e.g. many *Aeshna* and *Ischnura*).

(2) Release her so that she flies alone to the ovipositing area (usually close by). He then hovers or perches nearby and acts as a sentry, intercepting intruding males who try to mate with her. This is called 'non-contact-guarding'. For dragonflies it is less energy-costly than the next strategy but suffers from the increased risk that imperfect guarding could lead to losing a mate to another male, before she has deposited all her eggs. It has the advantage that the male may be able to mate with a second female while on guard; also, he can still maintain a watching brief over his territory.

(3) Stay with her and engage in assisted or 'tandem oviposition'. This is called 'contact-guarding'. Dragonfly males do much of the work during aerial tandem-dipping, making this the most expensive strategy in terms of energy. It is also costly in the amount of time spent with the female, during which all other mating attempts have to be forsaken. Its great advantage is that it almost always keeps the female free from interference until she has laid all her eggs. A disadvantage for territorial species is the need to leave the 'monitoring post', with the result that the territory may be lost to a rival. However, for damselfly species in which the probability of securing more than one mating per day is low (either because receptive females are scarce due to the time taken to mature each batch of eggs or because females are widely scattered in freely available oviposition sites), then contact-guarding is the only sensible tactic. Unlike in dragonflies, contact-guarding in damselflies consumes little energy; the male just sticks vertically up above his mate as she oviposits, attached by his anal claspers. His mere physical presence tends to prevent interference from other males. By contrast, non-contact-guarding in damselflies often embroils the male in costly physical chases and fights with intruding males intent on mate theft.

▼ In most damselflies, the male remains connected to his mate after copulation, as in these large red damselflies, *Pyrrhosoma nymphula*, from Europe, and contact-guards her as she lays her eggs.

ZYGOPTERA Damselflies

Mating strategies in territorial species

In most damselflies courtship is absent and the male technique of 'pounce and grab' is almost universal. However, courtship displays are found in some species, most notably in the Calopterygidae. Many damselflies are territorial. Males of the American *Calopteryx maculata* defend oviposition sites in turbulent streams (Waage, 1973), practising resource-defence polygyny. An 'owner' may manage to hold on to a territory for 1–8 days. A female generates the owner's 'cross display', during which he points his abdomen skywards while curving up its end to flaunt the pale undersides of segments 8–10. Simultaneously, he bends his hind wings forwards but leaves his forewings in their normal position folded over his back. If already perched on an oviposition site (e.g. a partly submerged tree root) he stays put for his display. Otherwise he heads out to meet the female and escorts her back to the oviposition site before giving the display. Courtship consists of an aerial 'dance' before the perched female, the male's wing-beats being greatly accelerated and the degree of flexion of the wings reduced so that they appear as a constant blur. The female may communicate rejection by using the 'wing-clap'. This involves snapping open the wings while angling the abdomen upwards. Similar 'keep-off' displays are found widely in damselflies and are utilized by both sexes. They vary from family to family, differing mainly in whether the abdomen tip is curled upwards (Platycnemidae, Calopterygidae, *Ischnura elegans*) or downwards (Coenagrionidae). In the latter family, there is also a brief accompanying vibration of the wings. As a threat display, the wing-clap can be very effective in encouraging an intruder to back off without physical contact; it has been found to be 100 per cent effective in *Ischnura verticalis*.

In *C. maculata*, sperm-charging occurs while the pair is perched in tandem. Copulation is relatively brief, averaging less than 2 minutes. After splitting up, the male flies back to the oviposition site, followed by the female a minute or so later, whereupon he repeats the cross display. As she lays her eggs, the female remains in full view, while her mate stands nearby, expelling invading males. Expulsion is not accompanied by the long flight common in territorial encounters but with a truncated chase aimed at doing its job while not leaving the female unattended for too long. The 'guarding dividend' can be considerable; in one study, guarded females averaged 12–15 minutes of undisturbed egg-laying, while unguarded individuals were discovered and interrupted by cruising males after only 1–2 minutes (Waage, 1979a). Females find certain oviposition sites especially attractive, giving the 'owners' the status of 'super-studs' who mate with a succession of females. Although this means temporarily neglecting any egg-laying females in order to mate with a newcomer, the potential benefits are great and a 'super-stud' may eventually end up presiding over a whole string of ovipositing females (Alcock, 1979). He seems able to recognize former mates and allows them to oviposit in his private domain, while quickly picking out and expelling any 'cheats' who try to lay their eggs without 'paying the entry fee' by first mating. Any non-contact-guarder is always faced with this problem once he has left his mate, but there appears to be some fairly reliable mechanism by which males are able to identify a former mate, even after an interval of several minutes (see also *C. splendens* below). As a territory-holder ages he becomes less able to expend the energy needed to maintain his position against his younger adversaries. Now he may become a 'sneak', surreptitiously attempting to hijack females from within the territories of his youthful rivals (Forsyth & Montgomerie, 1987).

'Sneaking' behaviour is also found in other damselflies and in dragonflies such as *Nannophya pygmaea*, as well as in other insect groups such as bees. Sneaking is generally a 'make-the-best-of-a-bad job' response to deteriorating personal circum-

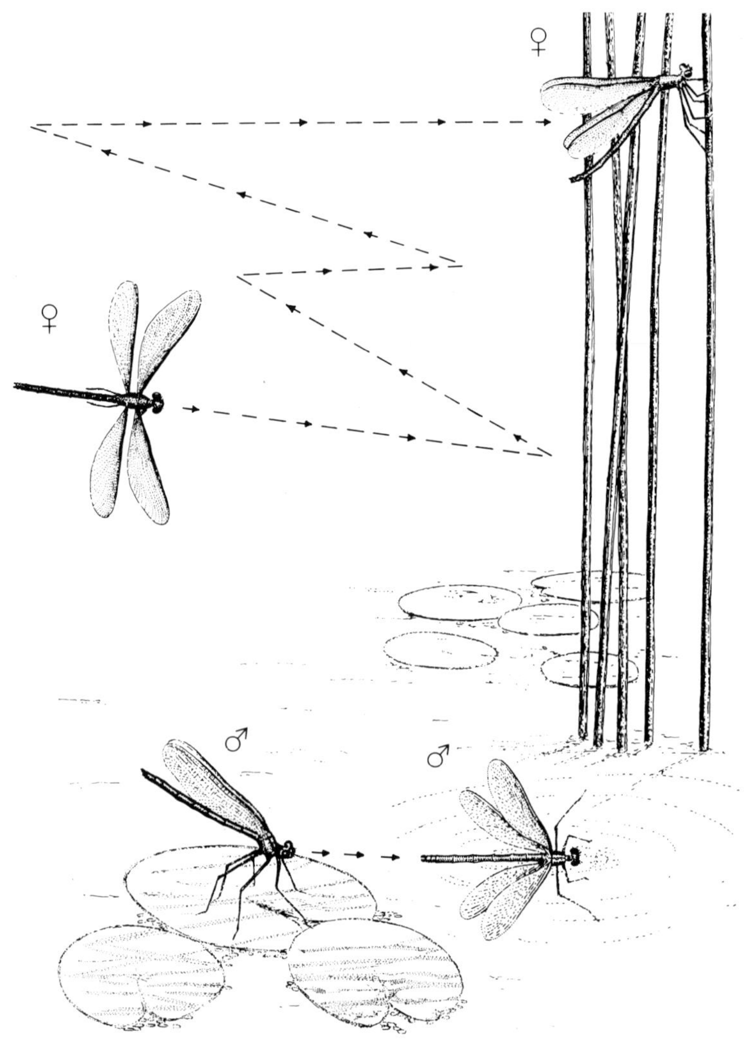

► Part of the complex display repertoire of the European damselfly, *Calopteryx splendens*. Upon seeing a female entering his territory (and she is probably using a 'Lolita'-style soliciting flight), the male jumps off his lily leaf and into the water in his 'belly-flop' display. This 'invites' the female to mate and also indicates the spot chosen by the male as being a high-quality oviposition site. The female 'accepts' his invitation by performing a zigzag flight above him; this also enables her to memorize the spot on which the male is displaying and to which she should return to oviposit. She flies to a nearby rush stem and adopts an acceptance posture. The male performs a brief hovering courtship beside the female before actually mating. Should she delay in following him back to his chosen oviposition site, he will go and fetch her, indicating the way back with a characteristic wing-whirring flight low over the water; he will also float on the proposed spot and waggle his tail to encourage her along. As she gradually submerges to lay her eggs, he perches nearby and defends her fiercely against intruders. If all else fails, he will perch on her back, thus denying her to his rivals until she is safely submerged. This is a highly coveted territory, with its lily pads sheltered from the wind by the nearby clump of rushes. The 'owner' of such a property is virtually guaranteed 'super-stud' status. He will have been obliged to fight long and hard to win ownership of such a prized location. During aerial dogfights which may last two to three hours, males engage in protracted helter-skelter flights and 'height battles' during which one male may dive-bomb his opponent, forcing him into the water. Under conditions of severe overcrowding the whole territorial system breaks down (perhaps only temporarily) to be replaced by a free-for-all system of scramble-competition, which entails severe drawbacks for the females.

stances, although it probably never reaps anything like the same rewards as the high-cost, high-yield strategy of territory-holding. The European counterpart of *C. maculata* is *C. virgo*, which has very similar male displays allied to above-water oviposition by the females. A number of species, such as *C. dimidiata* in the USA and *C. splendens* in Europe, also employ male displays but go in for submerged oviposition.

Other wing displays

Courtship involving wing displays is not restricted to the Calopterygidae; it is also found in another damselfly family, the Protoneuridae. In the Indian species *Chloroneura quadrimaculata*, males flaunt dark sepia-toned spots on the wings in a courtship flight beside the female. During their tandem flight, approaching males will be warned off by the female, using the 'abdominal-curving' rejection display. The males are contact-guarders, attached in an upright posture to the ovipositing female by the anal claspers. This prevents take-overs, but may also be of more direct utility to the female herself. During oviposition, the either partly or wholly submerged females of many damselfly species often become quite waterlogged, and it is common to see a male hoisting a bedraggled female out of the water, possibly saving her life. Such apparently unselfish behaviour benefits the male as well, for she may well have more eggs (fertilized by his sperm) still to lay, and he also retains the chance of mating with her again. Contact-guarding males may exhibit a remarkable degree of behavioural flexibility when 'rescuing' females which have been grabbed and pulled under by aquatic predators. If a spell of tugging mightily with whirring wings does not pull her free, the male may relinquish the grip with his anal claspers and attempt to lift her bodily with his legs. This is not a misdirected attempt at copulation, as is evident from his failure to try to re-establish a hold with his anal claspers.

Leg displays

In some damselfly families, the males display brightly coloured parts of the legs. These are normally broader than in the female, thus increasing the visual effect. In the European white-legged damselfly, *Platycnemis pennipes* (Platycnemidae), the white legs are drooped down in front of the female during a courtship flight. In *Platycypha caligata* (Chlorocyphidae), an African streamside species, the males execute one of the most dramatic of displays. The upper surface of the male's robust abdomen is mostly an eye-catching bright blue; while the rather oar-like tibiae are a combination of red and white (Robertson, 1982). The female is black and brown, with unadorned legs. The males maintain territories based on high-quality oviposition sites such as tree roots and snagged logs, only holding sway for a maximum of 3 consecutive days before being deposed. Owners regularly inspect and sometimes 'tread' their oviposition sites, presumably to check the quality of the wood, which must be in the barkless state preferred by the females. This is similar to the regular morning aerial inspections of submerged water plants by male *Calopteryx splendens* and, when males defend an oviposition resource, they probably always check its condition on a regular basis.

Non-territorial males constantly trespass in an attempt to evaluate potential oviposition sites. If such an interloper lands, he is quickly confronted by the resident, who briefly flashes the white areas on his legs. If the interloper flies up, the resident reacts with a bout of abdomen-waggling, accompanied by a flourish of the red marks on his legs, which are angled tightly upwards to expose the coloured areas. A 'no-touching' contest may then ensue, in which waggling by both males is combined with varied but stereotyped flight patterns. This may go on for several hours before one of the contestants gives way and retreats.

When a female flies in, the territory-owner briefly exposes his white tibiae, flies around her and then tempts her towards his oviposition site with a low-speed flight, during which he droops and waves his abdomen to make the most of its blue upper side. Once the female has landed, the male launches into his spectacular courtship, brandishing his outstretched legs and showing off the white tibiae in a vibrating blur during an aerial display beside her. A receptive female flies to a nearby stem and they mate, after which the male leads her in tandem back to the oviposition site. He briefly courts her again until she begins to oviposit and then deserts her, showing no specific guarding behaviour other than the normal reaction to intruding males. Such apparent carelessness is surprising, given that females are willing to mate several times a day (up to five times). However, most matings seem to be with territory-owners, who are effective in guarding their general air-space, so presumably chances of hijack are rare.

Mating in *Hetaerina vulnerata*

Similar mating systems, in which the males select and then defend a likely oviposition site, are common in odonates. However, in at least one species, the American calopterygid *Hetaerina vulnerata*, the male is willing to accompany his mate in tandem out of his territory, in a flight directed by the female to locate an oviposition site which meets with her approval (Alcock, 1982). The need for such an expedition is somewhat puzzling, as the males defend a carefully selected territory overlooking a fast-flowing stretch of a stream – just the kind of spot normally favoured by females. Yet females seem to be dissatisfied with the resident's choice in at least one third of matings, hence the out-of-territory tandem quest. Many good potential sites are inaccessible due to interference from the local territory-owners, but if the female signals interest by 'slamming on the air-brakes' the male will quickly respond by landing and relinquishing his hold, enabling his mate to back down into the water for a quick investigation. Even though the females undertake submerged oviposition, the males still guard them. When this takes place within his 'own' territory, the male is keen in his defence of both mate and air-space. By contrast, guarding within a 'foreign' territory is a deliberately low-key affair, designed not to attract any hassle from the rightful owner. Mate-guarding is probably necessary to ensure re-possession of a female should she 'change her mind' about the suitability of a locality and re-emerge to try another – a common happening. In *Hetaerina americana*, males have often been seen forming hybrid pairs with *Calopteryx maculata* females, and proceedings may go as far as actual attempts at oviposition. *H. americana* males seem to be regular 'offenders' in this way, and up to 1.25 per cent of all pairings can be of this type. Such cross-species pairings are also found in other odonates, although it is surprising that the female ever allows matters to proceed very far. She must be receiving all the wrong messages – both visual and tactile – and should not react by raising her abdomen and coupling.

Mating strategies in non-territorial species

Most of the common damselflies which lay their eggs in still water are non-territorial. Densities in some species are often extremely high, with males greatly outnumbering females near the water's edge. Coming out on top in the scramble to couple with a receptive female when she arrives at the waterside is therefore not easy, and males may be forced to adopt alternative strategies to optimize their reproductive success. *Enallagma hageni* from North America is probably typical of species which have to confront this situation. The answer is to opt for two alternative strategies: either beat all the other males in the quest for females skulking in the vegetation at the edge of the pond (where they spend the bulk of their time) or wait around near the (frequently clumped) oviposition sites and try to shanghai a female there (Fincke, 1985). The latter option is less energy-consuming than searching, but more difficult, as the gatecrasher has to reckon with the males who stand on guard near their mates as they submerge to oviposit. Getting past a guard in the early stages of oviposition would be difficult. However, some guards fail to notice a female who emerges prematurely, while others simply fly off anyway after their mate has been under for 25 minutes or more. Both eventualities give an alert loiterer the chance to apprehend the female as she surfaces. When seized in this way, she will usually re-mate again immediately, although the number of eggs remaining to be fertilized and laid may be small, making this a rather hit-or-miss strategy.

Faced with two rather dubious options, most males use them both, thereby maximizing their reproductive chances. They often switch to the 'waiting' mode in the afternoons, when more females will be available on the water than in the adjacent fields. Such dual strategies are possible when the females are willing to mate repeatedly on a single day (as occurs in most damselflies), until they have completed laying the whole of their current batch of eggs. Only then do they refuse a further copulation, so there is always a reasonable chance that either of the two strategies will be profitable.

Female *Ischnura ramburi*, a small damselfly from lakes and ponds in North America, use alternative tactics to avoid sexual harrassment during oviposition (Robertson, 1985). Two different female morphs occur. One is drably coloured and basically cryptic (heteromorph) while the other (andromorph) is brighter and closely remembles a male. During encounters with males, the 'female-in-male-clothing' andromorph reveals that the masculine resemblance is not merely skin-deep by acting like a male, hovering and circling in the kind of 'face-off' conducted between two males. This tactic almost always succeeds in duping the male, who simply moves off without making any attempt to mate. This would suggest a major drawback – that the andromorph female will be at a disadvantage when she actually needs the inseminatory services of a male. However, this is a notably ephemeral species, so that during her brief life of only a few days, a single mating probably provides all the sperm required to fertilize a female's lifetime output of eggs. Finding and 'seducing' just the one male needed for this should not be too difficult, and is probably accomplished by adopting a 'female' style of flight.

The great dividend of the andromorph's male-like behaviour is its ability to reduce the number of unnecessary copulations to which the andromorph female has to submit (about half the number undergone by heteromorphs), resulting in a generally quieter life and undisturbed oviposition. She is spared the constant sexual molestation suffered by the heteromorphs, who waste a great deal of time and energy in unwanted matings. This is an unpleasant side-effect of spending all their lives in crowded conditions near the water. The ability to avoid repeated pairing is especially valuable in this species, for the duration of copulation is exceptionally long, averaging 3 hours. This may represent a form of pre-emptive mate-guarding. By monopolizing the female for such a long period, the male makes it too late for her to take another mate when she goes to oviposit, a task she carries out unguarded. Andromorphs are found in many other species of damselflies, and in the dragonfly *Erythrodiplax umbrata*, although the exact function of the behaviour in these other species is uncertain.

ANISOPTERA Dragonflies

Patrollers

Dragonflies follow many of the same strategies as damselflies, but with certain differences. Males of most of the large 'hawkers', such as the European *Aeshna cyanea*, establish territories over water, circling around for anything from 10 to 40 minutes, after which they leave and are replaced by new arrivals. Such temporal partitioning of a pool is probably standard in these large hawkers, being dictated by the amount of energy they burn up while in constant and expensive flight during their tenure over the water. Neighbouring males usually veer away from one another at their common boundary in a 'meet-and-turn' routine without any physical contact. Constant aerial vigilance is necessary both to repel invading males and to be sure of spotting any females who show up.

A female can adopt one of two strategies when arriving at the water, depending on her condition. A female who has previously mated needs to concentrate on the protracted task of laying her eggs, without wasting time in a superfluous and lengthy (often an hour or more) copulation with a resident male. An *Aeshna cyanea* female therefore keeps low, often skulking near the bank, searching for decaying tree roots and logs above the waterline in which to lay her eggs. While thus absorbed, her mottled green and brown patterning is very cryptic, protecting her from the prying eyes not only of predators but of her own interfering males. Females of other *Aeshna* and *Anax* species are similarly cryptic. An ovipositing *A. cyanea* who is dragged from her post by a male's anal claspers will seldom co-operate by raising her abdomen and coupling. Instead she struggles to tear herself free as they race in tandem around the pond. If successful, she will quickly return to her interrupted task, often closely harried by the insistent male, who nevertheless is usually forced to give up in the end (pers. obs.). In complete contrast, a female who arrives at the water requiring the services of a male to fertilize a newly matured batch of eggs does not play at all hard-to-get. In fact she actively solicits a mate. Her first move is to make a few circles over the water. If she is not quickly grabbed by a male, she makes herself more conspicuous by switching to a series of bobbing, soaring 'Lolita' flights, gaining more height so as to be visible from further away. When she is finally taken in tandem she immediately brings her abdomen upwards and couples, without any struggle (pers. obs.).

Mate-guarding in the Libellulidae

Most libellulids undertake some form of mate-guarding. Males of *Orthemis ferruginea* from Trinidad are territorial and defend their patch of water against rivals with mixed success, depending on the length of the contest (Harvey & Hubbard, 1987). Short disputes always go the way of the resident; longer contests put both contestants at almost equal advantage, with only a slight bias towards the resident. As in all libellulids which mate on the wing, copulation is brief, averaging around 11 seconds. During oviposition, the female flips her abdomen to and fro, using the inflated tergites of the seventh and eighth abdominal segments as a ladle to flick on to the bank small droplets of water containing eggs. Meanwhile her mate hovers nearby and chases off intruding males intent on making off with the free-flying female; despite the guarder's best efforts, such intrusions do succeed about 10 per cent of the time. Strangely enough, the male often seems to be more dedicated to the completion of egg-laying than the female – when only part-way through her exertions she may suddenly fly off into the forest. Her mate sets off in close pursuit and ushers her back to the pool to finish the job. This makes an interesting parallel with the 'showing-the-way-back' behaviour of male *Calopteryx splendens*. It seems likely that, in view of their high vested interests in their mates, many male odonates (including, of course, contact-guarders) probably have suitable methods for 'insisting' that a female completes the task. As the act of copulation in *O. ferruginea* (and many other libellulids) is so brief, the male may be able to intercept and mate with a newly arrived female as he hovers on guard. He then oversees both females until they have completed their egg-laying.

Some libellulid males practise resource-defence polygyny. In the American *Paltothemis lineatipes*, the defended resource comprises small areas of smooth sand beneath shallow water, scattered among the otherwise stony beds of mountain streams. As in other odonates, the male is skilled at identifying sites of greatest value to the females (Alcock, 1990). Territorial defence takes several modes. Often an intruder is simply repelled after a brief horizontal chase. This may escalate into a joint, steeply soaring flight up to a height of 10–20 m, after which the interloper makes off and leaves the resident in charge. The most serious disputes involve swift dashes up and down the stream. This tends to be risky for the residents, who lose their territories to the rival in a quarter of all engagements. Copulation takes place in a rapid to-and-fro flight above the oviposition site (such a flight is common in libellulids having aerial copulations). Most females then 'play the game' by using the male's pre-selected site, although a few try to look elsewhere. The male encourages them back by hovering above the spot with his abdomen tipped upwards. The female oviposits alone, dipping up and down to wash her eggs off into the water (a common method) while the male hovers nearby and chases intruders.

Sympetrum 'darters' exhibit a wide range of post-mating oviposition behaviours. Contact-guarding is the typical strategy, although it may be mixed with non-contact guarding, *Sympetrum* males being behaviourally flexible and capable of adapting their tactics to circumstances. In the non-territorial European ruddy darter, *S. sanguineum*, all males resident on a pool compete actively for females (scramble-competition polygyny). As in many small dragonflies, egg-laying takes place in tandem, with the male doing a large part of the work of lifting the female up and down in a bobbing-and-flipping action. In the ruddy darter, this propels the eggs into dense vegetation, such as rushes, on the waterside. Such a habit is unusual in dragonflies but it does mean that ovipositing pairs are not as conspicuous as those laying eggs over open water. In low-density situations, where the male is at little risk of losing his mate to a hijacker, he will often release her and hover or perch nearby, using less energy-consuming non-contact guarding (Convey, 1989). The 'decision' by the male 'to release or not to release' therefore seems to be related to the degree of harassment suffered during tandem oviposition, i.e. 'lots of hassle = keep hold of her', 'little or no hassle = let go'. Females may also manage to arrive surreptitiously and spend a period ovipositing alone before they are noticed and seized by a male.

The same dichotomy in behaviour also occurs rather less frequently in *Sympetrum striolatum*, a successful species which seems adept at responding appropriately to

◀ **Behavioural flexibility is the hallmark of the male ruddy darter, *Sympetrum sanguineum*, from Europe. His 'decision' about whether or not to remain with his mate during oviposition (as seen here) is geared to the degree of pressure from rival males. This species is unusual in dropping its eggs into dense vegetation on the waterside, rather than into open water.**

changing circumstances. In fact, under extreme conditions, the males' reaction to disturbance may even be the reverse of that seen in *S. sanguineum*. On small ponds, a single dominant alpha male may monopolize the prime position next to the pool, excluding rivals who are forced to perch further away. The alpha male therefore enjoys the major access to females, but while he is away copulating and engaging in tandem-oviposition, he loses his position, at least for that day, to a beta male who gains promotion to the alpha position.

When conditions enforce unnatural overcrowding by shortening the reproductive period, the situation can change completely. This happened in Britain in 1976, when a long (highly unusual) drought led to dried-out pools until late in the season. The first appearance of standing water provoked an orgy of activity in *S. striolatum*. As both sexes were present around the water in approximately equal numbers, and all females were receptive, the males were in the unusual position of being able to mate more or less at will. Because of high densities of tandem-ovipositing couples over the limited area of standing water, interference was intense. Yet some males still relinquished their mate part-way through and left her to finish ovipositing alone, despite the very high risk of a take-over by another male (pers. obs.). Such a strategy may possibly be explained by the deserter 'knowing' that he had an equally high chance of being able to re-mate almost at once, as females often accepted a new male within seconds of being released. Under these abnormal conditions, 'deserting' could maximize the number of different females inseminated within a short period; the male is in effect 'putting fewer eggs into more baskets'. The numerous repeat matings by the females may have been due to unusually high egg loads accumulated during the long absence of water. Similar flexibility in male response is also seen in *Sympetrum depressusculum*, in which overcrowding prompts males not only to conduct full contact-guarding as standard, but also to conduct an extended period of pre-copulatory tandem-guarding.

Male *Sympetrum* (and other darters) often fly alone above the water and make dipping movements similar to those used when ovipositing in tandem. The function of this is uncertain. It may serve to assess the quality of the egg-laying site; it could be a display designed to attract females by showing that the male is fit and ready to render them the service of assisted oviposition (in return for mating); or it could be that males who have lost a mate become confused and 'go through the motions' by themselves. Male assistance is certainly of benefit to the female, as it increases her egg-laying rate, compared with a solo performance, while virtually eliminating take-overs which would waste valuable time.

In *Plathemis lydia*, the males respond to increasing density at the water by resorting to a system of 'temporal leks'. The dominant alpha male will allow up to five other conspecific males to share his territory, but only as long as they do not challenge him and always exhibit submissive behaviour towards him (Campanella & Wolf, 1974).

Courtship in dragonflies

Courtship is absent in most dragonflies, but species with patterned wings may make use of them in a pre-copulatory display. In both sexes of the pretty little African *Palpopleura lucia lucia*, the wings are decorated with conspicuous inky blotching. The drab brown female benefits from the full impact of the male's patterned wings and powder-blue abdomen during a hovering and fluttering flight which he performs before her. Both sexes of *Palpopleura sexmaculata* from India also bear spotted wings, but the male's spots are more prominent; males are blue, females brown. The males defend miniature territories adjacent to small patches of open water in vegetation-choked streamlets (Miller, 1991). A newly arrived female has a 'self-advertisement' display at her disposal if her initial provocative perching position some 20–30 cm in front of the male fails to raise his interest. Faced with no response, she either flicks her wings or makes a brief upward flight, before landing closer to the male. He may respond by whirling around her in a series (2–4) of tight circles, first one way, then the other, then back again, all at breakneck speed, so that the whole drama only lasts some 5 seconds. Occasional variations on this theme include a zigzag display in front of the female and a hovering flight with upraised abdomen, designed to flaunt its bright blue finish.

BLATTODEA Cockroaches

Most cockroaches are nocturnal, carrying out their courtship and mating activities after dark, so it is not surprising that the main elements in courtship are based on scent, taste and touch. Females emit pheromones which act primarily to release male courtship efforts, while pheromones produced by the males release the female mounting and feeding responses described below. Tactile contact is also combined with pheromonal communication to produce the various sexual behaviours. However, in one major group of cockroaches it is sound – another medium which does not depend on light for its effective function – which plays the leading role.

Latiblatella angustifrons from Honduras employs a courtship procedure which, with numerous minor variations, is found in many other cockroaches. The female often starts by taking up a 'calling' stance, the tip of her abdomen heavily drooped to expose a narrow air-space between its upper surface and the underside of the wings (Willis, 1970). It is thought that this 'calling' posture is designed to pull in males by creating ideal conditions for the dissemination of a male-attracting female pheromone. 'Calling' by females has been observed in three of the five families of cockroaches, and is capable of tempting in males from distances of 10 m or more. In *L. angustifrons*, the emphasis on pheromonal stimulation switches to the male when he has finally made contact with the calling female. Upon meeting, the pair engage in some mutual antennal fencing which communicates a cocktail of stimuli via the senses of taste, smell and touch. Sexually receptive males often respond with side-to-side oscillations of the body before adopting a special courtship posture. Turning his back on the female, the male curves his abdominal tip downwards, droops both head and thorax, and flicks his closed wings and their cases upwards at an angle of about 60°. This lays bare a pheromone-producing gland or 'excitator' on his seventh abdominal tergite (near the tip); this was formerly masked by his wings.

The odour emanating from this 'excitator' usually proves strongly attractive to the female, who steps forward and nuzzles the gland's odorous lobe, which is adorned with a bunch of short bristles bathed in a secretion. Feeding by females on male 'excitators' is also seen in millipedes, harvestmen and malachiid beetles. In the first two, and in cockroaches, the feeding is aimed at tempting the female to adopt a specific stance relative to the male, and to hold her position for the required amount of time. In cockroaches, this is probably long enough to induce the necessary gaping of the female genitalia and maturation of the male's spermatophore. This sexually

▶ **Two stages in the courtship of the cockroach *Latiblatella angustifrons*. (Top) The male performs the full-wing-raising display which 'invites' the female to mouth his tergal gland. (Bottom) She walks forward on the male's back, mouthing the top of his abdomen and feeding on the 'excitatory' secretion.** (After Willis, 1970).

stimulating pheromone is found in many male cockroaches and, as a result of its function, it has been called 'seducin'.

In *L. angustifrons* (and other cockroaches), the lure of the excitator prompts the female to step right on to the male's back. She edges forwards, nibbling away along the male's back, before wrapping her front legs around his underside. As she nears his elevated wings, the male senses her position. Now he shoves himself backwards underneath her in order to grip her gaping genitalia with his own extended organs. As coupling is achieved, the female is standing atop the male, looking closely at the back of his head. Immediately afterwards, she steps sideways off his back and turns around to face in the opposite direction, at 180° to her mate. This bodily rotation after coupling is typical of most cockroaches. Many aspects of this procedure are common to most cockroaches, notably the antennal fencing, the upraising of the male's wings, the 'feeding' on the male's exposed gland, the edging forward on his back and the post-connection rotation. The main components of cockroach courtship can thus be summed up as:

(1) Contact.

(2) Courtship and male 'full-wing-raising display'.

(3) Mounting and feeding by the female.

(4) Backing-up by the male.

(5) Genital connection and about-turn.

Category 2 must obviously be absent in wingless cockroaches, such as *Eurycotis floridana* from the USA. Both sexes are wingless, a relatively unusual situation, as in cockroaches the males tend to be winged even when the females are not. *E. floridana* is also aberrant in having females who actively (rather than passively, as in 'calling') initiate courtship by rushing up to the males. In fact, a female may even try to mount a stationary male who is exhibiting no visible interest (Barth, 1968), although such apparently impulsive behaviour is probably prompted by a male pheromone. A receptive male responds with antennal body-stroking which is returned in kind, this being the prologue to a bout of male sideways body twitches similar to those of *Latiblatella*. Several of these spasms are interrupted by shifts of position which place the male's rear end near the female's face. This induces her to quiver and mount, after which she gradually nibbles her way along the length of his abdomen, feeding for far longer than any known winged species. The male does not back up as she nears his head, but mating in the standard end-to-end position still ensues.

Sound as a stimulant

In view of seducin's function in coaxing the female on to the male's back, it is not surprising that in one group of cockroaches lacking pheromonal communication, females do not mount the male at all. In the 'hissing cockroaches' from Madagascar, sound seems to have replaced scent as a species-specific identifier and stimulant. The males of many species, such as *Gromphadorhina portentosa*, are larger than the females and have heavily armoured pronotal shields. This makes them very much the battletanks of the cockroach world. They throw their weight around during highly bellicose encounters with other males. When two males blunder into one another, they engage in a preliminary skirmish with some frenetic antennal fencing, and then prepare for real battle by hunching down to bring the twin-humped pronotal battering-rams into play. Then, perhaps to the accompaniment of some aggressive hissing (produced by expelling air through the second pair of abdominal spiracles), the two heavyweights ram one another, often making contact with an audible thump (Barth, 1968). In the ensuing pushing match, the loser is the one who fails to stand his ground and ends up being propelled backwards. Victory is often celebrated by a barrage of hissing and a volley of abdomen-slapping. This is often against the ground, but is also inflicted upon the loser's shiny abdomen. He may suffer further ignominy by being chased and harried off the battlefield. Not surprisingly, the biggest pugilist usually wins, and his victory may earn him a place at the top of the dominance hierarchy, which is thought to govern priority of access to females. This would not be surprising, as these large cockroaches spend the day in hollow logs, a scarce commodity frequently leading to multiple tenancy. Under such crowded conditions, the evolution of a dominance hierarchy would make sense, as it should reduce the amount of disruptive combat between rivals. Low-ranking males seem to be able to avoid punishment from their superiors by adopting a submissive posture, hunching down low to the ground.

Conjugal relations are less contentious, and the initial antennal fencing lacks the militant animation seen in male–male encounters. In fact, the female wields her antennae in a delicate caress upon the male's body, spurring him into parading around her with his 'chin' held high, venting frequent hisses and reciprocating the female's caresses. This sexual hissing is much less strident than in aggressive encounters or defence. After a minute or so, the male extends and depresses his rear, trying to hook up with the female's genitalia by wiggling his abdomen from side to side as he backs forcibly into her. In so doing, he achieves the back-to-back position in a single step, unlike other cockroaches with their complicated sequence of mounting, mouthing and rotation. At the moment of shoving backwards to connect up, the male delivers an exceptionally dense train of hisses. These induce the female to adopt a special posture facilitating smooth coupling (Nelson & Fraser, 1980). Artificially muted males fail to achieve coupling, while muted males whose efforts are accompanied by playbacks of their hissing enjoy a normal degree of success. In these noisy heavyweights, hissing parallels the role of seducin in other cockroaches; both are indispensable keys which unlock certain essential female responses.

A nuptial gift in *Xestoblatta*

In *Xestoblatta hamata* from Costa Rica, the male offers the female a substantial post-nuptial meal by raising his wings, extending his abdomen and gaping the genital chamber widely, disclosing a white secretion composed of uric acid. The female feeds on this greedily, often guzzling away for long periods. The urates are absorbed and make a contribution to the development of her eggs, the degree of benefit depending on her own nutritional state at the time of mating (Schal & Bell, 1982). This action thus constitutes a form of parental investment by the male and represents large-scale post-copulatory nuptial feeding. The additional nourishment also has the advantage of accelerating the onset of oviposition, thus giving greater assurance that the donor will retain sole paternity (she is less likely to resorb her eggs through inadequate feeding, or to mate with another male). Males probably acquire most of the necessary urates by feeding on bird- and reptile-droppings in the rainforest. The male commands a long period of copulation (4 hours), probably to ensure that, by the time they separate, the female will be able to consume the bulk of the urate gift. This is doubly necessary, as the male stores urates in such quantities that failure to get rid of them can, in time, lead to premature death through toxicosis. As females spend the early part of the night feeding, the crop is usually full when they seek out a mate, usually around midnight. The prolonged mating time allows the crop to empty, making it ready to receive the full complement of urates on offer.

Defence of mate and mating chamber

Female promiscuity is probably the rule in most species, although the pest cockroach *Periplaneta americana* only copulates once. Males of the American wood roach *Cryptocercus punctulatus* endeavour to assure paternity of repeated egg batches by cohabiting with a female in a special mating chamber inside a rotten log (Ritter Jr, 1964). Intruding males meet with a hostile reception, being fiercely rebuffed in a no-holds-barred fight. Finally one roach retreats, although the winner is not always the original resident – if he loses his home, then his mate goes with it and is claimed by the victor.

'Transvestite' behaviour

Cockroaches often live in dense aggregations, especially in bat roosts in caves. Under such crowded conditions there may be considerable competition between males for the relatively few females who will be receptive at any one time. Consequently there is a premium in securing a scarce resource, and this has probably led to the evolution of the 'pseudofemale' or 'transvestite' behaviour seen in *Blaberus, Byrsotria* and *Archimandrita*. The 'transvestite' inserts himself surreptitiously into an ongoing courtship sequence between a rival male and a highly prized (by both males) female. Once the initial antennal sparrings are over, and the suitor has adopted the 'full-wing-raising posture', it is the 'transvestite' who accepts the implied invitation and rushes forwards on to his rival's back. The lowermost male is now very vulnerable to a sudden assault from his 'transvestite' rival, whose sexual confidence trick has earned him a position in which he can unexpectedly deliver a vicious series of bites. With luck, the surprise and stress of such an assault will ruin the whole courtship sequence, thus increasing the 'transvestite' assailant's chances of appropriating the waiting female for himself (Wendelken & Barth, 1985).

ISOPTERA Termites

In termites, which are highly social insects, the production of sexually active adult stages is a carefully contrived process whose general *modus operandi* is shared by most species. Unlike the workers, the sexual forms develop wings with which to carry out their sole task of dispersal. These winged 'alates' cluster together in special chambers awaiting the date of their release from confinement. The actual time will be decided by the workers who tend them and usually coincides with the start of the rains after a prolonged dry period. As the correct conditions can be detected equally well by the inhabitants of many nearby nests, there is normally a synchronized mass release of alates, ensuring abundant out-breeding

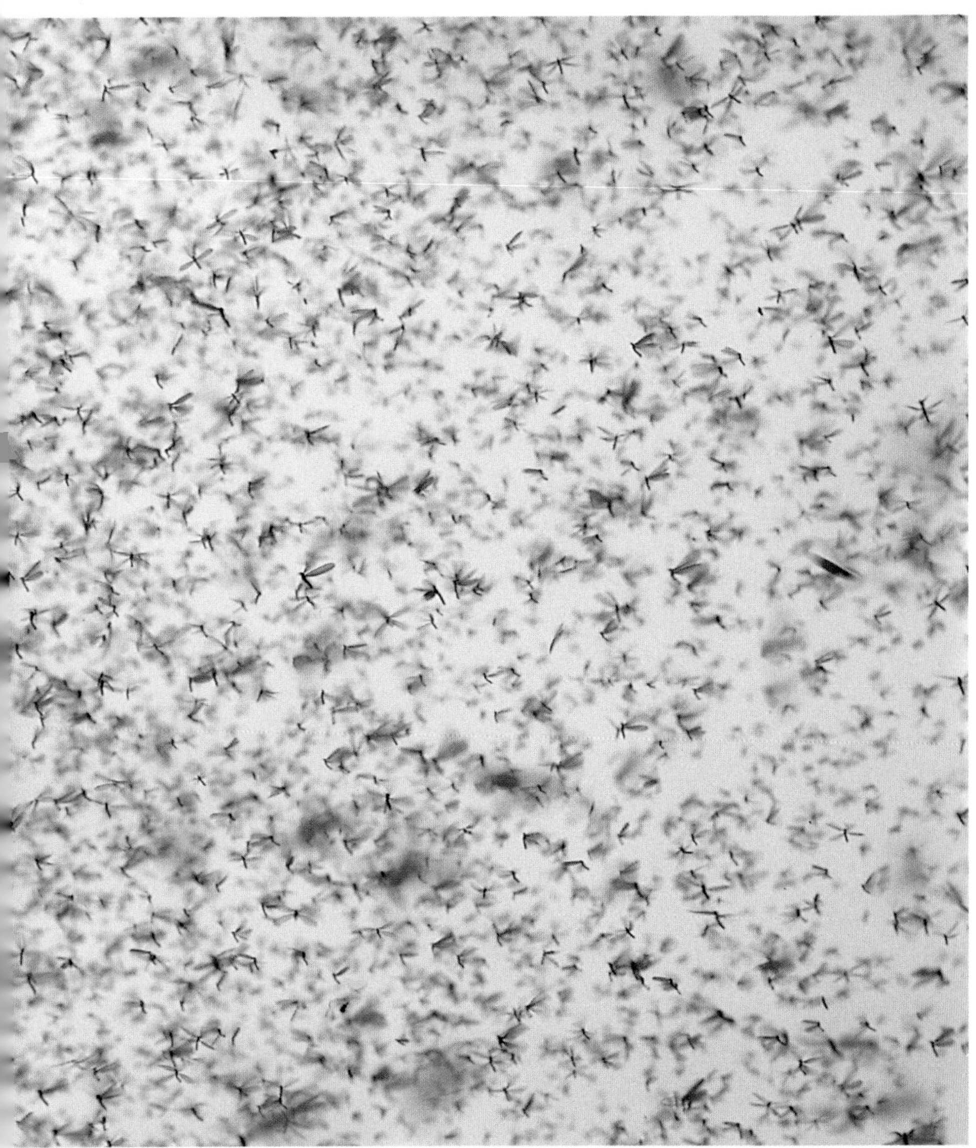

▲ Some termites form daytime swarms around landmarks such as trees. This is a Brazilian species, which releases its winged males and females on the first day of rain after the long dry season.

◀ Shortly after emerging, the females of this *Odontotermes* species (probably *O. obesus*) take up a 'calling' stance low down on a plant or even on a stone which just protrudes above ground level. Calling never takes place from a higher position, nor from the ground itself, but always from a height of about 2–6 cm. While calling, the female flaps her wings vigorously and arches her abdomen outwards to disseminate a pheromone to which the males orientate. Immediately after being joined by a male the female sheds her wings, her suitor quickly doing likewise, and they begin the 'tandem run' (male following female). If a female loses her mate after shedding her wings, she will resume calling by merely arching her abdomen. Nepal.

between nests. A colony's entire stock of alates is not normally released in one go, some being kept back for future departures. As the 'big day' approaches, the workers tear a number of holes in the mound's surface, through which the alates will make their exit. In some species, such as *Armitermes euamignatus* from the Brazilian *campo cerrado*, special launch-towers cover the surface of the mound. It is only necessary to pierce a hole in the top of these and the alates stream forth, while a protective army of soldiers and workers musters on the adjacent surface. The nuptial flight is often brief and the females land first, discarding their wings via a line of fracture near their bases. Bereft of her wings, the female now pauses, raises her abdomen in a 'calling' stance and releases a volatile pheromone which will attract a male. When a male shows up, he too sheds his wings and begins to shadow the female nose to tail in the so-called 'tandem-run', tapping her rump with his antennae as the search for a suitable nest site proceeds. If a good spot is found, they co-operate to excavate a hole in which they will establish their new colony, nourishing the first few offspring from their own bodily secretions.

Some species gather in swarms to perform a nuptial dance. *Cornitermes cumulans*, which builds earthen mounds in the Brazilian *cerrado*, releases a multitude of alates on the first day of heavy rain in September. The workers prepare for the event by constructing special wedding-day 'exit ramps', narrow strips about 4–12 cm long placed near the top of the mound. These ramps are initially roofed in with a temporary covering of soil. This is dismantled by the workers at 'zero hour' as the alates stream forth, protected by an army of soldiers and workers milling around over the mounds's outer surface. Such saturation defence is not available to a species of *Anoplotermes* which is also common in Brazil, as the soldier caste is absent. This species also builds release-towers, whose tops are snipped away just before use. This leaves a hole which, with the exception of the gap required for the egress of the alates, is blocked by a circle of workers' heads forming a living 'door'. The alate launches itself skywards from amidst this circle of protectors, some of whom may succumb to marauding ants. This is the cost which has to be paid to protect the alates. Males and females from several nests congregate around a conspicuous landmark, such as a solitary tree, and use it as a swarm-marker, dancing up and down in thousands before settling on the trunk or ground to mate. As light is needed to see the marker and establish the swarm, the dispersal flights are launched in daylight, usually when the rain is just easing off after a heavy downpour; most other species fly only at night.

In *Hodotermes mossambicus*, the respective roles of the male and female in attracting a mate and finding a nest site are reversed (Leuthold & Bruinsma, 1977). This species is also unusual in timing the emergence of its alates to coincide with a period of sunshine following rain, mainly during the hotter part of the afternoon. The alates soon drop to earth after covering less than 100 m; like other termites, they quickly shed their wings. However, in this species it is the *male* who now starts 'calling' by strolling around with his abdomen raised, exposing his massive sternal pheromone-releasing glands. He also starts digging the nest hole, but during his labours he still manages to keep his abdomen tilted upwards in the 'calling' pose until a female finally arrives. The females increase their ability to detect a 'caller' by shinning up a stem to 'sniff out' the nearest male. A female navigates towards a male's scent-beacon with remarkable accuracy, tapping him upon the abdomen with her antennae upon arrival. He turns around and checks on her identity by mouthing her abdomen, receiving the necessary 'switch-off-calling' message upon confirming her gender and species. If at this point a female is experimentally taken from the male, he re-instates his calling behaviour. In some cases the female manages to rendezvous with the male during his initial 'calling run', in which case she assists him with his digging right from the start.

MANTODEA Praying mantids

For over 200 years, it has been stated that a normal feature of conjugal relations in mantids is for the female to graze away on her mate's head – while his rear end is still busily doing its inseminatory duty. The discovery that the vigour of the male's copulatory movements may even be increased by decapitation lent unfortunate weight to this idea. This even led to the notion that decapitation was in fact necessary for successful copulation by removing 'inhibitors' in the male's brain. His restriction to but a single mating as a result of this cranial sacrifice (instead of the maximum possible number usually sought by male insects) was said to be balanced by the benefits of 'offering' his body as a bizarre form of 'nuptial gift'. He would thus make a single yet substantial posthumous contribution to the development of the one batch of eggs which he would ever fertilize. Females of the Asian mantis *Hierodula membranacea*, which were artificially maintained on a low but not starvation diet, were more likely to cannibalize their mates. They subsequently produced egg batches that yielded larger numbers of babies, giving an apparent pay-off in terms of increased numbers of offspring for the male's self-sacrifice. However, we do not know how many times female mantids mate under natural conditions. If they mate more than once, then any self-sacrifice made by an earlier male would be to the benefit of his later rivals, assuming that last-male sperm precedence operates. Although it is still difficult to be certain, it is likely that the 'sacrificial gift' view is quite false, at least in regard to many – perhaps most – species. One of the problems in establishing the facts is the scarcity of observations on actual *courtship* in mantids under natural conditions; even mating pairs are seldom encountered. During 5 years of searching for insects in the tropics, the author has found only six pairs of mating mantids. In five of these, circumstantial evidence pointed strongly to 'non-fatal' conjugal relations as being the likely norm, while the probable importance of female 'calling' in ensuring that affairs stay peaceful was also indicated.

However, in at least one species of mantis, the European *Mantis religiosa*, it seems that males do indeed lose their heads on a regular basis (Lawrence, 1992). Observations made by Dr Susan Lawrence under natural conditions in the wild established that males were decapitated in about one quarter of all copulations.

The males are normally far smaller than their mates, which might seem to increase the risks, but in fact probably reduces them considerably. The males tend to fly well and are more agile than the females. This probably enables them to 'drop in' on unsuspecting females in the hope of securing co-operation once they are 'in the saddle'. Such 'Lone-Ranger' tactics are common among male insects as a short-cut, replacing the protracted alternative of mounting a complex courtship display (e.g. as in some grasshoppers or tephritid flies) Once he is in place, the male mantid's small stature means that his head is placed well back, forcing his mate to make a special effort to take the first exploratory bite.

The supposition that male mantids could just 'drop in' on females unannounced is backed up by the author's experiences in Kenya. On just a single day, two separate examples of males clinging 'hopefully' to the wrong females were discovered. One pair comprised a male *Polyspilota aeruginosa* with a female *Danuria* twice his length; the other was also a *P. aeruginosa* , but this time firmly clasping a female *Tenodera superstitiosa*, also almost twice his size. Both males appeared to have their genitalia engaged, although it is unlikely that this could have led to the transfer of any sperm. In neither case did the oversized female appear to object to the presence of the interloper, who was still safely in place more than an hour later. The significance of these cross-matings lies in the males' ability to mount the females and then remain in place in complete safety *without* being 'invited' aboard. After all, any courtship performed by the male *P. auruginosa* would be unlikely to resemble at all closely that of the 'correct' males, making it unlikely that they could have induced a co-operative state in their respective mates. In neither instance were the 'partners' in the same genus and, in one case, they were not even in the same family! *P. aeruginosa* is in the Mantidae, whereas his *Danuria* 'mate' is in the Vatidae.

▲ Intergeneric coupling in mantids: this *Polyspilota aeruginosa* male is seated in perfect security atop the much larger female of *Tenodera superstitiosa*. Such happenings indicate that a correct form of courtship may not be important in enabling male mantids to mount females under wild conditions; it also hints that female cannibalism of the males is probably not at all common, contrary to 'popular' belief. The use by the males of a surprise leap – the 'Lone-Ranger tactic' – probably explains how such a misguided mating occurs.

An abrupt 'Lone-Ranger' style of mounting seems to explain the above events but raises an additional question. How did the males come to make such a time-wasting and unproductive mistake in the first place? The answer could be that female mantids normally advertise for mates by 'calling'. The pheromones released may not be sufficiently species-specific (or even family-specific) to preclude the occasional luring of 'foreign' males. 'Calling' has indeed been well documented in at least one mantis, *Acanthops falcata*, and is likely to occur in others. Support for this stems from some further observations by the author: in only four normal matings seen in the wild, no fewer than three involved the attendance of a second male. In two of these, the mating pair was in a well-hidden location, reducing the likelihood that two males just happened to arrive by chance. It is more probable that both were homing in on a deliberately released female pheromone. In one pair, again involving *Polyspilota aeruginosa*, the female had already eaten her original mate's head and was absorbed in working her way down his body. The second male clung to the more normal position on her back, having probably pushed the incumbent's body to one side in a scuffle which had presumably provoked the female's attack by breaking the conventional chain of behaviours.

Cannibalism need not inevitably follow such a gross interruption. In Trinidad, an *Acontista* pair was mating on a leaf when a second male flew down and landed directly in front of them. Such precision indicated the likelihood that he was attracted by female pheromones. He seemed to sum up the situation instantly, for he clambered up over the female's raptorial front legs, across her head and on to her thorax. This was not accompanied by any threatening moves on her part. What then ensued was a pushing match between the two males upon the female's back, during which her head served as a support for the attacker's straining back legs. She made no attempt to take advantage and seize the attacking male, despite his vulnerable position. Nor did he show any signs of 'nervousness' about being so close to her potentially life-threatening front legs. In the end he gave up and flew off. Similar take-over attempts are common in insects, but are seldom successful unless the attacker is considerably larger than the incumbent (for examples see conopid and scatophagid flies, longhorn beetles, brenthid weevils). In a second instance, two *Isomantis domingensis* males were perched on a female beneath a leaf. This in itself hints that a pheromone had attracted them. Neither male was *in copula*, and both were ignored by the female, despite the fact that all three heads were very close together. This trio was still present 3 hours later, with no change.

'Calling' by the female

Observations on female 'calling' behaviour in the Neotropical *Acanthops falcata* indicate that the males are immune to attack because they come visiting 'by invitation only' (Robinson & Robinson, 1979). Both sexes are superb mimics of dead leaves. The females are wingless and resemble a bent shrivelled leaf, whereas the males are fully winged and more like a flattened leaf. The male therefore faces two major problems. Firstly, he must somehow detect females who are not only widely dispersed in the forest but who also resemble their surroundings to such an extent that they are virtually impossible to distinguish. Secondly, he must find a female who is willing to mate, and must manage to mount her without being seized and eaten (however slight that possibility may be). In practice, both problems are solved in advance by the virgin females, who woo the males from the surrounding forest by adopting a special 'calling' posture. They minimize the risks of losing incoming suitors to predators by only 'calling' for a very brief period each morning, just after dawn when the flying males are less at risk from hungry birds. When 'calling', the female hangs beneath a twig, raises her wing cases and curves her abdomen downwards, exposing paired black protuberant glands on its upper surface. Males may be attracted

▲ A newly arrived male (left) of an *Acontista* species mantis in Trinidad responds to finding a male already in place by trying to push him off the female. He follows up by biting at his genitalia (as seen here); anything goes in the fight for reproductive success. Despite the struggle taking place on her back, the female at no time exhibited any cannibalistic tendency; the assailant eventually withdrew. Attempted take-overs in insects are rarely successful unless the attacker is considerably larger than the defender.

◄ Judging by the angle of the mounted *Polyspilota aeruginosa* male's wing, quite a struggle ensued when he interrupted an already mating pair. This kind of gross disturbance of the normal peaceful mating sequence could have caused the female to take a bite at her original mate's head; his headless body projects at an angle to his mate. His rear end will still continue to perform its inseminatory task, even when the female has finally consumed the rest of his body.

in some numbers, landing nearby and walking towards the female with a rocking gait (an identification display?) before taking a flying leap on to her back. In numerous such pheromone-initiated copulations (observed in cages), not a single case of sexual cannibalism was noted and mating only ever took place just after dawn.

ORTHOPTERA Crickets, locusts, grasshoppers

Sexual communication in the Orthoptera is mainly acoustic, although there are numerous exceptions. 'Singing' is typical of the males; in females it is chiefly confined to brief answering calls in a few species. Male acoustic mating systems fall into two broad types: 'male call' and 'male search'. Male callers attract females by the regular emission, usually from a fixed point (e.g. a leaf in Tettigoniidae and some Gryllidae; a burrow in Gryllotalpidae and many Gryllidae), of a species-specific acoustic signal, the calling song. This carries over a considerable distance (sometimes augmented by the use of specially built 'amplifiers') and depends on the active homing response of a receptive female to the acoustic beacon produced by the calling male. Male searchers (most Acrididae, many Gryllidae, some Tettigoniidae) roam in an active quest for females, broadcasting their identity and intent by producing intermittent calls which advertise their location at any one point. In many grasshoppers (Acrididae), the males do not sing at all but locate females purely by searching, often on a specific food plant where they are likely to be present. Courtship in such species depends on visual, tactile and possibly pheromonal cues.

The method of sound production is mainly by stridulation (i.e. the rubbing of one body part against another). However, certain kinds of grasshoppers, such as *Gamarotettix* raphidophorids, drum the abdomen against branches and leaves, or against stones (one toad grasshopper), while others tap against leaves with the legs. The European oak bush cricket, *Meconema thalassinum*, has long been known to tap out his amorous message in this way. One hind foot drums rapidly against the leaf while the other leg is stretched out to the rear to act as a brace. A remarkable volume of sound is thus produced, audible at a range of several yards. A vegetation-transmitted method of communication called 'tremulation' has also been described in a number of tettigoniids (see page 49).

STENOPELMATIDAE King crickets

In the few stenopelmatids so far studied, courtship seems to be lacking. Males of the New Zealand tree weta, *Hemideina crassicruris*, appear to assemble harems of up to nine or more females within refuges in hollow branches (Moller, 1985). There is considerable competition for these marriage-quarters, as high-quality examples with the right characteristics tend to be scarce. There is also much demand for the resident harem, whose 'master' may spend part of his time acting as sentry at the entrance, blocking access to any passing males. (This probably constitutes one of the relatively rare examples of harem-defence polygyny in insects.) However, a male intent on commandeering a refuge can often evict its occupant by grasping a back leg and hauling on it. If the victim happens to be a female, she will usually turn and run as soon as she is exposed in the open. But the males are equipped to stand their ground and make a fight of it, having greatly enlarged, shiny head-capsules and formidable horny mandibles. These are intimidatingly brandished in a wide gape as an initial threat gesture between an assailant and the male which he has just dragged out of his home and released. The pair may sit for several minutes in a gaping 'face-off' before one of them tries to make for the hole, possibly being intercepted half-way.

If neither male backs down at this point, then conflict will commence. This is a straightforward affair in which the opponents deploy their bulbous heads as battering rams, making a simultaneus charge whose impact may send one of the combatants spinning off the tree towards the ground. It seems that, in tree-weta society, one head-butt is enough to settle the dispute. The vanquished male makes off, while the winner may announce his newly acquired status by stridulating briefly as he enters the gallery with its coveted harem. Copulating pairs have been seen out in the open on branches but, in *H. femorata*, mating mainly takes place in the privacy of the galleries, where the females can be kept away from rival males.

Sexual cannibalism in Jerusalem crickets

In the American Jerusalem crickets, *Stenopelmatus intermedius* and *S. nigrocapitatus*, the nuptial rites are so similar that they can be treated as one (Tinkham & Rentz, 1969). In the wild, a male probably attracts females by drumming his abdomen against the sides of his burrow. The initial confrontation between the drummer and a receptive female who has answered his call does not appear to augur well for his future success. A clearly audible clicking of the large gaping mandibles presages their use as pincers to grip one another's legs, as if to inflict mortal damage. Yet no such thing occurs and the couple engage in several of these brief but harmless grappling matches before the male mounts the female. Facing towards her rear, he seizes her hind legs in his mandibles and curls his abdomen to connect with her genitalia. Shortly afterwards, he extrudes a large spermatophore along with a quantity of fluid in a copulation lasting only 5 minutes. Within a minute or two, the female begins eating part of the spermatophore (for details and function of this see page 49). Unlike in tettigoniids, the female Jerusalem cricket may supplement this post-nuptial snack with a much larger meal – the male himself, who is quickly killed and eaten. Such sexual cannibalism might conceivably be frequent in the wild and it would certainly account for the surprising lack of male specimens in collections compared with the number of females.

TETTIGONIIDAE Bush crickets, katydids

Male leks and fighting

The chirping and buzzing calls of male katydids (Tettigoniidae) are a familiar night-time sound in warmer countries. The calling of numerous species, each with its own species-specific song, can lead to the problem of acoustic interference, which is often solved by so-called 'duetting', in which the songs of different species tend to alternate. Exceptional overcrowding of the air-waves may even force some species to alter the time of their singing to avoid competition. In Panama, acoustic interference from three related species causes the cone-headed katydid, *Neoconocephalus spiza*, to switch its main calling activity from the preferred period after dark to the daylight hours (Greenfield, 1988). If the 'noise' level is artificially lowered by removing the competitors, *N. spiza* reverts to nocturnal singing.

Some male katydids appear to sing in regularly spaced groups which appear to constitute leks. As such leks are intended to attract large numbers of females by creating a salient focal point, a considerable level of competition might be expected for the privilege of membership. Internal disputes might also be anticipated, designed to disenfranchise competitors and maintain the lek's density at an optimal level by keeping the spacing of singers at a regular minimum distance. Such inter-male rivalry does indeed seem to be present in several species of katydids, notably in the North American *Orchelimum* meadow katydids, in which outright belligerence seems to be an endemic characteristic. Each male executes a brief song, comprising a number of ticks broken by a buzz but, as numerous singers are operating in close proximity (average distance apart 1.7 m), the whole display resembles a choir rather than a solo performance and has the continuity of a concert (Morris, 1971 & 1979).

The competitive element becomes manifest when a singing male *O. gladiator* suddenly quits his serenading and forsakes the relative security of his song-post to zero in on the song of one of his neighbours. The result is a vicious fight during which both opponents try to inflict as much damage as possible with their powerful mandibles. Although the males' heavily clawed feet are mainly devoted to clinging on to the vegetation, they also do duty as additional weapons, being raked savagely across the opponent's abdomen. The victor 'celebrates' his triumph with a short burst of song as the loser makes off. Should the latter try to re-establish a nearby song-post by pausing in his retreat to stridulate, the victor will quickly respond by dashing across and completing the rout, usually without any further need for violence. However, some losers are reluctant to admit defeat and a cycle of singing, chasing, fleeing and singing again may be continued for some time before the loser is finally exiled to the edge of the lek, where his mating success will be limited. Alternatively he can remain within the group – but only as a sexually inactive mute.

In most such contests, it is the attacking male who wins, suggesting that, before embarking on his enterprise, he must have accurately estimated his chances of victory. He probably gauges his opponent's size, and therefore his prowess in fighting, by the volume of his song (smaller males sing less loudly). The high level of aggression in this species, as well as the need for leks, may be explained by the likelihood that the females only mate once. This would make virgins a scarce commodity, as males always try to mate as often as possible, while males and females usually occur in approximately equal numbers. Ease of attraction therefore becomes pre-eminent, hence the leks, but undue levels of competition would be correspondingly unwelcome, hence the tendency for the largest males to cut out the competition by making pre-emptive strikes against their smaller neighbours.

Tremulation

One method of reducing acoustic competition, while simultaneously reducing the chances of being detected by enemies, is to engage in 'open-skies' singing only for as long as absolutely necessary, and then change to a less dangerously overt tactic (tremulation) for close-in communication. This is the strategy adopted by several species of cone-headed katydids which inhabit dense tropical rainforests, where insectivorous bats are thought to use the normal songs of katydids as acoustic beacons, resulting in the instant snuffing out of the caller's song.

The male of *Copiphora rhinoceros* from Costa Rica establishes his calling platform beneath a large leaf, often a *Philodendron* (Morris, 1980). From here, he leads off into a two-part advertisement, comprising bouts of 'typical' buzzing stridulation alternating with periods of tremulation – but never both together. During tremulation, the caller lifts his body rather higher off the leaf, and in a more head-down stance than while stridulating, and initiates a triple bout of oscillations. These send vibrations through the feet and into the leaf, where they tend to be amplified by the leaf's angle with the stem. Several bursts of tremulation transmit a string of pulses through the plant and any connected vegetation. A receptive female navigates actively towards a calling male and may alight on the leaf directly above him. Once he senses her arrival, the male instantly switches his calling mode from the wide-band 'open-skies' method represented by stridulation to the 'private line' of tremulation. The female may now also tremulate occasionally, but in a less vigorous manner than the male. Continued bursts of male tremulation coax her to join him on his side of the leaf. To repeated bursts of tremulation he now backs very gradually down the leaf's midrib, only resorting to the odd round of stridulation if required to urge on the female when she shows signs of dawdling.

The male's goal is to coax the female towards the vertical stem of the leaf. Here, they will be reasonably well concealed from predators by the leaf arching out above. Also, the vertical alignment enables the male to engage his genitalia while sitting vertically above the inverted female. Sometimes several males may call from within a small area, inviting females to choose between them in a possible lek. When a female is able to detect tremulations from more than one male, she probably chooses the source of the most powerful transmissions. This is because the strength of the vibration pulses is directly related to the weight of the male; and this is related to the size and weight of the spermatophore which he will be able to deliver.

The katydid spermatophore and its role

Spermatophore size is of considerable importance to a female katydid. Shortly after the male has departed, she will bend forwards so that her body forms a curve, enabling her mouthparts to contact the spermatophore, which the male has just attached to the rear of her abdomen. The spermatophore consists of a small sac (ampulla) containing the sperm, surmounted by a much larger protein-rich gelatinous appendage (often bilobed) called the spermatophylax. The female tucks into this bulky spermatophylax first, so that by the time she gets around to mangling the ampulla, its vital load of sperm will all have been safely transferred into her body. The size of the spermatophore varies greatly from species to species, representing only 2 per cent of male body weight in some and

▼ On a damp night in a Brazilian rainforest, a female katydid (Tettigoniidae) eats the large spermatophore donated by the male during copulation.

up to a massive 30–40 per cent in the European *Ephippiger bitterensis*. In view of such an immense physical investment, it is hardly surprising that *E. bitterensis* males do not begin calling again for some 3–5 days after mating, using the interim to channel resources into the production of another bulky spermatophore.

By contrast, males of the Nearctic *Neoconocephalus ensiger* transfer only a small spermatophore during a prolonged copulation and then are able to resume singing within minutes of separating from the female (Gwynne, 1977). Presumably a fresh spermatophore is forming inside the male while he copulates, so he is ready for more sexual action as soon as he finishes. The small spermatophore in this and related species may serve an additional function as an internal mating plug, prohibiting further matings by the female for a while. In this respect, it is similar to the sphragis of some butterflies. The male *N. ensiger* thus makes a small investment in a large number of females, while conversely *E. bitterensis* makes a huge investment in only a small number of mates.

The main function of a huge spermatophylax, representing such a significant proportion of the male's bodyweight, is the subject of some dispute. Experiments with the Australian *Requena verticalis* have indicated that the large spermatophylax comprises a nuptial gift, which exemplifies a considerable degree of parental investment by the male (Gwynne, 1988). The availability to the female of a substantial nuptial gift has been found to increase fecundity, or to be beneficial to her eventual offspring, in a number of diverse animals (e.g. the spider *Pisaura mirabilis*). In *R. verticalis*, the products of the spermatophylax have been shown to pass into the female's eggs. As a result, these are larger than in females who have been denied access to a spermatophylax. As large eggs result in more rapid nymphal growth, male offspring will mature more quickly and enjoy an increased chance of being at the head of the eventual queue for females (who mature later). The father's contribution to the future success of his male progeny via his spermatophylax could therefore be substantial (Gwynne, 1988a).

In contradiction to this, the eggs of the European wartbiter, *Decticus verrucivorus*, do not appear to be significantly larger in females who have had access to a spermatophylax, nor is their fecundity increased (Wedell & Arak, 1989). In this species, the main function of the very bulky spermatophylax appears to lie in delaying the onset of the female's destructive feeding activities. By the time she has chomped her way through the bulwark of the spermatophylax (3 hours), the contents of the ampulla will have been safely evacuated. The spermatophylax would therefore serve mainly as a sperm-guard rather then a direct male investment in his progeny (sperm-guarding delaying devices, but of a different nature, are also seen in crickets; see page 51). Full transfer of the ampulla's contents is vital, as the amount of sperm received by the female has a direct influence on the number of eggs laid and the length of her post-mating refractory period, during which she will not mate again. Thus, in katydids, the sperm itself could be said to replace the active male mate-guarding behaviour seen in many other insects, in which there is no female refractory period between mating and oviposition.

Sex-role reversal

Whatever its prime function, the sheer size of the spermatophylax donated by many male katydids represents a vastly greater allocation of resources than is conventional in most male animals. Normally, the weight of the gametes lost during mating constitutes an insignificant proportion of total body weight. Males can therefore be expected to behave in a generally promiscuous and unfussy way, mating indiscriminately with any female on the basis that they have little to lose by selecting a 'poor' mother. On the other hand, a male katydid who invests generously in a weighty spermatophylax could be expected to apply a different set of rules to govern his selection of a fitting mate. Rank promiscuity is impossible anyway due to the period (several days) needed to regenerate a new spermatophore. As he is 'betting heavily' on just a few females, he could be expected to be rather fussy, choosing carefully between partners.

Indeed this seems to be exactly the case in at least one species, the Mormon cricket, *Anabrus simplex*, an inhabitant of the sagebrush deserts of the American west (Gwynne, 1984). The spermatophore represents 30 per cent of the body weight, and only well-fed males with swollen abdominal glands (perhaps comprising only a small proportion of the males in a given habitat) indulge in sexual activity. With such a desirable bounty on offer, each male need only call for a few minutes in the early morning for one or more females to come running. If two would-be courtesans arrive simultaneously, they will fight for the privilege of access to the plump caller (Gwynne, 1981). This is an extraordinary state of affairs, as normally it is *males* who do battle over access to *females*. The winning female clambers on to the male's back and they link genitalia. This represents 'decision point' for the male. Instead of proceeding to pass across his large succulent spermatophore, he may well disengage his abdominal claspers and walk away in an unmistakable 'not today' gesture. Such fickleness is rare in male animals, although common among females. However, in Mormon crickets, it appears that the males 'weigh' the mounted females, being capable of distinguishing plump, egg-laden, well-fed 'successful' females from skinny undernourished 'losers'. Small females probably die while still spinsters, possibly after numerous abortive mating attempts, while the heavyweights enjoy a lusty sex-life with a whole string of matings. However, in richer habitats, where all males are well nourished, the females revert to being choosy. A similar reversed-role mating system is also found in an Australian *Metaballus* sp.

Male body parts as nuptial gifts

Nuptial feeding takes rather a bizarre direction in the Russian *Bradyporus tuberculatus* – the female nibbles away at the male's back and laps up the oozing blood. A variation on this vampire-like behaviour is found in the North American *Cyphoderris buckelli*. This is a member of the very small and 'primitive' family Prophalangopsidae which pre-dates the more 'modern' Tettigoniidae and Gryllidae (Morris, 1979). Consumption of a spermatophore could perhaps be considered a specialized form of sexual cannibalism *after* the appendage has left the male's body; in *C. buckelli*, however, the female starts to gnaw away at part of the male's body – his wing tips – while they are still in place. This is apparently not merely opportunistic 'free-loading' but an intrinsic part of the mating process. The male's wings are unusually fleshy, and the female derives considerable benefit from devouring both the tissue itself and the haemolymph which oozes forth upon injury. She is able to feed in comfort because she sits on the male's back (a reversal of the male-under-female position of the Tettigoniidae). The male invites her to take a bite by raising his wings into a position where she can easily bring her mouthparts to bear. The male is thus offering a version of parental investment in terms of wing tissue ingested by the female. In theory he could stand to lose a great deal by allowing 'feed-and-leave' females to make a living out of repeated gastronomic encounters, but each time leaving 'without paying the bill' in the form of accepting a spermatophore. Such unproductive and destructive liaisons are neatly prevented by the male's ability to force the gourmandizing female to stay put until he has transferred the spermatophore. At the tip of his abdomen there is a so-called 'gin-trap' device consisting of two opposed sets of curved hooks. These firmly clench the female's abdominal tip and impound her until copulation is complete.

GRYLLIDAE Crickets

Courtship in the Nemobiinae

Courtship methods have been well studied in the subfamily Nemobiinae and vary from the simple to the extremely complex (Mays, 1971). The American *Hygronemobius alleni* stands at the simpler end of the spectrum, relying merely on a long bout of antennal lashing across the female's body, preceded by some brief chirping. After several minutes of this, the male produces a spermatophore, but about another 20 minutes pass before this is ready for transfer. He gives a 'ready-now' sign with chirping and some brief antennal stroking, then droops his wings as the female climbs on board. With a lightning movement, he shoves the spermatophore into her genital opening. Having dismounted, the female walks around for a few minutes with the spermatophore attached to her rear end, before rubbing it off and eating it. Two more bouts of courtship and copulation then follow, with the average male usually managing three couplings in a single session. Such rapid multiple matings are probably linked to the small size of each spermatophore. By transferring his sperm in three small packages, the male can fully inseminate the female without the need for the delaying tactics (nuptial feeding or post-mating guarding) required with larger single spermatophores. The process in the European wood cricket *Nemobius sylvestris* is similar but more drawn-out and complicated.

Nuptial feeding on male body parts. Nuptial feeding on some specially adapted part of the male's anatomy is also found in many crickets. Indeed, it is an intrinsic part of copulation in all Oecanthinae. Its purpose is to spin out the period during which the female is mounted on the male's back until the contents of the spermatophore have been fully transferred. This delaying tactic is necessary because, once the female dismounts, she will usually rub off and eat the spermatophore within a short time. The males' nuptial secretions therefore act as sperm-guards. Crickets which do not practise 'in-the-saddle' feeding, but instead make a rapid transfer of the spermatophore, achieve the same end with an extended period of post-copulatory guarding by the male, who physically prevents the female from eating the spermatophore. This means that there is often a direct relationship between mating time and the duration of post-copulatory guarding behaviour.

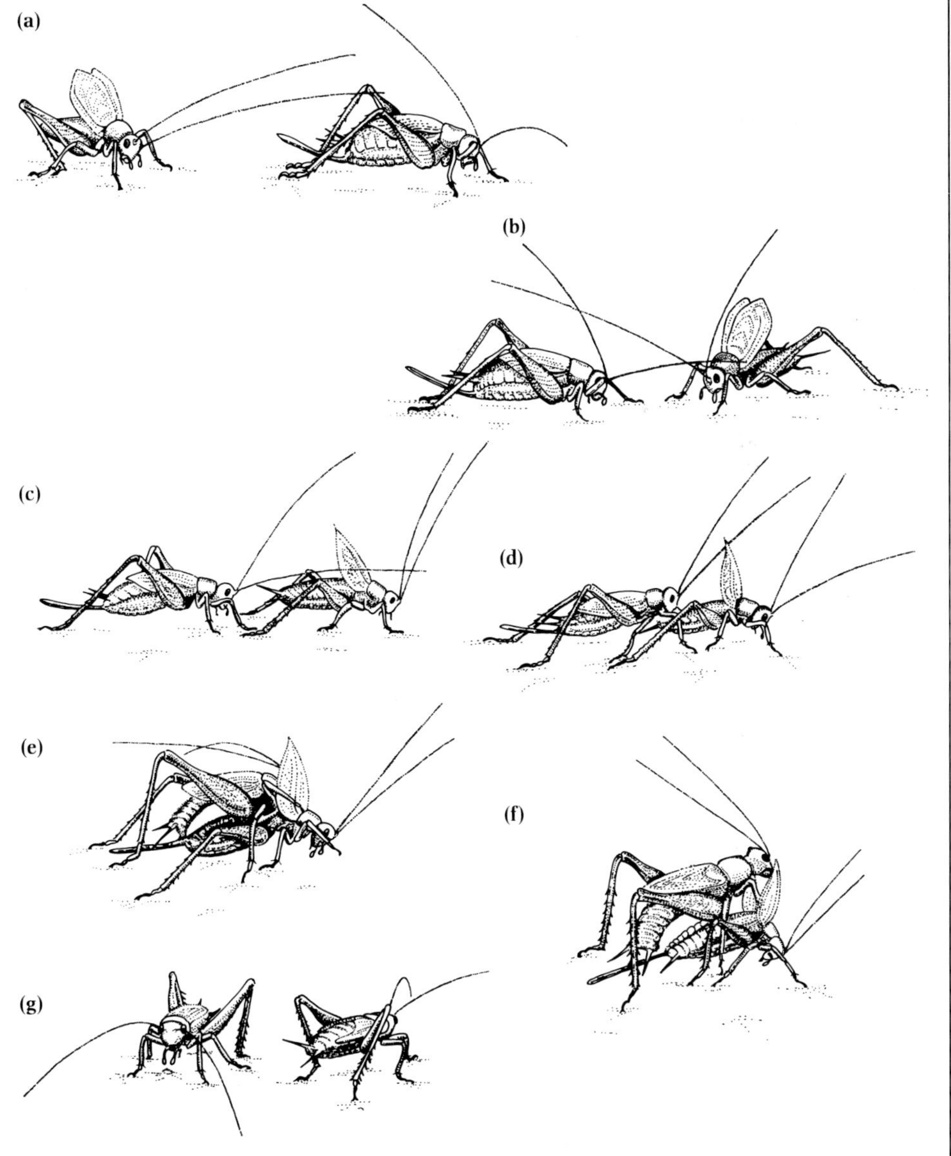

▶ Stages in the courtship behaviour of the brown bush cricket, *Hapithus agitator* (Gryllidae). (a) Upon spotting a female, the male (left) raises his wing cases. (b) If she does not run off, the male (now right) produces a spermatophore and approaches her. (c) He turns his back on her and 'invites' her to mount. (d) She moves forwards and starts to mouth his back. (e) She moves forwards, all the while mouthing his back until her head is buried at the base of his wing cases. (f) The male 'hooks' the end of her abdomen; simultaneously, the female raises her head and starts to chew on the ends of the male's wing cases. (g) The frayed state of the male's (right) wing case after copulation. (After Alexander & Otte, 1967).

In the North American genus *Pteronemobius*, the males are unique in serving up yet another form of bodily 'nuptial dish' – spines on the inner surface of their hind tibiae. These are chewed upon by the female during a rather lengthy period *in copula* (Mays, 1971). It is likely that the hollow spines are connected to glands which feed a secretion through them. This secretion then passes out at their tips, oozing out through a hole gnawed by the female. As in most crickets, courtship in *Pteronemobius ambitiosus* involves antennation and chirping, although, strangely enough, males who are unable to stridulate mate just as often as normal males. In the latter, chirping escalates into a non-stop trilling accompanied by a to-and-fro rocking as the male invites the female to mount, backing in to her with lowered wings. As their genitalia meet, the two crickets erupt into a brief bout of pitching and tossing before the female springs off. Only now, after this brief yet hectic 'pseudo-copulation' does the male form the spermatophore, continuing his courtship at a low level for up to 40 minutes before he again trills and rocks. They now remain *in copula* for 20–30 minutes, during which time the female is kept busy chewing away at the male's tibial spines. He facilitates this by closing his back legs up against his body (a most unorthodox posture for a cricket). The female eventually eats the spermatophore, while the male may also launch into a bouncy 'dance' shortly after she dismounts. The function of the preliminary 'pseudo-copulation' is unknown. What is certain is that, without this, the male will not come up with a spermatophore. Once a female is in place, nuptial feeding on the tibial spines seems certain to be a delaying device, buying enough time for the secure attachment of the spermatophore.

In the North American decorated cricket, *Gryllodes supplicans* (Gryllinae), the 'delaying device' seems to be the same as in many tettigoniids – a large spermatophylax attached to the sperm ampulla. Immediately after mating, the female usually tears off the spermatophylax and eats it, which occupies her for some 40 minutes, long enough for the ampulla to have evacuated the bulk of its contents (Sakaluk, 1985).

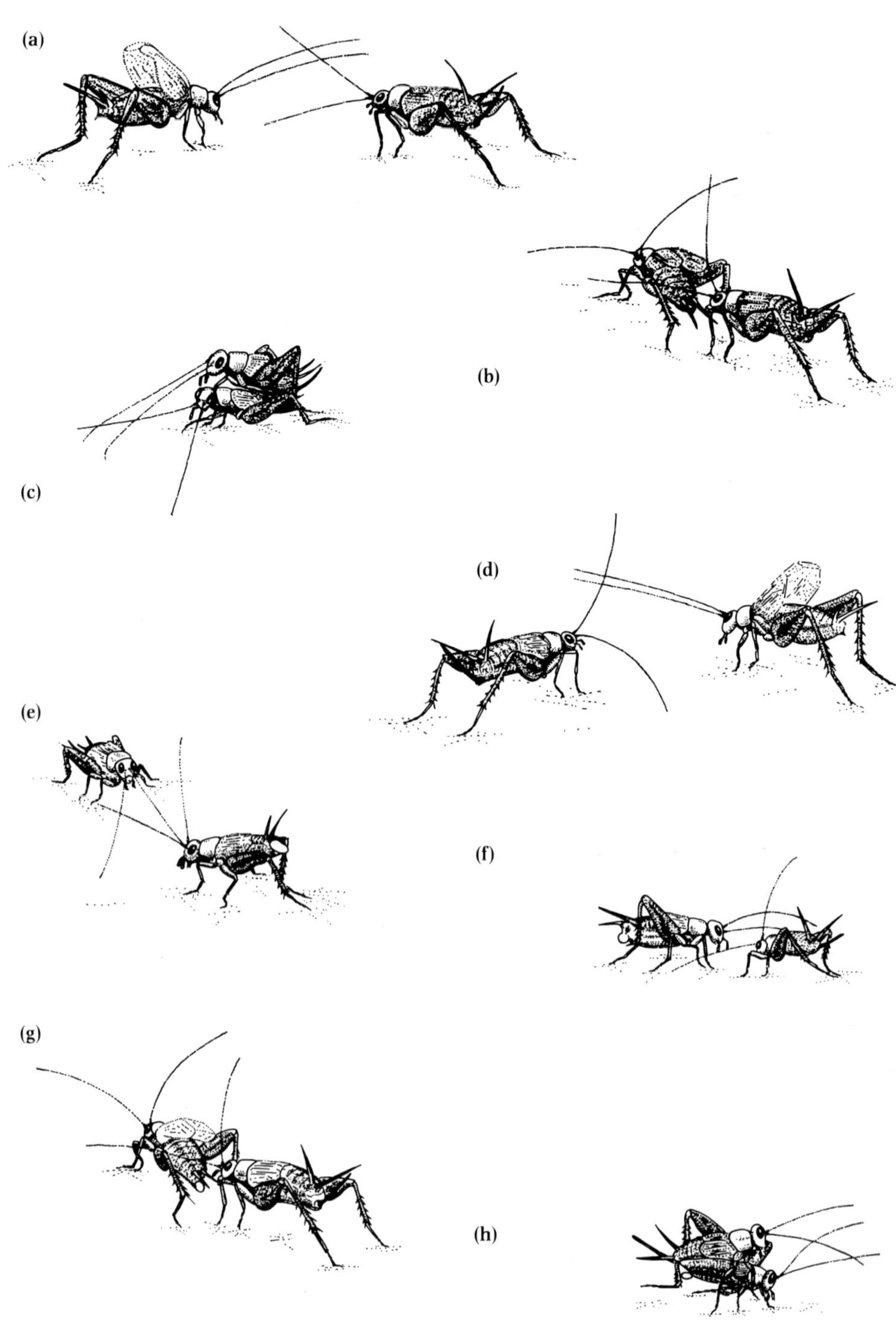

▲ The courtship sequence in the cricket genus *Pteronemobius*. (a) Antennal lashing during initial courtship (male on left). (b) The male backs and trills. (c) The female briefly mounts the male without a spermatophore having been produced. (d) Further courtship. (e) The male (right) produces a spermatophore. (f) Courtship takes place with the spermatophore now mature. (g) The male backs and trills. (h) A much longer session of mounting, during which the female feeds on the male's tibial spur while he transfers the spermatophore. (i) The female eats the spermatophore while the male 'dances' nearby. (After Mays, 1974).

Mate-guarding and ecology

In three species of Australian crickets, the males' varied sexual strategies may be at least partly moulded by the exigencies of their different habitats (Evans, 1988). The highly mobile male of *Balamara gidya* devotes a great deal of time to a mainly silent quest for receptive females, followed by a drawn-out courtship requiring great persistence. He eventually donates a very large spermatophore, which also does duty as a form of non-contact mate-guarding, constituting a mating plug which renders the recipient unreceptive until she has laid her eggs. This is an efficient system for a cricket which lives among dense tangled vegetation, hardly the easiest circumstances in which to carry through a long period of contact-guarding with a female who may be determined to run away.

The male *Bobilla victoriae* differs substantially by employing an elaborate system of calling and searching for a female, and then enticing her with a succession of spermatophores which she is permitted to eat shortly after they are received. Such is the attraction of this multi-course feast that, if more than one female is present, they will fight for the right to copulate and receive their nuptial gifts. If a female tries to cheat by leaving as soon as she has eaten the first spermatophore, the male chases after her and persuades her to stay by vigorously swaying his body, emphasising his intent with an 'aggressive' song.

The mode of persuasion employed by the male of *Teleogryllus commodus* in the same circumstances is to try and maintain contact with his mate by laying his antennae across her body (Loher & Rence, 1978). If she behaves properly he just maintains a watching brief, but if she shows signs of restlessness or tries to eat the spermatophore – he is especially sensitive to the latter eventuality – he brings her to heel by lashing her with his antennae and rocking his body, sometimes accompanying this by chirping. The total post-copulatory guarding time is around 1 hour, a lengthy period which is only viable because they mate within the natural confines of the male's burrow. By attracting her into his home with his vigorous calling, the male secures

Song, territory and male rivalry. Crickets have long been famed for the quality and volume of their singing. Males who use a call system based on a burrow to attract females may utilize the physical qualities of the burrow itself to maximize their acoustic output; louder calls attract more females. Suitable structural modifications may yield increases in volume to the point of being almost painful to the human ear. The Asian *Gymnogryllus elegans* constructs a horseshoe-shaped earthen wall around the mouth of its burrow. Soon after nightfall, each male stands with his back to the open section and gives vent to a truly stentorian call, this being powerfully amplified by the 'megaphone' formed by the burrow-wall sound system.

Although being the loudest singer does generally pay dividends in attracting the most females, it also has drawbacks which tend to dilute the total amount of advantage. Male field crickets *Gryllus integer* call loudly at night from suburban lawns and gardens in the USA (Cade, 1979). As in many crickets, the males fight savagely for top spot in a dominance hierarchy and the right to establish a prime-site calling burrow. The biggest males rely on their large jaws to bully their way to the top of the dominance hierarchy. They increase their own reproductive success still further by seeking out copulating competitors and disrupting their mating activities. Such bullyboy tactics are a logical extension of the need to maximize lifetime reproductive output but the possession of a top-notch burrow, and the powerful voice to go with it, has disadvantages for the heavyweight occupant. The lustiest singers risk attracting not only females but also less welcome visitors – previously silent males of their own species. These home in on a stentorian rival, using the loudness of his voice as a measure of the likely value of his burrow, and then try to evict him and take over his property. Satellite males try to cash in on a burrow-owners's acoustic output by 'sneaking' illicit matings with females on their way to a sexual rendezvous in the caller's burrow. By keeping mum, except for a brief and rather soft courtship song when he spies an approaching female, the satellite greatly reduces his chances of being attacked by the parasitic flies, which normally home in on calling males and whose larvae eventually kill them. But, by being soft-voiced, the male also reduces his chances of actually getting to mate, as most of the females are too fast or too perceptive to allow themselves to be appropriated by such a 'soft-spoken' suitor.

Satellites have not so far been observed in the European field cricket *Gryllus campestris* in which owners seem able to maintain an 'exclusive acoustic zone' around their burrows. This zone varies in size according to the population density and the likelihood of attracting females (Hissmann, 1991). When females are scarce, the larger males often make raiding sorties from their burrows to expel singing neighbours and create an enlarged 'exclusion zone'. Sneaks appear to be absent, possibly because such potentially unproductive behaviour is only worthwile where pressure from parasitic flies makes 'playing dumb' an acceptable alternative to singing.

▲ On Mount Kinabalu in Borneo, a male *Nisitrus* sp. cricket (Gryllidae) serenades his mate with a song produced by rubbing together modified areas at the base of the wings. He has raised the latter into a position where they act as resonators; unlike the deafening chirps produced by some crickets, the sound produced by this species is almost inaudible to humans.

for himself a reasonably safe period of *post-copula* contact-guarding, a luxury which would be impossible for a wandering species. The importance of his alert supervision is clear when he has finally left the female. She soon bends forward so that she can nip off the spermatophore and eat it. She will do this much earlier – with dire results for its cargo of sperm – if the male is not present to prevent it.

Female choice in crickets

Female 'choice' is always difficult to demonstrate. In so many instances, females do not appear to exercise any 'choice' whatever between competing males; in others they 'choose' apparently inferior mates while sometimes they are so inconsistent that no obvious pattern can be established. Nevertheless, female 'choice' does indeed seem to play a large part in the nuptials of the European field cricket *Gryllus bimaculatus*. Females prefer to mate with males who sing from burrows, in which oviposition can also take place. Males fight ferociously for the possession of burrows, with the largest males usually winning. So, by choosing to mate with burrow-owners, females automatically choose also to mate with the larger males (Simmons, 1986). However, the females have an additional method of ensuring that only the sperm from the largest (and presumably fittest) males manages to fertilize their eggs. Although a female may 'agree' to mate with males of various sizes, she normally eats the spermatophores from small males, thus effectively using them as an unintended source of nuptial food. However, the spermatophores from the largest males are left in place to transfer their full cargo of sperm. In addition, the female may also accept multiple matings with a large male, presumably to ensure that any sperm remaining from matings with smaller 'inferior' males will be thoroughly flushed out by the overwhelming influx from the 'superior' male. The female's pay-off for being so choosy is the production of nymphs which grow faster and can therefore enter the 'mating game' earlier than smaller nymphs. This maximizes the chances that her offspring (especially her sons) will enjoy increased mating success.

Sound-baffles in tree crickets

In the tree crickets (Oecanthinae) males generally attract mates by calling from a platform on a leaf. In at least three species of South African *Oecanthus*, the singer exploits the intrinsic capacity of a leaf to act as as an amplifier. *O. burmeisteri* chews out a neat pear-shaped hole in which he sits head-down. His head, thorax and front legs protrude from the upper side of the leaf, at the tapering 'base' of the pear, while his abdomen hangs out on the other side. The wing cases are raised at right angles so as to fill in the rest of the gap, closely contacting the rim with their outer edges. Chirps are produced at 2-second intervals, although they do sound rather subdued to the human ear. Even so, the output of sound produced in this manner is nearly four times higher than is possible without the aid of the foliar amplifier.

Nuptial feeding in tree crickets

When a calling tree cricket finally 'pulls in' a female, he quickly switches to a close-up mode of communication. The black-horned tree cricket, *Oecanthus nigricornis*, draws upon a mixture of signals involving sound, tremulation, scent and taste (Bell, 1979). A characteristic feature of all oecanthine males is the occurrence, just behind the wing bases, of a relatively large metanotal cavity bathed in a secretion from an adjacent gland. During copulation, the female stands on the male's back and laps at the secretion, but retains her hind (and often middle) feet on the leaf. The male's provision of a nuptial meal is a delaying device aimed at keeping the female in place until the spermatophore has emptied. If she leaves prematurely, she often turns and eats the spermatophore. A female who loses her appetite and tries to shuffle backwards off the male is encouraged to return to her meal with an imperative burst of forceful stridulations and vibrations. Should such exhortations fail and she dismounts, her place may quickly be taken by a second female who has been waiting near by. This opportunist is now in a position to defraud the male of his nuptial offerings without herself being inseminated (Bell, 1980). Such kleptoparasitic behaviour is operable because the males require at least 30–60 minutes to form a fresh spermatophore. The importance of nuptial feeding as a delaying device has been demonstrated in the European *Oecanthus pellucens*. A female who has been prevented from feeding on the male's metanotal glands removes and eats the spermatophore after only 2 minutes – 4 to 5 minutes less than the time required for all the sperm to be transferred. The male can

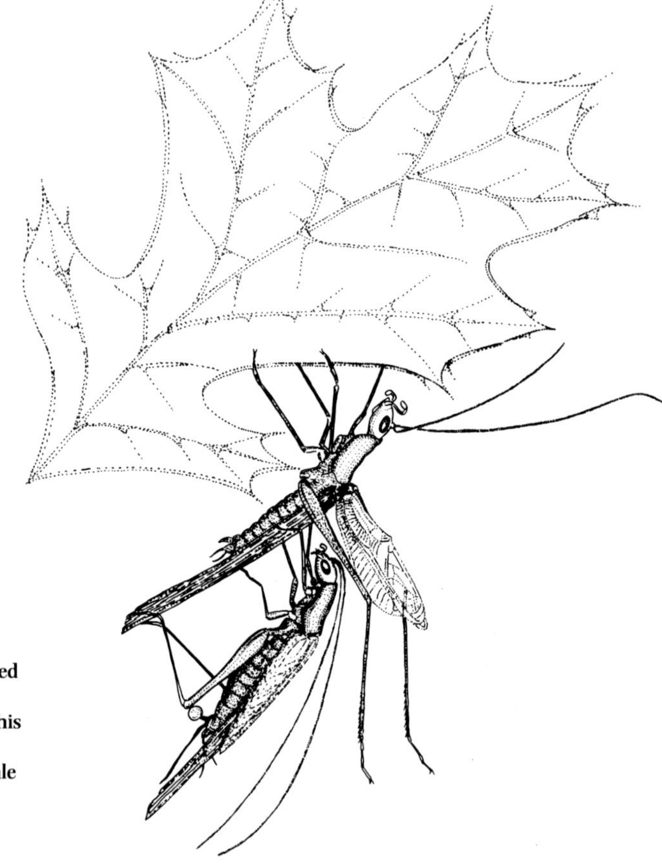

▶ **In tree crickets (Gryllidae: Oecanthinae) the female normally feeds on the male's metanotal gland (situated in a cavity at the base of the upraised wings), once he has passed across the spermatophore. In this two-spotted tree cricket, *Neoxabea bipunctata*, the female is just about to commence feeding.** (After Walker, 1978).

only manage a single copulation each day, but the female often stays with him and mates repeatedly, such a pair often remaining together for up to 6 days.

Sperm-scooping in a cricket

By guarding his mate so as to guarantee a full uptake of sperm, thereby inducing a refractory period, a male reduces the chances that his mate will copulate again in the near future. However, in some species, the females seem to accept repeated matings with different partners in quick succession. Under such competitive circumstances, guarding behaviour would need to be too prolonged to be practical. Males of the Japanese tree cricket, *Truljalia hibinonis* (Podoscirtinae), maximize their individual mating success by using penile sperm-scoops, as in many damselflies (Ono *et al.*, 1989). The male crickets probably suffer from high levels of sperm competition due to the females' habit of mating repeatedly on the same night. This promiscuous behaviour is spread over a considerable period before they finally get around to laying their eggs, which by now is quite late in the season.

In response to the pressures of keeping pace with such a sexual marathon, the males use a system which contributes to the high level of energy reserves needed for the cycle of repeated matings. They simultaneously 'disarm' and recycle their rivals' most vital resource – their semen. As soon as a male has finished copulating, he bends his abdomen forwards (as in a female katydid) and feeds on the semen which is adhering to his genitalia. Experiments have shown that this is not the diner's own semen, but has been deposited by earlier suitors and removed by the male during copulation. It is probably flushed out of the female's spermatheca by the new supply of sperm. Such a method of removal is possible because this cricket is unusual in injecting liquid sperm, rather than attaching a spermatophore. This ensures particularly high rates of sperm precedence for the last male to mate. (Semen-flushing, but not with subsequent oral ingestion, has also been suggested in the migratory locust, *Locusta migratoria*, in which the last male fertilizes 86 per cent of the eggs). *T. hibinonis* males invest large amounts of time, energy (in singing) and body substance (metanotal-gland secretions) in securing numerous copulations, most of which will not achieve the desired aim of fertilizing any eggs. It therefore seems that harvesting the investment made by other males by acting as 'sexual kleptoparasites' on their sperm is a good way of keeping the total losses down. A similar situation exists in many small soil arthropods, in which males recoup some of their investment (in leaving numerous spermatophores lying around willy-nilly) by eating those of their competitors.

GRYLLOTALPIDAE Mole crickets

Mole crickets broadcast their soft churring calls from burrows in the earth. The architecture of the burrow assists the caller in transmitting his message to as many females as possible by functioning as a sophisticated amplifying device, able to convert the maximum percentage of physical effort into productive audible output. In the European mole cricket, *Gryllotalpa vinae*, the 'loudspeaker' part of the burrow takes the form of a letter Y, with the exponentially expanding paired horns at the top reaching the surface as twin orifices (Bennet-Clark, 1970). When calling, the male faces inwards with his head and thorax just inside a bulbous chamber placed near the base of the Y. His raised wing cases are placed against the ceiling, just before the bulb. By sitting thus, the singer utilizes the bulb as a filter to minimize sound loss into the burrow. The length of the bulb equals a quarter of the song's wavelength. In *Scapteriscus vicinus* and *S. acletus*, the horn is single and the caller performs with his wings raised to touch the roof of the bulb, instead of in front of the bulb, as in *Gryllotalpa* (Nickerson *et al.*, 1979). Because the bulb is less than a quarter of the wavelength of the song, a position in front of the bulb would be ineffectual, and the caller's wings can serve to block access to the rest of the tunnel system. When his singing brings in a flying female, the caller responds by rocking from side to side with drooped wings, to a constant churring accompaniment. If she accepts him, she climbs on to his back and he transfers a spermatophore.

ACRIDIDAE Grasshoppers

Unlike most other orthopterans, the grasshoppers are active almost exclusively in daytime. This means that courtship and mating take place under conditions in which it is possible for the participants to see one another. It is not therefore surprising that courtship, where it exists, depends heavily on visual displays involving various parts of the body. Even mate location may be via a visual search by the male. Males of many species sit and stridulate to entice females to come to them, or actively search while occasionally announcing their location by stridulating. Males also engage in acoustic warfare by utilizing the so-called 'rival song'. This differs from the normal calling song in various parameters, such as the volume or the speed of stridulatory strikes. The closer the two rivals spatially, the more distinctly their rival songs depart from their normal calling songs, reflecting an intensification of aggression with increasing proximity – but expressed acoustically rather than by actual combat. However, many subfamilies are almost or completely mute, so the percentage of species deploying sound at some point in the courtship process is relatively restricted.

Purely visual courtship

The tropics abound with brilliantly coloured grasshoppers, most of which are mute. Courtship often seems to be absent – the male simply jumps on top of the female in a 'Lone-Ranger' strategy and tries to avoid being kicked off. The female can always make it impossible for him to couple his genitalia by raising the tip of her abdomen, but male persistence often pays off and she eventually gives in. In some cases, it seems possible that males might improve their chances through having gained the acquaintance of a female over a protracted period of 'engagement'. In the green and gold solanum grasshopper, *Drymophilacris bimaculata*, from Costa Rica, the sexes locate one another visually. Because of its poisonous food plant, the bright colours could normally be expected to serve a 'warning' function, but this is not so. Instead, the gaudy uniform appears to be a crucial 'identification badge' in pair formation and the subsequent maintenance of a loose pair bond. Males and females wear a similar pattern, in which a key aspect seems to be a pair of gold spots contrasting with the black rump. The males search out females on their food plants, announcing themselves by drumming on the leaves with their back legs. A nearby female will answer with a similar vibratory signal, but from then on visual cues are paramount as the two follow each other around, keeping in touch by orienting to the black 'back-marker' spots.

Visual flight displays

A combination of highly visual in-flight wing displays often coupled with crepitation – a loud cracking sound generated by the wings in flight – is the typical mode of sexual advertisement in many band-winged grasshoppers (Oedipodinae: e.g. *Oedipoda, Sphingonotus, Trimerotropis*). In these very cryptic grasshoppers, the hind wings are often red, orange, yellow or blue, being invisible beneath the forewings until the insect makes a brief publicity flight. Generally the males perform, it being likely that crepitation serves to bring them together into rather open groups which are more likely to attract females. Females do not move around much and seldom crepitate. However, it seems probable that, in some species, newly moulted virgins use crepitation as a solicitation display, the more promptly to secure their first mating. This seems to be so in *Trimerotropis agrestis gracewilyae* in Utah (Willey & Willey, 1970), as well as in several other American oedipodines and acridines. It appears that most female grasshoppers are highly receptive only for the first few days after moulting into adulthood. They are easily mated at this time, but play hard to get and are susceptible only to intensive courtship thereafter.

No courtship – the 'Lone Ranger' male

Some grasshoppers manage to mate without preliminaries, while others do so only by 'seducing' the female with an extraordinarily complex repertoire of signals. *Melanoplus tequestae* from Florida merely leaps straight on to the female's back 'Lone-Ranger' style (Bland, 1987). An unreceptive female who wishes to deter such an attempt signals her rejection by turning her back and raising her hind legs, confronting the male with the yellow inner faces of the femora. This appears to be very effective at stopping the males in their tracks, and no mounting attempts are made upon females exhibiting this 'not-now' display. Even when a male does manage to mount, he has to earn the female's compliance through a demonstration of his strength and stamina. This operates on similar principles to a 'bucking-bronco' event; a male who manages to hang on to the wildly cavorting and kicking female for more than three to six jumps proves his worth and wins the female's co-operation. She spreads her rear femora slightly apart so that he can couple more easily. However, he has to prove himself at the first try; if he is thrown off quickly, no amount of further inveigling will secure acceptance. The females apparently do not want males who cannot make a quick conquest, so there seems to be a considerable degree of female choice in this species. This whole 'trial-by-strength' system may be essential because the female must avoid being monopolized by a 'wimp' male during the extremely long copulating period – up to 36 hours.

Simple courtship

Members of the Cyrtacanthacridinae, such as *Melanoplus sanguipes* from the USA, are mute, and the male apparently locates a female by some intensive foot-slogging. Courtship is simple – he lowers his antennae and points them at her, while swaying his head a little from side to side (Pickford & Gillott, 1972). After these meagre preliminaries, he jumps on to the female's back. Experimental evidence shows that sight plays the major role, as no copulations occur with temporarily 'blindfolded' males. In many grasshoppers, pheromones are probably of little importance in the sexual context, but *M. sanguipes* males are adept at distinguishing immature females from adults, and virgins from non-virgins.

Complex courtship

The acme of complexity is reached in two North American species of *Syrbula* (Otte, 1972). Males of *S. admirabilis* demonstrate the use of 18 distinct motor patterns, involving no fewer than seven parts of the body; *S. fuscovittata* comes a close second with 17 separate 'displays'. However, the 'decision' on whether even to mount a display seems to depend on circumstances. When a female *S. admirabilis* responds positively to a calling male by heading towards him, she will accept him after no more than his short 'approach' song – a less strident version of his regular call. Females may also stridulate to attract males, who can therefore guarantee a positive reception. In both instances, the female is confirming her desire to mate, even exercising a certain degree of choice by running to earth a specific caller. Complex courtship by the male is therefore superfluous. However, the context is not always so advantageous. A male who simply spots a potential mate must take the role of supplicant and his reception is liable to be cool.

To win her, he must bring into play his entire portfolio of courtship manoeuvres, which can be divided into successively more complex phases.

(1) He begins by standing alongside and jiggling one lowered antennae before her eyes. There is a hint of stridulation and he angles his drooped abdomen in the female's direction. With a fore-and-aft pitching he starts to intensify the vibrations of his femora, finally kicking out abruptly with both rear tibiae; leg vibration then ebbs quickly away, the whole sequence lasting a mere 5–10 seconds.

(2) This phase lasts from 12 to 46 seconds and is typified by stridulations of increasing intensity, initiated by the femur nearest the female which is held slightly lower than the other, the two acting slightly out of phase. Meanwhile the antenna next to the female is waved up and down through 10–20°. The other antenna remains immobile at first, but gradually joins in to describe an arc of 90°. The palps also join in the performance, being simultaneously waggled up and down during the more vigorous volleys of stridulation.

(3) This phase (16–120 seconds) is marked by out-of-phase movements of the rear femora. He pumps the one nearest the female up and down at 4–7 strokes per second, while the more distant one, which he holds rather higher so that the female can watch it across his intervening body, oscillates rapidly at 32 strokes per second, giving forth a soft buzzing.

(4) He appears to be ready to mount, but droops his abdomen as he walks up to the female, curves it towards her and stridulates. Now he does a rather peculiar thing – he makes rapid dabbing movements with his front nearside leg against the female's leg, as though trying to step up on to her. If she is receptive she 'presents' by elevating her front end, then spreads her rear femora slightly and strongly depresses the tip of her abdomen.

The courtship of *S. fuscovittata* employs many of the same elements, though applied in a different species-specific way. There is no 'tapping' with the nearside front femur (phase 4 above), but there is a wing-flick. In both species, nearby males may interrupt proceedings and try to muscle in on the action. 'Sneaking' occurs when a non-stridulating male tries to waylay a female responding to a rival's call. However, more often, up to four males form an audience near the courting couple, observing events intently but not drawing attention to themselves by moving. In *S. admirabilis*, one or more of this expectant congregation will dive in and try to mount as soon as the female displays her receptivity, thus acting as parasites on the suitor's courtship exertions. In *S. fuscovittata*, a similar band of sexual muggers may actually strong-arm the performer during the final phase of his courtship and pre-empt his attempt at mounting. Unfortunately it is the very length of the proceedings which is the suitor's undoing, there being plenty of time for nearby males to realize what is going on and take steps to cash in on a rival's enterprise.

The only other species anywhere near as resourceful as this are: *Gomphocerus rufus* (11 movements, 6 body parts) and *G. sibiricus* (11 movements, 5 body parts), both from North America, and the mottled grasshopper, *Myrmeleotettix maculatus*, from Europe (10 movements, 5 body parts). As in other acoustic grasshoppers, *Syrbula* males consistently interact with one another via their songs. In *S. admirabilis*, these combine to produce choruses, each male being stimulated to start singing by the song of a rival. Choruses therefore tend to be punctuated by lengthy periods of silence, until one male 'decides' to start singing and the whole thing begins anew. In *S. fuscovittata*, there is more of a tendency for a male to start singing before a rival has finished, a common trait in grasshoppers.

Resource-defence polygyny in a grasshopper

Ligurotettix coquilletti bases its sexual behaviour on resource defence, and the inevitable competition for the prime resource sites may lead to actual physical combat – rather an unusual phenomenon in grasshoppers. The cryptic male establishes a territory on creosote bush, *Larrea tridentata*, in the Sonoran Desert system. He singles out a bush from its neighbours for the low incidence of defensive phenolic compounds in its leaves (Otte & Joern, 1975). Bushes with lower levels of chemical defences are also preferred by ovipositing females, so a male 'owning' such a desirable property is more likely to attract a mate than a male with a less estimable bush. The males call more or less throughout the day. Even so, this may not deter an intruder intent on hazarding a take-over. After an initial 'squaring-up' – during which they raise and shake their hind legs threateningly – the two males get to grips with mandibles and legs, kicking and biting with such abandon that they often fall struggling to the ground. 'Owners' tend to win such fights but, in areas of high population density, even the most aggressive of males may be forced to share his bush with a rival. Under these circumstances, the two males co-exist by ignoring one another's presence, although both sing actively.

Satellites of singers

Goniatrum planum is almost restricted to southern blackbrush, *Flourensia cernua*, in the neighbouring Chihuahuan Desert. As in *Ligurotettix coquilletti*, there is normally only a single singing male on each bush but he may be accompanied by up to three 'satellites', who lurk silently, waiting to intercept a female attracted by the 'owner's' song. By parasitizing the owner's song in this way, they save the energy expended in singing and are less at risk from predators. Whether they enjoy the same degree of sexual success as the singer is, however, unknown; it seems unlikely that they do.

Sperm precedence and mate-guarding

Male grasshoppers usually leave their mates to oviposit alone. During her quest for a suitable egg-laying site, the female may be waylaid and mated by a second male and it is he who will gain the benefit of fertilizing all or most of her eggs. The percentage of eggs fertilized by each of several matings varies from species to species, often depending on the female's internal anatomy. In the migratory locust, *Locusta migratoria*, a male who mates with a female close to ovipositing can expect to father 40 per cent of her offspring, with an equal cost to her original mate. In the desert locust, *Schistocerca gregaria*, being last to mate pays a 100 per cent dividend. The males use a female-defence strategy, seeking out a female who is about to oviposit. The idea is to guard her until she has finished, for it is only immediately after laying one of her several batches of eggs that the female becomes receptive to further copulations.

Little is known about patterns of sperm precedence in grasshoppers generally, but the behaviour of the male elegant grasshopper, *Zonocerus elegans*, from Africa, suggests that he stands to lose everything to a successful usurper. Once the females have mated, they have to delay oviposition until suitable conditions arise at the onset of seasonal rains. Then they gather in egg-laying aggregations which are a powerful lure for unmated males. A female will mate as soon as she is mature, so, with the subsequent delay until egg-laying begins, it is not surprising that the mating system involves an extremely prolonged period of mate-guarding.

When a male first comes across a virgin, he jumps on her back and hangs on tightly until she has finished bucking. A receptive female does not use her powerful back legs to kick him off, and she is then stuck with her rider for anything from a few days to several weeks – at least until the rains come (Wickler & Seibt, 1985). During this long period 'in the saddle', the male mates repeatedly, so that by the time the eggs are laid he has donated numerous spermatophores. Although individually small, these add up to a considerable investment in his future offspring. Efforts to unseat the male reach a crescendo when the pair finally arrive at the crowded oviposition site, which is patrolled by dozens of sexual bandits bent on mate theft. This is by far the most dangerous juncture for the guarding male, as he risks losing his entire investment in time and spermatophores by allowing himself to be displaced at the last moment. So he remains in position as his mate probes into the soft earth. He greets a rival by raising his back legs and pedalling them slowly, warning of the powerful kick to follow if the gatecrasher presses home his attack. This often comes from the front, precisely in order to avoid the legs. With so much at stake, the intruder often pitches in with everything at his disposal, biting and scratching at the consort and trying to pull

▼ Standing up high on his front and middle legs, a male African elegant grasshopper, *Zonocerus elegans*, starts to bicycle with his back legs in a warning to an intruding male to stay away. It is at this point, as the female is beginning to lay her eggs in the moist earth, that her mate could lose his entire investment of the past few weeks if he concedes her to a rival male.

him down. This is not easily done, as the rider retains a very powerful grip on his mount with his forelegs. The chances of making a successful take-over are therefore very slim. Yet it must be worth the effort in view of the enormous reward – the chance to fertilize 100 per cent of an egg batch which has been partially nurtured – at great expense – by the original male's repeated input of spermatophores.

PHASMATODEA Stick insects or walkingsticks

Courtship generally seems to be absent. Males probably locate virgin females through a pheromone which they release when sexually mature. The male simply walks straight on to the female's back where, being much smaller than his mate, he often looks like a different species. He is carried around by the female, who acts as though he were not there. In some species, the male's entire adult life may be spent riding on a single female, although, during such a protracted consortship, actual copulation is very intermittent. The male Indian stick insect *Necroscia sparaxes* may spend as much as 79 days mounted upon a single female, a record for the insect world. Such extended ridership is probably a form of long-term mate-guarding, preventing rival insemination of the female until she has laid her eggs. Egg-laying often takes place over an extended period, which obviously influences the amount of mate-guarding which will be necessary. In the Australian *Extatosoma tiaratum* (among others), females engage in multiple matings and males are able to displace almost all a rival's sperm (Carlberg, 1987).

PLECOPTERA Stoneflies

Courtship largely consists of drumming against the substrate with the underside of the abdomen. *Siphonoperla* species from Europe vary the pattern with an extremely-high-frequency vibration of the abdomen, which makes no further contact with the substrate. The females generally sit around waiting for drumming males to reach them; only virgins respond in kind to the males' calls. In the European *Isoperla grammatica*, the females also take an active part in finding a mate. In some species, a two-way 'discourse' develops, with the male and female drumming to one another until they eventually meet. Drumming is coded and species-specific. (Szczytko & Stewart, 1979).

THYSANOPTERA Thrips

The most interesting aspect of reproductive behaviour in thrips is the fighting and guarding behaviour in the males. Within the suborder Tubilifera, there is a great range of male size and armament, from heavyweight 'oedymerous' males boasting large robust front legs tipped with sharp chitinous points, which act as stabbing blades, to small wimp-like 'gynecoid' males, which are built more like females, virtually bereft of the large males' martial hardware. Between these two extremes lies a range of males of varying stature and with varying degrees of foreleg development. A male's sexual prospects are largely ruled by his size

▲ An enormous difference in size between males and females is typical in phasmids (walkingsticks or stick insects). Courtship is absent and the males typically ride the females for a very extended period, in a form of mate-guarding. This is *Ctenomorphodes tessulatus* from Australia.

◀ A male African elegant grasshopper, *Zonocerus elegans*, clamps on tightly to his mate as a large male intruder makes a determined effort to hijack the female shortly before she lays her eggs. Despite his superior size, the assailant failed in his endeavour, mainly because the males of this species are exceptionally well able to cling very firmly to the females.

and armament. Even so, being a midget does not necessarily exclude him from the mating game, as he has various alternative tactics at his disposal, such as 'sneaking', which may secure at least some access to females at low personal risk. Conversely, being a well-armed bruiser certainly brings its reproductive rewards, but the price to be paid – in certain hard-fighting species – is the increased expectation of an early death due to the risks involved in combat.

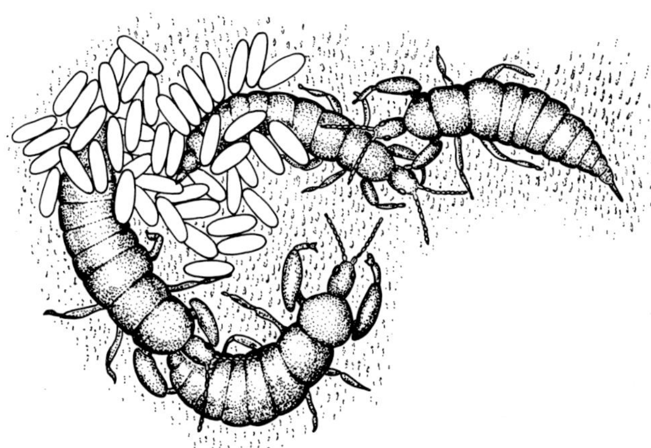

► Male–male rivalry at an egg-mass area in the thrips *Hoplothrips karnyi*. The 'owner' of the egg mass (the large male at bottom left) stabs a large rival, while a smaller 'sneak' (top right) takes the opportunity to try and mate with the egg-laying female. (After Crespi, 1988).

Combat is one of the necessary costs incurred in maintaining access to receptive females. In the European *Hoplothrips pedicularius*, which breeds on bracket fungi on trees, the males set up guard-posts next to communal egg masses. These are regularly visited by females arriving to deposit more eggs (Crespi, 1986). Smaller males bidding to sneak a mating with these females are driven off by the guarders. Matters are not so easy if the intruder is as large, or nearly as large, as the guarder. This results in an escalated contest to decide who will remain the dominant figure at the egg-laying station. During these fights, the opponents use their pointed front tarsi to stab one another on the abdomen from behind. They also wag their bodies in an effort to swipe each other aside; wagging also makes stabbing attempts more difficult to prosecute. Some of these fights become remarkably prolonged – up to 10 minutes – with the outcome apparently being decided in favour of the male who lands the most stabs. The winning count seems to be around 26 stabs, after which the loser gives in, with little or no real damage being suffered.

The behaviour of *Elaphrothrips tuberculatus*, which lives on mouldy oak leaves in the USA, is broadly similar. However, the dominant 'mega-males' of another American species, *Hoplothrips karnyi*, have raised their personal stakes considerably by opting for a 'fight-to-the-death' policy when opposing a challenge (Crespi, 1986 & 1988). They are apparently following the classic 'hawk' strategy which advocates 'escalate until injured or victorious'. This strategy is fairly rare among animals generally, but is also found occasionally in mites, spiders, fig wasps, gall aphids, bees and beetles. In *H. karnyi*, there is no question of restricting contact to a few harmless skirmishes, followed by 'sensible' retreat when the going gets too rough. This is probably because of the enormous reproductive benefits in being a guarder, compared with an alternative which, in terms of reproductive success, is not much better than being dead. The dominants sit on guard near the communal egg masses, where they are well positioned to meet incoming females. These always allow the guarders to mount and mate at will. Resistance is noticeably absent, presumably because the females 'know' that guarders are guaranteed to be top-notch 'fit' males. Guarders therefore obtain far more copulations than non-guarders, who have to resort to brief liaisons with females before they arrive at the egg-laying site. If last-male sperm precedence applies in this species (as it almost certainly does) then such a

Aggressive mate theft. Combat during take-overs of mated females is generally rare in stick insects. However, there are exceptions, such as *Diapheromera veliei* from North America. The male is equipped with robust hooked 'fighting spines' jutting out from the mid-femora (Sivinski, 1978). These spines are present, but in a much less well-developed form, in the females and in many other phasmids. The events preceding actual battle are initially given over to attempting to settle the issue without resorting to conflict. The consort's reaction to an approaching rival is to protect his interests without risking the dangers involved in coming to blows. By using his anal claspers to turn the tip of the female's abdomen inwards, he places her genitalia out of reach of the intruder. This gambit is usually doomed to failure especially when the challenger has the weaponry to make a fight of it. As he forces himself in, a fight becomes inevitable. The combatants use their spines as picks, raking them across the opponent's body and sometimes drawing blood. The battle takes place with the combatants suspended by their anal claspers from the female's rear end. This leaves both front 'fighting legs' unhindered, something which is only possible because the anal cerci are modified into powerful claspers. This character is not universal in the phasmids and may, like the well-developed 'fighting spines', be linked to the pugilistic habit. Consorts tend to be rather poor at defending their property, losing the fight more than 50 per cent of the time, after an engagement of anything from just a few seconds to several minutes.

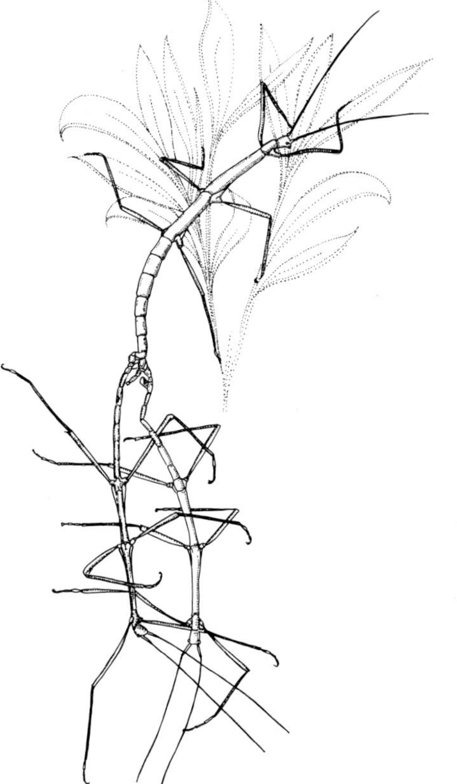

▲ Male–male combat in the stick insect *Diapheromera veliei*. Both combatants are only attached to the female by their rear-mounted clasping organs, giving them free rein to lash one another with their spined legs. (After Sivinski, 1978).

tactic would be of little merit, as the female is sure to be re-mated by a guarder when she goes to lay her eggs.

A second alternative strategy is also available – to become a 'sneak' within the 'danger zone' around the egg mass. To be successful, the sneak must be able to intercept incoming females, while at the same time being constantly ready to flee – he must avoid having to challenge the guarder. Sneaking really comes into its own when the guarder, as sometimes happens, is confronted with an excess of females – up to eight may turn up to oviposit at the same time. The guarder then runs the risk of falling short of semen, allowing a competitor to obtain productive fertilizations away from the egg mass. He virtually eliminates this eventuality in an interesting way. After climbing on to a female, he turns around and partly embraces her abdomen, apparently 'measuring' its size in order to decide whether she might already have laid her eggs. This enables him to mate preferentially with females who have yet to oviposit, ensuring that none of his semen is wasted.

Guarders do not win their sexual laurels easily. They have to beat an existing guarder for the privilege, and then they have to keep on battling away to defend their position against challengers aiming for higher status. These are drawn mainly from the ranks of large males who have recently pulled off a mating with a female in the egg-laying area. The ferocity of takeover battles tends to vary according to the relative sizes of the contestants. Well-matched opponents generally fight harder and longer, as both are 'convinced' that there is a good chance of winning. The accompanying stabbing matches are serious, in particular the stab-and-hold tactics which inflict nasty wounds, leading to the loser's death within 36 hours or so. A guarder never backs off from a challenge, always doing whatever is necessary to maintain his status, even if that means fighting to the death. It is not only his position which is at stake – if he succeeds in mortally wounding his opponent, then he has removed one more competitor from the ever-present pool of sneaks and potential challengers.

HEMIPTERA True bugs

In both suborders (Homoptera and Heteroptera), stridulation probably plays a major role in bringing the sexes together, but only cicadas are audible to humans.

Cicada songs

In cicadas, it is the males who sing and the females who respond by flying or walking to join them. In the 17-year periodical cicadas *Magicicada septendecim* and *M. cassini* from North America, the males' synchronized songs – their 'common calls' – draw in individuals of both sexes, so that a group of trees radiates a tumultuous sound beacon far and wide (Covalt Dunning *et al.*, 1979). In-coming males quickly add their own contribution to the clamour, singing one or two phrases of their calling song before fluttering to a nearby twig for another short burst. This peripatetic mode of singing is continued for a while and, in *M. cassini*, the males may co-ordinate their song-phrases and movements, so that the trees are a constant bedlam of song and movement. (This synchronized singing, forming a composite sound beacon, parallels the synchronous flashing of *Pteroptyx* fireflies, which constitutes a composite light beacon). Real courtship only starts when a male settles down and sings from a fixed position, using the same song but in an unbroken form. Any mute cicada is likely to be a female, so a singer approaches any silent companion, often using a slightly modified version of his song with shorter gaps between phrases. A receptive female allows him to stroke the ends of her forewings, at which point he shortens his song-phrases to build an up-beat tempo, and mounts. Once coupled, he relapses into silence.

In the Australian tick-tock cicada, *Cicadetta quadricincta*, the usual male–female roles are reversed and it is the females who are sedentary while the males move around. A receptive female responds to a male's song by flicking her wings, generating a characteristic sound which the male recognizes during his roving method of self-advertisement. He manages to cast a wide acoustic net for scattered females by emitting only 30-second 'sound bites' from each perch before flying on to another. This itinerant strategy appears to entail much higher risks for the males, who fall prey to predators such as web-building spiders far more often than the static females (Gwynne, 1987).

Song in hoppers

Many of the smaller homopterans, such as leafhoppers (Cicadellidae) and planthoppers (Delphacidae) also employ stridulation, which functions in several ways. As a 'common song', it advertises their presence; as a 'courtship song', it initiates sexual contact; as a 'pairing song', it is used after copulation has started; and as a 'rivalry song', it is employed between males of the same species.

Dalbulus males (Cicadellidae) from Mexico and Guatemala all employ a similar courtship routine (Heady *et al.*, 1986). As in cicadas, the male's common call accelerates somewhat during initial courtship, but the females are not mute and reply with their own song in a duet which brings the pair close to one another. Such duets have been recorded in many other species of planthoppers and leafhoppers. The male now launches into his 'song-and-dance' routine, darting jerkily around the female, occasionally patting her with various parts of his body while varying his song-line with the odd burst of wing-buzzing. An effort to couple may be met by a hefty kick from the female's hind leg, but a receptive female allows him to stand beside her and connect his genitalia. The females only mate once, which leads to considerable competition for their favours. When two rivals confront one another in the presence of a female, they will settle the issue of access to her with a 'singing contest' rather than physical combat. The specific 'rivalry song' now comes into play, but just how the contestants 'know' who is the 'best' singer and therefore 'top dog' is not known.

Female solicitation in treehoppers

Treehoppers (Membracidae) often live in dense aggregations on twigs. The opportunity therefore exists for a receptive female to solicit several suitors and allow them to compete for her. This is what seems to happen in the thorn bug *Umbonia crassicornis* in which two types of mating strategy are apparent: 'male search' and 'female solicitation'. In 'male search', a would-be suitor wanders around the food plant in a quest for females. If successful, he may just sit beside her or, if she is already walking, he will trail along behind her, fanning his wings with an audible buzz (Wood, 1974). In 'female solicitation' the female appears to tout for males by fanning her wings. It could be the resulting buzz which is the main attractant, or possibly a pheromone which is blasted far and wide by the wings. Either way, a gaggle of males soon congregates around the buzzer, often climbing on her back and remaining for several days as she continues to walk around fanning and buzzing. Eventually one of the riders succeeds in mating, although how he beats his rivals to the draw is unknown. A solo male will also seek out and stay with a female in this way and protracted 'riding' seems to be an essential preamble to copulation, common in many tropical membracids.

Stink-bug singing and 'calling'

In many stink bugs, the males 'call', releasing a pheromone which attracts females, and often other males as well. *Eurygaster integriceps* 'calls' by making rhythmic up-and-down movements of the upraised abdomen (Zdárek & Kontev, 1975). This disseminates a pheromone which smells strongly of vanilla. Males of the green stink bug, *Nezara viridula*, release an aggregating pheromone which serves to assemble numerous individuals of both sexes. However, scent only fixes up the rendezvous and, in later interactions, the males employ seven discrete songs, including a 'duet' with rival males and two different call–answer scenarios with females, who can themselves produce three different songs (Harris *et al.*, 1982). Song routines in *Nezara viridula* from the USA differ somewhat from those in a southern European strain, suggesting the evolution of 'dialects' in this widespread species. Stridulation alone, or in combination with pheromones, is thought to assemble the dense mating aggregations of bugs such as those seen in the European *Eysarcoris fabricii*, which may contain hundreds of individuals. These can be very enduring, lasting several weeks. In the European pied shield bug, *Sehirus bicolor*, the males are extremely persistent, spending the entire day chasing after females on their food plant. The males' powers of discrimination are deficient and they will butt against anything remotely resembling a female, including juvenile wolf spiders (pers. obs.). Butting against the female's side is a common element in stink-bug courtship. There are three male courtship songs and a single female one. An unreceptive female responds by leaning over towards the male, with her nearside touching the leaf, preventing him from levering up her carapace to expose her genitalia.

The procedure in *Murgantia histrionica*, a pest species in the USA, is probably fairly typical for many pentatomids (Lanigan & Barrows, 1977). With vibrating antennae, the male runs up to the female, at which point she either 'squats down' on the leaf or makes a run for it. If she squats, he launches into a prolonged stimulation of her body with his antennae. The female signifies assent by tilting her rear end upwards, enabling the male to connect his genitalia end on. Behaviour in the cryptic American *Brochymena quadripustulata* incorporates one rather unusual component (Gamboa & Alcock, 1973). After grabbing a static female with his front legs and hanging on tight, the male's next step, if he is not carried off head-long, is to position himself above her head and perform a crabwise 'quick-step' to and fro across the front part of her body. This highly unconventional 'caress' encourages a receptive female to elevate her abdomen. After some 20–30 side-steps the male moves to one side, strokes her with his antennae, and then sticks his head beneath her rear end and

► In many stink bugs, the male butts against the female's sides. She responds by tilting over, preventing him from shoving his head in beneath her. This is *Sehirus bicolor* from Europe, a species whose males are poor at discriminating between females and other small invertebrates. He is likely to head-butt the wolf spider on the leaf by mistake, something he can do with impunity; this bug is distasteful to wolf spiders and immune to attack.

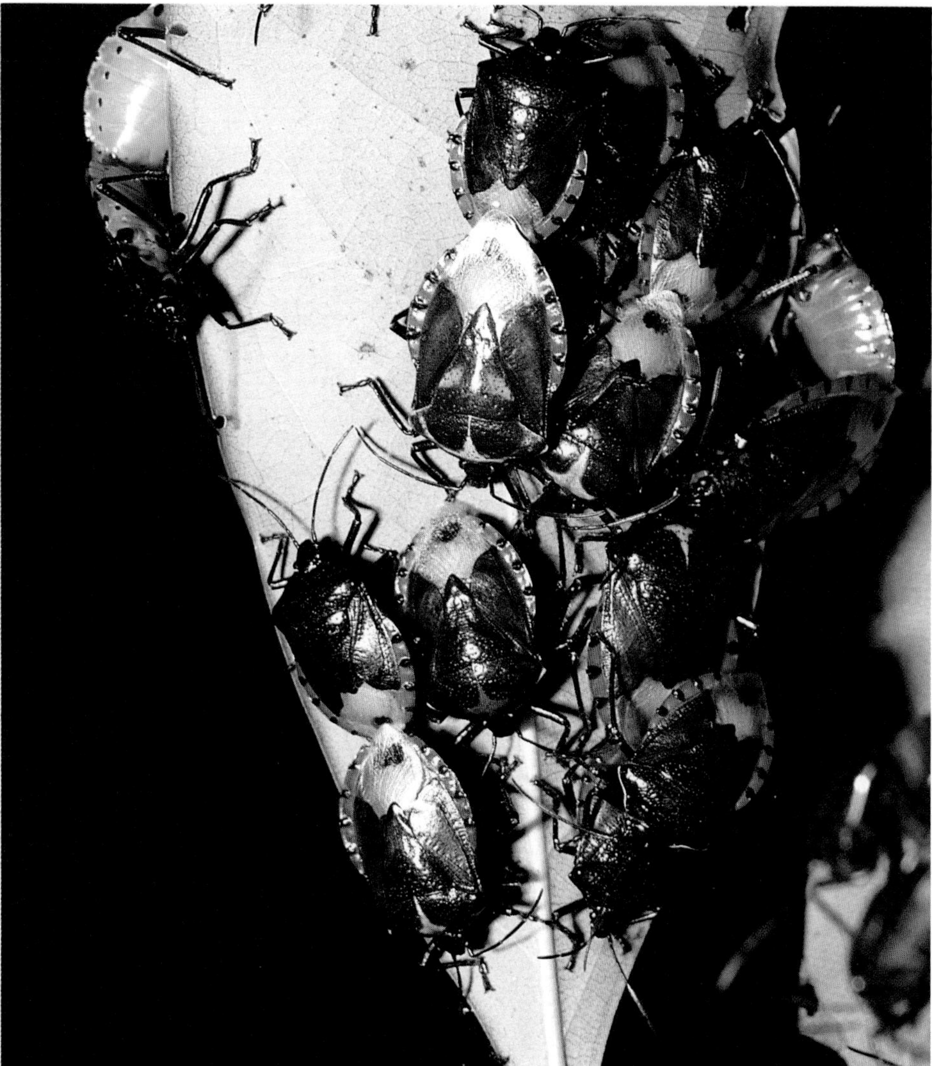

▼ Mating aggregations in stink bugs may be remarkably long-lived. Mating pairs of these *Vulsira violacea* (Pentatomidae) from Costa Rica were present on this same leaf for at least 2 weeks.

beats a rapid tattoo against her underside with his antennae. Genital linkage is achieved end to end, although in other pentatomids it may be initiated with the male above the female and facing in the same direction (e.g. *Brachymena, Dolycoris, Calidea*), or with the female on top (e.g. *Podisus*.)

In-copula mate-guarding

In bugs it is not unusual for mating times to last several hours. However, in the highly promiscuous circumstances of the North American ragwort seed bug, *Neacoryphus bicrucis* (Lygaeidae), the males tailor the duration of copulation to the density of the population, and hence the degree of likely competition for their mates once they have released them. By hanging on to his mate for as long as possible, a male reduces her chances of re-mating that day and supplanting his sperm through sperm competition (McLain, 1989). In the equally promiscuous European *Lygaeus equestris*, mating times can be anything from half an hour to 24 hours (Sillén-Tullberg, 1981) and sperm displacement is 90 per cent effective. Small wonder that the longest matings occur when a male is seeking to maintain control over a heavily gravid female. Her soon-to-be-laid eggs will be fertilized by his sperm – but only if he can keep her away from his many rivals.

Courtship in lygaeids is quite interesting, with male stridulation. In *Ligyrocoris diffusus* from the USA, the males stridulate as part of a courtship 'dance' (Thorpe & Harrington, 1981). The male see-saws his body to and fro, regularly thrusting his abdomen downwards to produce sounds from opposed stridulatory mechanisms on the abdomen and back legs. Stridulation is found in at least 18 genera within the family, although in *L. diffusus* it appears to function as a species-separating mechanism rather than a 'seductive overture' aimed at arousing the female. Muted males mate just as successfully as singers, but the females react irritably and negatively to the songs of the wrong species.

The male African *Stilbocoris natalensis* presents the female with a nuptial gift – the only hemipteran known to utilize this form of courtship. The male spears a fig seed kebab-style on the tip of his rostrum and injects it with saliva to pre-digest its contents. If he presents this to the female, he is usually accepted – but males who come minus gifts are rejected. This is in line with other insects which present nuptial gifts prior to mating, and 'no gift – no copulation' appears to be the norm.

Leg-fights in a coreid

In many coreids, the males patently have much fatter and spinier rear legs than the females. Despite the 'obvious' suggestion that these must be weapons for use against rivals (and therefore evolved by intrasexual selection), this has only quite recently been confirmed. *Acanthocephala femorata* males defend territories on the stalks of their food plant (the sunflower, *Helianthus annuus*) to which females may be attracted to feed (Mitchell, 1980). A territory-owner reacts to a trespassing rival by splaying his hind legs widely apart and turning his back on the interloper. If he does not like the look of the owner's gladiatorial hardware, the intruder may just turn tail and run, instead of taking up the implied challenge. If he 'decides' that he has a chance of winning he stands his ground and strikes the same pose, back-to-back with his rival with their rear legs touching. In a trial of strength, each combatant strives to enclose the other's abdomen in the gin-trap of his hind legs, making good use of the formidable femoral spines. Well-matched contestants may strain away in this fashion for quite a while. In a poorly matched contest, the smaller male usually yields early on, doing his best to get himself off the hook unharmed. Actual wounds seem to be rare, despite the lethal appearance of the femoral spines and the amazing power with which they are squeezed home.

Harem-defence polygyny in a coreid

Examples of the defence of several females by a single male are rare in insects. In the winter cherry bug, *Acanthocoris sordidus*, from Japan, males use their 'fighting legs' to establish ownership of small 'harems' on the food plant (Fujisaki, 1981). Females tend to congregate naturally while feeding, provoking intense competition between males for ownership. Once established, an 'owner' wins all subsequent take-over fights unless the challenger is substantially larger. Males do not defend plants without females and soon abandon a plant when the members of their harem disperse to lay their eggs. In addition, the bigger the harem, the bigger the male who defends it, indicating an ability to assess its value before contests over ownership. Oviposition follows soon after dispersal, leaving little or no chance for the females to be found and inseminated by haremless males.

Surface waves and resource defence in water striders

In the calmer backwaters of certain Australian streams, the male of the water strider *Rhagadotarsus* cf. *kraepalini* attracts females to a pre-chosen mating and oviposition spot by generating coded surface waves with his legs (Wilcox, 1972). The waves consist of two types: large amplitude, constituting a calling signal designed to lure females; and low amplitude for close-up courtship. When summoning a female with his calling signal, the male clings with his front legs to some suitable oviposition material, such as a floating stick. He switches to courtship waves when he detects the female and she responds with her own courtship ripples. He relinquishes his oviposition resource to the female during mating, which only lasts a minute or so. The male then floats nearby and generates post-copulatory signals (synonymous with the courtship signals), presumably as a 'reminder' to his mate to stay and lay her eggs within his personal piece of wood. Males often respond to one another's emissions and attempt to take over a calling resource, provoking a specific aggressive ripple from its owner, followed by physical assault if the intruder fails to back off. If the males can monopolize the best of the very patchy egg-laying substrates, the females have little choice but to mate in exchange for guarded oviposition.

NEUROPTERA Lacewings, mantispids

Green lacewings (*Chrysopa* in the broad sense) communicate with one another via substrate-borne vibrations (Henry, 1979). Each species has its own special 'calling code' which probably helps to prevent time-wasting interactions between closely related species from the same habitat. The low-frequency vibrations are not produced by direct drumming against the leaf but are generated by rapid vertical oscillations of the abdomen, which transmit vibrations down through the legs into the leaf. In *Chrysopa carnea*, perhaps the commonest species in both Europe and North America, the male sits beneath a leaf and broadcasts in volleys lasting only $\frac{1}{3}$ second at 1 second intervals. A sexually receptive female within around 15 cm of the caller will reply with a similar call, striking up a duet. Still transmitting, the male finds the female and bows low before her while folding his antennae back beside his wings. She responds by quivering her antennae against his head and they both twitch their bodies vigorously. The female now walks forward to 'kiss' the male, during which action food or pheromones may be exchanged, as certainly happens in many flies. Suspended by their front legs, they now swing downwards, using their interlocked mouthparts as a hinge. The male produces a final burst of vibrations and they couple, end to end. Other *Chrysopa* species exhibit similar procedures (Henry, 1980), with a few variations such as the absence of a female response call. In some European *Meleoma*, courtship is more elaborate and the female feeds from a cavity situated on the male's head, while his antennae are provided with hooks for grasping the female.

Pair formation in mantispids has rarely been observed, but has been studied in a few species. *Climaciella brunnea* from the southwest USA is one of a number of American mantispids which are superb mimics of different *Polistes* wasps. As visual mate location could therefore be unreliable, the use of species-specific pheromones has evolved. To avoid possible confusion between his own females and their model (the wasp *Polistes fuscatus utahensis*), the male *C. brunnea* broadcasts an olfactory message which can only be deciphered by the correct recipient (Batra, 1972). He produces a very powerful pheromone which is easily detected from as much as 3 m away. This presumably accounts for the aggregations of females sometimes noted around males. During courtship, the male merely raises and lowers his spread wings while 'rowing' with his front legs. Copulation may last as long as 24 hours, during which he attaches a white spermatophore to the female. She carries this around for anything from 24 to 36 hours, during which time it is gradually absorbed into her body.

COLEOPTERA Beetles

Complex pre-mounting courtship is relatively rare in this vast family. Males more often simply mount the females without any preliminaries, although not necessarily without some difficulty. In order to 'capture' a mate, the male ground beetle *Pasimachus elongatus* from the USA has to outrun her, before hurling himself on to her back and clamping his jaws around the rear of her thorax. Then he deftly flips her over on to her back – with himself underneath (Cress, 1966). After adroitly transferring his jaw-clamp to the space behind her eyes, he begins a rather belated courtship by kicking her abdomen with his back legs and probing with his genitalia. Once she opens her genital pore, he copulates for a minute or so, then unclamps his jaws and gives her the push-off – literally.

Among the vast hordes of rotund-bodied more lumbering kinds of beetles, nuptial relationships are less feverish. The male clambers slowly up on to the female and rocks from side to side, trying to couple his genitalia with hers. This probably provides tactile stimulation and is often backed up by stroking with the legs and antennae. The male may also delicately nibble the female's thorax or head with his mouthparts. She often rocks violently from side to side in an effort to dislodge her unwelcome rider, while pushing him vigorously backwards with her rear legs. Male persistence is characteristic and seems aimed at wearing the female down until she gives in. However, persistence is not always possible, nor advisable. In *Pterostichus lucoblandus* (Carabidae) from the USA, the female squirts an over-insistent suitor full in the face with a spray from the tip of her abdomen (Kirk & Dupraz, 1972). This instantly slows him down and immobilizes him within a few seconds, plunging him into a corpse-like state for anything from 1 to 3 hours. Only the males are affected in this way; the females are immune to one another's discharges. However, this does incur a cost for the female, who has to divert resources to replenish her chemical supplies; it would be better for her if the males were less pushy, one of the many examples in which male interests (to mate as often as possible) and female interests (to mate only when necessary, probably once only) do not coincide.

Bringing the sexes together

In many cases, sexual liaisons are engineered by pheromones, which probably play a greater part than so far confirmed. In some species, it is obvious when pheromones are being released because the beetle adopts a characteristic 'calling' stance, as in some lycids and scarabs. In the males, the antennae are often longer and sometimes conspicuously pectinate, increasing their

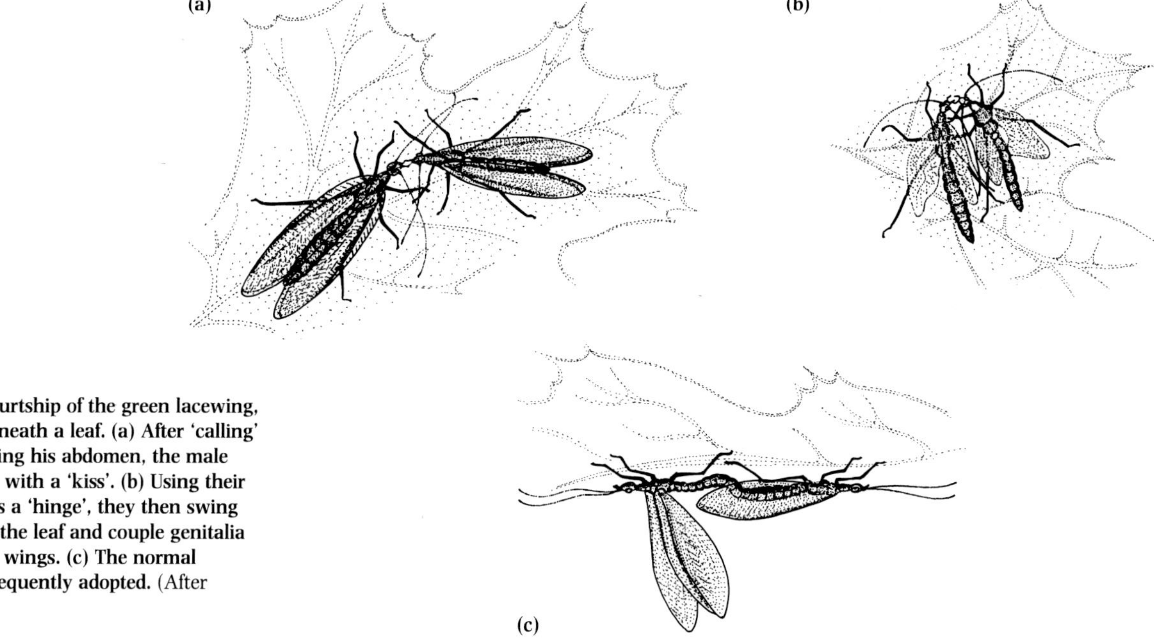

► Three stages in courtship of the green lacewing, *Chrysopa carnea*, beneath a leaf. (a) After 'calling' to a female by vibrating his abdomen, the male eventually meets her with a 'kiss'. (b) Using their locked mouthparts as a 'hinge', they then swing downwards beneath the leaf and couple genitalia while fluttering their wings. (c) The normal mating position subsequently adopted. (After Henry, 1979).

► In many beetles, the male does not perform any preliminary courtship; he simply mounts the female and rocks from side to side, attempting to couple. In this simple way, a persistent male often eventually persuades the female to accept him. Such behaviour is typical in weevils such as this giant *Entimus granulatus* from Peru.

sensitivity. Sometimes it is purely the female's physical appearance which acts as a releaser for the male's sexual activities. This can have awkward consequences for the males if 'counterfeit' females are present in the environment. In Australia, males of the buprestid *Julidomorpha bakewelli* are sexually 'hooked' on a local type of beer-bottle called a stubbie. The problem here lies in the preference shown by some male insects for mating with the largest 'female' available, so that giant-sized models of females may act as super-normal releasers for copulation. This seems to be the case with the stubbies, which have a shiny brown tuberculate strip around the base, closely resembling the punctured wing cases of the female buprestid. This fools the male into treating the whole bottle as a giant version of the female. So powerful is the attraction that the males will allow themselves to be eaten alive by ants, rather than check their frantic efforts to shove their genitalia into the unyielding glass (Gwynne & Rentz, 1983).

Sound plays a part in the sexual behaviour of many beetles. Stridulation occurs in a variety of families, including Cerambycidae, Scolytidae and Passalidae. Production of sound via percussive means is also found in a number of beetles. Males of the notorious death-watch beetles (Anobiidae) bang their heads against their gallery walls to attract a mate. Some *Psammodes* (Tenebrionidae) from southern Africa are locally known as 'tok-tokkies', an onomatopaeic word derived from the male's habit of noisily tapping the underside of his abdomen against the ground. The sound produced is so loud that it can easily be heard several metres away, so it probably serves as a fairly long-distance attractant for passing females.

Some beetles assemble in landmark-based swarms, this being common in some chafers. Swarming is often restricted to certain weather conditions. The trigger in the Australian scarabaeid, *Heteronyx obesus*, is showery or thundery humid conditions near dusk (Morgan, 1977). Large numbers of both sexes emerge from the soil *en masse* and swallow large quantities of air. This inflates the oesophagus and crop, presumably to help maintain the energetic flight during swarming activity. The swarm itself is a turbulent and disorderly affair, the members continually blundering clumsily into one another, while hordes of males descend in clusters upon females who perch. The South African scarabaeid, *Platychelus brevis*, carries out its swarming close to the surface of desert sand-dunes (Louw, 1987). Any female who alights on the sand and tries to dig herself in will be quickly covered in a knot of males, all frenziedly kicking and shoving in an effort to capture her. A ball of grappling beetles may roll down the face of the dune before one of their number manages to dig down behind the female and mate with her beneath the sand. Scramble competition on a similar mass scale, even including the formation of 'mating balls' is also seen in certain hunting wasps and bees.

The lure of light

Light emissions bring the sexes together in the glow-worms or fireflies (Lampyridae). Two signal systems are used:

(1) The females sit motionless on vegetation and emit a constant light, which attracts flying males on the lookout for such an invitation. In *Phausis reticulata* from the Appalachians, the male also signals while searching, stimulating a non-transmitting female into switching on her beacon when she spots him.

(2) The males attract the attention of resting females by flying around in darkness and displaying an amazingly bright light, either as a continuous glow or as a sequence of flashes. The light is flashed in a species-specific code which may vary in colour, length of flash duration, number of flashes in a given sequence, and time interval between flashes.

In the American *Photinus macdermotti*, the flashes occur in pairs about 2 seconds apart, each pair being repeated every 4–6 seconds or so. The female has her own specific reply, giving a single flash about 1.2 seconds after the male's second flash. Rival males may seek to interrupt this visual dialogue and hijack the female by interposing flashes of their own, although the way in which this operates varies from one area to another, with distinct local idiosyncrasies (Carlson & Copeland, 1988). Each firefly species tends to fly at its own particular height above the ground and restricts its main flight activities to a certain period of

the night, such as just after dusk. This helps to reduce visual interference and possibly time-wasting confusion where two or more species share the same habitat.

The main drive behind the incredibly varied and sophisticated multiple-code signalling systems used by many North American *Photinus* was presumably the need to outwit the capabilities of predatory female *Photuris* fireflies as visual tricksters. Such a complex signalling system is not found in Old World fireflies, which do not have to contend with the hazards posed by *Photuris* females (*Photuris* spp. are restricted to the New World). It is therefore especially fascinating that some *Photuris* males perform a kind of 'double bluff' by mimicking the signals of their normal prey. Some *Photuris* males are amazingly versatile, being able to mimic the signals of at least two other species of *Photinus* or *Pyractomena*, as well as having their own species-specific flash code. Just as important, they are capable of deploying their deception in a convincing manner, flying not only at the same height but also during the same period of the night as the species being mimicked (Lloyd, 1980). This is maintained on a seasonal basis, so that the appropriate behaviour is expressed according to seasonal changes in the availability of the models. During periods of the night when neither *Photinus* nor *Pyractomena* is active, the *Photuris* male can switch to using his own flash code, as he also does when the flight season for his models is over. With so many species involved, the whole situation is incredibly complex, and probably forms the most sophisticated system of behavioural mimicry so far recorded in any animal.

One problem remains: what are the *Photuris* males up to? They could be angling for meals-on-wings, as practised by their own females upon *Photinus* males (Lloyd & Wing, 1983). But it seems certain that their own females are the real target. Once a female has mated, she is likely to be on the prowl for a meal, rather than seeking further sexual liaisons. The *Photuris* male's mimicry of just the kind of meal she is looking for – a male *Photinus* or *Pyractomena* – is apparently designed to lure her up close. This strategy of drawing her in under false pretences is necessary because she would be unlikely to respond to one of her own males flashing a courtship invitation. A potential meal is another matter, and is likely to attract her attention. If a meal is what she expects, then this may not be just an extra copulation for the fraudster, but his last as well if the female (who is much larger and more powerful than he is) still insists on grabbing a quick bite afterwards. The possibility of this happening suggests a further reason for the mimetic male's signals. Perhaps old *Photuris* males who are sexually almost 'past it' switch to mimicking meals as a last-gasp 'kamikaze tactic' aimed at securing one last suicidal copulation. If the geriatric signaller ends up being eaten, so much the better, for his body may benefit the last offspring he will ever father. Offering himself as a parting 'nuptial gift' would therefore make sense.

In the Asian genus *Pteroptyx*, the males mount a synchronized light display in which thousands of individuals crowd into a tree and flash in unison. The impressive synchrony takes place not only within a single tree but also between individual trees spaced out along a riverbank, so that numerous beacons of light suddenly appear and are equally abruptly extinguished. *Pteroptyx* males may congregate in millions in shoreline trees. The flash rhythm is species-specific so that both males and females are attracted towards a given tree. In *P. tener* from Malaysia, the males vacate their daytime resting places around dusk and quickly establish their synchronous display in nearby mangrove trees (Case, 1980). While moving they give low-intensity flashes, changing to high-intensity display flashing from a prominent fixed position. Females, who also flash (though never in synchrony with the males), are attracted to the display in its totality, yet upon arrival may show interest in a male regardless of whether or not he is flashing. This is puzzling, as in other species the females zero in on flashing males and may 'choose' between competitors on the strength of their light output.

During courtship itself, the male sits on the female's back and vigorously taps her abdomen with his rear legs. He also twists his abdomen round so that its tip hovers above her eyes; then he blasts away with his light at point-blank range. The reason for this apparent optical overkill is uncertain, but it could be a way of 'blinding' the female to the enticing array of lights all around her. The close-up display may also stimulate receptiveness as, just before he makes each of his frequent copulation attempts, the male interrupts a 2- to 3-second dark period with a volley of exceptionally intense flashes. The female also flashes during courtship, often just before the male's dark period. Males also flash brightly when approached by intruding rivals. Courtship is therefore exceptionally complex, contrasting strongly with the habits of non-synchronizing species, in which courtship is absent and light emission rare once the pair has got together.

Lampyrids are not the only beetles which use light as a sex attractant. The so-called fire beetles, *Pyrophorus* (Elateridae), from South America bear two brightly glowing spots on their pronotum, displaying these in impressively rapid vertical ascents into the forest canopy.

Fighting in horned beetles

One aspect of beetles which has intrigued human beings for centuries is the apparent heavyweight 'fighting equipment', such as the enlarged mandibles of stag beetles (Lucanidae), so prominently carried by many of the males. Despite their 'obvious' utility as weapons, the precise function of these overgrown appendages has been much disputed. Observations in nature and under experimental conditions have now established beyond reasonable doubt that their main purpose is in combat between rival males to decide access to a mate. This can be either directly, in an argument over

▲ A sap-run on a rainforest tree is the perfect lure for both sexes of this Kenyan scarab *Chelorrhina polyphemus*. Although the males do not have the extravagant horns seen in some species, they do use their nasal adornment in fights with other males to maintain a territory on a such a sap-run (resource-defence polygyny). The 'owner' is then able to mate with any females who come to feed on the fermenting exudate.

a nearby female, or indirectly, to contest ownership of certain resources which will attract females e.g. oviposition sites. Thus two rival stag beetles coming face to face at a sap-run on a fallen tree will vigorously attempt to grip one another in their massive caliper-like mandibles, each straining to flip his opponent off the log.

The most strikingly adorned fighting machines are the rhinoceros beetles (Scarabaeidae: Dynastinae). The male *Golofa porteri* from South America is fairly typical, with a prominent forward-curving horn on his thorax and an equally formidable forward-jutting counterpart on the front of his head. The front legs are also greatly lengthened. He deploys all this offensive hardware in combination to dislodge other males from resources such as sap-runs or over-ripe fruit, to which females will also be attracted. During fights, the head-mounted horn is forced underneath the opponent, who is dislodged by a combination of a sudden vicious upward levering, using the horn as a crowbar, and raking actions with the long front legs. These latter are aimed at breaking the contact between the opponent's legs and the substrate (Eberhard, 1977). In the normal chain of events, both combatants struggle valiantly, ending in a confused tussle whose eventual outcome may be far from clear. As they toil away, both combatants may stridulate by rubbing a file on the upper side of the abdominal tip against a scraper on the underside of the elytra. The winner usually gives a 'victory chirp' immediately after a successful outcome.

Other dynastine males, though less impressively armed, make full use of their horns in disputes over the ownership of 'mating burrows' in plant stems. In *Podischnus agenor* from South America, the long facial horn is curved backwards, thereby opposing a short, stout, forward-jutting counterpart on top of the thorax. Both sexes often cohabit in tunnels in sugar-cane stalks and solitary males advertise for female company by 'calling' from the burrow entrance (Eberhard, 1979). Unfortunately, the strong-smelling pheromone released by the caller also attracts the unwelcome attention of other males intent on stealing both the burrow and any females. Small residents simply give way when faced with a larger adversary, surrendering their home without a fight. A more evenly matched pair will come to blows, firstly inside the burrow, where the resident's tough rear end can provide an effective block to further penetration. If he turns and faces the entrance, both combatants can bring their horns to bear. Each beetle strives to secure a clinch, using the opposing horns on head and thorax to hold the opponent's front end in a vice-like grip. Quite often they both manage to snap the 'vice' shut simultaneously, but sooner or later one of them allows his hold to slip, permitting the other to lift him off the stem and drop him to the ground. The winner 'celebrates' his victory with a brief burst of stridulation. Residents do not tend to enjoy any 'home-ground' advantage and a first-time winner may become the loser in a rematch. The only clear tendency is for larger males to win more often. These fights look ferocious but the rather short, blunt-tipped horns are incapable of inflicting any real damage on the opponent's shiny armour. In *Heterogomphus schoenherri*, the males' much longer pickaxe-like horns are able to puncture an opponent's integument during clamping operations, sometimes doing fatal damage.

Horns are also used for fighting by much smaller beetles, such as the forked fungus beetle, *Bolitotherus cornutus*, which inhabits bracket fungi in North American forests. Only the males are offensively armed, having an impressive pair of horns which project well out in front of the head. The length of the horns and the size of the body both vary considerably and the available evidence points to large-bodied, large-horned males as being able to win greater access to fungal feeding sites, and therefore to females (Brown, 1980; Brown, *et al.*, 1985). The horns are employed in two different contexts. Males often seek to remove potential competitors for vital but finite resources – such as food and females – by knocking them off the fungus altogether in butting contests. However, fights are not inevitable, as several males may be able to co-exist peacefully on a single bracket. When a receptive female is at stake, a fight is far more likely and attempted take-overs of mated females are common. The assailant strives to get a secure grip on the consort by clamping him with his horns and levering away until he is pushed backwards off the female. The forked fungus beetle employs a distinct, if rather simple, post-mounting courtship. Facing towards the female's rear, he strokes her head with his back legs and rocks from side to side. After anything from 10 minutes to several hours of this, he makes an abrupt about-turn and tries to connect his genitalia. Mating is brief but the male remains on his mate for several hours as a 'contact-guarder' (Conner, 1989). Unlike many beetles, the male cannot 'force' copulation, as he lacks any genital clasping equipment. So he is obliged to wait until the female voluntarily lowers her strongly armoured abdominal tip before genital linkage becomes possible.

In species whose females are in desperately short supply, fighting to the death may be the only sensible strategy to preserve contact with a female. In the phengodid *Zarhipis integripennis* the larviform females are much larger than the males and always in short supply, probably because of the longer time needed to grow to the larger size (Tiemann, 1967). The males possess knife-like jaws which they use to slash one another to pieces in fights over receptive females. Fatal fights are frequent because a male guarder will refuse to surrender 'his' female to a challenger and they both adopt the strategy of fighting to the death because there is little hope of ever finding another female, either guarded or unguarded.

Mate-guarding in longhorns

As longhorn beetles (Cerambycidae) are often concentrated in time and space on certain favourable tree trunks, there is fierce competition for females and attempted take-overs are common. In response to such pressure the males have

▶ Size and strength alone may not be the sole male assets that dictate success or failure in takeover attempts of mated females. Sheer *persistence* may also be needed, as demonstrated by these European soldier beetles *Rhagonycha fulva* (Cantharidae). In this species *in copula* mate-guarding for an extended period means that unoccupied and receptive females are always in short supply, making takeovers of mated females common. Here a large male is energetically tugging backwards on a slightly smaller rival in an effort to pull him off his mate. When this failed, the would-be usurper turned his attention to an intruding female, only to be easily rebuffed. Yet this same female then immediately succumbed to the more persistent overtures of the male seen mating in this picture. He, meanwhile, had perversely deserted the mate to whom he had clung so tenaciously only minutes before.

Weevil club-fights. In the Central American weevil *Macromerus bicinctus* (Curculionidae), the males possess greatly lengthened forelegs with conspicuously clubbed ends. These are used to belabour an opponent during disputes over access to females who are about to lay their eggs (Wcislo & Eberhard, 1989). Considerable numbers of these mini-pugilists may be found on fallen logs in rainforest. Females wander around on the logs, seeking somewhere to drill a hole with the rostrum; just a single egg will be laid in the resulting hole. The males scuttle around on the look-out for a drilling female. If she is alone, he will mount her in a brief unresisted copulation lasting only a few seconds. After dismounting he remains by her side, perhaps resting his front legs on her back, acting as a guard until she has finished drilling; only when she turns to oviposit does he depart.

The chances are that another male will show up and mount a take-over attempt on the drilling female. After some brief 'measuring-up' preliminaries, during which the two contenders tap each other gently with their front legs, they launch into the genuine '*slugfest*'. The technique is to raise the clubbed front legs vertically and then bring them down with considerable force on the opponent's back. The recipient may stand motionless and take this treatment for some time before meting out some of his own to his opponent, who now himself stands and takes it. Alternatively, they will slug away in a 'one-for-me, one-for-you' fashion. Larger combatants usually win and serious fights do not develop when potential opponents are unequally matched – the smaller simply fights shy of conflict and beats a retreat. His 'decision' is based on the larger opponent's habit of stamping his feet to emphasise his superiority. During clubbing matches, it is usually not the clubbed tip which strikes the opponent, but the curved section above it. The weighted tip probably serves to increase the leg's momentum. Despite its force, the clubbing is probably highly ritualistic and, though small, the combatants are exceptionally solidly built.

Males of various European leaf-rolling weevils (Attelabinae) attempt to dislodge one another from leaves by using their enlarged back legs as rams. They sit back to back, abdomens cocked upwards, and strain away with their rear legs until one of them gives way or falls off. Males of *Byctiscus populi*, on the other hand, rear up on their hind legs and grapple with their front legs and interlocking rostra. In the bizarre giraffe-necked weevil, *Trachelophorus giraffa*, from Madagascar, the display is probably purely ritualistic. Two males stand face to face and nod their incongruously elongated 'necks' up and down until one gives way and retreats.

▼ Club fights in the weevil *Macromerus bicinctus*. Beside a drilling female, two males battle it out. The one on the left raises his clubbed front legs (a) and brings them down upon his opponent's back (b). (After Wcislo & Eberhard, 1989).

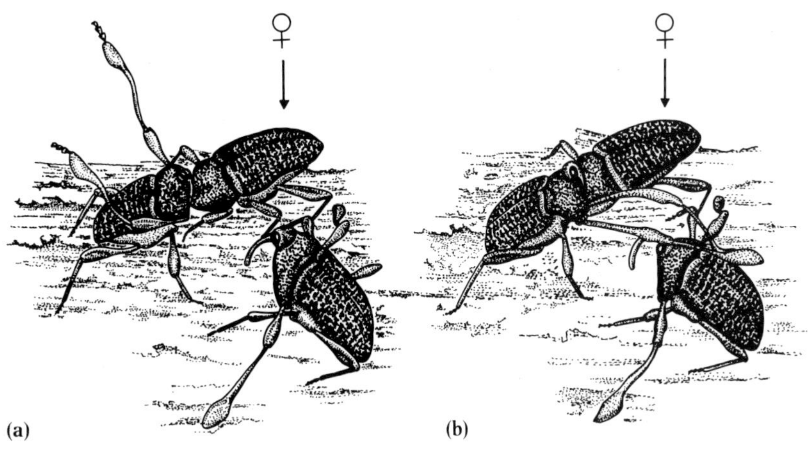

evolved a variation on contact-guarding by establishing a 'pair bond' during which copulation and oviposition are repeated several times. After he has mated, the male of the American white-spotted sawyer, *Monachamus scutellatus*, stands guard over the female in a 'half-mount' posture, clasping the tops of her wing cases with his specially modified 'non-slip' front feet; his rear legs remain on the trunk (Hughes, 1979 & 1981). He shields her as she gouges a hole in the wood with her mandibles, and then shuffles around behind as she turns to lay an egg in the slit. At this point he may re-mate immediately and then tag along behind as his mate lays a further egg. His antennae are exceptionally long, protruding well out beyond the female's, and enabling him to detect the arrival of an intruding male. As consort, he may be able to prevent an assault merely by making threatening jerks with his antennae, which can also serve as weapons with which to lash a rival. During an escalated contest, the consort dismounts and there is a face-to-face grappling match during which both contestants stridulate constantly. The consort usually wins, but does lose sometimes, presumably making mate theft a worthwhile strategy, given the shortage of receptive females.

If a female has started to lay her eggs by herself, any male who finds her frustrates her efforts by curving the tip of his abdomen beneath her rear end and levering upwards. Far from objecting to such cavalier treatment and trying to shake him off, she will immediately accept him as a mate. As sperm precedence should now assure him of paternity, the male establishes a pair bond and allows her to proceed with oviposition. It is in his interests to spin out the pair bond for as long as possible, as females who are permanently guarded seem to lay more eggs. The guarder can therefore increase his reproductive success considerably if he can

▲ The male of the African longhorn *Promecidus linearis* stands guard over his mate as she lays an egg in a crevice in the bark of a fallen tree. Only by doing this can he ensure paternity of the egg being laid, as rival males are abundant in the vicinity.

▼ Despite their lack of obvious weapons, such as horns or oversized mandibles, male longhorns are formidable fighters when engaged in an attempted take-over of a mated female. Here the assailant clamps his powerful jaws on one of his rival's antennae and yanks backwards on it, trying to pull him off his mount. He gave up after trying just about everything but, having no size advantage over the consort, never had much chance of succeeding. This is *Stenocorus meridianus* from Europe.

Courtship in longhorns. Preliminary courtship is usually absent and it is only after the male has somehow succeeded in mounting the female that he can begin to persuade her to accept him. As in many cerambycids, the female of the European spotted longhorn, *Strangalia maculata*, often strenuously objects to the male's presence, reacting by vigorously prising him off with her back legs; the same technique is also used for terminating copulation. Males generally resort to the 'Lone-Ranger' tactic of jumping on the female and trying to take her by surprise. He must have enough time to grab hold of one of her antennae in his mandibles before she realizes what has hit her and throws him off. After fastening his mouthparts around an antenna, the male *S. maculata* yanks back lustily as if reining in a horse. Notwithstanding his firm grip, his mount struggles so violently that she may still succeed in throwing or kicking him off. If he manages to hang on, he probes her ovipositor with his genitalia and may make a connection. In many Lepturinae, the male then performs a series of rear-end press-ups with his back legs, pulling the ovipositor out fastened to the end of his penis. When the ovipositor is fully out, the penis extrudes and the ovipositor is withdrawn.

In *Stenocorus meridianus* from Europe, the male devotes most of his mounted time to pulling out the ovipositor. This may possibly act as a continued stimulant, reminding the female not to boot the male off her back. In *Rhagium*, the females often viciously attack their suitors, even perhaps biting off a leg or antenna before a male succeeds in climbing on board. Once mounted, the male cerambycid often calms the female down by gently nibbling at her head, thorax or abdomen; he may also tap her with his head to keep her quiet.

Dismounting can be interesting. After copulation, the *Rhagium inquisitor* male removes just his hind legs from the female's back and rests them on the substrate. A little later, the female executes her 'round dance', turning in a circle and gradually 'wiping' the male off, his back legs being unable to keep up with her circling due to their position at the outer perimeter. His rear end gradually gets left behind until he is completely dislodged (Michelsen, 1962). Longhorn males often fight savagely over females, often biting off one another's antennae or legs.

▲ A pair of *Rhagium bifasciatum* longhorn beetles (Cerambycidae) mating peacefully on a wild rose in England. The male appears to be intact; he is lucky, as females of this genus often viciously attack their suitors, even biting off a leg or antenna.

manage to stick with his mate. She finally rids herself of her escort by jerking her abdomen upwards and stridulating vigorously. Pair bonds can be cut short in a number of other ways. The consort may 'jilt' his existing mate in favour of trying his luck with a passing female. Or he may lose her during a take-over attempt by another male. Most annoyingly, he may rush off to do victorious battle with an intruder, only to find upon his return that his mate has wandered off and he is unable to find her again.

Mate-guarding and fighting in the Brentidae

Brentid beetles breed in timber, often occurring in great numbers in a favourable habitat. Size varies greatly, especially in the males, whose elongated rostra often serve as weapons, which have apparently evolved by intrasexual selection through their use in take-over attempts of guarded or mated females.

The Neotropical *Brentus anchorago* possibly holds the record for any insect in respect of size differences. The males vary in length from 7 to 52 mm while the heaviest may weigh 26.5 times as much as the lightest (Johnson, 1982). Such vast disparities, plus the crowded conditions and heavy competition for females, exert a powerful influence over whether or not a male will ever manage to mate, or how often; large males mate most. Males contact-guard drilling females who are about to oviposit. A big male can provide a reasonable level of security just by resting his elongated snout across the driller's back. A small male has to make the best of a bad job by mounting her and using his whole body as a shield. All males usually copulate several times during the long process of drilling, so that the female's spermathecae are thoroughly swamped with the guarder's sperm by the time she finally lays her single egg. Guarding is often rudely interrupted by a gate-crasher. During attempted take-overs the combatants side-swipe one another with their elongate lance-like snouts, facing in opposite directions so that their rear ends receive the brunt of the attack. Blows are deftly parried with the hind legs, but the whole affair is more ritualistic than truly violent in nature, and is mostly concerned with estimating relative lengths. An interloper will also use his long snout to lever a rival upwards off the female, or he may use his snout to swat viciously at his genitalia.

A female will also try to wreck a rival's drilling operations by levering her upwards with the snout, or she may even employ her whole body as a lever by inserting it between the driller's rostrum and the tree trunk and then straining upwards. If the driller has a male guard, he will drive off the attacking female, proving his utility as a dual-purpose protector. Roving females will also try to disrupt copulating pairs, much as happens with males under similar circumstances. This whole complex scenario indicates an unusually intense level of competition in this beetle, not only between males for mates but between females for oviposition sites. This is linked to the frequent larval overcrowding, forcing females to improve the prospects for their own larvae by 'spoiling' rivals' drilling operations. The males always prefer to mate with the largest females, probably because they are more capable of defending their oviposition sites. As bigger females also lay bigger eggs, the bigger larvae which result are more likely to prosper against lightweight rivals when conditions are overcrowded. Because circumstances oblige males to lavish attention in guarding mates and fending off rivals, they end up investing nearly as much as the females. So a preference for a large 'winning' female is also to be expected. Similarly, females reject small males more often than large ones, presumably because the latter make more effective guards during attacks by 'spoiling' females.

In another Central American brentid *Claeoderes bivittata*, the 'top earners' in the sexual stakes are, as to be expected, the largest males but the outright losers tend to be the middle-sized males, not the runts. This is probably because the midgets are so small that their size becomes an asset, enabling them to 'sneak' by hiding away beneath a female while a large male is present and emerging to mate with her when she is unattended (Johnson, 1983). A similar tactic is used by midgets of the New Zealand brentid, *Lasiorhynchus barbicornis*, which also displays extreme size dimorphism both within and between the sexes. The largest males are physically able to mate with remarkably tiny females only a fraction their length. Midget males copulate with enormous females during 'mating windows', when the large males are too busy fighting one another to notice that they have been cuckolded. During such contests, these grotesque giants employ their incredibly elongated snouts to swipe at one another, trying to land a decisive blow which will knock the opponent off the tree (Meads, 1976).

In the gangling-legged zygopine weevils *Mecopus audinetii* and *M. bispinosus*, from Indonesia, a male seeks to gain the upper hand by thrusting his long snout beneath that of his rival; then with a mighty heave he can lever him off the tree (Lyal, 1986). 'Courtship' is unusual, in that the male bestraddles the female and traps her within a mobile corral of his long legs and rostrum. He then shadows her every move until she eventually gives in to this sexual kidnapping and tips forwards against the bark so that her abdomen points upwards at an angle. This is the cue for her 'jailer' to caress her head and thorax with his long front legs; then he copulates.

Spoiling tactics in a leaf beetle

'Spoiling' behaviour is also seen in the European leaf beetle, *Gastrophysa viridula* (Chrysomelidae). In parts of England, this beetle seems able to survive only in the ecological conditions near to major rivers. Its preferred food plants are the docks *Rumex obtusifolius* and *R. sanguineus*. When these become exhausted, the females will reluctantly oviposit on *R. hydrolapathum*, on which the larvae do not thrive. Larvae which have hatched from the first-laid springtime egg batches rapidly consume most of the locally available food plant. Further larvae hatching from subsequent egg batches could bring about starvation for them all, yet the females seem willing to mate repeatedly and lay more and more eggs. In response to incipient overcrowding and its dire consequences, certain males seem to exhibit behaviours designed to limit the further ruinous addition of yet more fertile eggs. They become 'spoilers' who act *co-operatively* to ruin copulations. With one spoiler up on the female's back, straining to push the consort off backwards, one or two 'assistants' stand on the leaf behind, trying to do the same by pulling at his legs or antennae. (Males also act alone to effect a dethroning.)

Any kind of co-operative effort among males is unusual (but see burying beetles, page 71), but once a female is available, the males could be expected to cease co-operating and lay claim to her. However, no such thing happens (at least when observed by the author). While an unseated male scuttles off, his tormentors ignore the female and trot off across the leaf to repeat the process with another mated pair. The spoilers do not operate as a band, but often convene by chance beside a mated pair, wreaking their sexual havoc with male after male, yet never attempting to profit by their actions by trying to mount the vacated females. Why do these 'spoilers' do it? It seems likely that they are males who emerged early and were the first to mate. It is therefore their offspring who have the

▼ 'Spoiling' tactics by a male pot-bellied emerald leaf beetle, *Gastrophysa viridula* (Chrysomelidae), in England. The male on the right is just about to pull a rival completely off the female. Instead of mating with her, the assailant will seek out another mating pair and do the same thing, probably to reduce future competition for his own larvae developing on the overstretched food plant.

most to lose if overpopulation wipes out the food plant. Further matings would not therefore be to their advantage – quite the opposite, as they would risk losing their *entire* investment as a result. It would serve their interests best to prevent any other males from mating, which is what they do. Spoiling behaviour would presumably be density-dependent, being triggered when damage to the food plant reached certain detectable levels. Here is a rich field for further detailed research.

Fighting and 'sneaking' in burying beetles
As resources, dung and small animal corpses have much in common – they are ephemeral, widely spaced and unpredictable in appearance. Competition for their ownership is therefore liable to be intense. However, in burying beetles, the circumstances of discovery may dictate the finders' initial mode of behaviour. If more than one male of the burying beetle *Necrophorus vespilloides* (Silphidae) congregates on a mouse corpse, they may co-operate peacefully to bury it. However, such good-natured mutual assistance only materializes while there is still no female present. If a female is not attracted by the smell of the corpse itself, then the combined pheromonal output of the males will draw one in, for they evince further collaboration by 'calling' as one. Once a female shows up, all collaboration ceases and the former allies become fierce rivals, fighting tooth and nail for possession both of the female and the corpse. The latter's value rockets with the female's arrival, for it is now guaranteed to nurture a brood, sired by whichever male comes out on top; so there is plenty to fight for.

After driving away his former allies, the victorious male takes up residence with the female in their subterranean nuptial and breeding chamber. From now on he mates frequently with his 'spouse', this being a necessary 'insurance' measure as she will almost certainly have been inseminated before her arrival. Only by frequent brief (20–90 seconds) multiple matings – just before and also during oviposition – can the consort ensure a high degree of paternity (up to 98 per cent). This has to be achieved through sperm competition within the female (Müller & Eggert, 1989), as unlike in many insects, the male cannot rely on the 'last in, first-out' principle of sperm precedence, in which the last male to mate fertilizes the bulk of the eggs. So he has to swamp a rival's sperm by repeated matings. It is absolutely critical to be sure of paternity as, in strong contrast to most male insects, he will devote considerable time and resources to rearing his offspring. Repeated matings are also advisable because he may not be able to guarantee exclusive access to the female. Sometimes one of his former allies will hang around in the hope of a 'sneak' mating (Bartlett, 1988). As in some other 'sneaks', he may have little to gain. At best he might perhaps secure one or two matings, fertilizing a paltry number of eggs, if any. However, for a very small male, even this trifling level of achievement may be preferable to the alternative – engaging in repeated co-operative 'calling' and burying sessions with larger rivals. This has one major drawback – he is always going to come out as underdog in the final battle.

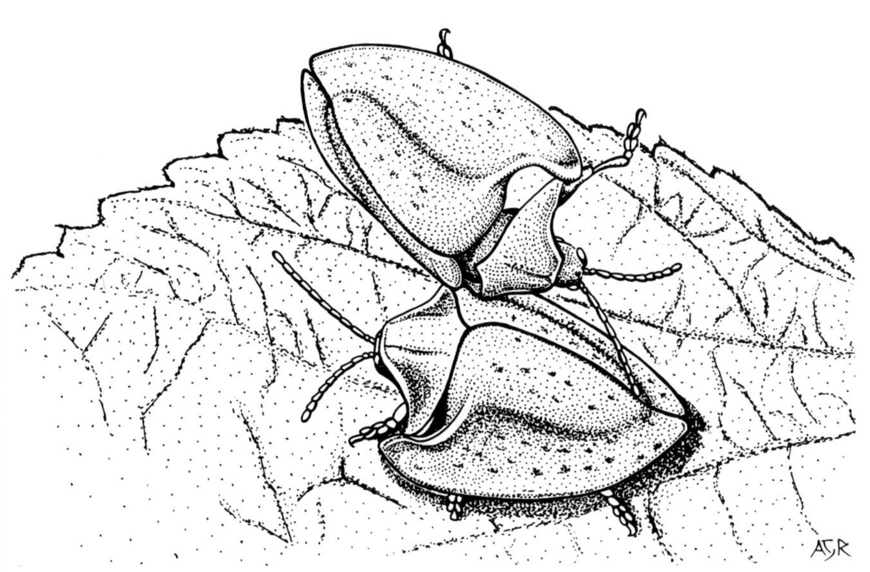

Resource defence in leaf beetles. A Central American chrysomelid (a *Doryphora* species closely related to *D. punctatissima*) defends an easily exhausted food plant even before mating takes place. A beetle tries to remove a potential competitor from a plant by levering upwards with a horn which juts forwards from beneath the sternum ('chest'). The fact that males and females both possess similar horns suggests that resource defence is its primary function (Eberhard, 1981).

Males of the Neotropical tortoise beetle, *Acromis sparsa*, practise resource-defence polygyny (Windsor, 1987). They station themselves at the fresh, young, growing tips of the vine *Merremia umbellata* (Convolvulaceae), to which the females are attracted for egg-laying. There is substantial competition for these mini-territories and clashes are frequent, during which the opponents strive to butt or flip one another off the leaf. One opponent may also contrive to neutralize his rival in a highly original way, by 'capturing' him in a socket in his carapace.

▲ Males of the Neotropical tortoise beetle *Acromis sparsa* (Chrysomelidae) have their elytra modified so that they can indulge in chasing, flipping over and thus immobilizing their rivals in the manner indicated. The male on the leaf has lifted his rival up in his elytral notch and is in the process of turning him on to his back. (After Windsor, 1987)

Sexual strategies in dung-rollers
Dung is another temporary resource and arguments over ownership may be frequent and fierce. In many dung-rollers (Scarabaeidae), it is the male who initially fashions the ball from the crude dung and rolls it to a nuptial chamber, usually with the female sitting on top. She is attracted visually to the sight of the male rolling the ball. This represents both a ready-made food source and an oviposition resource and she can acquire it in exchange for mating, sometimes gaining his assistance in nest-building as well. The possible importance of male assistance is hinted at in *Canthon angustatus*, for a lone female rolling her own ball will abandon it in favour of joining a male, even though his ball may be no better than the one she has just discarded. A female who has joined a

ball-rolling male will also abandon the ball if the male is experimentally removed, preferring to search for a new ball complete with owner rather than stick with the original ball as a loner (Peck & Forsyth, 1982).

The ball itself is the main visual stimulus for the female (lone males are not attractive), although in some species the male may impregnate his ball with pheromones to increase its powers of attraction. Within the subterranean nuptial chamber, the two beetles frequently copulate and gorge on the dung ball over a period of several days. Then they re-emerge to prepare a fresh ball, the 'brood ball' on which to raise their offspring. The first ball has merely done service to cement their relationship, being the sole dish in a private 'wedding feast', hence its name of 'nuptial ball'. Not all species make a nuptial ball, as sometimes the brood ball suffices for both purposes. In some species, mating takes place while the ball is being made, or at any time thereafter, as in *Kheper (Scarabaeus) platynotus* from Kenya (Sato & Imamori, 1986 & 1987). In this species, both male and female co-operate to make the ball, but it is the male who rolls it, with his mate riding on top, before he eventually buries it. It functions solely as a brood ball, and there is no nuptial ball.

In *K. aegyptiorum*, the nuptial ball may be made by a lone male, a male–female pair or a lone female (Sato, 1988). When a lone male has completed his ball, he puts on a special leg-bending display which attracts a female, and later mates with her alongside the nuptial ball in the bridal chamber; mating never takes place above ground. He may even exhibit 'choosy' behaviour, sending a female packing by flipping her away if she arrives before the ball is complete, although he may quickly accept a later (possibly larger) arrival. In some dung-rollers, it is the female who makes the first ball and attracts the male, after which mating can take place on the dung, or later in the nuptial chamber. Males often dispute the ownership of a ball, particularly if it is exceptionally large. The would-be thief suffers from having to attack 'up the hill' against the owner, who has the advantage of holding the 'high ground' atop his ball. From here he can often insert his spade-like front legs beneath his assailant and flip him on his back. Challengers are often remarkably loath to give up, and even a small male may eventually succeed in appropriating at least a portion of the coveted ball. It is puzzling why such energy-consuming contests happen at all, given the nearby presence (often mere centimetres away) of ample quantities of fresh dung, from which the challenger could construct his own ball without the need for a fight. However, if an owner has already saturated his ball with a female-attracting male pheromone, an undersized thief could presumably cash in on his larger rival's superior pheromonal output by stealing some of it. By subsequently adding his own pheromonal contribution, the diminutive thief could perhaps produce a ball of sufficient quality to augment his otherwise slender chances of attracting a female.

▲ There is often considerable competition for dung balls among dung-rolling scarabs. This pair of a South African *Kheper* species is having to parry a determined attack by another pair intent on stealing their ball.

Behavioural transvestism in a rove beetle

Sometimes it is not the dung itself which is of direct value as a resource but the insects – such as flies – which it attracts. Both sexes of the rove beetle *Leistotrophus versicolor* (Staphylinidae) from Costa Rica home in on vertebrate dung or corpses, preying on the flies which also gather there (Forsyth & Alcock, 1990). The male beetles practise resource-defence polygyny, fighting ferociously to exclude rivals from the dung and thereby winning sole access to the females, who are more receptive while on the dung than in any other situation. Some males are much larger than others, so the smallest are at risk of never mating at all. Large males sport enormous sickle-like mandibles which wrap around the head when closed. Small males have much shorter, thicker pincer-like mandibles which are not much good in a fight with a big-mouthed rival. During some initial skirmishing, two contestants snap at one another with their mandibles, and then one turns his back and swipes his opponent on the head with his cocked abdomen. A large male may make good use of his capacious jaws to clamp a smaller rival and heave him over on his back, persuading him to beat a retreat and urging him along by snapping at his heels. In response to the sad prospect of a non-existent sex-life, the smaller males have evolved a strategy which not only allows them to wander around on the dung without being bullied, but also grants them freedom to engage in sexual unions 'under the noses' of their larger rivals. They achieve this by adopting 'transvestite'

tactics – mimicking a female when approached by a larger male, saving them from having to cut and run, or to submit to a fight which they cannot possibly win.

In order to understand how transvestism works, it is necessary to know the train of events during normal courtship. A female 'minces' along in front of a suitor, brandishing the tip of her upraised abdomen and swaying it enticingly from side to side. The male taps the top of her abdomen with his 'chin' and she halts, lowers her abdomen and thus allows her follower to 'thread' his abdomen forwards between his legs and connect his genitalia with hers. 'Transvestites' adopt identical movements and may fool a larger male into prosecuting a 'homosexual' courtship for some time. At length, the 'suitor' may finally see through the charade and start a fight – but not always. 'Transvestites' are made, not born, evincing remarkable flexibility in behaviour. Thus, a middle-sized male may be playing a 'mincing' female role for the benefit of a large rival directly to the rear, while simultaneously dealing out punishment to a third male of inferior stature. 'Homosexual' courtship drags on longest when the 'transvestite' is substantially smaller than his duped 'suitor'. This is probably because a small male, hopelessly ill-equipped for combat, simply cannot afford to change tactics mid-way and switch to aggression. He just has to stick with his 'act' all the way, or at least until the menacing 'suitor' eventually loses interest and can be given the slip.

Even the largest males can opportunistically make the temporary switch to transvestite behaviour to 'get them off the hook' when challenged by an equally large rival. This normally happens when they are unable to go to battle-stations immediately because their fighting jaws are occupied with demolishing a fly or tapping a female's abdomen in courtship. Habitual transvestism pays off by allowing the practitioner to spend more time on the dung then he could if he had to fight to stay there. This means that he earns more opportunities to 'sneak' matings. He may even do so under the very 'nose' of a 'homosexual' suitor, who gullibly continues with his misdirected courtship while his would-be 'bride' is busy mating with the 'transvestite'. Such alternative strategies may be especially important in this species, due to the males' habit of expelling a mate from the dung immediately after copulation. This is probably done to deny them to other males, making the competition for incoming females even hotter.

(a)

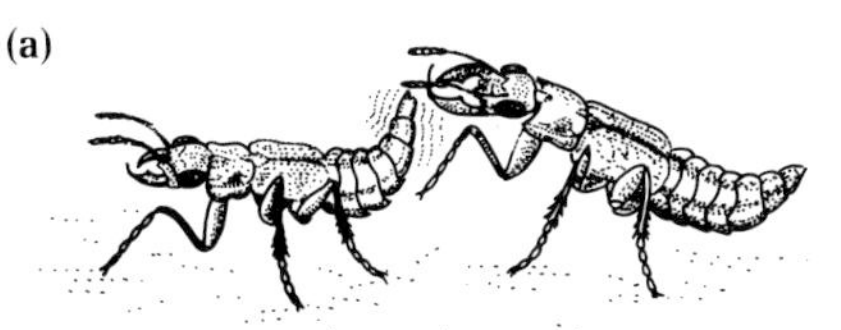

(b)

(c)

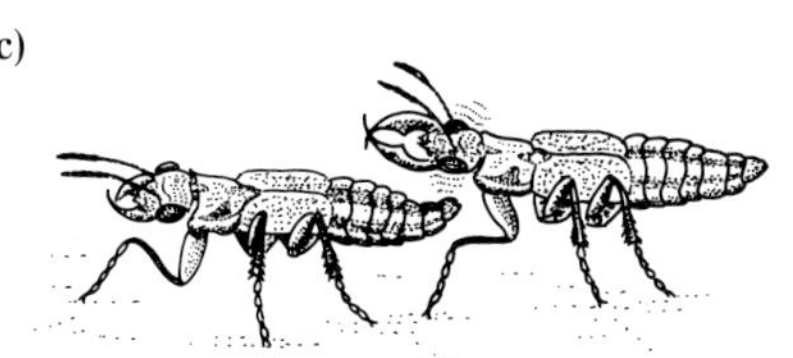

(d)

▲ **Normal courtship in the tropical rove beetle *Leistotrophus versicolor*. (a) A female 'minces' along in front of a male, brandishing the tip of her upraised abdomen and swaying it enticingly from side to side. (b) The male taps the top of her abdomen with his 'chin' and she halts. (c) The female lowers her abdomen. (d) This allows her follower to 'thread' his abdomen forwards between his legs and connect his genitalia with hers.**

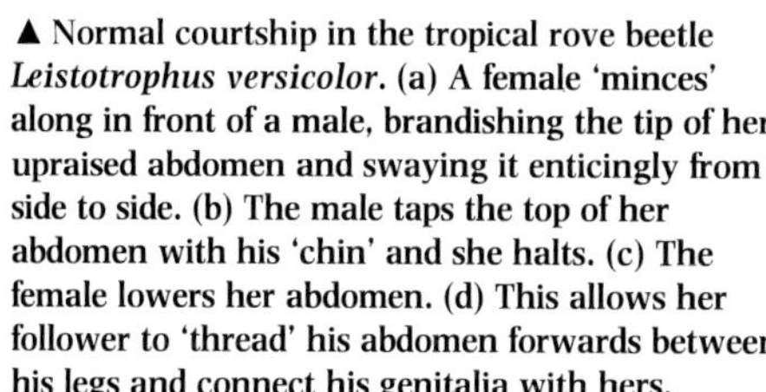

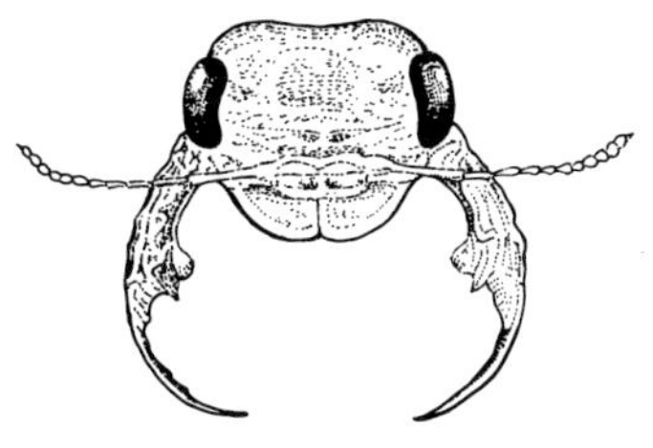

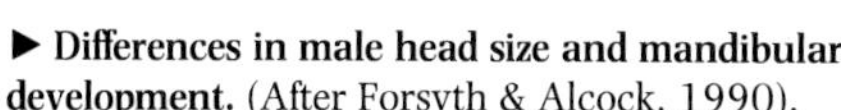

▶ **Differences in male head size and mandibular development.** (After Forsyth & Alcock, 1990).

Chemical transvestism in a rove beetle

Similar 'transvestism' is also exhibited by the European rove beetle *Aleochara curtula*, which congregates on carrion and feeds on flies which are attracted to it. Young males who are sexually immature, and older males who may have been sexually capable for some time, can become temporarily starved of the resources necessary to manufacture a spermatophore. They are able to remain on the carrion, and there fatten themselves on flies, only by secreting a female sex pheromone on the cuticle. This protects them from being driven off by any assertive mature males who are already present. The system does however suffer from the disadvantage that a chemical cannot just be switched off, as can the purely behavioural transvestite tactics of *L. versicolor*, which can be changed instantly to normal courtship. This means that female-mimicking *A. curtula* males are shunned by their own females.Thus although food may be available, sexual liaisons are 'off the menu' until after each male has assimilated sufficient nourishment and a spermatophore has been manufactured.

◀ (Top left) In several species of black tenebrionid beetles from the Namib Desert, males spend all day following unreceptive females. In this species, *Lepidochora discoidalis*, males are forced to traipse along like this because females are only receptive at long and unpredictable intervals. The smaller male at the rear would have little chance of beating his rival to the female if she were to become receptive.

◀ (Top right) In many desert tenebrionids, males vastly outnumber receptive females. This leaves the males with a problem; how to react when they do eventually run across a female. This male *Adesmia* is investigating a female who has already been claimed (but not mated) by a rival male. What should he do? He could try and push the rider off; he could sit on top of him and 'hope' to triumph in the struggle to mate if the female becomes receptive; or he could run off and try to find his own unattended female.

◀ (Bottom left) What he 'decided' on was to add himself to the pair, only to be surmounted a minute or two later by a third male who 'made' the same 'decision'. In this species no tactic is likely to yield a higher rate of return than any other; all are equally low.

◀ (Bottom right) A male *Onymacris rugatipennis* beetle (Tenebrionidae) pursues a female doggedly across the dunes of the Namib Desert. He needs to stay with her until she 'digs in' at dusk; then he can mate with her – as long as he can keep his many rivals at bay.

Female-defence polygyny in a desert tenebrionid

In the African Namib Desert, the daylight hours are marked by the sight of numerous large black darkling beetles scurrying around, either in pairs or trios, tagging constantly behind one another in apparently aimless pursuit. These sexual 'groupies' are often *Onymacris rugatipennis* (Tenebrionidae), whose males devote nearly 20 per cent of their day to the tenacious 'shadowing' of a female (Hamilton *et al.*, 1976). The aim is to stick close on her heels until the evening, when she will bury herself in the sand for the night. Only at this point will she uncritically accept any accompanying male as a mate. However, on some days she may hole-down by mid-day, in which case her pursuer will need to maintain ownership of the space above her until evening. This can be easier said than done, as receptive females are in short supply, and there is an enormous surfeit of males. Attempted take-overs are common both during the above-ground pursuit stage, and after the female has buried herself. In his capacity as follower, the male rarely if ever feeds, but has to stand around just behind the female as she stops for a nibble, ever at the ready to defend her against a hijacker. His response to an intruder is to deny him access to the female and claim ownership by performing a half-mount on her back. This is usually sufficient to discourage the challenger. Once the female has gone to ground, her follower has to switch to defending the space above her, plus an extended 'exclusion zone'. He has to do this just when she is liable to pull in the greatest number of rival males by at last becoming attainable, so she is very much a hot property. If butting does not deter an intruder, the defender will run over him and, if they are evenly matched, this may further escalate to wrestling.

Wrestling is an incredibly ritualized routine governed by certain 'rules' about who does what and when. One combatant duly shoves his head beneath his adversary's front end. The latter then tries to 'throw' him by embracing him around the body and straining to heave him off his feet, which are widely spaced to make this difficult. In a successful 'throw', the 'thrower' keels over backwards, pulling his opponent on top of him. Then they scramble to their feet and 'take guard' again, but this time the positions may be reversed, with the 'thrower' now taking the head-down position. The thrower's technique may be undermined by his opponents efforts to tip him off balance by kicking his feet out from under him. One combatant may also hasten the outcome by seizing a leg in his mandibles and doggedly refusing to let go. Biting is quite rare, possibly being a 'cheat' strategy. How the winner is decided after numerous 'throws' is uncertain, but if one beetle is too weak to lift his opponent, then defeat is certain and he leaves the scene. Larger males tend to win and gain most access to the females, although, given the high densities of beetles, one victory may merely lead to another immediate fight with a new challenger. The victor harries the loser's retreating rear and then executes a brief 'victory dance' on stilted legs, before reaping his reward by mating with the female in her subterranean boudoir. On the following day, the males again start afresh as equals and all females are 'fair game'.

Pre-copulatory mate-guarding in a tenebrionid

The black rotund males of *Physadesmia globosa* from the Namib Desert are faced with a thorny problem. They need to acquire females who are in short supply, never stop running (except briefly to feed), and seem disinterested in copulating except on rare and unpredictable occasions. It seems that the only answer is for the males to spend most of their lives in tenacious pursuit of the errant females in the hope that, one day, perseverance might pay dividends (Marden, 1987). As a female seldom stops moving, opportunities to climb on board and have a go at eliciting some interest are severely limited. Nor can the pursuer risk diverting his attention for long enough to feed in case he loses sight of his quarry; yet he has somehow to muster sufficient energy to fight off numerous take-over attempts. No alternative mating strategies are viable, mainly because there is no way of predicting when or even why a female may suddenly 'change her mind' and become receptive. Pre-copulatory mate-guarding, over whatever length of time is required, is therefore a better strategy than the only alternative – to run madly around in the slim hope of chancing across a female who just happens to be both receptive and unguarded.

Contact-guarding in a ladybird

Pre-copulatory mate-guarding occurs in the Australian ladybird *Leptothea galbula* (Coccinellidae), and is directed not at an adult female but at a female pupa yet to eclose; it is on the latter that the male takes up his guarding station (Richards, 1980). When he becomes mature, a male devotes his time to searching for pupae, which are attached to the leaf by their anal organs. Upon locating one, he strokes it with his antennae and palps, then climbs on top and settles down for as long as 4 days. Female pupae which are not 'ready' for a guarder, being a long way from maturity, are capable of resisting a male's exploratory probings by flipping up vertically, throwing him off. If he then prods at the exposed underside, he could be shot violently aside as the pupa snaps back on to the leaf. A guarder will have to repel repeated boarding attempts by rivals, and these are stepped up when the female is due to eclose; then a flock of attendant males may converge on

Mate-guarding in tiger beetles. Male *Cicindela* tiger beetles (Carabidae) are able to avail themselves of a simple technique to stay in contact with their mates. At the rear of her thorax the female has a groove known as the 'coupling sulcus'. This provides a secure fit for the male's mandibles, which remain clamped around this sulcus as he rides around on the female's back (Freitag, 1974). In most cases the sulcus is species-specific, so that only the correspondingly contoured mandibles of a conspecific male can obtain a snug fit. Interspecific matings, as frequently hazarded by the males, are therefore seldom successful, as the mandibles will be a loose fit, allowing the females to throw off an interloper. The males use 'Lone-Ranger' mounting tactics, so the tight grip permitted by the coupling sulcus in same-species matings is probably a boon to a hard-pressed male who has just endured a hectic chase after a fast-running female and succeeded in jumping on board.

Many species of tiger beetles inhabit brightly-lit deserts, where the males can presumably recognize their 'own' females at some distance before they make their final leap. However, the gloomy conditions at ground level in tropical rainforests could inhibit reliable intersexual recognition, so some of its denizens employ specific courtship signalling. In *Odontocheila mexicana* from Mexico, the male gesticulates with his prominent white feet from just in front of the female, while *Cicindela xanthopila* from Gabon repeatedly gapes his white mandibles from a similar distance (Pearson, 1988). As the white marks on the appendages are unique to the males, they presumably have a communicative function.

Once the male is safely mounted, he may stay in place for some time, although only a tiny fraction of this period is actually spent in copulation – in the American *C. marutha* only 2.3 per cent of the total duration (Kraus & Lederhouse, 1983). The rest is dedicated to contact-guarding until the female is almost ready to lay her eggs. The chance of retaining paternity is therefore greatly increased by an extended period of ridership, and success rates in attempted take-overs are probably low, due to the rider's firm jaw-clamp on the female. The females try to unseat their riders on a regular basis, but are usually beaten by their own coupling sulcus, which enables the male to retain a firm grip despite the amount of bucking and twisting going on. He may or may not manage to stay with her until she lays her eggs. In the European green tiger beetle, *C. campestris* (and probably many others), he often remains, reducing the degree of contact to the jaw-clamp alone, which leaves the female's abdomen free to perform its egg-laying duties unhindered.

▼ Male tiger beetles ride the females for long periods without any kind of genital contact; they are there purely as guards to counter take-over attempts by rival males. The rider's grip is a powerful one, facilitated by his mandibles' snug fit around the female's coupling sulcus. This is *Cicindela aurulenta* from Borneo.

the pupa. Its 'owner' is highly intolerant of such company and usually responds by taking the fight to them, mainly by butting and pushing; only occasionally does this mild jostling develop into a fierce fight. Chases and fights tend to proceed on a stop-go basis, due to the guarder's habit of breaking away at regular intervals to sprint back and touch the pupa, which is eventually mounted by the victor.

As the female begins to emerge, several males crowd in, while the guarder himself walks on to her back and secures himself in place by clinching the rim of her elytra with his tarsal claws. As she moves off, he has to brave the blitz of butting, shoving and biting from the throng of challengers, several of whom may stack themselves up on top of the 'wedding couple'. He has to resist for around an hour, until the female has dried her wings and is in a fit state to copulate. Despite all this scrimmaging, whoever happens to be guarder as the female emerges will usually manage to hang on and mate with her in the end.

Various escort behaviours

Take-over attempts reach high levels where beetles occur at high densities, or when the females release a pheromone designed to grab the attention of every male in the neighbourhood. It is the latter which probably accounts for the struggling balls of up to 200 males seen in the Japanese leaf beetle *Popillia japonica* (Scarabaeidae). In the middle of such a cluster is a female emerging from her pupa in the ground (Barrows & Gordh, 1978). Despite her embarrassment of suitors, the virgin female is exceptionally shy of accepting one as a mate. It seems that her strategy may be to 'play them off' against one other, reserving her virginity for as long as possible until the only male fit enough to stay the course eventually ends up as consort. Every male will strive to remain mounted for as long as possible, but pays a high price in energy needed to repel his numerous rivals. During these 'anything-goes' assaults, several assailants may simultaneously push, pull, lever, bite, scratch, ram and kick the rider, while he, hanging on desperately with his hooked feet, responds likewise to the best of his ability. As these brawls take place on the female's back, and she has to bear the continuous burden of one or more males, the costs to her must be considerable. The advantages of saving her virginity until just before she lays her eggs must therefore be substantial, and presumably outweigh the

◀ *Popillia* males (Scarabaeidae) often remain mounted but uncoupled for long periods atop unreceptive females. The female 'invites' strenuous competition among the males to be the one who will still be in position when she suddenly becomes receptive. This is *P. bipunctata* from South Africa, which often forms 'mating balls' on flowers. The male is preparing to fend off a rival (out of picture at left) with his back leg.

▲ The malachiid beetle *Malachius bipustulatus* is one of many species in the family in which the females 'feed' on 'excitators' on the male's body; in this case situated on the male's face.

▼ The ability of a male beetle to cling to his mate while being tugged and pushed simultaneously by several rivals is facilitated by his hooked tarsal claws. These are particularly obvious on the front legs of this South African chafer *Genyodonta flavomaculata*, hooked securely under the leading edge of the female's thorax.

costs. The most likely benefit is that, by 'choosing' a 'winner' to father her offspring, her sons may inherit some of their sire's staying power, making them more likely to prevail against the odds in next year's sexual marathon.

Like most beetles under siege, the male's ability to keep a firm grip on his mount is facilitated by his hooked tarsal claws. *Cicindela* males have a tailor-made jaw-clamp which is very effective in performing the same 'stick-tight' function. However, males of the lycid *Calopteron discrepans* employ a most bizarre method of fixing themselves in place – they bite a hole in the 'shoulder' of the female's right wing case and 'staple' themselves in place by plugging their sharp mandibles into her living tissue (Sivinski, 1981). Late-coming males stack themselves up on top of the original pair, each one plugging himself into a different part of the unfortunate female's elytra. Although hanging on tightly is in the original male's best interest, his method of doing so clashes with the female's need to stay healthy. Any reduction in egg-laying potential resulting from her wounds would constitute a debit to be deducted from any benefit accruing to the male through his behaviour. However, lycids are rather tough insects, so such rough treatment may not be too harmful.

Excitators in melyrids

Although pre-copulatory courtship is absent or minimal in many beetles, there are some notable exceptions. In at least two families, the Melyridae (also referred to as Malachiidae) and Meloidae, there is extensive use of 'excitatory' glands situated on the head, thorax or abdomen of the males. In *Malachius bipustulatus*, the 'excitator' consists of a glandular area on the face, just below the antennae, whose bases bear a row of yellow peg-like projections. Both the glandular area and the antennal projections seem to be bathed in a liquid which is extremely attractive to the females. When making an approach, the male holds his head low while pointing his antennae upwards just past the vertical. This exposes the glandular area on the face and the antennal pegs, which now project forwards. The female rushes forward 'eagerly' and starts nibbling at the 'excitator', while the male now lowers his antennae, pressing the yellow pegs against the female's face. As she feeds, the male often nods his head slightly, while repeatedly slapping her smartly on the sides of her head with his extended forelegs.

If she breaks off he quickly re-presents his face in an invitation to resume her nibbling – and she almost always accepts. These nuptial preliminaries may continue for an hour or more – and then it is the male who suddenly withdraws and flies off, without even attempting to mount the female! It seems that the female's full co-operation is essential for copulation to proceed and, if she fails to give a 'come-on' signal, the male simply pulls out once his 'excitators' have been licked dry. The 'excitators' rarely succeed in their job of rendering the female receptive, and copulation is almost never seen in the wild. Could it be that the males have no way of recognizing a virgin, so allow themselves to be 'parasitized' by feed-and-run females who have already mated?

Numerous other European melyrids exhibit a similar courtship, with eventual copulation also being but rarely seen. This lack of a result reaches its peak in the *Axinotarsus pulicarius* male in which the excitatory glands are situated at the tip of the body. After a quick face-to-face introductory antennation of the female, he turns his back and presents himself ready for her attentions. Despite her obvious enthusiasm for battening on his 'excitators', the level of stimulation again seems always to fall short of the minimum required to make her receptive (Matthes, 1962). Several observers report seeing hundreds of protracted courtship sessions – without noting a single copulation! Much the same lack of success applies to almost all the species studied so far.

Excitators in the Meloidae

Some male blister beetles also possess glandular 'excitators' – usually twin grooves on the face, into which the female's antennae are introduced during courtship. It is therefore via receptors on her antennae, rather than through her mouthparts, that the meloid female picks up the stimulatory male pheromones. Unlike in malachiids, whose females always seem eager to accept the offer of a 'free lunch', female blister beetles do not voluntarily press their antennae into the males' grooves; it is the male who does this, as the most vital part of courtship.

In *Tegrodera erosa* from the southwest USA, courtship is split up into three phases (Pinto, 1975): (1) the preliminary, (2) the display, and (3) the copulatory. In phase 1, the male aims to persuade the female to stay in one place long enough for phase 2 to begin, standing behind her and lightly tapping her rear with his antennae and mouthparts. He then moves around to the front to begin his display. If she still moves off, as she often does, he simply goes back to the rear and starts again. This may become a tiresome process, occupying the bulk of the total courtship time, before he finally either loses her or persuades her to stand still and allow him to begin his display. Facing the now immobile female, he stilts up high so that he overtops her head, curls his antennae around hers, which she deliberately presents in a vertical position, and then yanks them rapidly back and forth into his excitatory grooves. The female often makes off half-way through, but signals acquiescence by depressing her head and raising the tip of her abdomen. The male takes this as his cue to crawl up over her head and mate.

Not all *Tegrodera* males bother to carry through the whole of this tiresome procedure before having a go at mounting. In *T. aloga*, there is the 'jump-and-mount' tactic seen in some other meloids (e.g. some *Epicauta*). However, such 'short-cutting' is rarely successful, as the female can still kick the male off with her rear legs. She can also deny him her genitalia by pushing them down against the substrate. Perhaps in compensation, the males of *T. aloga* are blessed with exceptionally large genitals, almost twice the size of those in *T. erosa* and *T. latecincta*, both of which normally 'play it by the rules'. The *T. aloga* male is therefore amply equipped for attempting 'rape'. However, as in many beetles, a female who at first resists may eventually yield, even without the entire nuptial ritual. However, the female of *Pleuropompha tricostata* can cut short an unwelcome courtship in a most effective way (Pinto, 1973). As the male vibrates his antennae over her rear end, she up-ends and smashes the tip of her abdomen and her back legs under his 'chin'.

Antennal pulling is also seen in the subtribe Eupomphina, but the male stands on the female with his head above hers and pulls her antennae upwards and backwards (Pinto, 1977). In *Eupompha*, the courtship varies from virtually nothing (*E. viridis, E. edmundsi*) to antennal pulling while mounted, accompanied by stimulation of the female's mouthparts, which the male rubs with his front feet (*E. elegans, E. imperialis*). *Epicauta pennsylvanica* females on goldenrod, *Solidago* spp., in the eastern USA seem to choose their mates on a combination of size and courtship persistence. Males seldom have a chance to mate without first having come to blows with several rivals. This ensures that it is mainly the larger males who gain access to the females. Courtship lasts for at least 30 minutes before the female finally allows the male to mount without summarily kicking him off. His technique is to nuzzle the tip of the female's abdomen with his mouthparts. It is more a trial of endurance than sophistication; if he can keep going for 30 minutes, while also having the energy to beat off

▲ Sexual activity in this Indian *Mylabris tiflensis* oil beetle (Meloidae) seems to take place only during a brief period just before dusk, when virtually every individual in the localized populations typical of this species may find a mate and copulate within a few hectic minutes. During courtship the male (right) faces the female and touches her head with his rapidly vibrating front legs.

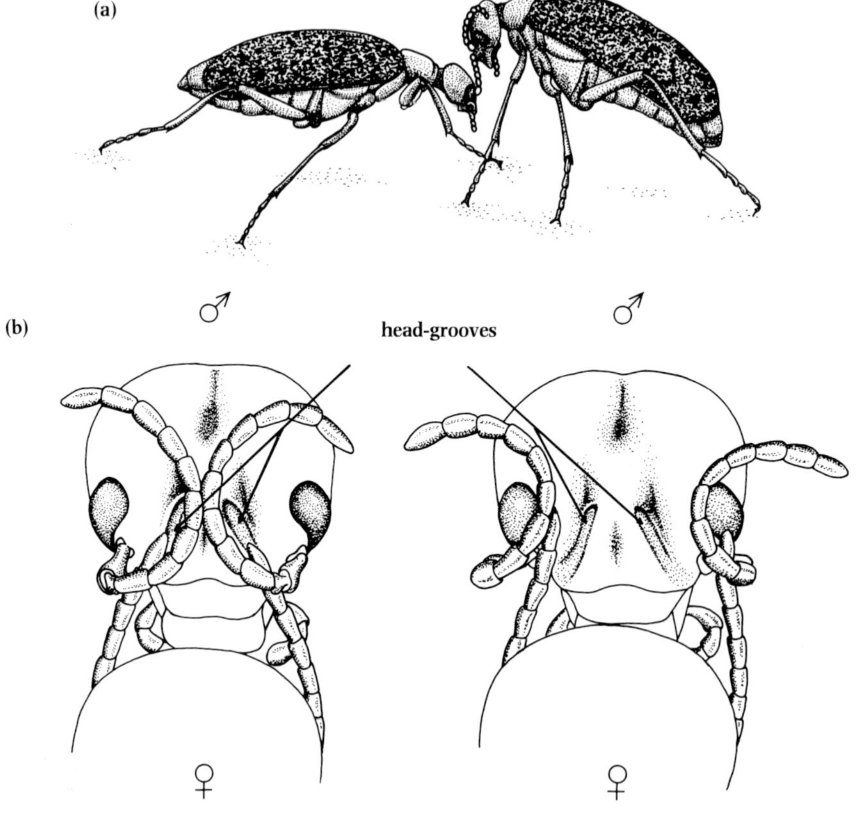

► In the blister beetle, genus *Tegrodera* (Meloidae), (a) the male repeatedly pulls the female's antennae into his head-grooves. (b) A close-up of the heads to show the process in *Tegrodera erosa*. (After Pinto, 1975).

intruders, he has 'proved' his qualities and is accepted. Such a female-operated selection system, favouring males who can 'prove' their competitiveness by their ability to carry through an extended courtship (in the face of frequent opposition) is similar to that of the fly *Physiphora demandata* (Otitidae).

However, a female may gain another advantage by ending up with the 'toughest' (and usually biggest) of the bunch. During mating, male meloids donate a large and highly nutritious spermatophore, which is 'digested' by the female and augments egg production. If larger males also produce larger spermatophores, this would be an added reason for the females to evolve a system which virtually guarantees a large mate. Males of the American *Lytta magister* also transfer a large spermatophore containing the unpleasant chemical cantharidin. This contributes not towards egg production but towards the females' chemical defences.

Choosy males

Playing 'hard to get' is a common trait in females. When males also act in this way, it is usually because they are raising their contribution above the norm, e.g. the huge spermatophore donated by certain male katydids. Male investment may alternatively be expressed in terms of time and energy spent in preparing a suitable 'home' for his mate and their offspring. The 'home-owner' could therefore be expected to be fussy about which females he invites to stay and share the fruits of his labours. Such home-based 'choosy' behaviour seems to be evident in at least three species of *Ips* engraver beetles (Scolytidae). The males commit considerable effort, over a day or so, to the construction of a 'nuptial chamber'. When a female shows up the owner may just flatly refuse to allow her entry to the chamber. Alternatively, he may 'test' her qualities by subjecting her to a protracted period of 'jostling' before finally vouchsafing her entry (Schmitz, 1972).

For the females of *I. confusus, I. paraconfusus* and *I. calligraphus*, the secret of gaining entry is to give the correct 'password' by stridulating. Silent females are stubbornly excluded, so standing outside the entrance and 'singing' is the only way to gain admission. However, even protracted stridulation, backed up by pushing against the entrance, may be to no avail if the occupier has already reached his 'quota' of three 'wives'. Home-owners can afford the luxury of being choosy, presumably because not all males manage to establish a nuptial chamber. This means that those who do so can take their pick from a pool of available females, usually collecting an average of three 'wives' in the nuptial chamber.

In bark beetles, initial colonization of a host tree can be carried out either by males or females, depending on the species. The first arrivals emit an aggregation pheromone which attracts members of the opposite sex, who usually announce their presence by giving a species-specific and sex-specific bout of stridulation. In the minute bark beetle *Cryphalus fulvus* it is the female who sets up home, sending out a pheromone which attracts males and females in about equal numbers (Sasakawa & Sasakawa, 1981). Males arrive to find a nuptial chamber ready to receive them, where mating takes place.

MECOPTERA Scorpionflies, hangingflies

SCORPIONFLIES

Courtship

The provision of some kind of nuptial gift by the males is characteristic of most members of the Mecoptera. Male *Panorpa* scorpionflies are able to utilize three separate mating strategies depending on circumstances. A prospective suitor lays the groundwork by tracking down the corpse of an insect. The most likely place to find such a scarce commodity is in a spider's web, which makes life rather hazardous for the searching male. Once a suitable corpse is found, the male attracts females by 'calling' and releasing a pheromone. Unfortunately, as in a number of other insects (e.g. the cricket *Gryllus integer* and the grasshopper *Syrbula fuscovittata*) 'calling' is a communications channel which his rivals can also tap into. These may fly in to deprive him of his corpse and appropriate any female who arrives. He cannot avoid this undesirable side-effect (of which he himself may take advantage on some future occasion), so he will have to defend his corpse against the intruders. Males are able to judge the size and value of a corpse, and the bigger it is, the longer and harder they will fight to own it.

If insect corpses are unavailable, the male resorts to his back-up tactic and 'manufactures' a wedding gift by secreting a rapidly-hardening, pillar-shaped gob of saliva on to a leaf. He attracts a female by 'calling' and invites her to feed by shaking his wings and abdomen; the meal keeps her occupied while they mate. This self-made gift represents no meagre contribution by the male. His salivary glands are relatively huge compared with those of the female, being quite a substantial proportion of his body weight. As with corpses, rival males will also try to steal a salivary gift and use it in their own nuptials.

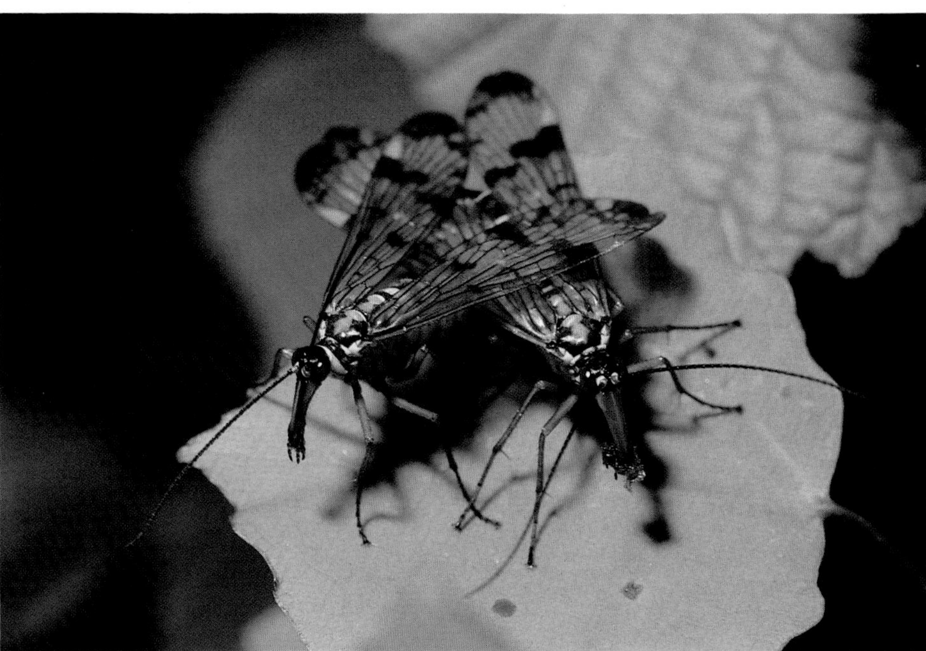

▼ A pair of European *Panorpa communis* scorpionflies mating. The female on the right is feeding on a salivary gift presented by the male.

Rape as an alternative strategy
If a male finds himself unable to compete for corpses, he has a problem. For a start, he will be unable to attract a female directly by acquiring a corpse. Secondly, because males also feed on corpses themselves, a starved male will lose his capacity to produce a large and costly salivary gift as a stand-in. In the absence of any better way, the only answer may be to take a female by force.

'Rape' is not necessarily a 'policy decision' which a male is doomed to follow for his entire reproductive life. Nor is it necessarily connected with failure – it may be quite the reverse. After a run of sexual successes brought about by the use of salivary gifts, the male may find his salivary reserves in a temporarily depleted state. If he cannot rapidly replenish them by gaining access to a sufficient supply of insect corpses, then turning to rape as a fill-in strategy could make sense. As the would-be rapist is fully 'aware' that his parsimony is going to invite a negative response from the female, he adopts the appropriate measures and takes any passing female by storm. His main weapon of subjugation is his large bulbous genital capsule with its efficient claspers, set at the tip of a flexible 'tail' which can be cast around until it 'hooks' its target (Thornhill, 1980a). If he lands a firm grip on one of the female's legs or wings, the male can manoeuvre her so as to obtain a more secure hold on her wing edge with his abdominally-mounted 'notal clamp'. A female always fights tooth and nail at first but, once she is wing-clamped, her only remaining option is to try and keep her genitalia out of his reach. If her attacker finally succeeds in capturing her genitalia he holds her firmly with his notal clamp until copulation is complete, perhaps after as long as 2–3 hours. The most rewarding of these strategies turns out to be to guard a large insect.

HANGING FLIES

Courtship
Most hangingflies (Bittacidae) also provide nuptial gifts, which are captured using the raptorial back legs. In some species both males and females are hunters; in others, such as the Australian *Harpobittacus australis* and *H. nigriceps*, the females do not appear to catch their own prey, but rely on obtaining free hand-outs from the males during the mating season, exactly as happens in some empidid flies (Bornemissza,

1966). Even when females can hunt for themselves, they can still be fussy when assessing the value of the nuptial gifts offered by males. The female *Bittacus apicalis* is sparing with her compliance when mating with a male who has only served up a small inferior gift, and she consents only to a short stint of copulation. Her treatment of a male with a large juicy gift is more generous and he is allotted a rather longer period of copulation (Thornhill, 1977). The females are able to terminate copulation before the male has transferred all his sperm, so it behoves him to keep her happy for as long as possible by providing a substantial meal.

The nuptial behaviour of *B. apicalis*, which can occur in large numbers in damp gloomy forests in the eastern USA, is typical of many hangingflies. The scene is thus set for intense interference competition between males, spurred on by the females' willingness to mate several times each day (Thornhill, 1977). Having caught an insect, the male hangs from a leaf with his front legs, probing the corpse with his mouthparts in a 'quality-control' exercise to secure a gift with maximum 'meal appeal'. If the prey turns out to have a repugnant flavour (it is e.g. a ladybird or lycid beetle), he jettisons it and looks for something better. Small insects with a volume of less than 19 mm^3 are also rejected, indicating an instinctive ability to select gifts which are both large and tasty enough to satisfy the fussy females.

With the corpse slung between his hind tarsi, the male now launches a series of short exploratory flights through the undergrowth near ground level. He interrupts each flight to hang briefly from a leaf and 'call', shivering his hind wings and curving his abdomen upwards to evert the abdominal vesicles which release the female-attracting pheromone. The wing-fanning may serve to spread the pheromone more widely. An answering female hangs likewise from a nearby leaf, and the male then ceases 'calling' and they meet face to face. The female now lowers her wings, 'inviting' the male to proffer his gift, which he does, but keeping it covered behind the tip of his abdomen, probably to thwart any snatch-and-run attempt by the female.

◀ Most male hangingflies (Mecoptera: Bittacidae) present the female with an insect as a nuptial gift. The size of the gift is related to the amount of time which the female will allocate to her mate for copulation: the bigger the meal, the longer the mating time. This *Harpobittacus* species from Australia has served up a plump fly to his mate.

Once she has buried her 'snout' in the gift and seized it with her hind legs, the male tries to link genitalia, usually without success. In common with such disparate gift-bearers as the spider *Pisaura mirabilis* and many empidid flies, he still retains an interest in the gift by holding it with his long hind legs. The female holds her genitalia well out of the way until she has thoroughly taste-tested the gift to assess its value. If it fails because it is too small or tastes bad she will reject it and fly off. Even if she does signal approval by offering her abdomen tip, the duration of her continued co-operation is still quality-dependent. A really good gift keeps her quiet for so long that it is the male himself who breaks off copulation after about 23 minutes. Her interest in a measly gift soon expires; she uncouples after a shorter period. Either way she may occasionally make off with the gift during a brief post-copulation tussle. If the male retains the gift, he again 'tastes' it before deciding whether it will make the grade for a second or even third round of courtship. If not, he drops it and looks for something better.

The ability to make an astute assessment of the gift's capacity to prolong copulation to full term is of vital importance for the male. It seems that only after male-terminated copulations of more than 20 minutes do females start to lay eggs and enter a 3-hour refractory period, during which they are unavailable to rival males. This means that a large gift functions as a form of 'proxy' mate-guarder, preventing the female from replacing her mate's sperm with that of a competitor, without the necessity for actual time-consuming physical guarding. Smaller males who can only catch inadequate gifts are therefore presumably penalized in the same way as small males (in other species) who can be easily supplanted by larger rivals during *in situ* mate-guarding. As a large *B. apicalis* male could possibly copulate with several more females during his first mate's 3-hour refractory period, the strategy of presenting nuptial gifts is a considerable time-saver compared with contact-guarding, enabling a male to squeeze in far more matings within a given period. The females also 'milk the system' for all they are worth by obtaining up to four free meals a day and laying around 14 eggs each. The daily input of free meals is sufficient to nourish an egg batch fully without needing to resort to any other source of food. It therefore comes as no surprise that, during the peak of the mating season, when males are serving up nuptial meals in some numbers, the females more or less give up catching their own prey and rely completely on the hand-outs.

Transvestite behaviour

Going through the laborious business of catching a gift is both time-consuming and hazardous, putting the hunter at increased risk from predators such as spiders. Some males have evolved a low-risk, high-yield behaviour aimed at reducing the intervals between matings by pirating gifts from 'calling' males. The technique is a subtle one but the 'pirate' does not rely on energy-wasting assault, which would invite robust retaliation. Rather, he banks on receiving a 'friendly' welcome through deception, by pretending to be exactly what the calling male is most expecting – a female (Thornhill, 1979). For his masquerade to pay easy dividends, the 'transvestite' only has to act the part convincingly by lowering his wings on arrival in 'invitation'. The 'caller' is well enough duped to hand over his precious gift in around two-thirds of all such encounters. This still leaves about one-third in which the calling male is presumably 'on the ball' and sees through the sham; or else the 'transvestite' fails to get it quite right and fluffs his performance. However, having secured the hand-over solves only part of the transvestite's problem. He still has to cope with the male's habit of retaining the gift while trying to make genitalic contact with the 'female'. The 'transvestite' keeps the tip of his abdomen out in the clear in a very female-like way, but this can only go on for so long. Eventually the 'suitor' rumbles that something is wrong and tries to repossess the gift. In tugging matches which ensure, the 'transvestite' manages to hang on to his spoils in about 22 per cent of encounters. Such a success rate makes this form of piracy particularly rewarding, especially as it is a low-risk, quick-result strategy. The major pay-off for the transvestite is the capacity to mate more often, due to the reduced time needed for obtaining a nuptial gift compared with capturing his own.

Sexual muggings

Some males try to achieve the same ends by straightforward 'mugging' of 'calling' males, but violence pays less well than deception, securing the gift in only 14 per cent of attempts. Copulating pairs also stand a good chance of being assaulted by a male with an eye both on the gift and the female. Such pairs apparently make almost irresistible targets for any solo male, who will almost always have a go at stealing the gift for presentation straight back at the same female. Two tactics are used: the intruder may simply attempt to 'bounce' the gift away by coming in hard against the pair, or he may try to wrestle it from the owner's grip. In dense populations, nearly 90 per cent of copulations are violently disturbed in this way, which is not surprising in view of the high yields for the perpetrators – the success rate in stealing the gift is more than 50 per cent, while mate theft is still a respectable 15 per cent. Late in the season, when they have already mated many times, the females also increasingly adopt fraudulent tactics, accepting the males' gifts but refusing to copulate in return.

Bittacus stigmaterus (Thornhill, 1978) from the USA and *Harpobittacus australis* and *B. nigriceps* from Australia all behave in a similar fashion. However, in the two Australian species, the female refractory period after mating seems to be absent and she will mate with a rapid succession of gift-bearing males. In these species, the males possess a long penile filament which could conceivably be used, as in damselflies, to scoop out sperm from previous matings.

DIPTERA True flies

The *modus operandi* of courtship within this large order is incredibly varied, making flies among the most fascinating of all insects for the study of sexual behaviour. It would seem appropriate to describe first the gift-bearing behaviour of the dance flies (Empididae), which permits easy comparison with procedures in the bittacids just dealt with on page 80.

Nuptials in dance flies (Empididae)

Males in the genera *Empis*, *Hilara* and *Rhamphomyia* are generally known to present a nuptial gift to the female as a prelude to mating. The nature of the gift varies somewhat; it can be eminently practical or completely useless and purely ritualistic. At least three aberrant species of *Empis* do not offer a gift at all. The flies usually assemble near a 'swarm-marker', which may be a bush, the branch of a tree, a gap in a hedge, a clump of nettles, a depression in a sandy road, a pond, a pile of logs or even human garbage. Swarms are usually originated by males, who advertise their possession of a gift by flying in a characteristic manner, e.g. in circles, in spirals or from side to side. Females join the males by coming down from above, whereupon the tricky job of handing over the gift takes place. In swarming species, the males generally have large holoptic eyes whose upper regions are furnished with especially large facets. These enable them to gain a clear picture of objects approaching from above, permitting rapid recognition of females.

The conditions needed for the formation of swarms vary even between members of the same genus. During the month of May in the British Isles, the large *Empis tessellata* only swarms from late morning to mid-afternoon, and only when the sun is shining. Even a passing cloud causes a retreat on to nearby vegetation. The slightly smaller *E. livida* is on the wing in July, when temperatures are generally higher and day length extended. It swarms during the long evenings, between about 19.00 hours and dusk, even when the weather is cool. *E. tessellata* males take a wide range of invertebrates as gifts, including caddisflies, mayflies and moths. However, the main prey always consists of various kinds of flies, such as large plump *Sarcophaga* flesh flies and *Calliphora* bluebottles. At certain times, the large black furry males of the St Mark's fly, *Bibio marci*, often bear the brunt of the attacks, probably because they too are occupied in aerial swarming, which makes them an abundant and easy target (*Rhamphomyia* males in the Arctic are known to recognize the swarm-markers used by mosquitos, and so habitually prey on swarming males (Downes, 1970)). Large *Tipula* crane flies are also much favoured, but they pose considerable handling problems. They struggle valiantly and their long legs become hooked on nearby vegetation, so it may take several minutes for the empidid to finish one off.

Most male dance flies, and especially *E. tessellata* and *E. livida*, will also prey on rival males and use them as gifts. In fact this seems to be a much favoured strategy, as a male will frequently 'make the rounds' of nearby leaves, pouncing on one rival after another. Once he has caught an insect the male usually hangs by one leg from a leaf or twig and 'prepares' it by probing it repeatedly with his proboscis. There is little evidence that he actually feeds on it himself, except perhaps in *E. livida*, whose males often probe an insect for long periods in an apparently non-sexual context (pers. obs.).

During May, as many as three species of British empidids may simultaneously carry out courtship in close proximity, without any obvious signs of interference, save that the smaller species may be served up by the larger as nuptial gifts. Thus, the large *E. tessellata* will be swarming at the top of a small tree, while the much smaller *E. opaca* is performing singly or in twos and threes only a metre above the ground. Although the *E. opaca* females can probably see the larger species flying around above them, they do not respond, indicating the existence of certain species-specific characteristics, such as size and flight mode. A similar separation exists between simultaneously courting swarms of *E. barbatoides* and *E. poplitea* in the USA (Alcock, 1973). In many aspects, these two species form the counterparts of *E. tessellata* and *E. opaca* in Europe.

The male enters a swarm with his gift slung beneath him, held by the tarsi of his middle legs. Males of *E. tessellata*, *E. livida* and *E. opaca* all then advertise their presence and the availability of a free meal with a rapid side-to-side 'dance' as if hung on the bottom of a pendulum. If, after a few seconds, a male finds no takers he moves a short distance and tries again. However, in stark contrast to the situation in most insects, a shortage of females is generally a dancer's last worry. One of the most notable characteristics of empidid courtship is the 'enthusiastic' participation by the females. Their 'eagerness' to couple with a dancer is so fervent that, when there is a temporary shortage of males, the females themselves will rally together in 'invitation' swarms. The members of such all-female swarms constantly fly up and touch one another as if 'checking out' a neighbour's gender. Such swarms of 'frustrated' females are common in several species, usually at the termination of the day's nuptial activities or at the end of the courtship season. Their occurrence seems to have given some observers the erroneous impression (e.g. in *E. livida*) that it is the females who habitually form swarms, rather than the males; in only one or two species do the females replace the males as the dancers.

The females sit around on leaves and fly up to join a male, often responding to the mere sight of his becoming airborne and before his dance proper has begun. Dancing often occurs in cycles, the sight of one male taking to the air encouraging others to join him. As the dancer spots a female above him he suddenly rises vertically; then they

▲ **This male dance fly *Empis tessellata* from Europe has presented his mate with a substantial nuptial gift, – a large *Tipula* cranefly. To capture such a rewarding gift may require a considerable investment in time and energy by the male *Empis*. A large *Tipula* will struggle for an extended period to escape its captor. When this happens the *Empis* male may deliberately ground them both in order to come to grips with the problem under more favourable conditions. For as much as half an hour the *Empis* may pin the squirming, kicking cranefly to the ground and probe its thorax with his proboscis in an attempt to finish it off, only to see the cranefly finally free itself and fly off, apparently unharmed. After a more successful conclusion to such a battle, the *Empis* male deftly hands over his victim's corpse to the female during a brief rendezvous on the wing. In this species, males form swarms above landmarks such as small trees; females mate many times each day and are 'eager' to make contact with the gift-bearing males. Previous writers have suggested that the male presents a gift as a safety device to placate a predatory mate and prevent her from attacking him during mating. However, in gift-bearing *Empis* species the *females* are *not* predatory; it is only the *males* who catch other insects. The gift must therefore serve some other purpose, such as a delaying device to ensure the maximum time *in copula*. Contrast this with non gift-bearing *Empis* such as *E. trigramma* and *E. stercorea* in which the females *are* predatory. In these species, the male injects a liquid gift which the female can only claim *after* the male has mated and left. This may, therefore, truly represent a safety device.**

briefly hover in a 'holding position' while the gift is handed across. With a rather wobbly flight mode, they now form a 'trio', the male at the rear, the female in the middle and her gift in the lead. In this strange fashion, they fly to a leaf, over whose edge the male now hooks his front legs, supporting the weight of the entire trio. This 'trapeze act' allows the female free rein to turn the gift over and over as she probes it with her proboscis. The only major departure from this posture occurs in a small unidentified British *Empis* in which the male often lies on his back on a leaf, with the female and gift heaped up on top of him (pers. obs.). Some of the smaller empidids mate entirely on the wing. 'Quartets' are formed when a male cannibalizes a rival who happens to be holding his own nuptial gift in his middle legs. The gift tends to remain in place, caged between the spiny legs, when the dead male is finally presented to a female, who holds them both out in front of her.

In all the hundreds of nuptial meetings observed by the author in *E. tessellata*, *E. livida* and *E. opaca*, the male's insect-gift has been accepted as adequate 'payment' for copulation. However, other authors report a much lower success rate, for a variety of possible reasons. Amongst others, these could entail rejection of a 'low-quality' gift by the female. It is possible that she makes a rapid evaluation of the gift before accepting it, as males with obviously unsuitable prey may be instantly rejected. However, this must vary from place to place, or according to a female's appetite at the time. For example, in many species, e.g. *Empis livida*, *E. tessellata* (pers. obs.) and *E. confusa* (Chvála, 1980), the gifts are re-used several times and may become very tatty, yet they are still snapped up by the eager females. One observer even noted that male *E. opaca* were presenting females with the inedible fluffy seeds of dandelions as a substitute for an insect gift (Hobby, 1932). It is possible that these were mistaken for insects while being blown by the wind – but unlikely, as empidids are far too sharp-sighted to be so easily deceived. It is more likely that this was a 'deliberate' attempt at deception, and it seems that it paid off handsomely, as the 'cheats' were able to mate just as often and as long as when serving up genuine gifts.

The success of such sharp practice in a species which normally presents an unwrapped insect is interesting, as it may throw some light on the evolution of the behaviour in some *Hilara*, in which the male hands over his offering gift-wrapped in a 'balloon' of silk. This packaging may indeed contain a genuine insect, but it is also frequently used to conceal an inedible object, such as a petal or a seed. Sometimes the 'balloon' may even be empty, yet in all cases the female vigorously probes it as though it were a juicy insect. The gift has therefore apparently lost its original purpose and has become a mere 'token', a functionally useless yet still essential part of a display which has become purely ritualistic in nature.

In *E. opaca*, the males can fool some females but not others when 'trying it on' with a fake gift. One male observed by the author spent a whole afternoon presenting nothing but the white fluffy seeds of a willow tree as his nuptial gifts. However, his subterfuge met with mixed success. He had to 'dance' for an unusually long time, suggesting that some females had already appraised the situation from previous contact, and were boycotting the male bearing the brilliant white ball. Even when he did attract a female, he was often abandoned after a brief aerial contact, suggesting a quick evaluation and rejection of the fake gift. Then, when he did manage to mate, his time *in copula* was exceptionally short. In these truncated matings, it was always the female who terminated copulation by struggling free. Females of *E. opaca* also break off relations when the male has served up a genuine but useless gift, such as the empty drained husk of an insect. However, when a fresh insect of reasonable size is presented, the female is usually preoccupied with it for long enough for the male to finish transferring sperm and make the break. Yet, interestingly enough, females who did fall for the fraud turned the fluff over and over and probed it with their proboscis just as if it were a juicy insect. Presumably, if he could find enough such dupes, the fraudster's strategy (if such it were) would pay off, at the expense of his rivals who would use up extra time and energy catching real insects – willow seeds do not fight back, nor fly away. However, in the present state of nuptial relations in *E. opaca*, fraudulent males can fool some of the females all of the time, but cannot yet fool all of the females some of the time, let alone attain the acme of success by fooling all of the females all of the time.

▼A possible short cut to sexual success for a male dance fly could be to present a useless but easily obtainable nuptial gift. This male *Empis opaca* from Europe has 'fooled' his mate into copulating in return for a 'gift' consisting of a ball of willow seeds. She turns it over and probes it with her proboscis as if it were a genuine insect; other females were less easily duped and rejected the male immediately upon contact.

One interesting aspect is the female's willingness to let the male reclaim the gift at the termination of copulation, there being no struggle over its ownership as seen in bittacids. The male often uses his back legs to retain a light contact with the gift throughout copulation. The female probably surrenders her meal so easily because she 'knows' how easy it is to obtain another one in very short order – she could well fly straight into the arms of another male and be probing a fresh insect within 30 seconds of leaving her mate. At the height of the season, females indulge in a day-long 'binge' of going from one free meal to the next, all for the low cost of accepting the donors as mates. This can go on for 2–3 weeks, which probably enables the females to build up the resources needed for egg production entirely through the nuptial gifts. Males also mate several times each day, but have to waste some time catching insects to serve as gifts, although some species keep this to a minimum by re-using the gifts, such that a large insect may be presented to a succession of females, before it is finally reduced to a shapeless pulp and discarded. Females of *E. livida* seem noticeably willing to accept these recycled gifts, even those in very poor shape. Given the fairly restricted activity period of *E. tessellata* and the extended duration of copulation (45–90 minutes), each female probably still manages to get through 4–7 free meals a day, giving her around 5–6 hours of total feeding time daily. Not bad, considering the minimal costs incurred – no wonder the females are so 'enthusiastic'.

In *Empis borealis*, there is a sex-role reversal; it is the females who form the swarm and the males who fly into it. Once surrounded by 'eager' females, the male acts in a 'choosy' way – after all, he holds all the cards, these being the nuptial gift which represents a substantial investment that is in considerable demand (Svensson & Petersson, 1988). If none of the females in the swarm satisfies him, after he has performed an ascending flight with each in turn, he may leave and try another swarm. The American longtailed dance fly, *Rhamphomyia longicauda*, exhibits an interesting variation on the above theme (Newkirk, 1970). The females swarm, performing an aerial ballet whose purpose is to make them as conspicuous as possible. While 'dancing', each female inflates her abdomen and displays her feathered legs alongside it, bobbing and weaving like a balloon at the end of a string on a windy day. The males are attracted to these prominent female swarms and present the usual nuptial gift. They sometimes mistake other similar-looking natural objects for females and have been seen approaching burred fruits by mistake.

In three other *Empis* species (all in the subgenus *Xanthempis*), there is a complete departure from any of the above procedures, showing that *advance* presentation of a nuptial gift is not a necessary prerequisite to mating within the genus. *E. trigramma* utilizes a much simpler and less hectic mating system. The male is attracted to a 'calling' female sitting below on a leaf, locating her by the pheromone which she releases when in need of a mate. While 'calling', she vibrates her wings, wafting the pheromone far and wide and possibly also providing a visual beacon from the glinting wing surfaces. The male lands 3–4 cm in front of her and agitatedly flicks his wings, holding them above his thorax. Then, to the flicking, he adds a pedalling with his front legs in front of his face. The female responds likewise and the gestures of both insects rapidly rise in intensity until they meet head-on and vigorously stroke one another's faces. Seconds later, the male leaps upon her back and instantly couples. He arrives empty-handed. He cannot present an insect gift, as the males of this species are not predacious, in contrast to the females, who are efficient hunters (Hamm, 1933; O'Toole & Preston-Mafham, 1985). Shortly after they break up, the female bends her body downwards into a curve, so that her long proboscis can contact the tip of her abdomen, from which there now appears a large glob of colourless liquid. She sucks this dry before resuming her normal stance, a behaviour similar to a female tettigoniid feeding on a spermatophylax. It seems therefore that, although the *E. trigramma* male arrives bearing no obvious gift, he does in fact inject some kind of nutritious liquid during the act of copulation, and it is this which the female squeezes out and eats. In *Empis (X.) scutellata*, the males form dancing swarms to which the females are attracted (Hamm, 1933). However, no gift is handed across but, after a pair has split up, the female again prepares to sup on an 'injected' gift. Her method for reaching her meal is quite different; she exudes the blob of liquid and then picks it up in her middle legs. These then deftly transfer it to her front legs, where she can feed on it in comfort, accomplishing all this without rupturing its surface. Neither sex is predatory.

◀ Courtship in an empidid with predatory females. In *Empis stercorea* it is the female who often initiates courtship. She is able to detect a male sitting invisibly as much as 20 cm away, and will quickly make her way towards him, presumably homing in on a pheomone that he releases. What happens next can be rather surprising. It is she who shows 'enthusiasm' by striking up her wing-flicking courtship in front of the male, and he who manifests disinterest by flying off after anything from one to thirty seconds. However, on other occasions it is the male who homes in on the female and initiates courtship (top), in which case it may be her turn to spurn his advances.

This variation in male–female roles is probably linked to the considerable contribution which the male makes during mating. He injects a transparent liquid, which the female imbibes immediately after the male has left (bottom left). It is probably the desire for this free gift which impels females to initiate courtship. However, as the females are fiercely predatory and pose a considerable risk to the males, who are often eaten (bottom right), it is not in the males' interests to respond rashly to a female's overtures.

It is possible that females are able to detect whether or not a male contains a suitable quantity of nuptial fluid via the substrate-borne vibrations induced by his courtship wing-flicking; heavier replete males would produce more powerful vibrations. Males who 'know' that their reservoirs are exhausted would thus be less likely to respond positively, in case they are snapped up by the female in lieu of the absent nuptial gift. Similarly, males may prefer to mate with heavyweight females who are laden with eggs, and therefore close to oviposition, making paternity more likely. Such a female would also be less in need of a large meal such as a complete male, so reducing the risks to him.

Courtship in robber flies (Asilidae)

Unlike empidids, all robber flies are exclusively predacious. Despite the male's ability to present a fly as a sop to a fierce mate, there is no proof that this ever happens. Reports of gift presentation by asilids are probably due to the males' frequent (but purely coincidental) habit of courting females who are already feeding on prey. Unlike in empidids, robber-fly females are not normally enthusiastic participants in the courtship process – they have no nutritional rewards to gain. Mating times are generally much shorter, sometimes only a few seconds and rarely more than a few minutes. Nor does the male seem to require a circumspect approach to his large and fearsome mate. Far from it – in the majority of species there is no courtship and the males jump on to any available female. Admittedly females are occasionally cannibalistic on males, but there is no evidence for this being in a sexual context.

The male's mode of approach can be divided into four categories:

(1) No courtship.

(2) Simple courtship while perched.

(3) Simple courtship while hovering.

(4) Complex courtship – either on the ground or in the air.

No courtship

Most asilids fall into category 1, but the pattern of locating and 'capturing' the female varies somewhat. *Efferia frewingi* males from the USA search for females by flying near the ground and diving on to any female spotted in flight (Lavigne & Dennis, 1975). In the ensuing struggle, they drop to the ground, where the male may succeed in forcing genital contact. During copulation, the male sits atop the female and gently caresses her abdomen with his hind legs, a habit also seen in *E. helenae* and *E. staminea*. The male announces the termination of copulation with a prolonged wing-buzz, then leans backwards and launches himself into the air. In *Efferia varipes*, the female 'turns off' an unwelcome suitor's interest in her by falling into a state of apparent thanatosis as soon as she is grabbed. She continues to 'play dead' until the male loses interest, due to the lack of the necessary feedback, and flies off (Dennis & Lavigne, 1976). This seems to be a more effective female rejection display than the abdomen-raising 'sky-pointing' stance employed by most other species, during which the males often press home their attack regardless and struggle to engage their genitalia.

Stichopogon catulus from Arizona employs similarly crude techniques, but is one of many species which only pay attention to females who are resting on the ground (Weeks & Hespenheide, 1985). It is interesting that females occupied with prey are less likely to perform their 'sky-pointing' refusal display.

Simple courtship while perched

In *Nannocyrtopogon neoculatus* from California the male lands near a female on a log (Hespenheide, 1978). His precise landing site – off to one side – seems to be chosen, because if she sees him coming she will make a bolt for it. Once safely in place, he steals around to face her, slips his front legs under hers and pumps them up and down as in a hearty double handshake. On occasion, a second male will join in, sitting on top of the first! The small male of *Ablautus rufotibialis* from Colorado performs a dance before the female (Lavigne, 1972). Stilting up on high, he waves his front legs alternately up and down, flaunting the tufts of black and white setae on his feet, while simultaneously rushing back and forth in an arc. He is very persistent and, if the female flies off, he will shadow her closely until she lands and immediately resume his efforts (most male asilids with any kind of courtship do the same).

Simple courtship while hovering

Leg displays are often performed with the male hovering in front of a perched female. Even within the same genus, there are variations in the degree of courtship. Thus, while *Stichopogon catulus* simply jumps upon a female with no preliminaries, *S. trifasciatus* (Weeks & Hespenheide, 1985) flies back and forth in front of the female, drooping and shaking his legs before dropping on to her for attempted copulation. However, as in a number of other insects, such as grasshoppers, he may sometimes dispense with preliminaries and simply pounce on a female.

In many asilid males, the front legs are distinctively coloured and ornamented, and courtship techniques are designed to display them to best effect before the perched female. The male *Heteropogon stonei* from the western USA (Alcock, 1977) begins his aerial approach with his legs drawn in, but then extends and splays them widely at the very last second as he almost skims them against the female's face. He finishes off by perching near her, then has another go. In between every four or five frontal approaches, he darts around the back and brushes his legs against the rear of her abdomen. As the female detects the imminence of this manoeuvre, she prepares to kick him clear by 'cocking' her hind legs beside her abdomen, while also spreading her wings so that the crimson apex of her abdomen can blatantly proclaim its 'not-today' message. The male is usually reluctant to take 'no' for an answer and persists in his efforts, perhaps squeezing in more than 350 flights over half an hour, yet having nothing to show for it at the end.

Such cold-shouldering by the females is typical of asilids. In fact, a female will often dart out and snap up an insect – right in the middle of the male's display. This could be because such females only ever mate once, so a meal to nourish developing eggs would be more important than another superfluous mate. However, multiple mating is present in some species. Strangely enough, 'Lone-Ranger' take-by-storm tactics seem more productive than the most complex courtships, which seem to reap the lowest rewards in terms of copulations. Despite their predatory habits, females rarely chase the males away, but usually either steadfastly ignore them or fly off, ducking and weaving in an attempt to lose their pursuer.

In *Heteropogon maculinervis*, the male brandishes the white patches on his foretibiae as he flies to and fro in front of the female (Lavigne, 1970a). The flight of *H. wilcoxi* is similar but slower, while in *H. lautus* the male also swings from side to side. Females of the first two react to the male's frontal approach by spreading their wings. Then they stick their rear legs up in the air and vibrate them rapidly in a rejection display. Foreleg-waving is also seen in the diminutive *Cerotainia albipilosa* from the eastern USA (Scarbrough, 1978). The males often start to display at other insects, such as other asilids (a bit risky this), hover flies and tephritid flies, as well as males of their own species. Such apparent incapacity to distinguish females from similar-looking objects is typical of most male asilids, although 'mistaken-identity' courtship tends to be short due to the lack of the correct response. In *C. albipilosa*, prolonged courtship flights tend to attract other males, resulting in dual courtship. However, the first male usually 'claims ownership' by driving off intruders in a downwardly spiralling flight, usually also barging and wrestling with the interloper. The 'owner' usually wins and returns to finish his courtship, although his victory may avail him little, courtship success being very low.

In the Central American *Mallophora schwarzi*, the male is unusual in approaching from the rear (Shelly & Weinberger, 1981). He bobs to and fro with his fore and middle legs tucked up against his thorax and his hind legs splayed widely apart. Now and again, he swoops right in, hooks his hind tarsi under the female's wings and rapidly flips them open. Why he does this is uncertain, but in *Mallophora* the wings do protrude well beyond the rather stubby body, obliging the male to spread them in order to gain access to the genitalia.

Complex courtship

The height of courtship virtuosity is found in *Cyrtopogon*, which typically rest on logs and rocks (on which courtship takes place) in mountainous areas of North America and Europe. In some localities, two species may occur together, e.g. *C. auratus* and *C. glarealis* in Wyoming (Lavigne, 1970). The latter has the simpler nuptial repertoire – just an alternate raising and lowering of the front tarsi, combined with a see-sawing of the abdomen. The success rate seems to be abysmally low (one mating out of 75 observed encounters) This may account for the males' tendency to court anything remotely like a female, including other males, both sexes of *C. auratus*, various large-bodied flies, such as blow flies, and even hunting wasps.

The arena for the elaborate nuptial choreography of the male *C. auratus* is a fallen pine log; but only where the sun strikes the wood. As such spotlit stages are not abundant, several pairs may cavort simultaneously within a sunspot only 30 cm or so in diameter. Raising both front legs slightly off the log, the male quivers them slowly in unison, showing off their silvery tarsi, then raises them above his head and palpitates them at a greater rate, but this time in an alternating mode. The performance may finish with this 'Act I', but usually the male must perceive some sign of interest, which sparks off 'Act II'. Still quivering his upraised front legs, he now points his abdomen upwards and starts to swing it to and fro, showing off its glossy black and gold upper side. To this spellbinding performance he adds regular flicks of his spread wings. After about a minute of energetic prancing, he 'bounces' on to the female in a quick 'test' and back again. He is doggedly persistent over many hours, yet the rewards for this most enterprising of displays are incredibly low (one copulation out of 125 observed encounters).

The rate for *Cyrtopogon ruficornis* in the European Alps must be similarly disappointing, at least in mid-August. During his courtship dance the male prominently parades the shiny black tip of his abdomen as he waves it around in the air. He also raises his body as high as possible on his middle and rear legs, using his front legs in a rowing action until he comes in close, when they turn to stroking down across the female's face. Now he also extends his shiny black proboscis and dangles it before her gaze. As proboscis length could possibly be correlated with hunting prowess, he may perhaps be demonstrating his 'fitness' as a mate. However, the males' rate of success is extremely low. Yet, in *C. marginalis* from the eastern USA, 50 per cent of 12 courtship attempts were seen to meet with success (Lavallee, 1970), bucking the abysmally low trend in the genus. As this was on an early date – in May – virgin females may have been more abundant.

Scramble competition and mate-guarding in dung flies

Males of the common yellow dung fly, *Scatophaga stercoraria* (Scatophagidae), assemble in some numbers on a freshly deposited cow pat, where they can be guaranteed to meet females hurrying in to lay their eggs before the surface dries out too much. Females are always greeted by an excess of males, giving rise to intense rivalry which entails considerable drawbacks for the females who, being smaller, suffer possible damage and certain delay as a result of the males' aggressive take-over tactics (Parker, 1970a).

As soon as a lone female lands on a pat she is deluged by scuffling males, one of whom will win her and mate. Females sometimes give body-shaking rejection displays, but the powerful males seldom take any notice and just force a mating anyway. Once copulation has finished, the male remains astride his mate and contact-guards her as she lays her eggs. On fresh pats, the density of unmated males is so high that disturbance adds an average of 12 minutes to the 'normal' oviposition time of 45 minutes (Parker, 1970). There is much to be gained from a take-over, as the last male to mate with her fertilizes 80 per cent of her eggs. Being small, it is the females who come off worst during take-over attempts, particularly if several males are piling on top of her simultaneously (up to seven at once). Even with just a single assailant, the female will probably be pressed into the mire, often face-down, her body twisted and turned, her wings bent and scuffed by the bristly legs of the heedlessly battling males. On the very rare occasion, she may be drowned, or her abdomen may even be torn apart and the eggs – for

▶ The most complex courtship in robber flies is found in the genus *Cyrtopogon*. This male *C. ruficornis* from the French Alps 'dances' in front of the female (left), showing off the black tip of his abdomen as he bobs it vigorously, gradually advancing to stroke her face with his front legs. Note the different colours and pelage of male and female, a typical character in asilids having complex courtship designed to show off special male adornments.

whose paternity the males are so fiercely competing – scattered uselessly across the dung. Despite all the fuss, only 1.7 per cent of take-over attempts actually succeed.

Females appear to show a preference for mating with large males, probably because they will be guarded more effectively and be less likely to suffer serious delay and injury (Borgia, 1981). A take-over attempt can be pre-empted altogether if the mounted male can prevent the intruder from actually *touching* the female. The males are not very good at distinguishing between males and females by sight alone. As a result, the males spend much of their time jumping on and off one another in a brief examination of gender. When 'jumped' by a rival, a contact-guarding male therefore performs a stereotyped push-up with his front and hind legs, while his middle legs are thrust vertically upwards. This 'shrugs off' the intruder before he contacts the female and realizes that it is worth making a real fight of it. This 'shrug-off' is effective over 90 per cent of the time, but is obviously more likely to do its job when performed by a large rather than a small male. As a successful 'shrug-off' prevents a scrap which could damage the female, her preference for larger males would make sense. Larger males are also better able to air-lift the female to safety if the pat is threatened, e.g. by a cow lying on it. A female cannot use her wings while burdened with a male, so he has to carry out any flying which is required. Small males are not much help in this, often being unable to lift their mates at all. This may explain why, if 'captured' by a small male, a female may shake her body, inciting a take-over by larger rivals standing nearby.

Mate-guarding in tipulids

In many crane flies (Tipulidae), freshly emerged virgin females release a pheromone which attracts males to them for mating; the females then oviposit alone. However, in at least three American species, *Dactylolabis montana*, *Limonia simulans* and *Antocha saxicola*, males congregate around oviposition sites on the edges of streams and attempt to intercept egg-laying females (Adler & Adler, 1991). Competition is intense, partly because of the excess of males over females, and partly because females will mate with any males who grasp them, if necessary with two or three partners in rapid succession. Assuming last-male sperm precedence, there is great pressure on males to secure a copulation while the female still has eggs left to lay.

► A male dung fly *Scatophaga stercoraria* successfully parries an attack from a rival.

▼A large male dung fly *Scatophaga stercoraria* has just successfully pushed a smaller male (left) off the female, which he has been guarding, and is about to mate with her himself. Females prefer large males, probably because they are better able to resist such interference, which is often detrimental to the female herself.

A male will guard his ovipositing mate by 'caging' her with his legs, these being longer in the males than in the females. The male *A. saxicola* also boasts heavily clawed prehensile tarsi, which hook on to the female and make him difficult to dislodge. His ability snappily to re-couple with the female as soon as he senses a nearby rival is also of great protective value by denying access to her genitalia. The males therefore make effective guardians, managing to fend off gatecrashing males about 85 per cent of the time.

As guards, the other two species are comparatively ineffective, losing their mates over 65 per cent of the time. However, guarding behaviour still pays off, even in such apparently dire circumstances. This is because an intruder who is locked in battle with a guarder is not yet in a position to copulate with the guarder's mate, who continues to lay 'his' eggs. His guarding therefore gains valuable time in terms of extra eggs laid, even though he eventually loses his mate.

Resource defence in the cactus fly

A rot-hole in a cactus is the resource defended by the cactus fly *Odontoloxozus longicornis* (Neriidae). Males establish territories on areas of necrotic tissue (high-quality egg-laying sites for the females) on large columnar cacti such as the saguaro, *Carnegiea gigantea*, in the USA and Mexico (Mangan, 1979). Only the largest males are able to hog the prime spots, leaving their smaller rivals to adopt a 'drifter' strategy, wandering around in the hope of chancing across a female. Under ideal conditions this would be an inferior mate-finding tactic but, in the high desert temperatures, any prime spots which occur in full sun may become 'too hot to handle' for the territory-owners. They are forced to adopt the 'drifter' role themselves, poking around for females in cool nooks and crannies. However, territorial behaviour is probably the best bet, as the 'owners' are able to ensure their paternity of any eggs deposited within the territory. The 'owner' is also more likely to secure a receptive mate, as the females are fussy and tend to shun 'drifters', although not invariably. An 'owner' by contrast needs only to leap upon a female to be instantly accepted. He may well copulate immediately, although he will also sometimes allow her to lay a few eggs before he mates, possibly to check that she really is gravid and therefore a rewarding prospect. He then engages in repeated bouts of

Enigmatic behaviour in sepsids. Sepsids are known for their habit of forming vast swarms on vegetation, which becomes covered in a seething carpet of flies emitting a clearly detectable odour. Sexual activity is absent, and the flies incessantly, and apparently aimlessly, parade up and down. The purpose of these swarms has been the subject of much speculation. One clue to their function may be their habit of only appearing in late summer and early autumn. Secondly they can be remarkably enduring, perhaps lasting in the same spot for several weeks and then re-appearing in the same place a year later. It has been suggested (Pont, 1987) that the characteristic smell may be due to a marker pheromone. This is laid down by the females to assemble the swarm and enables males and females to re-locate at the same spot for mating after hibernating as adults through the winter.

The species involved in these swarms is invariably *Sepsis fulgens* (often misidentified as *S. cynipsea* in the past). The mating tactics of the latter species are also something of an enigma, as females arriving at fresh cow pats to lay their eggs invariably seem to have mated. They are met by males who have established territories on the pat in order to intercept egg-laying females. These then ride on the egg-layers' back as pre-copulatory contact-guarders (Parker, 1972). This is the reverse of the situation in dung flies, as is the existence of territorial behaviour. The 'riding' male is well able to resist the regular take-over attempts, due mainly to his own 'stirrups' – a spine-and-notch 'lock' on the front legs, which clamp firmly to the female's wings, making it difficult to unseat him. When the female finally walks off into the grass, her rider has a chance to mate, although after his long expenditure of time and effort, he may still be unseated by an unreceptive female.

▼ A swarm of *Sepsis fulgens* containing thousands of individuals. Such huge gatherings are somewhat puzzling but may function to lay down a long-lasting 'marker' pheromone, which enables overwintered adults to find one another and mate next spring.

mating and oviposition, possibly to guarantee a high rate of paternity, as females may have already mated with 'drifters' before coming to oviposit. The 'drifters' thus risk ending up not fathering any offspring at all, as they are not in a position to guard the female during oviposition, for this would mean challenging a territory-owner.

Resource defence in a dryomyzid

Dryomyza (Neuroctena) anilis males defend small animal carcasses (Otronen, 1984). A female arriving to oviposit will mate instantly with the 'owner'. He is not so easily satisfied, having to fight hard to control a commodity scarce enough to attract several females. So he becomes choosy, a luxury he can well afford as he can discriminate between heavily gravid 'good investments' and less rewarding immature or egg-depleted females. He checks out a female's fecundity by probing her abdomen with his hind legs, making a correct 'judgement' more than 70 per cent of the time. If she passes the test, they copulate.

Resource defence by a dolichopodid

Scramble competition is typical of most dolichopodids. The beautiful males of *Argyra diaphana* from Europe are an exception, defending small puddles of water, such as water-filled ruts, in shady woodland. The females will lay their eggs only in this patchily distributed habitat and, as population levels are very low for a dolichopodid, territorial control is economically feasible.

The male is a gem with an iridescent abdomen which flickers as he changes position; the females are smaller and quite drab, looking like a different species. The male's sparkling colours are somehow important in male–male rivalry rather than in nuptial relationships, as courtship appears to be absent (in other dolichopodids, the males scissor their wings frantically in front of the female, or hover before her and dangle ornamented front legs). A territory-owner perches beside the water and intercepts passing insects of a similar size (pers. obs.). When the intruder is another male, the 'owner' engages him in a breakneck circling flight just above the water; then they zoom upwards in a spiral flight which takes them away from the territory. (Similar ascending 'spiral flights' are common in territorial encounters in insects as diverse as dragonflies, butterflies, bees and hunting wasps.) Usually it is the 'owner' who returns some 10–30 seconds later, making several low passes over the territory before landing. If this 'inspection' turns up a fresh intruder, who has settled in the owner's absence, a fresh 'duel' breaks out. An intruder bent on making a serious challenge eventually refuses to be put to flight and sticks tight. The 'owner' switches tactics and now 'dive-bombs' the intruder, striking his back repeatedly. A really obstinate challenger is unmoved, forcing the 'owner' to adopt a third tactic: settling nearby and watching his rival's every movement. Now it may be the interloper who 'throws down the glove' and initiates a chase-and-soar flight, after which he may at last prevail. Females are allowed to potter around and lay their eggs for as long as they like without being molested, although the territory-owner will regularly, but only briefly, check out their identity (presumably he can recognize females with whom he has recently mated).

Complex courtship in a sarcophagid

Wing-scissoring is an integral (and often the main) ingredient of sexual communication in several families of flies, particularly the Sepsidae, Dolichopodidae, Tephritidae and Drosophilidae. As any kind of courtship is rare in the Sarcophagidae, it is interesting that wing-scissoring forms part of the complex courtship ritual practised by *Phrosinella aurifacies* (Miltogramminae) from the eastern USA (Spofford & Kurczewski, 1985). In fact, this is truly outstanding in being the first known example of a sophisticated courtship in *any* calypterate fly (calypterates include generally fat-bodied flies such as Tachinidae, Muscidae, Calliphoridae etc). The sexual rendezvous is made near aphid colonies, where abundant honeydew is available, on which the adult flies feed. The males arrive first and, while not busy fighting, manipulate honeydew-coated sand-grains between the proboscis and front feet (tarsal brushes). After casting a grain aside, a male then proceeds to lick his feet. It is possible that sand-grains so treated are being 'marked' with a male pheromone so that the whole area becomes attractive to newly emerged females from the surrounding countryside. This would make it a scent-marked mating station rather than the landmark stations employed by most other species of calypterate flies.

During courtship, the male first stands and shows off his golden face, a uniquely male character and possibly the only example within the family of mate recognition through a sexually dimorphic visual 'badge'. He then scissors his wings from various angles around the stationary female, who usually responds by flying off, trailed by the male who resumes his scissoring as soon as he can (shades of asilids here). Simple follow-my-leader trailing flights are broken by so-called 'yo-yo' flights, entailing a rapid rise and descent to the ground, and by the aptly named 'roller-coaster flights' in which the participants join in long undulating chases. After a few such bouts, the female either finally gives the male the slip or else settles on the ground and watches one final bout of scissoring before she succumbs and copulates. The chases appear to be an integral part of the sequence, as without them copulation does not take place.

Techniques in tephritids

No courtship

Wing-waving is especially characteristic of the Tephritidae, although not necessarily always in a sexual context; males and females both engage in spontaneous waving while sitting alone. Even when highly patterned wings could be expected to play an important role in courtship, this may not always be so. Strongly marked wings are found in both *Chaetostomella undosa* from the USA (Steck, 1984) and *Trypeta stylata* from Europe. Yet the male always seems to use the 'Lone-Ranger' approach, making a sudden leap and then trying to hang on desperately as the female shudders violently in an attempt to throw him off. If he 'passes the test' by holding fast for more than 10 seconds or so, the female's antics gradually subside and eventually she extrudes her ovipositor and allows him to mate.

Male pheromones

In some species, the female arrives in a complaisant mood, having followed up a pheromone message released by a 'calling' male. Examples of such 'calling' are known in a number of tephritids, including the Mexican fruit fly *Anastrepha ludens*, in which the male 'calls' from a territory beneath a citrus leaf. He 'switches on' his pheromones by elevating his abdomen and inflating pouches situated on either side of, and in, the rectal area (Robacker & Hart, 1985). (Eversion of abdominal sacs is found in several other tephritids as well as the chloropid *Thaumatomyia notata*.) Periods of sitting motionless in this stance are broken by 1-second bursts of rapid wing vibrations, dabbing of the rectal pouch against the leaf and on-the-spot circling, all of which actions probably promote the dispersal of the pheromones. If a female arrives, the 'caller' mates after a brief and rather variable courtship. Females of several species are known definitely to be attracted to pheromones released by male abdominal pouches. These include *Dirioxa pornia*, *Anastrepha suspensa* and *Toxotrypana curvicauda*.

Singing and nuptial gifts

In many species, alternate waving of the patterned wings is the main courtship device, although it is possible that the wings' movements may be conveying acoustic as well as visual information. In *Anastrepha suspensa*, wing-fanning produces sounds which play an important role in sexual communication. There are two different kinds of output. The 'calling song' is produced by fanning the wings and simultaneously emitting a pheromone. While doing so, the male defends a mini-territory in association with other males, thereby forming a lek. A 'pre-copulatory song' immediately precedes mating and may continue thereafter. This song has been shown to influence male mating success, and may bear upon female mate choice by indicating a suitor's vigour, which is linked to his size (Sivinski *et al.*, 1984). The male *A. suspensa* thus offers the female a cocktail of stimuli, which may also be typical of many other tephritids.

In *Aciurina mexicana*, a patterned-winged American species, the male employs a combination of visual, olfactory and gustatory stimuli. He stands up high on his legs and inflates his abdominal pouches, dispensing an odour which is clearly detectable by the human nose (Jenkins, 1990). A wing-flicking, body-swaying dance in front of the female blends with his scent to induce an answering bout of wing-waving, whose gradually increasing intensity indicates her growing mood of co-operation. The male recognizes this as his cue to start investing in physical resources, as well as time and energy, by serving up a nuptial gift. He exudes a clear liquid from his mouthparts and dabs it on to the leaf. As soon as the female steps forwards and starts to feed, he turns off his display and moves away a little, causing her to cease her own wing-waving, so allowing him to mount. If she rejects his initial attempts to mate, he may resume his wing-flicking and increase the 'bribe' by enlarging the nuptial gift, possibly making several additions until the female is 'satisfied' with the size of his investment and becomes absorbed in her feeding, allowing him to mate. In several tephritids, the nuptial gift consists of a frothy mass deposited on the leaf of the food plant. *Eutreta sparsa* from the USA secretes a tall, cone-shaped mass on to the leaf of its goldenrod (*Solidago*) food plant (Stoltzfus & Foote, 1965). The male stands near his gift and tries to attract the female's attention by darting from side to side and occasionally flicking his wings. The females are usually slow to respond and need considerable encouragement before they creep forwards to feed, but if they do so, they also allow the male to mate. Rival males seem to value the froth highly, and will try to evict the owner and dine on it themselves. Frothy gifts are also produced in the widespread North American *Stenopa vulnerata* (Novak & Foote, 1975). The male also blows bubbles from his mouth, possibly to disseminate a close-range pheromone to attract a mate, although it may also serve to broadcast an enticing whiff of the tasty nuptial gift on offer later. This can be excreted either as one large mass or sometimes as a row of small ones placed across a leaf. The male flicks his wings to draw a female's attention to the froth, and mating takes place only when she has started to feed.

The most complex nuptial-feeding behaviour is found in *Schistopterum moebiusi*, which occurs from South Africa to the Middle East (Freidberg, 1981). Each male defends a territory on a leaf, chasing off rivals or engaging them in rear-up, face-to-face grapples. A female is treated to a bout of courtship wing-scissoring as the resident male rushes around her. If she stays put he squirts out a mass of white froth from his proboscis. This initially forms a vertical pillar, but then he converts it into a mushroom shape by applying extra material to the summit and then tilting it over to one side. The female watches all this with obvious anticipation, and the male may have to employ some furious wing-scissoring to prevent her from barging in prematurely. Her very first taste seems to transmit an instant message to extend her ovipositor to its full length, allowing the male to mount and mate. He often repeats the whole process several times, adding more and more froth to the pillar, interspersed with repeated copulations. The main function of the gift in these froth producing tephritids remains uncertain. The male may be making a considerable nutritive investment in his offspring. Alternatively (or additionally), the froth may serve to keep the female in one place long enough for successive matings to transfer sufficient sperm. It seems that the froth also induces a receptive state in the female, but perhaps only after she is satisfied that the remuneration is high enough by reserving her co-operation until the gift exceeds a certain minimum size. Small males with small gifts would therefore be penalized and find it difficult to secure long enough copulations, or even copulations at all. Similar nuptial gifts also occur in the Drosophilidae, Platystomatidae, Sciomyzidae and Micropezidae.

In some tephritids, the nuptial gift is passed across directly by mouth-to-mouth contact (as in many Drosophilidae). The male *Paracantha gentilis* from California attracts females by 'calling' from beneath a thistle leaf, during which his orange abdominal pouches are inflated (Headrick & Goeden, 1990). He may have to 'call' for several hours before a female shows up. Then, after some quick wing-waving, he progresses into his 'bubble-gum' routine, extruding his proboscis and puffing up the lower surface to such an extent that the underside now faces upwards. From the top of this grotesque 'bubble', he secretes a droplet of whitish foamy liquid, which he then 'vibrates'. This may help to accelerate the diffusion of an attractive odour across to the female, for she quickly steps forward and sups on the liquid. She signals her readiness to mate by extending her ovipositor, causing the male to break the 'kiss' and position himself behind her. He must still wait for the final 'all clear' as the female droops her wings and telescopes her ovipositor to its maximum length.

In the Eurasian fleabane gall fly, *Spathulina tristis*, nuptial feeding is unusual in occurring *after* copulation has finished (Freidberg, 1982). Donating a nuptial gift *after* copulation has occurred is highly unusual because the female stands to lose by it. She cannot assess a male's fitness by the quality or quantity of his gift until after she has committed herself to accepting his sperm. She also risks being defrauded by males who mate with her and then run off without making a donation. However, walking out on a mate is probably not in the male's interests either, as extensive research has established that the only likely function of the kiss is to increase very slightly the number of eggs laid by 'kissed' versus 'unkissed' females.

In some tephritids, the males only bother with full courtship soon after female

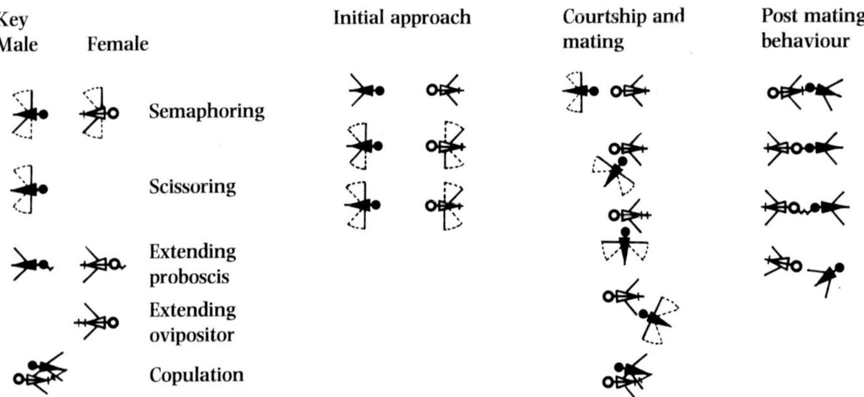

▲ **The courtship display sequence in the fleabane gall fly, *Spathulina tristis* (Tephritidae).** (After Freidberg, 1982).

▼ **During oviposition, the female *Trypeta tussilaginis* (Tephritidae) from Europe is reluctant to move once she has inserted herself between the dangerous hooked burs of her food plant. If a male finds her thus engaged, he will remain on her back until she has finished; then she will almost always allow him to mate with her, cutting out the need for a lengthy and often futile courtship.**

emergence. Once egg-laying begins, the males often take advantage of the females' preoccupation and dispense with courtship. Early in the season, males of *Rhagoletis mendax* from the USA conduct their full nuptials on the leaves of the food plant, but later transfer their mating attempts to females busy ovipositing in the berries (Courtney Smith & Prokopy, 1982). Mating success is particularly high when females are temporarily immobilized while the ovipositor is pushed home. Males of the European *Trypeta tussilaginis* do likewise.

Territoriality in tephritids. Males of *Paracanthera gentilis* exhibit considerable territoriality on thistles. This is not unusual in itself, as most tephritid males are intolerant of their rivals, lunging and butting at one another with proboscis and wings extended. However, the territorial behaviour of *P. gentilis* is more complex. The opponents employ purely visual 'jousting' using stereotyped wing displays, but they also come together, proboscis against proboscis, and 'wrestle'. There is even an acoustic 'battle', comprising a 'call–answer' sequence of buzzes produced by rapid vibrations of just a single wing. Exactly how the loser is determined is uncertain, but as usual it is the intruder who most often gives way and is seen off the premises by the occupier.

Like red deer stags in the rut, the males of several species of *Phytalmia* from the Australasian region contest territorial ownership using antler-like headgear. The various species are very large for tephritids and occur on fallen tree trunks in rainforests. Several males are usually present on a

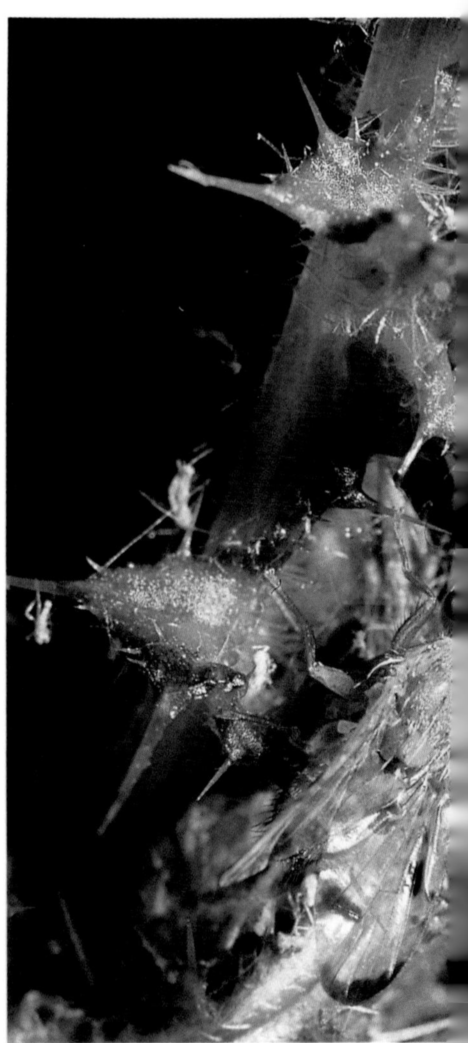

► **Ritualistic 'fighting' in male *Phytalmia megalotis* (Tephritidae) from Queensland.** (After Moulds, 1977).

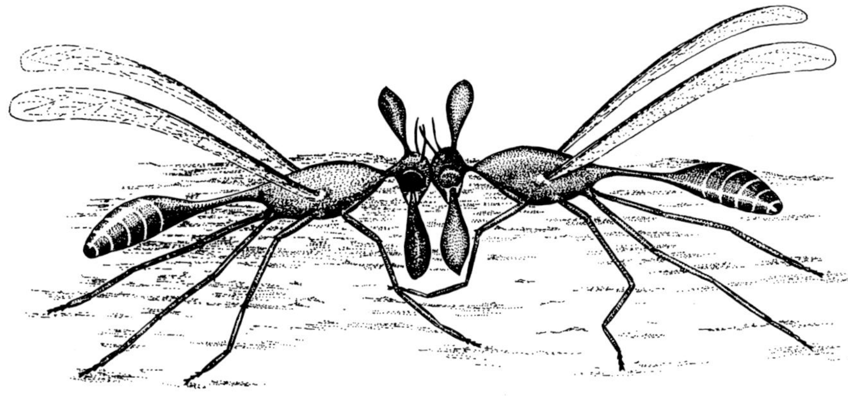

favourable trunk, on which they parade up and down within their territories (Moulds, 1977). 'Prime' territories have a good supply of crevices in which females can deposit their eggs – but only after the 'owner' has mated with them. As an example of resource-defence polygyny it has similarities with the cactus fly. When two males meet they almost always engage in a ritualized contest whose duration depends largely on the relative sizes of the opponents. *Phytalmia* males differ very noticeably in stature and the mere sight of a superior opponent sends a smaller rival running for cover. A more evenly matched pair will edge speculatively forwards before finally engaging their long cheek projections or 'antlers' and embarking upon a pushing match. As they strain away, they gradually rise up on stilted legs until their flattened faces come to take the full force of their efforts. They can only keep this up for a few seconds before one combatant takes a tumble and scuttles off.

A mounted male secures a firm grip on the female's wing bases with a series of spines on his forelegs, remaining on board as a guard while she lays her eggs, although he will dismount to engage in head-to-head battle with an intruder. A large male will use his

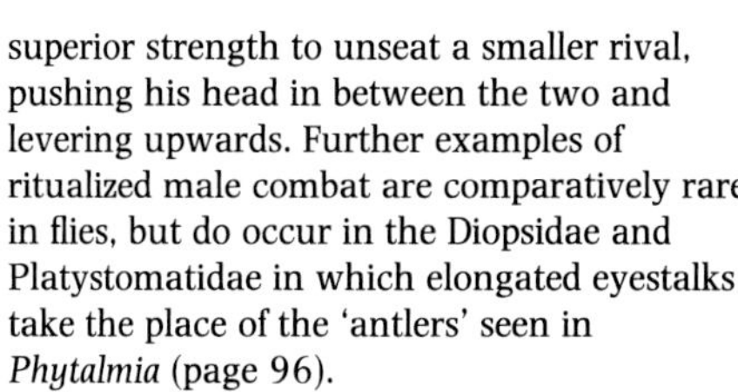

superior strength to unseat a smaller rival, pushing his head in between the two and levering upwards. Further examples of ritualized male combat are comparatively rare in flies, but do occur in the Diopsidae and Platystomatidae in which elongated eyestalks take the place of the 'antlers' seen in *Phytalmia* (page 96).

◄ **Mating procedures and rivalry between males in the European tephritid *Xyphosia miliaris*. This species develops in the unopened capitula of thistles, mainly marsh thistle, *Cirsium palustre*, and creeping thistle, *Cirsium arvense*. In general, only one male is present per plant, and he sits immobile for long periods and pumps the labellum on his proboscis. This presumably releases a pheromone, as females often alight directly in front of the 'calling' male. A receptive female approaches the male face to face, waving her wings. He responds by adopting a peculiar crouching stance with his wings, which are not waved, but held up at an angle above his back. If she continues to indicate her receptivity by coming closer he will jump on to her back, facing towards her rear. Then he does a quick about-turn and rapidly strokes her ovipositor sheath with his back legs. This induces her to extrude her ovipositor so that he can connect his genitalia. Females who are not receptive are able to dart out of the way in an instant as soon as the male pounces. Mating therefore only seems to take place when a female actively responds to a calling male.**

When a second male tries to trespass on a calling male's thistle, a fight often breaks out. The 'owner' of the plant usually adopts a posture similar to that taken up during courtship (far left, fly on left) while the intruder often seeks to defuse the implied aggression and continue his trespass unmolested by adopting female-like wing-waving behaviour (far left, fly on right). Occasionally, the intruder will use this behaviour to sneak up on the 'owner' and launch an attack without warning (top left). Scuffles are usually brief but vigorous, and sometimes end up with one male seated atop the other (bottom left). When this happens, the uppermost male seldom makes copulatory movements but just sits there, sometimes for long periods. This could possibly represent a form of dominance behaviour.

Lekking

A number of tephritids – mostly species which utilize a range of food plants – assemble in leks. A *Procecidochares* species from the USA is an exception, forming galls solely on *Chrysothamnus nauseosus* but migrating to *Atriplex canescens* for lekking purposes (Dodson, 1986). While 'on parade' the males engage in regular head-butting and grappling contests. The sight of two males having a fight seems to attract females who may mate with the winner, usually the larger of the two. This indicates that females may directly choose 'fit' mates by watching the outcome of male fights within the lek.

Courtship in fruitflies (Drosophilidae)

Lekking and fighting

Courtship in tephritids and drosophilids has much in common, save that sound plays a larger role in drosophilids – the 'love-songs' of fruit flies having been known for some time. On the other hand, 'calling' behaviour using inflated abdominal pouches seems to be absent.

The largest and most flamboyant species are endemic to the Hawaiian group of islands, where the family has evinced a spectacular tendency to speciate vigorously. There is a notable degree of sexual dimorphism, typified by modifications to the wings, abdomen, legs, mouthparts or antennae of the males. Research into sexual behaviour has increasingly elucidated the sexual role played by these modifications, their main purpose being to aid in discrimination between co-occurring species, of which there is an exceptionally large number in Hawaii. Hawaiian drosophilids tend to form lekking or 'jousting' displays on feeding or ovipositing areas, or on leaves nearby (Ringo, 1976). 'Jousting' males use up to 13 separate movements or poses, often in combination. *Drosophila grimshawi* males form up in rows in which neighbouring flies face in opposite directions (Ringo, 1974). The 'jousters' than position their abdomens at various angles, manoeuvre their legs and wings in certain stylized ways and jostle against one another whilst circling. They also rub their front legs and hind legs up and down against each other and may also exude a drop of liquid from the anus, smearing it over a large area by dragging the abdomen on the substrate. It is likely that the odour emanating from a multitude of such droplets may serve as a powerful attractant to the females by mimicking the yeasty odour of decaying vegetation, in which females would lay eggs (Steele, 1986). By combining their efforts, the males would, in effect, be drawing in females under false pretences by creating a 'false resource'. Jousting behaviour has been noted in several Hawaiian species, including *D. crucigera*, *D. formella*, *D lasiopoda* and, in a less complex form, in *D. pullipes*.

Most Hawaiian drosophilids are highly secretive and select very private lekking arenas among thick cover. *D. heteroneura* and *D. silvestris* differ by congregating prominently on leaves in the shrub layer (Spieth, 1981). Up to ten males may gather on a favourable tree-fern, each parading up and down in his particular sector. Male wing-waving attracts females by visual rather than olfactory means and, as several males are more conspicuous than one, this presumably explains the reason for the lek. Positions within the lek are maintained with a purely ritualized contest. The contestants perform press-ups with their front legs as they advance, until finally their bodies are pointing upwards at an angle of around 80°. Just before they touch, the middle legs are moved out in front of the forelegs to act as a brace. Each fly then turns his head and raises his forelegs to embrace his adversary. Battle now commences, each combatant using the *side* of his head and the underside of his thorax to push against the corresponding areas of his opponent. The winner is the one who can force his opponent to his 'knees'. The larger males intimidate smaller rivals, who sense their inferiority by not being able to stand up as high during the preliminary face-off. With the unfavourable odds being blatantly obvious, the smaller of the two males backs off and retreats before he gets into a fight which he cannot possibly win.

In *D. heteroneura*, combat is so vital in deciding status that the males have developed 'fighting mouthparts' – greatly enlarged maxillary palps which interlock and clamp the two opponents firmly in position as they engage in a face-to-face trial of strength. When mounting, *D. silvestris* and *D. heteroneura* shove their heads violently up under the trailing edges of the female's wings, their entry eased by species-specific contours of the males' head and antennae, whose plumose aristae exert a cushioning effect. While mounted, the male stimulates the female by stroking her with his hairy feet, a character restricted to the males.

Drosophilid love-songs

The males of six species of Hawaiian drosophilids produce a 'clicking' noise very different from the low-frequency sounds produced by other fruit flies. Even the other Hawaiian species employ a significantly different repertoire, such as 'low-frequency' singing generated by large-amplitude wing oscillations, or the 'purring' sounds produced by abdominal vibrations in certain species (Hoikkala *et al.*, 1989). In 'clickers', such as *D. lineosetae*, the sound is remarkably loud, being audible to the human ear up to 50 cm away. This makes an astounding contrast with other drosophilid 'songs' which require amplification to be audible. Instead of 'singing' by wing-semaphoring, with brief bouts of vibration, as in most drosophilids, the *D. lineosetae* male produces his amazingly loud clicking by vibrating his wings slightly in their 'stowed' position. As he sings, he grips the female's ovipositor with his mouthparts and touches it with the long hairs on his front feet. In *D. glabriapex* and *D. fasciculisetae*, the male touches his front legs to the female's abdomen during clicking sessions, while *D. disjuncta* and *D. affinisdisjuncta* males stroke the female's abdomen with their hairy legs, beating time with their wing oscillations. It is probably no coincidence that all these 'clicking' males actually touch the female during singing. High-frequency sounds are transmitted more efficiently via actual contact and this is probably augmented by the highly developed rows of hairs on the males' front legs.

The importance of sound in fruit-fly nuptials can be judged from the fact that virtually every species investigated in *Drosophila*, *Zaprionus* and *Scaptomyza* produce songs, which form a quick-and-easy 'proclamation of identity' at an early stage in the courtship process. The songs consist of a series of pulses produced by wing vibrations. The nature of the pulses and the intervals between them vary according to species, giving a 'signature' enabling females to identify their own males (Kyriacou & Hall, 1982). In a few *Drosophila* species, the females answer the males' songs in a duet, although this is far more typical of *Zaprionus*. *D. americana* is unusual in that the female is the first to sing after being stroked by the male (Donegan & Ewing, 1980). Her song is of a more regular pattern than the male's, which would make hers the more efficient as a species-identifier. Later on a duet ensues, and the male is also unusual in not trying to mate until the female indicates her readiness by splaying her wings in a raised V-shape. If he fails to take rapid notice, she spurs him on by breaking into song again. He also keeps singing after mounting, a habit shared with *Zaprionus*, in which genus the males employ two different types of song, one for courtship and the other as a post-copulatory exercise. They may also sing momentarily during interactions with other males (Bennet-Clark *et al.*, 1980).

Nuptial gifts

In a number of *Drosophila* species, the male provides his mate with a nuptial gift. In *D. subobscura*, he invites the female to dine by elevating his wings into a V-shape. At this signal, the female approaches and 'kisses' him, quickly draining a droplet of fluid from his tongue (labellum). This droplet is derived directly from the contents of the crop, and females who have been short of food (their normal state in the wild) derive considerable benefit from this nuptial meal in terms of increased and accelerated egg production (Steele, 1986). If the female scorns his offering, the male does not waste it but sucks it back in. If she does take it, he circles around to her rear end and attempts to mount, usually being successful whenever the female has indicated approval by accepting the gift. However, any approval may be purely conditional; females can 'choose' between mates merely by refusing to stay still long enough for a male to copulate if he has dished up an undersized or dilute watery drop of little nutritional value. However, only undernourished females need to discriminate in this way, so, in a situation of abundant natural foods, all males would probably enjoy equal mating success. However, as females in the wild are often short of food, males who serve up large yeasty gifts are in a good position to reap the highest rewards (Steele, 1986a).

Similar nuptial gifts are donated by the males of several other members of the *obscura* species-group. The *Drosophila pseudoobscura* male disgorges a substantial proportion of his crop contents on to the substrate in plain view of the female. If she takes up the offer, he tries to mate: if she walks off he quickly slurps his gift back up, then follows her and resumes his vibratory courtship. Nuptial gifts also figure in other species-groups, but differ in being derived from the anus rather than the mouthparts. *D. nebulosa* curves the tip of his abdomen sideways towards the female's face and exudes a droplet of anal liquid. He then fans one wing behind his offering as if to waft a 'mouth-watering' scent across to the female. Although she often feeds on this droplet, while the male circles around and mates, she also seems frequently to ignore it.

A number of Hawaiian species in the picture-winged *adiastola* subgroup 'kiss', although no transfer of liquid has actually been observed. Other members of this group seem to use air-borne scents as stimulants. The *D. pectinitarsus* male regularly moistens a comb-like appendage on the front tarsus by wiping it between the labellar lobes on the proboscis (Spieth, 1966). *D. tendomentum* drums his similarly moistened front tarsi against the substrate while facing the female. In *Idiomyia clavisetae*, the 'wafting' procedure is quite bizarre. The male proffers his enormously enlarged labellar lobes to the female whilst simultaneously arching his abdomen right up over his head, so that its apex finally hangs just above his eyes. The rectal tip is then everted to constitute a conduit from which volatile scents can be steered towards the female, their progress augmented by his long abdominal hairs.

Liquid gifts in Micropezidae and Sciomyzidae

A nuptial gift in liquid form also figures in the conjugal rites of at least three other families. The male stilt-legged fly, *Cardiacephala myrmex* (Micropezidae), from Central America, climbs on to the female, reaches over to dab a droplet of liquid on to her probosics and almost simultaneously connects his genitalia.

The 'snail flies' (Sciomyzidae) do not generally preface mating with any kind of courtship. However, there are some notable exceptions in the genus *Sepedon*. The male *Sepedon fuscipennis floridensis* from the USA responds to the nearby presence of a female by bending his abdomen on to the substrate and expelling a blob of watery fluid (Berg & Valley, 1985). He then fans with outspread wings, wafting across the tempting odour arising from the fluid. When he detects the female trotting closer, he moves to position his body as a barrier, cutting off her access to the 'gift' and obliging her to stoop low and creep beneath him in order to make contact with it. As she bends down to lap at the droplet, he walks up over her front end, does a quick about-turn and mates. *S. f. fuscipennis* adopts similar techniques, but with the difference that the male allows the female free access to the gift and then mounts her directly from behind. Instead of issuing his own droplet, the male often searches out some nice smelly object, such as a dead snail. He advertises his ownership by bobbing his abdomen, waving his front legs and vibrating his outstretched wings (shades of behaviour in *Panorpa* scorpion flies). The *Sepedon* male safeguards his investment by only allowing females to reach the 'gift' if they signal receptivity by answering with the correct wing movements. The 'owner' raises his forelegs to deter intruding males, or even females if they fail to give the correct signals. If his leg signals fail to put them off, he will rush at them to drive them away. In the Asian *Sepedon aenescens* the source of the droplet is the anus rather than the mouthparts, making a striking parallel with the drosophilids.

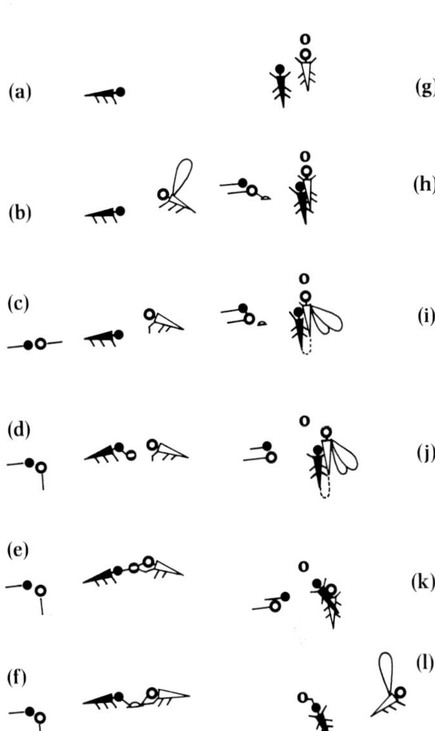

▲ *Asteia elegantula*, a widespread Palaearctic member of the Asteiidae, whose males form leks on leaves, also utilizes a nuptial gift in liquid form (Freidberg, 1984). The courtship display sequence of *Asteia elegantula* (Asteiidae). (a) A male seated on a leaf. (b) A hovering female approaches the male. (c) The male sways. (d) The male regurgitates a droplet of fluid. (e) The female rushes at the droplet. (f) Droplet placed on substrate by male and female. (g) The male detaches from the droplet. (h) The male mounts his mate. (i) The male caresses the female, their genitalia now coupled, and the female ceases feeding. (j) A period of extended copulation follows. (k) The male dismounts from the female. (l) The male feeds on the remains of the droplet while the female flies off. (After Freidberg, 1984).

Post-mounting courtship

Nuptial gifts in the Platystomatidae

In-copula nuptial feeding is found in some members of the Platystomatidae. After first brandishing his black front legs at the female, the male of *Euprosopia subula* from Australia mounts from the rear and embarks upon what seems to be a form of post-mounting courtship (McAlpine, 1972). He lays his front legs out flat across the female's eyes and stimulates her directly by vibrating them. He then leans round to place his head adjacent to her face, sticking his nearest front leg down on to the leaf as a prop. His mate responds by briefly touching his outstretched proboscis with a front foot; then she turns her head so that proboscis meets proboscis. After some spirited mutual tongue-massage a blob of liquid oozes from the male's proboscis and the female takes her fill. Several more bouts of nuptial feeding then follow. Similar regurgitative post-mounting feeding is also seen in *Rivellia boscii* and in the common European *Platystoma seminationis*.

The nuptial behaviour of *Euprosopia anostigma* is a complete enigma, as it is the female who produces a droplet of anal fluid. She does so after the male has provided suitable stimulation, stroking her wings with his front legs and rubbing her rear end with his proboscis. The male feeds on the fluid and attempts to mount, but is not often successful. Once mounted, he grips the female's head in his front legs and rocks it from side to side. He also taps his proboscis energetically on the front of her thorax; simultaneously they both sway to and fro. The function of the female's anal droplet is puzzling, but it may conceivably provide the male with information about her reproductive or nutritional status. He could possibly use this to decide whether or not to press on with the lengthy, and often futile, post-mounting courtship.

In *E. tenuicornis*, the mounted male uses the well-developed claws on his feet to comb the female's downy abdomen; he combines this with foreleg vibrations over the female's eyes.

Post-mounting courtship in stratiomyids
Post-mounting courtship occurs in *Himantigera nigrifemorata* from Costa Rica (Eberhard, 1988), after the male has ambushed the female in a mid-air scramble. Once mounted, he raises each leg one at a time high above his head and slashes it briskly down across the female's head, where her antennae bear the brunt of the blow. During breaks in this 'caning' activity, the male's legs 'twitch' rapidly as they lie, on contact, across the aristae of the female's antennae; he also shuffles his legs and sways back and forth.

Post-mounting courtship in a scathophagid
The European *Hydromyza livens* breeds in the leaves and stems of water lilies. Males spend the day slowly quartering lily pads, searching for females (pers. obs.). They will jump on anything which remotely

▼ Ritualistic 'fighting' in male *Achias australis* flies (Platystomatidae) from Australia. While standing thus, the two opponents can accurately gauge the width of one another's eyestalks. As head width is a function of total body size, and this is related to fighting ability, this could be a form of 'assessment' behaviour designed to avoid possibly injurious fights with overwhelmingly superior opponents (After McAlpine, 1979).

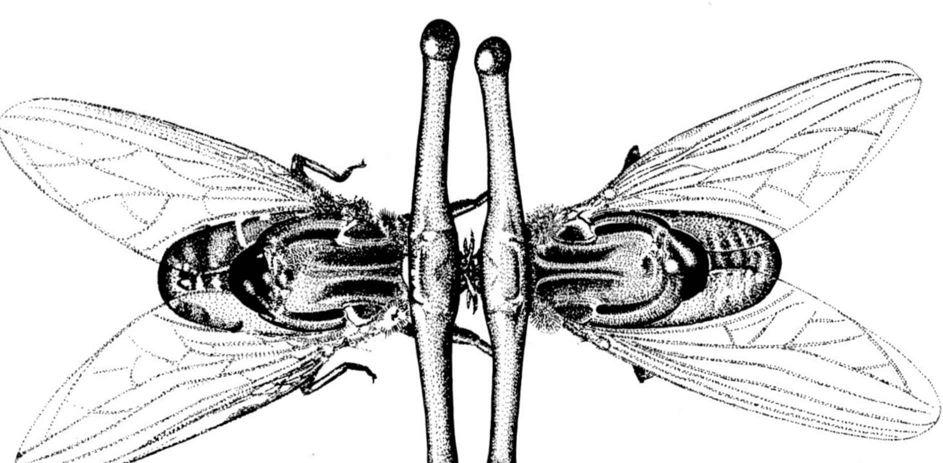

Mate-seeking in hover-flies. Hover-fly males often utilize very flexible dual mate-seeking strategies. Many species divide up the day, spending part of the time guarding resources which will attract egg-laying females, and the rest searching for females in feeding areas such as flowery rides in woodlands. During the morning, males of the American *Mallota posticata* and *Spilomyia decora* patrol near flowers, then spend the afternoon near wet rot-holes in trees, to which ovipositing females are likely to come (Maier & Waldbauer, 1979). The European *Helophilus pendulus* can adopt alternative tactics – either searching at flowers, or resource defence. Searchers demonstrate a limited capacity to recognize their own females, diving on to every insect on a group of flowers and violently striking them with the legs and the tip of the abdomen (pers. obs.). Insects thus 'inspected' include brown *Eristalis* hover flies, *Calliphora* bluebottles, bumble bees and, in particular, hive bees. None of these remotely resembles the black and yellow striped female of *H. pendulus*, although they are all in her approximate size range.

The alternative strategy is to wait near a marshy oviposition site. The territory-owner may spend several days defending his 'patch', flying up to inspect every like-sized insect which comes near. The respective rewards of each strategy are difficult to evaluate, but at least the territory-owner can be reasonably sure of fathering any eggs laid in his patch of mud. If an in-coming female should already have mated with a 'searcher', his sperm will probably be replaced by that of the last male to mate – and that will be the territory-owner. The furry, bee-like bulb fly, *Merodon equestris*, locates females on flowers and employs a helicoptering courtship which includes a series of 'bounces'. During these, the tip of the suitor's abdomen is bent downwards on the downstroke and, simultaneously, the female on the flower bends the tip of her abdomen upwards, probably in a gesture of refusal. This scenario may be broken momentarily as the male makes the odd rapid circling flight before resuming his hovering. The first contact is not made until after 10 minutes or so, when the hoverer drops down on to the female. In a mad scramble, they couple genitalia and retreat into cover below; copulation takes about 15 minutes (Colyer & Hammond, 1968).

► In a number of hover-flies (Syrphidae), the male hovers above the female in courtship and emits a high-pitched whine. This is *Eristalis nemorum* from Europe; females are rarely receptive and mating seldom seen, despite this being a very common species.

resembles a female, including grey blotches on the leaf, water-strider bugs (Gerridae) and dolichopodid flies. When two males meet, they sometimes interact with slow-motion 'fending-off' actions of the front legs, following this up with a strange 'cheek-to-cheek' contact, although no sign of pushing is evident. It is possible that this behaviour is ritualistic and establishes a dominance hierarchy among the males.

'Homosexual' mountings also occur and may last up to 20 minutes, despite the lowermost male's strenuous efforts to unseat his rider. It is possible that this is not a mistaken attempt at copulation, but an expression of dominance by the 'rider', who neutralizes his rival by making it impossible for him to leap on to an arriving female. A receptive female does not struggle much when mounted by a male, but parts her wings to an angle of 45°. This enables the male to insert his long front legs beneath her wing bases and out horizontally beside her face. By keeping his back legs on the leaf, the male can now use his front legs to rock his mate vigorously from side to side, by lifting each wing alternately. Copulation lasts some 10–15 minutes and then the female walks to the edge of the leaf, kicks the male briefly with her back legs, and then simply walks straight off into the water. Her brief kick is the signal for the male to slip neatly off her back, which he usually does obediently and exactly on cue.

If the female quickly re-emerges on to the leaf (10–40 seconds later), the male will immediately jump on board and attempt to re-mate. However, this time she will have none of it and will shake her body wildly while kicking out strongly with her back legs. By keeping her wings firmly shut she also deprives the male of his main foothold, and he is soon thrown off. It therefore seems that post-mounting 'rocking' is only feasible if the female is receptive anyway. If so, then why is it necessary at all? One possibility is that small males make poor 'rockers', and are therefore only allowed short copulations before the female walks off into the water.

Landmark mating systems and the right context

Why do male and female hover flies (and other flies) rubbing shoulders on some favourite flower ignore one another? The reason why requires some explaining. Take the common drone fly *Eristalis tenax*. Males and females often feed side by side on flowers, but never interact at all. This is

▶ **Rubbing shoulders on a small thistle flower, a male and female of the hover-fly *Eristalis tenax* (Syrphidae) completely ignore one another. Why does the male not take advantage of her proximity and try to mate? The answer is: this is not the right context. Females 'visit' males specifically for copulation at mutually recognizable 'landmark' sites, where they can guarantee acquiring a 'fit' mate.**

because they 'know' that the *context* is inappropriate. Mating does not take place on flowers but at a mutually recognized, landmark-based rendezvous point, such as a clump of leaves on an oak tree overhanging a path. Because the 'owner' of such a site can only remain in charge by constantly defending his position against numerous rivals, the female is guaranteed a 'fit' mate when she eventually makes her way there with the deliberate and sole purpose of copulating. The landmark-owner 'knows' why the female has come and mates with her in mid-air. This is why any old male on a flower just simply will not satisfy the female *E. tenax*. He cannot 'prove' his fitness to father her offspring just by eating at the same table, and he 'knows' it. Similar systems are operated by many other flies, including the very similar-looking *Eristalis pertinax*.

In both *E. pertinax* and *E. tenax*, some males spend the day hovering, a very energy-consuming occupation. The hovering station is usually located in a flyway leading to a 'landmark' perch, so it indicates that expensive hovering may be an 'alternative' strategy designed to intercept females on their way to mate with a landmark-based territory-owner. As females are 'expecting' to mate at a landmark, this implies that they may mate less readily with a hoverer (pers. obs.). Hovering is also adopted as a 'fall-back' tactic by *Criorrhina asilica*. The primary strategy is resource defence of a rot-hole in a tree, carried out partly by perching on the tree and partly by hovering near it. Males who cannot appropriate a rot-hole patrol back and forth in sunny glades. However, hovering is used as a *primary* strategy by a number of very common hover flies, such as *Episyrphus balteatus* and *Volucella pellucens*. These hover singly, but the small males of *Melanostoma scalare* form hovering leks about a metre above the ground, usually in a shaft of sunlight in woodland. Each male maintains his own discrete air-space within the lek, only interacting with his neighbours when nearing his invisible boundaries.

This is similar to the much more spectacular aerial lekking in horse flies (Tabanidae), such as the North American *Tabanus bishoppi*. These large flies can form leks of up to 75 males, which produce a loud humming like a swarm of bees audible from 50 m away (Blickle, 1959). The site is highly specific – a forest clearing forming a natural arena surrounded by tall-canopied trees. A female will be instantly pursued by several males until one of their number seizes her and mates. We now see one of the major differences between male behaviour in a lek and in a swarm. In a swarm, the successful male would usually have to fight off a cluster of competing males, all desperately trying to prise him off the female – classic scramble tactics. In a lek, there is more discipline and the female's 'choice' is accepted by the losing males, who duly return to their hovering. Nor do they chase a departing female who has just finished copulating. She probably avails herself of a special 'just-mated' mode of flight which extinguishes any further male interest in her.

'Singing' in a chloropid

Lipara lucens is one of a number of chloropids in which the males and females 'sing' to one another. The acoustic component of this 'song' is of doubtful utility. It seems more likely that the vibrations transmitted through the substrate provide the main communications channel, as in lacewings (*Chrysopa*) and tremulating katydids (Tettigoniidae). *L. lucens* causes galls on common reed, *Phragmites communis* (Mook & Bruggemann, 1968). The male is faced with two problems in locating a virgin female. Firstly, galls in which females develop are very widely scattered. Secondly, newly hatched females do not move far. This forces the males to devote most of their short lives to finding a mate, not an easy proposition for a small fly which spends most of its time walking and flies only sparingly. However, the male's searching time is probably reduced considerably by his ability to announce his presence every time he lands on a reed stem. He does this by vibrating his body for a few seconds; if a female is sitting on the same stem, she replies by vibrating in a female-specific manner. He can thus head more or less straight to her, although he may have to call her up once or twice more in order to run her to earth. Although he still needs to announce himself on hundreds of stems, he is spared the impossible labour of minutely searching up and down every one. 'Singing' may also be useful in keeping co-occurring species separate. In Europe, at least six species of chloropids breed on common reed and, in each of them, the males have their own highly specific song-phrases, although those of the females are all remarkably similar.

Olfactory-cum-visual leks in an otitid

Males of *Physiphora demandata* (Otitidae) form combination olfactory/visual leks on vegetation. They 'call' by emitting a pheromone from a droplet of fluid held on the extruded anal gland. This olfactory element is alternated with periods during which the banded front legs are raised and lowered (Alcock & Pyle, 1979). This virtually cosmopolitan fly breeds in manure and rotting vegetation. The males do not start to establish their lekking sites until around 13.00 hours. When a female arrives and lands near a male, one of two courtship procedures may ensue. In the initial 'shortened' version, the male quickly performs an about-turn, presenting the female with his rear end. If this arouses her interest, she places her outstretched proboscis on the tip of his abdomen and then 'drags' him backwards for several centimetres in a peculiar spiral walk.

She may possibly then mate immediately, but is more likely to refuse to spread her wings and allow the male to mount. If so, he opens up with his full repertoire of movements. This involves various moves which can be executed in several different combinations, so that each display is always followed by a different one, although not necessarily in the same order each time. He may *drum* on the female's head and thorax with his front legs. He may then *shake* his whole body so rapidly that it appears to vibrate. Next might come the *wing-wave*, during which he tilts his rear end upwards and rapidly vibrates the wing nearest the female, while moving it back and forth close to her head. This can be followed by the *wing-flick*, when one or both wings are flicked out rapidly to his sides. The *leg-wave* is a repeat of that already given during lekking, but the *middle-leg salute* is unique to this species, the middle leg farthest from the female being raised in a stiff vertical salute. Finally comes the *shuttle dance* when he backs away quickly and then returns. If she is 'impressed' by all this, the female allows him to mount and copulate. This is of exceptionally long duration – up to 2.5 hours – which probably prevents the female from having time to mate again that day, thus comprising *in-copula* mate-guarding. Numerous courtship episodes are interrupted by intruding males, usually leading to the loss of the female, although the intruder may very occasionally end up by claiming her himself. The females may be demanding such a lengthy procedure in order to select a male who can demonstrate his superiority. This arises through his ability to intimidate neighbours to such an extent that they dare not interrupt his protracted courtship.

Landmarks and leks in the calypterates

Lekking is probably found in many other flies, although it often appears to be landmark-based and males do not actively attract females by 'calling' or signalling. Some form of landmark-based system involving single males, assemblies or leks seems to be almost standard procedure in many calypterate flies, such as anthomyiids, muscids, calliphorids, sarcophagids and tachinids. Males either perch singly or in groups on bare areas such as tree trunks, piles of logs, fences and gates, rocks or leaves. The females are presumably attracted by some distinctive visual quality of the site. Male flesh flies, *Sarcophaga carnaria*, space themselves out across logs and make a swift aerial interception of arriving females. As their aerial assault almost always seems to result in copulation, it seems likely that the females are attracted to the sites for mating rather than for basking (pers. obs.).

In common with many other similar-looking calypterate flies, the males of *Hylemya alcathoe* (Anthomyiidae) congregate in large numbers on leaves in the herb layer of temperate woodlands (Alcock, 1982a). Each male establishes his own territory on a leaf from which he intercepts most passing insects. Some sites are more popular than others, probably because prominent tree trunks and overhanging foliage occur nearby. Males of *Hermetia comstocki* (Stratiomyidae) establish territories on *Agave* plants in Arizona (Alcock, 1990a). Large males tend to hog the best places, chasing off rivals with a typical rapidly ascending flight. Receptive females are attracted to the very plants for which the males have fought the hardest. This indicates that females as well as males are able to assess landmark qualities, an essential attribute in any landmark-based system if the sexes are ever to find one another. Each female demonstrates that her arrival is no accident by soliciting one of the resident males, circling the favoured agave plant and immediately mating with the first male to seize her. This resembles the procedure in *Mydas ventralis* (Mydidae), whose males practise 'hilltopping' in Arizona (Alcock, 1990) (by way of contrast, *Mydas heros* in Brazil practises resource-defence polygyny, defending an *Atta* leaf-cutting ant mound against other males and mating with females who come to lay their eggs in the nest). Several species of bot flies (Cuterebridae) also 'hilltop' (Alcock & Schaefer, 1983). The deserts of California play host to aerial lekking aggregations of *Lordotus pulchrissimus* bee flies (Bombyliidae). The same well-defined areas are used year after year by different generations, a feature also seen in many other insects which use landmark-based mating systems (Toft, C. A., 1989; Toft, S. 1989).

Lekking is just one of a number of different mating strategies found within the Bombyliidae. *Bombylius* males hover singly in open woodland glades, usually not far from flowers which will attract females. Males of different species hover at their own preferred heights above the ground. They are highly territorial, expelling other males which trespass in their air-space and intercepting females in flight. In *Exoprosopa*, the male is often decorated with a metallic 'badge of identity' such as a silver tip to the abdomen, this being absent in the female. He shows this off to best effect during an aerial courtship dance before the female.

Parading and ferrying in a phorid

In the tiny black *Puliciphora borinquenensis* (Phoridae), it is the females who go on parade in order to attract males. This is for a good reason, the females being wingless and requiring assistance in reaching fresh egg-laying sites. The necessary help is provided by the males, who are fully winged and able to transfer their mates to high-quality oviposition sites, such as dead insects. These are located by the males in advance of any ferrying operations (Miller, 1984). However, not all males are so helpful, and even a single male might not invariably prove helpful, as various strategies come into play according to the age of the male and the relevant circumstances. The males can employ up to four routines, any of which may reward them with frequent matings during their brief 3- to 4-day lives. In common with dung flies, this species goes in for explosive breeding on an ephemeral medium – a perfect example of scramble-competition polygyny.

In the late afternoon, the females sally forth from their hiding places on dead insects and 'parade' with abdomens pulsating. This presumably disseminates a pheromone and so comprises a 'calling' routine. A male can respond in one of four ways:

(1) He may jump on to a parading female and mate with her in a 'quickie' routine which secures him a whole string of different mates within a short period: an eminently rewarding strategy, but its drawback lies in the probable need for the females to emigrate to new oviposition sites. Being wingless, they cannot go far by themselves and would be fortunate to find another insect corpse within easy walking distance. So if they cannot lay eggs, there is not much point in mating with them.

(2) He may lurk around near an oviposition site, waiting for the arrival of males who are using strategy 3.

(3) He may deliver females to the resource, giving the lurking 'strategy-2' male a chance to dart in and attempt to mate when he has left. This is not easy, as 'strategy-3' males stay on and stand guard near their ovipositing mates, presumably because of the need to ward off pouncing 'strategy-2' males. The transporter-guarder 'strategy-3' male thus assures both paternity of his eggs and a good start in life for his offspring on a fresh corpse. He may also transport females who have already mated with 'strategy-1' males, a perfectly sensible tactic as he always mates with them himself, gaining paternity through last-male sperm precedence. Males who ferry consume more time and energy. They also take a different approach to copulation, using an 'on–off' routine, alternately mounting and dismounting several times before finally copulating. They then make a running take-off and lumber into the air in a rather laboured flight. Take-offs are sometimes aborted when a heavyweight female proves too much for a lightweight male. Some males may ferry numerous females in rapid succession to the oviposition site, while others may bring just a few before switching to strategy 2.

(4) He may air-lift a female, but then abandon her after failing to reach a suitable resource. Presumably this is a strategy of desperation rather than 'deliberate' policy.

Disporting on flowers

In the wasp-like *Conops quadrifasciata* (Conopidae) from Europe, males locate females on flowers by constant searching. Courtship is absent – the male just dives at a female and pins her down with his legs until he can scramble on top of her thorax. Once he is in place, she will cease struggling and ignore his subsequent presence. During copulation, the male's abdominal claspers reach down and under the tip of the female's abdomen and pull it upwards into position. This first copulation lasts about 14 minutes, but as his mate will continue to feed in full view until dusk – and will remate with any male who can subdue her – the male has no choice but to ride her as a guard for the next few hours. A female seems oblivious to her rider, and allows him to re-copulate every 50 minutes or so throughout the day (pers. obs.) These additional copulations vary in length from 2 to 14 minutes. Longer than that, and the female starts to shake violently in an attempt to throw the male off; he always gets the message and releases his claspers.

In his guarding capacity, the male sets himself up as a sitting target. His numerous rivals spend their day patrolling flowers, pouncing briefly on insects which resemble females, such as conspecific males, males of *Conops flavipes*, *Vespula* wasps, and wasp-striped *Syrphus* and *Helophilus* hover-flies. They also investigate honey bees and large bristly *Tachina fera* flies which are nothing like their own females in shape, size or colour. Take-over attempts are therefore frequent – every 16 minutes on average. However, the intensity of the assault does vary. Small intruders pounce momentarily on the rider's back, sense his superior size and withdraw immediately. When rider and challenger are evenly matched, the latter pounces quickly and positions himself atop the rider's thorax to form a trio. He can then take one of two courses, basing his 'choice' on whether or not the rider is currently engaged in one of his regular 'repeat' matings. If he is not, then the female's genitalia are potentially accessible to the intruder, who tries to grab them with his anal claspers. The rider responds with a 'tail-lift' which keeps the attacker's claspers well away from the female's rear end. After several fruitless attempts to connect, the intruder usually gives up and flies off.

If the rider is mating when the intruder pounces, this seems to be recognized immediately and the attacker modifies his behaviour accordingly. He clings to the

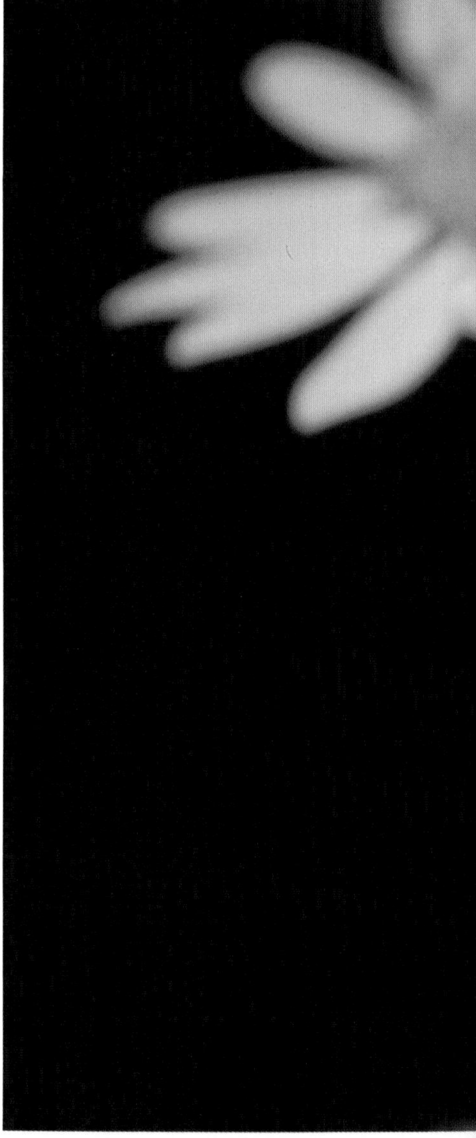

female alongside the existing male, waiting for him to uncouple his genitalia. Intruders may spend many minutes as extra baggage in this way, but rarely seem to persevere, usually flying off suddenly before the female's genitalia fall free. Perhaps the chances of finding an unmated female or non-copulating pair by resuming his patrol make the costs of waiting around simply too high. The reasons for the different position adopted during 'waiting' are also far from clear. Maybe the intruder is playing a long-term game, forcing the guarder to 'outstay his welcome' by staying *in copula* too long. This would eventually elicit efforts by the female to throw him off, possibly giving the intruder a chance to compete on equal terms to re-capture her.

◀ If an intruding male *Conops quadrifasciata* arrives when the consort is uncoupled, he sits on top of him and tries to reach over and join his genitalia with those of the female. The consort prevents this by jerking his abdomen sharply upwards, lifting his competitor bodily off his feet.

▼ If an intruding male *Conops quadrifasciata* makes a take-over attempt of a female while the consort is actually copulating, the assailant sits on one side of the female and waits for them to uncouple.

A small male is almost certainly going to be deposed. His capacity to hang on is severely reduced by a patroller's apparent ability to 'spot a midget' during his final approach. The sight of an 'easy' target seems to encourage the intruder to ram the pair with such violence as to knock them off the flower and on to a leaf. This alters the female's behaviour in a significant way. On a flower, she will continue to move around and feed, largely ignoring events taking place on her back. This makes it difficult for an intruder to force his way in. Once she has been knocked on to a leaf she stops moving, allowing the intruder to pin her down with his legs. This causes her to 'hunch down', and she will not now budge until the 'succession' is over. Now the larger intruder can gradually force his way in beneath the smaller guarder's body, eventually dislodging his grip. As soon as this happens, the guarder flies off and the winner takes his place on the female. Apparently the female recognizes what has happened, as she resumes her normal stance and calmly allows her new consort to lift her abdomen and mate. Females always seem to be immediately receptive to a victorious replacement. They do not show the same generosity towards an existing guarder after an unsuccessful take-over, refusing any attempt at an 'extra' mating within the normal 50-minute interval. It appears that females 'know' when their mates have changed. A patrolling male also remembers the location of a pair which he has already assaulted unsuccessfully, and next time he passes he presses home his attack with less conviction. In the end, if the pair remains on the same plant for an hour or more (as they often do), a passing patroller does little more than briefly touch the male before flying on. It seems that experience teaches a patroller not to bother with certain pairs in which the male has proved to be unbeatable. Attempted take-overs also decrease in severity as the afternoon wears on, probably because there is less to gain by appropriating a female so late in the day. By then, there is insufficient time to squeeze in more than a single 'repeat' mating, which is probably inadequate for effecting full fertilization or, perhaps, for displacing sperm from earlier matings.

By the end of each day, every female will have acquired a large mate, but not necessarily the largest available. The evoluation of this mating system is fascinating. Other conopid females (e.g. *Conops flavipes, Physocephala rufipes*) also feed on flowers, but when *in copula* they fly to nearby vegetation and mate inconspicuously on leaves. Females of *Conops quadrifasciata* presumably pay a penalty in terms of energy requirements for being lumbered with males on their backs all day, although the continuous availability of nectar balances this because they are able to feed without any interruption.

Courtship in some 'primitive' flies

Swarming behaviour has already been described in detail in the Empididae but, as practised in that family, it is so specialized that it has been treated as a separate subject. Rather specialized forms of swarming, which probably constitute lekking, have also been described in the Syrphidae and Tabanidae. The less complex habit of engaging in scramble competition by assembling in simple but highly competitive groups over some swarm-marker is found in many families of flies, but is perhaps most typical of the 'primitive' families, such as the Culicidae, Chaoboridae, Ceratopogonidae, Chironomidae, Mycetophilidae, Bibionidae, Simuliidae and many Tipulidae and Stratiomyidae. As in the Empididae, however, the males in these other swarming families are characterized by their holoptic eyes, which have fewer larger facets in their upper part and more numerous smaller facets below them. In the Culicidae, Chaoboridae and Ceratopogonidae, the males' beautiful plumose antennae with their specialized Johnston's organs are sensitive listening devices which can pick up the flight tones of approaching females. The swarm-marker can be virtually any natural feature, such as a tree, bush, lawn, a gap in a hedge, an opening in the tree canopy, a lake-shore, a road, a grazing mammal, a human being standing still, a chimney-pot or even a single cow pat. How swarm-markers are evaluated and recognized by members of both sexes so that they rendezvous at the correct spot is unknown, but it seems to be an efficient system of bringing males and females together from within widely dispersed populations. This is particularly important in species which do not feed as adults, as these cannot meet up at food sources, and in species whose females oviposit over a wide area in a freely available medium, such as water.

Swarming in bibionids

Swarming is typical behaviour in the Bibionidae. These are mostly black hairy flies in which the males, as in many swarming insects, have large holoptic eyes and can be instantly distinguished from the females. Certain of the larger species are very common and swarm in large numbers in springtime in both Europe and North America. The courtship behaviour of the American 'lovebug' *Plecia nearctica* typifies the intense nature of the 'scramble-competition polygyny' typical of the family. With their hind legs dangling downwards in a characteristic manner, the males hover at low altitude over female-emergence sites, usually in grassy meadows (Thornhill, 1976). The individual members within these swarms are not randomly distributed, but tend to be separated into distinct vertical layers. The largest males fly at the lowermost levels, where they are best placed to be first in line to grab any emerging females. Above them fly the middle-sized males, less well positioned but still in with a chance of mating. The topmost level is occupied by the smallest males, with little chance of intercepting a lone female unless she has somehow been missed by the 'heavy mob' lower down.

Frequently an emerging female gets no further than crawling up a grass stem above her pupal site before she is seized by a male. Any female who is not picked up at this early stage flies up into the male swarm above and is immediately ambushed by one or more males. The first male to secure her in his clutches is immediately under great pressure to obtain a firm grip with his genitalia, firm enough in fact to be able to resist a concerted attempt to prise him free by a scuffling gang of up to eight other males. Take-overs involving pairs who have previously engaged genitalia are only liable to be successful if the pair has not been *in copula* for long. This is because a tubular structure is gradually secreted by the male during copulation, and this eventually extends far enough into the female to resist any external pressure (i.e. from other males) which might force a disconnection. During any fight it pays to be big, as the larger males win most of the females. The actual act of copulation is incredibly prolonged and couples remain attached for more than 2 days. During this time, it is usually the female who gets to feed should the pair visit a flower. Only on certain types of flowers might it be possible for the male to feed as well, if he can bend himself around far enough to contact the pollen or nectar – not an easy task in the awkward end-to-end posture in which he finds himself. The cost to the male of such prolonged copulation may be the failure to secure a second mate; but the benefit is that he is assured paternity of at least one egg batch.

Behaviour of mosquitoes

Swarming

Swarms of mosquitoes (Culicidae) are typical examples of a scramble-competition system in which it is every male for himself as soon as a female enters the swarm. As a consequence, she will probably be deluged in a welter of males. An intense degree of competition is to be expected, as the females generally only accept a single mating. The female betrays her arrival by emitting a species-specific 'whine' with her wings, which the males detect using the Johnston's organs at the bases of their feathery antennae. There is no preliminary courtship and a receptive female couples during flight. An unreceptive female may be captured in a 'basket' of males' legs and she will have to fight hard to keep the tip of her abdomen out of the way. If she lands, she can obtain a breathing-space, during which her suitors continue to hold her caged in their legs, ready for her to become air-borne again and to continue the struggle to win her. *Eretmapodites chrysogaster* exhibits considerable discrimination in not mating with a 'starved' female who has yet to take a blood meal. Instead he follows her to a host and may sit near her as she feeds, copulating with her soon after she has finished.

The predatory females of the ceratopogonids *Bezzia* and *Palpomyia* adapt their normal hunting behaviour to their practical benefit during their nuptials. They are able to recognize the swarm-markers used by other insect species, and enter swarms of small flies and mayflies to pick off the members, usually males. When they enter a swarm of their own males they behave in a similar fashion, except that in this instance the meal makes itself available voluntarily in the guise of an ardent male. As he sinks home his genitalia in her abdomen so the female sinks home her mouthparts in his head. As they mate, she drains him dry, leaving behind a fragile husk which breaks away, leaving the tip of his abdomen still attached. This is one of the very few known examples within insects in which the females cannibalize the males as a normal part of the mating procedure. It may have been derived from the female's acquired ability to capitalize on males in swarms of other species.

Complex courtship

Courtship is generally rare within the Culicidae. In fact it has only definitely been confirmed in but a single species, *Sabethes cyaneus* from Panama, a beautiful iridescent creature shot through with a purple sheen, its middle legs extravagantly ornamented with large feather-like 'paddles', truly a 'bird of paradise' among mosquitoes. Other members of the subgenus *Sabethes* are similarly resplendent. The paddles play a role in the remarkably elaborate courtship, but their function remains strangely enigmatic (Hancock *et al.*, 1990). Males fly along searching for females, concentrating their scrutiny to beneath horizontal sticks from which the females hang by their front and middle legs. When a searcher finds a female, he hovers beside her with his paddled middle legs stretched out. The nearest leg casts around for a grip on one of the female's wings as, once secured, this will serve to swivel him around and bring him face to face with her. This is a tricky manoeuvre requiring considerable expertise, so it is not unusual for the male to fluff several attempts before finally latching on. Once in place he can begin his courtship. He waves his one free middle leg to and fro in front of the female, following up this introductory gesturing with body-bounces, enticingly flashy oscillations of his long, glinting, purple proboscis, and hind-leg swinging. The female reacts by lowering her abdomen, but to satisfy the male's requirements, it needs to be lowered almost to the vertical. If it is not, he launches once more into his courtship. Once he is satisfied with her stance, he tries to link his genitalia with hers, at which point he is more than likely to be 'given the boot' and kicked away by an unreceptive female. If she is receptive, he initially makes just a temporary link with her genitalia, using just his claspers.

If she does not try to kick him free he finally feels 'confident' enough to relinquish his hold on her wing. This gives him free rein to deploy his hind and middle legs synchronously in wide, flowing, 'rowing' movements, showing off his paddled middle legs to best advantage by giving them the 'starring role' of recording a broad and elegant arc; complete engagement of the genitalia then follows. During copulation, the male shakes his body and waves his ornamental legs until suddenly he kicks himself free and flies away. It might be thought from this performance that the male's elegant paddles comprise the most vital ingredient. Yet if these are artifically removed, the suitor is just as successful as if he still had them! Yet if a female's paddles are removed, she is almost completely ignored by passing males, indicating that the female's paddles form an important visual target for the males to home in on. The males possess paddles and what their precise role is therefore remains an enigma.

Pupal-guarding

Deinocerites cancer (Culicidae: Culicini) and *Opifex fuscus* (Aedini) adopt similar and highly unconventional strategies. The latter species inhabits pools on rocky coasts in New Zealand (Slooten & Lambert, 1983). The male spends almost his entire adult life loitering on the surface of the water, hunting for pupae from which adults are about to emerge. When he finds one, he sticks his head under the water and clutches the pupa, grabbing hold with his front legs whose large tarsal claws give him a non-slip grip. Next he arches his abdomen around and plants his genitalia in a special channel in the pupal case. His genitalia fit snugly and securely so he can now redeploy his front legs on the surface to keep himself and his submerged cargo in place. After about 10–20 minutes, the thorax of the newly emerging mosquito begins to appear from the gradually splitting pupal case. This is the cue for the waiting male to thrust his abdomen down inside the case, establishing genital contact shortly before the callow adult has completed its emergence. It is only at this late stage that the male discovers whether or not he has been wasting his time and energy, for only now can he tell whether the pupa is a completely worthless male or a valuable virgin female. In the former instance, the occupant is allowed to walk free as soon as genital contact is made; if the latter, the guarder mates with her shortly before she has finally emerged. His underwater tasks are facilitated by the numerous hydrofuge hairs which coat his front legs and eyes, preventing him from getting wet and possibly fatally waterlogged in the saline water.

Pupal guarding in the Mycetophilidae

In the fungus gnat *Leptomorphus bifasciatus* (Mycetophilidae) from the eastern USA, most females are mated as soon as they emerge from the pupa (Eberhard, 1970). The pupae are suspended on silken threads beneath logs in forest, where they are found by the long-legged males during bobbing search flights underneath the logs. Only female pupae which are about to eclose evince any interest from the male and, if he locates one in this state, he hangs beneath it and awaits developments. There is great competition for pupae and a 'guarder' usually has to stave off regular take-over bids. As the female emerges, her head and thorax are supported on the male's underside. When the end of her body comes free he immediately clasps it with the tip of his abdomen, moves the two of them into a vertical position and then delicately rotates her through 180° to face him.

A female pupa of the New Zealand mycetophilid *Arachnocampa luminosa* responds to a male alighting on her by glowing, so that by the time she is ready to emerge there may be as many as three males hanging on tightly. The first male to win the scramble to copulate when she emerges will therefore probably have to withstand several take-over attempts. If a female pupa is somehow missed by patrolling males, the adult female lures in a mate by switching on her light (Richards, 1960).

TRICHOPTERA Caddisflies

The males generally form swarms either just above the water or up near trees which act as swarm-markers. Most species swarm near dusk or dawn, either mating on the wing or flying to the shore to mate on vegetation. The males generally have much larger eyes than the females (as in swarming flies), emphasising the importance of being quick to spot a female arriving in the swarm. Speed is of the essence, as several males usually scramble for possession of an incoming female, and she will generally accept the first one who grabs her.

LEPIDOPTERA Butterflies, moths

MOTHS

Pheromones are the main channel of communication. The system is basically a simple one of 'female calls – male answers'. This is effective over remarkably long distances – up to several kilometres. Courtship is either a simple 'confirmation of identity' by the male or else non-existent. It is probable that male pheromones almost always come into play during close-up interactions. However, males also attract females using scent, sound or a visual display, or even a combination of these during lekking.

Male moths are remarkably good at locating 'calling' females some of which, such as wingless bagworm females, do not even leave the larval case in which they matured. In the African wattle bagworm, *Kotochalia junodi*, the male flies strongly in daytime, attracted to the pheromones released by the 'calling' female inside her bag. Perching on its exterior, he delves inside with his long telescopic abdomen and copulates with the female as she sits inside (Skaife, 1979). Females of the day-flying

▲ The burnet moths are exclusively day-flying yet still communicate via scent. This pair of *Zygaena lonicerae* in England are mating beside the female's pupa. Sometimes males find the females and mate with them when they have scarcely pulled themselves clear of the cocoon.

burnet moths (Zygaenidae) are also often mated as soon as they emerge from their cocoons, indicating a pheromone-based location system. In some other diurnally active moths, such as the red-belted clear-wing, *Aegeria formicaeformis* (Sesiidae), the female sits on a leaf, everts the tip of her abdomen and 'calls' until the male arrives.

In most nocturnal moths, the females call singly, often near their place of emergence, and their first flight is often postponed until after they have mated. The pheromone is not necessarily released in a constant plume, but may be pulsed by a 'throbbing' action of the abdominal tip, as typified by the American arctiid *Utethesia ornatrix* (Conner *et al.*, 1980). Arctiids are especially interesting as they exhibit some of the most complex nuptial behaviour found in moths and the males of at least three species foregather in leks.

As night falls across the east coast of North America, males of the saltmarsh moth, *Estigmene acrea*, gather in the gloom to form small assemblages (3–22 males). Each male perches at the tip of a plant and inflates a pair of long, curved, air-filled hairy tubes, called coremata, from the tip of his abdomen (Willis & Birch, 1982). These are thought to disseminate male pheromones which attract females to the lek. Coremata are found in many arctiids, although they vary enormously in size, and are akin to the inflatable abdominal 'pouches' of tephritid-fly males. By means of his 'calling', an *E. acrea* male attracts more males, who join in by inflating their own coremata, strengthening the downwind flow of pheromones so that females are more likely to be attracted from the maximum distance. When a female arrives, she flies into the lek and mates immediately with the first displaying male with whom she makes contact. Male display behaviour reaches a climax just after dark and then gradually declines. Then, 3 or 4 hours later, any unmated females also begin 'calling' on their own account in a fall-back strategy which should secure them a mate. There are therefore two different strategies, one male and one female, each adopted at different times on the same night.

Similar behaviour is also found in the Asian arctiids *Creatonotus gangis* and *C. transiens*. However, in these two species the coremata are known for certain to contain the pheromone R(-) hydroxydanaidal, derived from consumption by the larvae of pyrrolizidine alkaloids in the food plant. Both the amount of pheromone and the size of the coremata are influenced by the quantity of alkaloid ingested as larvae. The coremata may sometimes be huge – up to 1.5 times the body length – although at present the advantage of having giant versus small coremata is unknown. *Utethesia ornatrix* males use a pheromone derived from these alkaloids to provoke wing-raising in the female by thrusting the everted coremata towards the head of a 'calling' female. Only by raising her wings does she expose her abdomen and make it possible for him to mate and, unless the male throws the necessary pheromonal 'switch', she stubbornly refuses to co-operate. It has been suggested that the use of pyrrolizidine-based pheromones by the males of certain moths and butterflies enables the females to discriminate between potential mates on the strength of their chemical protection. However, this has yet to be proven. At least two other arctiids, *U. pulchelloides* and *U. lotrix* from Australia, also inflate noticeably large coremata.

Phragmatobia fuliginosa and *Pyrrharctia isabella* from the USA also evert coremata during the courtship flight, but only momentarily as they close in on a 'calling' female. The pheromone released by the coremata of each species differs somewhat in its chemical composition, and their respective females show a species-specific preference in response, being more sensitive to the pheromones of the 'correct' males. Strangely enough, although females of both species signal that they have picked up the pheromone by fluttering their wings and producing clicks of ultrasound, that seems to be as far as it goes. Male courtship success from that moment onwards is not affected by the presence or absence of the pheromone. However, in the dogbane tiger moth, *Cycnia tenera*, males can only enjoy maximum mating success after releasing a pheromone from their coremata or by producing ultrasound from their micro-tymbals.

Courtship in the swift moths (Hepialidae)

The formation of male aggregations is quite common among the swift moths. In this family, there is probably a definite link between differences in male and female coloration, the occurrence of scent-brushes on the legs of the males, and the habit of forming leks. In species which obey the 'conventional' rules, i.e. males fly in search of 'calling' females, the sexes are very similar and scent-brushes are normally absent.

The Eurasian ghost moth, *Hepialus humuli*, is a typical lekking species, the male being white on the upper side and dark on the underside and the female a mottled dark brown. The male's prominent coloration probably plays a central role during formation of the lek and in its subsequent ability to attract females. Around dusk, the males start to hover just above the tops of grasses, forming groups in which the males buffet one another, possibly to set up a dominance hierarchy. This could decide in which order the lek's members would be 'allowed' to mate with incoming females (Turner, 1988). The 'dancing' males face upwind and release a volatile pheromone from the everted scent-brushes on their hind legs. They hold their bodies in an almost vertical stance, exposing only the dark cryptic underside to an upwind observer (such as an enemy), but flaunting the white upper side to any females, who always advance from downwind. The pheromone released by the males' scent-brushes acts as a long-distance lure, and then the bobbing, white, ghost-like forms confirm the lek's exact position as the female draws close (Mallet, 1984).

Females exhibit definite choosiness, being

given full scope to do so by the lek's members, who just go on with their dancing and leave the choice to her. How she 'decides' that one male is superior is uncertain, but maybe something about each performer's demeanour informs her who is the 'hot-shot' male, that won alpha status in the dominance hierarchy. She 'chooses' one of the dancers by leading him towards a nearby grass stem, or she may even bump into him to indicate her choice. He probably perceives her wishes by recognizing a special 'receptive' flight mode which she adopts once she has 'decided'. Once she has made her selection, she sticks to it so insistently that she 'chases up' a male who is slow to follow or who temporarily loses her as she lands among the grass. She goes back to fetch him and then leads him to the spot which she has picked for mating. *Hepialus hecta* males also form leks in which they hover, loop and hover, or merely sit around. Their hind legs are modified into censers whose sole role is to disseminate pheromones (however, a female-attracting role for these pheromones has yet to be positively demonstrated). These male displays attract females, who also aggregate before flying into the lek. In *H. sequoiolus* from the USA, the males perch on leaves and display their hind tibial brushes. This summons other males to form an aggregation, which has the capacity to draw in more females than is possible for just one male operating alone.

Scent in noctuids

In the huge family Noctuidae, the males often possess eversible scent-brushes situated on the legs, thorax or abdomen (usually on the latter). In most cases, these brushes are restricted to showering the female with scent during the final close-up stages of courtship. However, it is possible that the use of hair-pencils in 'calling' behaviour might prove to be quite common. It has already been confirmed that females of *Trichoplusia ni* respond to pheromones released by hair-pencils situated at the tip of the male's abdomen. They follow the odour trail upwind in a zigzagging flight, but do not respond likewise to the products of the abdominal brushes located on the male's third to fifth abdominal segments. As in some arctiids, this species has a dual system, with males homing in on 'calling' females early in the night, while females respond to males throughout the night, but reach a maximum towards daybreak. In many other noctuid males, the hair-pencils are only everted during the contact phase with the female. Their function and effect, however, seem puzzlingly confused, as in some cases their products appear to increase sexual success, while in others their influence seems to be nil.

In the European angle-shades moth, *Phlogophora meticulosa*, the male flies around the female and gives her a quick 1- to 2-second burst from his scent-brushes, dousing her in a strong distinctive odour. This seems to operate partly as an arrestant, preventing her from flying away, and partly as an aphrodisiac, persuading her to accept him as a mate (Birch, 1970). The pheromone may also play a role in sexual isolation. In many moths, the males are unfussy 'flirts' who will happily court females of closely related species. It is left to the females to prevent hybridization by responding only to the pheromone released by males of their own species.

Complex courtship in moths. In some of the smaller moths, there is a genuine courtship combining olfactory, visual and tactile elements. Males of the Indian meal moth, *Plodia interpunctella* (Pyralidae), usually approach a 'calling' female, initially coming in from the rear (Grant & Brady, 1975). They flutter their wings, fanning a blast of pheromone, released from wing glands which are only exposed during courtship, in the female's direction. In her typical 'calling' stance, the female's abdomen is arched upwards, and the male nudges against it with his head and antennae. She responds by turning to face him, while lowering her abdomen to its normal position and withdrawing her protruded ovipositor and scent gland. This is an important step, as she has now withdrawn her olfactory request for males, leaving the field clear for her suitor to push ahead with little risk of interruption. A receptive female stands still and permits the male to shove his head in beneath her 'chin'. As he does this, her antennae flick forwards on to his wing glands, which his otherwise enigmatic 'burrowing under' has now rendered accessible. Receptors on her antennae can now pick up pheromones which increase her receptivity. To signal acceptance, she sticks her abdomen up between slightly parted wings, and this acts as the releaser for the male to initiate copulatory action. He arches his abdomen clear over his head in a 'copulatory strike'. Without the male's wing-gland pheromone, the female will not raise her abdomen to signal acceptance. Similar courtship has been observed in a number of other pyralids, including the almond moth, *Cadra cautella*, and raisin moth, *C. figulilella*.

▼ Four stages in the courtship of the moth *Plodia interpunctella* (Phyctidae). (a) The male (left) stands facing the female, whose abdomen is still concealed beneath her wings. (b) The male pushes his head underneath the female and raises his wings. The female brings her antennae forwards to contact the pheromones on the male's exposed wing glands. (c) Top view of last but the female has now raised her abdomen in the acceptance posture. (d) Male performs 'copulatory strike'. (After Grant & Brady, 1975).

(a)

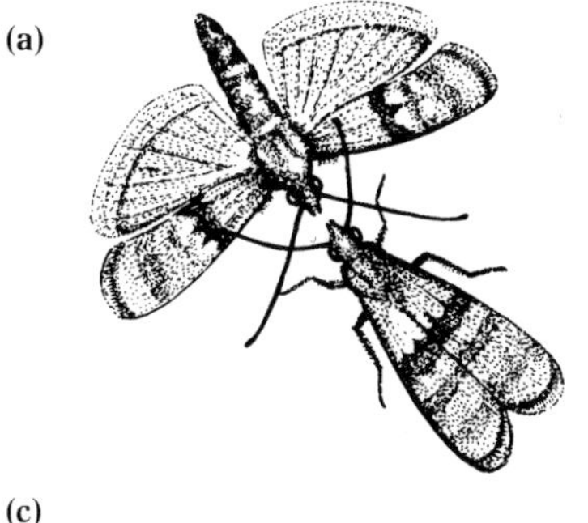

(b)

(c)

(d)

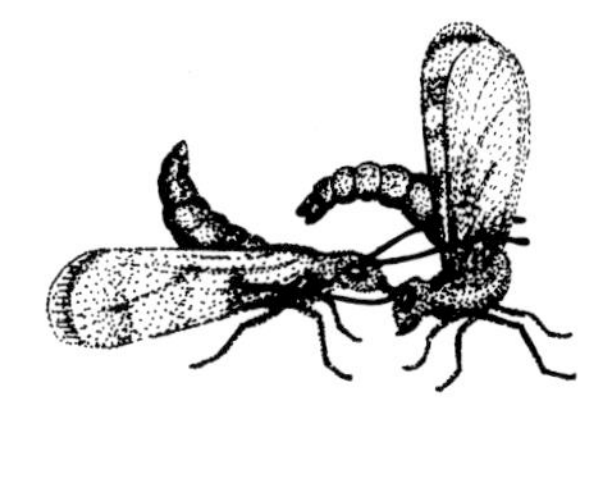

Courtship using sound

In some pyralids, ultrasound is important during courtship. The male of the greater wax moth, *Galleria mellonella*, uses a pair of small tymbals to generate short ultrasonic pulses as he flutters his wings (Spangler, 1985). A virgin female responds by fanning her wings, generating a low-frequency sound. When the male picks this up he releases a flood of pheromone, spreading it around by vigorously fanning his wings. The female can now home in on him, guided not by his initial interrogatory stint of ultrasound emission, but by the pheromone released in response to her positive answer. This constitutes a very neat 'question-and-answer' system of communication, ideally suited to short-range use within the confines of the moth's home within a bee-hive.

A long-range sound-based system seems to be operated by the Australian pyralid *Syntonarchia iriastis*. The males perch in prominent positions at the tips of foliage, spread their wings and emit a stream of ultrasound from a file-and-scraper mechanism situated on their genitalia and abdomen. The sound frequency is low enough to carry some distance, making a female-attracting function seem likely. However, no females have as yet been seen responding.

BUTTERFLIES

Courtship in butterflies

Courtship in the exclusively day-flying butterflies involves an exchange of information of a visual, tactile and chemical nature. The visual component is obviously important, but perhaps not to the extent which was formerly thought. This even applies when it seems 'obvious' that colours must be of supreme importance because they differ markedly between the sexes. Yet colour differences do not necessarily mean very much, as evidenced by the Neotropical nymphalid *Anartia amathea*. In an experiment, the bright red bands on the males' wings were concealed with black paint. The females, who have no red on them at all, found these dullards just as acceptable as

► Male long-horned moths (Adelidae) generally form daylight swarms around a visual marker, such as oak leaves. They only perform their aerial 'dance' while the sun is shining, enhancing the metallic sheen on their wings and their long white-tipped antennae. This is *Adela viridella* from Europe.

normal males (Silberglied, 1984). Set beside this, pheromonal stimuli are proving to be quite important, sometimes to the same extent as in moths. The overwhelming importance of chemical identification and stimulatory clues has already been established in certain species of *Colias* (Pieridae) and this will probably prove true in many more butterflies.

Territorial behaviour

Male mate-locating strategies can basically be divided into two types: territorial or patrolling. A patrolling male has a regular 'beat' up and down an area where females are likely to be resting or feeding. Two very common European pierids, the orange-tip, *Anthocharis cardamines*, and brimstone, *Gonepteryx rhamni*, are typical patrollers, regularly dipping down to investigate pale dead leaves or vaguely female look-alikes. In some species, patrollers may be territorial males who have been temporarily or permanently displaced; they then adopt patrolling as a less satisfactory alternative strategy.

The best-known example of a territorial butterfly is the European speckled wood, *Pararge aegeria* (Nymphalidae: Satyrinae), whose habits have received considerable attention. The male establishes his territory in a sunspot in woodland – just the kind of place to tempt a female. Sunspots are probably attractive because, in temperate woodlands, cool days are frequent, so flight can only be maintained in an area struck directly by the sun. The territory-owner perches on a leaf about half a metre from the ground (Davies, 1978; Wickman & Wiklund, 1983). If a female flies in, he chases her and she usually lands on a leaf. He alights beside her and woos her with a few jerky movements, a very simple style of courtship which is common in butterflies, particularly in the Nymphalidae. A receptive female presents her abdomen tip to the male as he curls his own abdomen around to connect and they mate. Virgin females probably almost always mate with territorial males, and are adept at avoiding physical contact when chased by males who have resorted to patrolling as a second-best tactic. Such peripatetic males are always on the look-out for a chance to take over a vacant territory. When a male trespasses within an occupied territory, the 'owner' evicts him and a spiral 'spinning-wheel' flight, lasting as long as 7 minutes, may take place before the intruder withdraws. The winner in such a contest is always the resident, but when he arrives back at the sunspot, he may find that it has been appropriated during his absence by another male. As the latter found the territory vacant when he took possession, he firmly 'believes' it to be his own. Yet the returning victor 'believes' likewise. The scene is thus set for an exceptionally drawn-out wrangle, involving 'spinning-wheel' flights and chases. These can last as much as ten times as long as 'normal' contests in which one of the males suffers from the handicap of 'knowing' himself to be a trespasser.

Although 'owners' always win disputes with known trespassers, the outcome involving males who have claimed 'vacant possession' is less clear. In one study, it was the original owner who always won; in the other, the newcomer prevailed. In general, the intensity of the disputes diminishes as the season wears on and the forest warms up. The value of a sunspot as a resource thus gradually diminishes from the cool days of early spring to the warm days of early summer, making them less worth fighting for as summer advances. Also, as more and more males emerge, population densities may reach the point where it is not worthwhile, or even practicable, to try and maintain exclusive use of a territory. When this point is reached, several males may co-own a large sunspot.

The outcome of territorial disputes in some tropical butterflies provides some interesting comparisons. Males of *Heliconius sara* and *H. leucadia* mount defensive patrols up and down 10–15 m long territories in sunny forest-rides. They are always successful in evicting intruding males, even when the latter have established themselves during the owner's temporary absence (Benson *et al.*, 1989). Resident *Eueides tales* are 95 per cent successful at executing evictions under similar circumstances. Intruding *H. sara* males often engage in a low-level circling flight which appears to represent a direct 'throwing down of the gauntlet' to the owner, but may switch to a noticeably sluggish flight when they eventually start to retreat in defeat. By substituting a slow non-aggressive 'I'm-going-anyway' flight mode for his previous provocative style, the loser apparently 'placates' the owner, who would otherwise harry the loser from the field of battle by 'snapping at his heels' as he withdraws.

Helter-skelter spiral flights soaring high into the air are common in nymphalids such as the small tortoiseshell, *Nymphalis urticae*, peacock, *N. io*, and comma, *Polygonia c-album*. In the peacock, the male who manages to achieve the topmost position in the spiral for two consecutive bouts wins the territory. The loser retires gracefully and accepts defeat (Baker, 1972). However, such aggravated contests usually break out only when an intruder has settled in a temporarily vacant territory. Under normal circumstances, the mere sight of the 'owner' flying up to meet him is sufficient to persuade an intruder to veer away. The actual territory is only occupied for a single afternoon, and is chosen for its position on a likely female flight line, rather than for any contained resources. When a female dashes through, the 'owner' abandons his temporary abode and rushes after her; he will need to stick close until she goes to roost, when he will get his chance to mate.

In the small tortoiseshell, the male also sets up a temporary territory, but this is resource-based, next to a bed of nettles, the sole food plant. He chases off intruding males from the general area until a female shows up. With such a priceless asset virtually within his grasp he cannot now afford to leave her unless he has to. So he switches to a form of very close non-contact-guarding of the female, perching just behind her. At regular intervals, the male snaps his antennae sharply down and raps the female smartly against her splayed hind wings. The clubbed tips land with sufficient force to be clearly audible. This probably 'reminds' the female that she has been 'claimed', and should not fly off. It would also be interesting to know whether the 'weight of antennal punch' has any influence on whether or not a female stays or flies, as this could indicate the size and weight of the guarder.

If another male rolls up, the guarder's main objective is to lure him away from the female before he has a chance to spot her. He does this by setting off in a rapid flight, which decoys the intruder away from the female for a distance of 100 m or so. He then contrives to lose his pursuer by suddenly turning tail and speeding back directly to the female. As soon as he returns, he strikes her with his antennae, but this time she responds by accompanying him down among the nettles, where the intruder, when he finally finds his way back, will not be able to spot them. It would seem that this 'let's-hide' wing-strike must be different in some way, as it elicits the opposite response from previous strikes.

Towards dusk, the female darts down among the nettles to roost, and her guard needs to be quick-witted not to lose her at this last hurdle. Once she is sitting quietly among the nettle bases, she will at last allow him to copulate. It is tempting to speculate whether the female 'insistence' in refusing to mate until evening may be a 'trial' of male capabilities, designed to secure a mate who can 'stay the course' against all the odds throughout an afternoon. As such, it would have much in common with the 'prolonged courtship in the face of adversity' system employed by the otitid fly *Physiphora demandata* (Alcock & Pyle, 1979) and a number of other insects.

Resource defence of food plants is found in numerous other species and has the additional advantage of being close to sites from which virgin females may be emerging. The desert hackberry butterfly, *Asterocampa leila* (Nymphalidae), aggressively defends the larval food plant, desert hackberry, *Celtis pallida*, against intruding males (Rutowski & Gilchrist, 1988). The 'owner' does not even allow a rival to remain within several metres of the food plant. Some hackberry trees can muster more territorial males than others, and these same males also enjoy the highest rates of mating success; this may reflect the quality of the plant as an oviposition resource. Males of the swallowtail, *Papilio indra*, also defend the larval food plant, but in a fiercely combative way, using their wings as weapons. These 'samurai' butterflies wreak havoc with one another's wings, and even antennae and legs may be casualties of the ferocious battles which frequently rage over territorial ownership.

Hilltopping and lekking

'Hilltopping' males defend an area of high ground devoid of any resources of any practical use to either sex – save as a place to pick up a mate. To the extent that hilltops are recognizably different from surrounding flatter areas, hilltopping comprises a typical landmark-based mating system. It seems to be especially characteristic of desert insects; in the North American deserts 'hilltopping' is found in butterflies such as the great purple hairstreak, *Atlides halesus*, as well as in bees, wasps, beetles and flies. The North American black swallowtail, *Papilio polyxenes*, also practises hilltop territoriality, chasing away any males which stray into a strict territorial zone of about 75 m² which contains neither flowers nor oviposition sites (Lederhouse, 1982). As territories tend to be clumped on some elevated point, such as a ridgetop, it could be that neighbouring males actually form a kind of lek. Holding a territory certainly seems to pay dividends, as nearly 80 per cent of matings take place either within a territory or in an adjacent vacant area. In addition, the best-quality territories, for which the males fought longest and hardest, also receive the largest number of visits from females. This indicates that both females and males are equally capable of evaluating the relative values of a 'landmark' site.

In the chryxus arctic butterfly, *Oeneis chryxus* (Satyrinae), the lek is formed on bare flat patches of sand and gravel among open forest (Knapton, 1985). Males defend their own private space within the lek for up to 11 consecutive days by tackling an intruder with the usual 'spiral flight'. In common with many other landmark-based mating systems, visits by females are puzzlingly rare and, when a virgin female does pay a call, she 'advises' the lek's members of her receptiveness by performing a 'Lolita' flight. If this is lacking, she will be allowed to enter the lek and even sunbathe without being chased – or even apparently noticed – by the males.

'Lolita' flights by other satyrines

The increased mating success derived from being a territory-holder makes this much the preferred strategy in the European small heath butterfly, *Coenonympha pamphilus* (Nymphalidae: Satyrinae). 'Drifters' who cannot establish and maintain a territory but wander in search of females are by comparison left out in the sexual cold, enjoying few copulations (Wickman, 1985). This is because virgin females undoubtedly seek out territories with the 'intention' of mating with the 'owner' (Wickman, 1986), inviting his rapid attention with a solicitation or 'Lolita' flight'. After mating, the female's subsequent behaviour is governed by her monandrous life-style. It could be expected that once a female has lost her virginity, she will give up playing 'Lolita' and even avoid trouble by not overflying territories. Instead, she should switch to a skulking way of life, aimed at ensuring the maximum amount of undisturbed time for her one vital remaining task – laying her eggs in the most suitable places possible. The small heath female does indeed make this sudden shift in attitude, becoming a skulker and avoiding 'covetous' eyes by veering suddenly away from obvious territories.

The female of the European ringlet, *Aphantopus hyperanthus* (Nymphalidae: Satyrinae), behaves likewise. A virgin flies into a grassy area where males may be expected to abound and 'poses' conspicuously with her wings spread (Wiklund, 1982). As soon as a male passes nearby, she makes doubly sure that he does not miss her by flying up and mating with him after a brief chase. Non-virgins exhibit the opposite behaviour, sticking tight when a male flies overhead and keeping their wings firmly shut.

Similar evasive tactics are also employed by some polyandrous species, such as the checkered white, *Pieris protodice*. However, in these multiple-maters the 'keep-low' tactic changes when the skulker's sperm supplies become depleted; then she again changes tack and actively solicits a further mating (Rutowski, 1980).

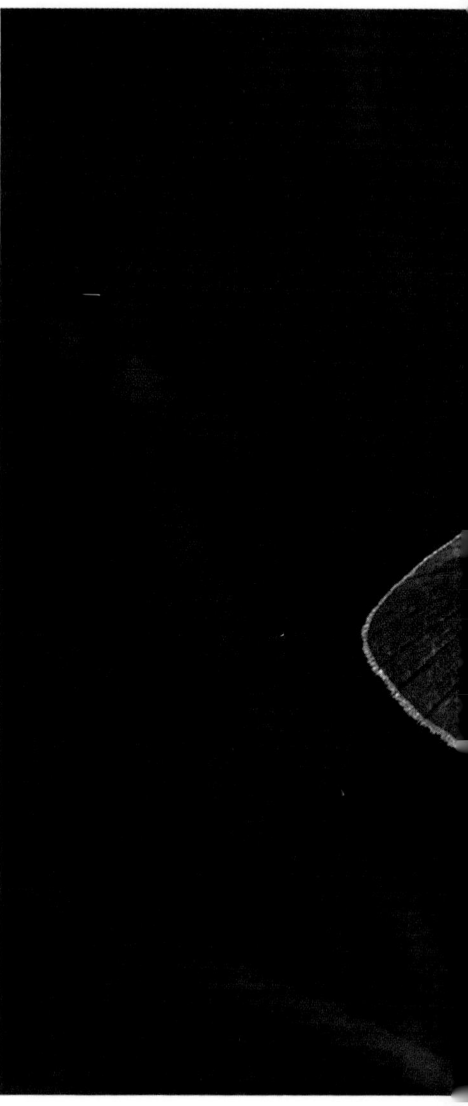

Courtship in some pierids

Courtship techniques can vary considerably within even a single genus, especially when similar-looking species fly together in the same habitat. This happens with the small sulphur, *Eurema lisa*, and barred sulphur, *E. daira*, over considerable areas of their range in Mexico and North America. The nuptials in *E. lisa* are fairly simple: the male simply buffets a perched female with his wings until she extends her abdomen and allows him to mate (Rutowski, 1978). Nothing could be simpler, and the whole 'courtship' lasts a mere 2 seconds, time enough for the brief tactile stimulus, plus pheromones released from structures near the trailing edge of the male's wing, to induce receptivity in the female. An unreceptive female signals refusal right at the start by fluttering her wings until the male leaves. If she is intercepted by a male while on the wing, she will go into an ascending flight and try to lose him. Such ascending flights are a common tactic used by female butterflies to signal rejection, although really persistent males may nonetheless occasionally succeed in mating.

◄ **Females of the European ringlet butterfly, *Aphantopus hyperantus*, gear their behaviour to whether or not they 'need' a male. Virgin females actively solicit males; non-virgins shun them.**

▼ **The 'wing-wave' of two male barred sulphurs, *Eurema daira* (Pieridae); the female continues to probe the flower, oblivious of her suitors.**

In sharp contrast to this rather stark procedure, the sympatric barred sulphur stages one of the most attractive courtship displays of any butterfly and the mechanics of its operation are unique. A male alights beside a female on a flower and suddenly unleashes his forewing and hinges it upwards in a static 'salute' beside the female's head (Rutowski, 1983). Next, he begins to use it as a flag, sweeping it up and down, slowly and gracefully on the upsweep and faster on the return. During its upward course, the trailing edge grazes the female's antenna, probably passing on a contact pheromone present on the wing. (The same end is probably accomplished by the comparatively clumsy wing-buffeting seen in *E. lisa*). A receptive female performs 'wing-flicks' during this performance and then signals her final assent by protruding her abdomen from between her wings in a 'coupling flourish'. Courtships in which 'saluting' is absent never succeed, apparently because the female fails to receive the stimuli which induce the coupling posture. During the display the ultraviolet-reflecting upper portion of the male's front wing is fully exposed before the female's eyes. *Colias eurytheme* females are known to prefer males whose wings reflect strongly in the ultraviolet range, so it seems likely that this reflection, plus pheromones released from the wings, act as releasers for the *E. daira* female's acceptance posture.

This distinctive wing display probably serves to maintain reproductive isolation between *E. lisa* and *E. daira*; and even more so between the latter and the dainty sulphur, *Nathalis iole*. The females are so alike that the males sometimes become confused and court them both. However, hybridization is unlikely to occur because the dainty sulphur male employs a distinctive display, spreading his wings in front of a perched female. This 'full-wing flourish' prominently exposes the exclusive red sex brand on his hind wings (Rutowski, 1981). Males only employ this display in about one-fifth of courtship attempts. They are significantly more likely to do so if the female has already performed a rejection display in response to the 'basic buffeting' courtship. The 'full-wing flourish' would therefore seem to function as an additional stimulus designed to 'change the mind' of an unreceptive female.

Female refusal displays. The 'ascending flight' is a common female refusal display. Most pierid females also employ a characteristic refusal posture, which has been dubbed the 'not-today-dear' gesture. She droops her wings downwards below the horizontal and raises her abdomen, often shivering as she does so. This posture seldom if ever repels a male immediately; if anything, it goads him into increasing his buffeting of the displaying female. On the odd occasion, she might finally acquiesce but more often he either gives up or the female tries to lose him by flying rapidly away. The nymphalid refusal posture is rather similar, save that the wings are held open and slightly upwards, while the abdomen is held almost vertically. This prevents the male from grabbing it with his questing terminal claspers, frustrating any attempt to resort to 'rape' to overcome her objections.

▼ The nymphalid mate-refusal 'not-today-dear' posture: as a male *Heliconius isabella* flutters low over a female, scattering her with androconial particles, she signals her unreceptive 'mood' by raising her abdomen almost vertically.

Chemical communication during courtship

Much male courtship is geared to the need to get essential pheromones to the right place at the right time. In many species, the suitor accomplishes this by flying back and forth just above the female, occasionally 'dive-bombing' her and sprinkling her body with scent-enriched androconial particles. These specialized scales, often called 'love-dust', are derived from highly modified areas on the wings. The scales fracture at pre-determined lines of weakness to form a barrage of minute 'scent-bombs'. In the Danainae, the androconial organs consist of a pair of eversible hair-pencils situated at the tip of the abdomen. During courtship, many male danaines bombard the females with scent-rich cuticular particles from these hair-pencils. They adhere to the females' antennae with a viscous terpenoid 'glue' and transfer their sexual message. Three related compounds – danaidone, danaidal and hydroxydanaidal – are the main chemical constituents of these androconial secretions. As they are found in a number of different *Danaus* species, they are unlikely to function as species-separators, but do seem to play an essential role in courtship. Other so far unclassified components of the hair-pencils may serve a species-isolating function. The danaines do not manufacture the dihydropyrrolizine chemicals themselves but synthesise them from pyrrolizidine alkaloid precursors obtained by feeding on withered plants, especially species of heliotrope (*Heliotropum*).

Courtship in *Danaus*

Methods of communicating chemical messengers to the female vary greatly and form an intrinsic part of the courtship procedure. In the aerial courtship of the North American queen butterfly, *Danaus gilippus berenice*, the male overtakes a female and dusts her head with the products of his expanded hair-pencils (Brower *et al.*, 1965). This aerial hair-pencilling is intended to transmit a 'land-immediately' message via the danaidone component which functions as a flight-arrestant. Once she has landed, the male hovers overhead, drenching her with androconial emissions; this is the ground hair-pencilling stage. This concentrated chemical bombardment induces the female to fold her wings, allowing the male to land and couple his genitalia. The products of the androconial organs are essential to success: no 'love-dust', no mating.

The American monarch, *D. plexippus*, has almost dispensed with chemical methods of persuasion. Instead, the male flies up to a female, who is sailing past with her characteristic languid flight, and nudges her from behind (Pliske, 1975). This makes her put on a spurt and try to outpace him, with the male close on her heels. Eventually he manages to dip down on to her back and capture her with his legs; this is the so-called 'aerial take-down'. By clamping his legs beneath the female's wings, he makes it difficult for her to continue powering away, so they glide gradually to earth, sailplaning comfortably on the male's outspread wings. Once they have landed they couple up, although usually not without something of a struggle. In some courtships, the male flies in front of the female and employs aerial hair-pencilling. A young healthy female can easily outpace a weakly-flying male, thereby ensuring that she weeds out 'inferior' suitors. She ends up with a 'fit' mate who can also manage the tricky manoeuvres involved in the aerial take-down and the subsequent rough and tumble of attempted copulation. Yet while the females undoubtedly act in a 'choosy' way, by playing exceptionally hard to get, males too are not necessarily always easy to please. Sometimes they prove that 'any old female in a hurry' is simply not good enough by prematurely breaking off pursuit. This 'fussy' behaviour is probably based on detecting the degree of vigour in the female's escape dash. If this is found wanting, then her desirability rating as a mate plummets to zero. Of course the alternative could also be true – if she is too fast, then she is probably carrying no eggs, with the same negative result on her desirability. From this, it seems certain that monarchs assess one another's qualities physically, rather than primarily by the chemical means used by the majority of other danaines.

◀ In many butterflies, the male flies back and forth above the female, occasionally brushing her with his wings and sprinkling her with scent-enriched androconial particles. This is *Phyciodes actinote* (Nymphalidae) from Mexico.

▼ Male danaines and some moths acquire their 'aphrodisiac' scents by feeding on withered heliotrope (*Heliotropum*) plants. These are *Danaus chrysippus* males in South Africa.

In the European grayling, *Hipparchia semele* (and probably in many other butterflies), the male pheromones are applied by direct contact between the male's wing-mounted androconial organs and receptors on the female's antennae. This intimate encounter cannot be arranged until after the male has landed in front of the female and flapped his wings in a characteristic way. This probably inhibits her from flying away immediately. He can now hood his folded forewings closely around her antennae, which can now contact his androconial patches (Tinbergen, 1941). An unreceptive female reacts by fluttering her wings, which repulses the male; a receptive female raises her wings, allowing the male to couple.

The male silver-washed fritillary, *Argynnis paphia*, a common European butterfly, also takes the female's antennae between his wings; then he claps them repeatedly, presumably to liberate larger quantities of pheromone and give her antennae a thorough dousing. This 'ground phase' of courtship follows an aerial 'dance' during which

the male repeatedly swoops up and under the flying female (Magnus, 1950). A similar wing-clap display is seen in most courtships by males of the gulf fritillary, *Agraulis vanillae* (Rutowski & Schaefer, 1984). For the male of the New Guinea birdwing, *Troides oblongomaculatus papuensis*, to perform a similar wing-clap he has to execute some slick loops around the fast-flying female in order to come upwards in front of her face – not just at the right place, but with his last wing-beat. This enables her antennae to slot into his exposed wing pouches which he opens at precisely the right moment (Parsons, 1983). An unreceptive female dives for the ground but this may just invite her pursuer to resort to his alternative tactic; he hovers overhead, buffeting her with his hind wings. This sheds copious quantities of hair-like androconial scales on to her antennae, where they adhere tenaciously in feathery tassels. Only unreceptive females seem to merit the full repertoire of male moves; virgins play easy to get, being satisfied with some brief hovering by the male. The proceedings in *Ornithoptera priamus* are similar, although the male pheromone saturates a fringe of long hairs on the hind-wing margins; these are applied directly to the female's antennae with a brushing action.

Chemical cues in ithomiines

Two conflicting roles have been suggested for the hair-pencil products in some members of the Ithomiinae, based on two independent sets of observations. The hair-pencil products normally consist largely of lactone, which is probably derived from pyrrolizidine alkaloids; ithomiines, like the danaines, feed on withered heliotrope. The hair-pencils are really hair-fringes, which can be everted vertically along the leading edges of the hind wings. One researcher (Haber, 1978) concluded that the purpose of this 'calling' behaviour is to attract males of several different lactone-using species to form mating aggregations. These can eventually contain several hundred butterflies belonging to 20–30 species. Such groups may take several days to come together, and may last several months. This might explain how large numbers of ithomiines and other butterflies (often in the same mimicry ring) meet together in dry-season assemblies in damp gullies in the forest. Here mating can take place in relative safety under the 'protection' of the overall mimetic 'badge'. If this were the case, then it would explain why Haber's observations and conclusions are at odds with those of another observer (Pliske, 1975a) who, under stringent experimental conditions, recorded that male ithomiines were *repelled* rather than *attracted* by the hair-pencil emissions of other males. These experiments were not conducted under incipient dry-season conditions, so males could be acting territorially at a different season, having nothing to gain from aggregating.

The hair-pencils are also unfurled during aerial courtship. *Mechanitis* males fan a stream of pheromones towards the females, dancing around and buffeting them as they enter a temporary 'hovering' phase of their chase (Pliske, 1975). As in *Danaus gilippus*, the pheromone's function is apparently to arrest the female's headlong flight and induce her to land. Once this happens, the hovering male can more thoroughly drench her with androconial emissions, while watching closely for her to signal compliance by exposing her abdomen and twisting it to one side so that he can couple. *Methona confusa psamathe*, by contrast, does not employ hair-pencilling, but instead drops on to the back of a flying female in an 'aerial take-down' similar to that seen in *Danaus plexippus*.

Female pheromones

Female butterflies also secrete pheromones which attract males. The number of known cases is currently small but the phenomenon is probably more common. The release of an attractive scent may not even be delayed until after the female has hatched and flown. In certain species, female pupae can be so irresistible that groups of would-be suitors cluster around

◀ This male *Mechanitis lysimnia* (Ithomiinae) in a Brazilian rainforest is 'calling' – he has raised a fringe of hairs on the leading edge of his hind wings. This disseminates a pheromone, whose precise purpose is currently disputed.

them before the females finally emerge. This is found especially in certain species of *Heliconius*, most notably *H. erato* and *H. charitonia*. Males of these two species regularly patrol the area near the rainforest floor where the female pupae are located, checking out their condition to be sure of being on hand during emergence. They will also try to pre-empt the event by trying to 'rape' the pupa itself. In both these pupal-mating species, the female pupa is furnished with a pair of horn-like 'handles' at its free end. These probably serve as foot-holds, enabling the callow female to hang on tightly, despite being loaded down with one or more males vigorously struggling to copulate with her. The males are sometimes attracted to pupae of the wrong species, and even contrive to copulate with them, possibly with fatal results to the immature female. Males are also initially attracted to conspecific male pupae but the allure fades before final eclosion, so that the adult male is able to emerge without harassment (Gilbert, 1976). Female pupae also attract clusters of males in the lycaenid *Jalmenus evegoras*, but here the allure is purely pheromonal, while in *Heliconius* a visual element also plays its part.

Circumstantial evidence also points to the likelihood of 'calling' behaviour by females of the European silver-washed fritillary, *Argynnis paphia* (Treusch, 1967). A virgin female reacts to the aerial approach of a male by 'tracking' him with the tip of her abdomen. She arches this upwards like an anti-aircraft gun and everts two small glandular pouches at its tip. These are said to contain a volatile female pheromone which attracts the opposite sex.

'Chastity belts'

The chemical method

Given the common occurrence of last-male sperm precedence in butterflies, it is in a successful suitor's interests to prevent the female from mating again until she has laid her eggs. The physical presence of the male as a guard is simply not a practical proposition in butterflies – the females roam so widely in search of food plants that they could never be viably guarded. Male butterflies therefore avail themselves of at least two *in absentia* strategies for ensuring their mate's subsequent chastity. *Heliconius erato* 'blights' the female by depositing a volatile 'anti-aphrodisiac' upon her during mating. This acts as a powerful 'turn-off' to other males and seems to be remarkably long-lasting, perhaps enduring for several months (Gilbert, 1976). This off-putting odour is always present in mated females, but not in virgins, and is dispersed from two organs known as 'stink clubs' on the female's abdomen. These 'stink clubs' fit snugly inside a gland-lined pouch in the male's clasping organ, and this is probably how the substance is transferred from a reservoir inside the male's body. This whole scenario does, however, raise an intriguing question. Why should any would-be Lothario take any notice of a rival's smelly 'keep-off' label? After all, if pressed hard enough, the female might well accept a second mating. The key possibly lies in this species' pupal-mating system, and the method by which male pupae – at a certain point – suddenly lose their allure for other males. This 'switching off' of the pupal attraction is probably brought about by the very same anti-aphrodisiac pheromone. Now, in the case of a male pupa, it is definitely in the interests of both parties not to waste time in a prolonged male–male mating attempt, so the evolution of an effective anti-aphrodisiac label which prevents this happening is perhaps inevitable. Its subsequent 'cheat' application to females as a form of durable chemical mate-guarding was perhaps only to be expected.

The physical method

A substantial physical block to further mating attempts, called the sphragis, is placed inside the female by the first male to mate in a number of butterflies. These are mostly in the Nymphalidae (Acraeinae and Danainae) and Papilionidae (Parnassiinae). The sphragis initially consists of a viscous liquid produced by the male's accessory glands

▼ In acraeines, the male leaves behind in the female a mating plug known as a sphragis. This prevents further matings for some time, although it may eventually fall off, when the female could mate again if she so wished. However, by this time she should have laid a batch of eggs fertilized by her first mate. This is *Acraea oncaea* from South Africa.

during copulation. This almost fills up the female genital cavity, where it soon dries to form a hard brown plug. In some species, such as *Cressida* and *Parnassius* (Papilionidae) and *Miyana* (Nymphalidae), the sphragis remains firmly fixed in position and is so big that it protrudes externally. It therefore provides an effective physical barrier to further matings. However, in the Australian papilionid *Cressida cressida*, the sphragis can be so big (half as long as the female's abdomen) that the mere sight of it is sufficient to deter further investigations by a prospective suitor (Orr & Rutowski, 1991). The female seems to connive in maximizing the effectiveness of the massive sphragis as a visual 'too-late' signal by taking the appropriate course of action when approached in flight by a male. As she sees him coming, the female changes to a flight mode in which the forewings do all the work, enabling the hind wings to be folded up over the back like a tent. By curving her abdomen strongly upwards, the female can now brandish her sphragis at the advancing male. In most cases, this causes him to veer away, although certain exceptionally persistent males still insist on grabbing her and trying to mate. In general though, the sphragis confers upon its carrier considerable benefits similar to those afforded to females by the presence of a guarding male – both methods reduce the cost of resisting unwelcome mating attempts.

In genera such as *Acraea*, in which the sphragis is universal, it may be less durable, eventually falling off and permitting multiple matings if the female is willing. A large sphragis, as donated to the female *Cressida cressida*, represents quite a substantial physical input by the male. The extent of his pay-off is uncertain; he certainly ensures paternity of the egg batch. However, if females are able to feed without interruption for longer periods, because they do not have to undergo lengthy enforced matings, he may possibly earn additional benefits in terms of increased egg number and possibly egg size.

Some male butterflies make a substantial direct investment in the form of a large spermatophore. In the chalcedon checkerspot, *Euphydryas chalcedona*, from North America, this may represent as much as 12 per cent of the male's body weight (Rutowski & Gilchrist, 1987). This may not be much compared with some katydids but it is still a significant investment which probably has a direct pay-off by increasing female fecundity.

HYMENOPTERA Sawflies, wasps, ants, bees

Courtship techniques in the Parasitica

Wing-fluttering beside the female is the main component in the courtship of the male braconid *Apanteles melanoscelus* (Weseloh, 1977), and this simple ingredient is also important in most members of the Pteromalidae. In the majority of these tiny wasps, a pheromone released from the female's abdomen is responsible for eliciting male courtship. Mistaken courtships between males are consequently very rare or absent. Male swarms have been recorded in several encyrtids and pteromalids. *Bothriothorax nigripes* (Encyrtidae) form swarms on and around ridgetop boulders in what appears to be an example of hilltopping behaviour. Females landing among the throng of males soon inspire courtship once a male is within about 1 cm. Courtship and copulation times are typically brief in all these small wasps – about 40 seconds for courtship in *B. nigripes*, during which time the male waggles his body some eight to ten times while standing in front of the female. Then he mounts from the rear, rocks from side to side and couples his genitalia as the female responds by opening her genital pore. Virgin females are apparently attracted to the swarm specifically to find a mate, as they always respond positively to male courtship. Once mated, the female's attitude changes and she will repel any further advances as she escapes from the swarm.

Swarms can be remarkably long lasting; one, composed of males of a *Trichomalus* species pteromalid, was present in an English forest for over 2 months (pers. obs.). This swarm was linear in nature – clouds of males were flying or perching on the foliage of trees for hundreds of metres along several rides in the wood. The swarming height was very restricted, at between 2 and 3 m. This could have been a pre-hibernation mating swarm of a type which has been frequently recorded in small parasitic wasps.

The North American *Pteromalus coloradensis* mates on thistle heads which are infested by its tephritid-fly host, *Paracanthera gentilis* (Headrick & Goeden, 1990). The male rocks from side to side and fans his wings, producing vibrations of a species-specific pulse rate and frequency which induce the female to permit mounting by the male. Such acoustic courtship has also been recorded for other pteromalids and may be widespread in the group. The mounted male stands with his front legs on the female's head and leans slightly forward, allowing her to beat the upper surface of her antennae against his closed mandibles; simultaneously he strokes the opposite surfaces of her antennae with the tips of his own. 'Singing' is also found in the Chalcididae, as for example in *Brachymeria intermedia* from the USA, which employs a tripartite 'song' consisting of three different auditory displays in a fixed series: notably rock, wing-quiver and buzz (Leonard & Ringo, 1978).

In some other pteromalids, such as *Nasonia* and *Eupteromalus*, it is the male who takes the sole active role in mouthing the upper surface of the female's antennae (see below). Male antennation of the female's antennae is, however, common within the Pteromalinae. The signal used by the female to indicate acquiescence is also widespread: she droops her antennal tips (flagellae) downwards so that they almost touch her face, while at the same time raising the last four tapering segments of her abdomen to expose her genital aperture. Courtship is very brief at some 5–6 seconds, but mating lasts around 30 seconds. In *Nasonia vitripennis*, the male evokes the female's receptive signal by nodding his head up and down while mounted (Van den Assem *et al.*, 1980). With each first nod of a series, he uses his mouthparts to dab the top of the female's antennae with pheromones. These are presumably responsible for actually inducing receptivity, as she always signals acquiescence at the beginning of a series when the pheromone has just been freshly applied.

Fighting over females

Nasonia and *Melittobia* males fight savagely over ownership of sites, such as exit holes in a fly puparium or a batch of host eggs from which virgin females will emerge. *Melittobia* males regularly fight to the death to become master of a harem of females; they have little alternative. The probability of locating an unguarded mass of the widely-dispersed eggs of the host is more or less nil. If a male's sole path to reproductive success dictates that he do battle with a series of rivals, then he loses nothing by following the principle of 'victory or death'. However, males of the scelionids *Trissolcus bodkini* and *Phanuropsis semiflaviventris* do not seem to adopt this principle when seeking to monopolize the parasitized egg batches of their stink-bug host in Colombia (Eberhard,

1975). A dominant male will be in a position to mate with all the females as they emerge from the host's eggs. During fights, the males use their mandibles as weapons, locking them together until one of them is forced backwards off the egg mass. Usually it is only in evenly matched pairs that battle actually commences – small opponents wisely back off after watching an intimidating demonstration of jaw-gaping by the larger male. The ability to monopolize an egg batch is vital – the females only mate once, usually just after emerging. However, the host is often abundant, which may make fighting to the death unnecessary.

Courtship in blind male eulophids

In the eulophid *Melittobia chalybii*, the normal practice of males homing in on female pheromones is reversed (Evans & Matthews, 1976). Instead the male, who is eyeless, goes on parade in a 'calling' stance with his antennae spread wide, his wings elevated and his abdomen directed upwards a trifle. He probably releases a pheromone as he displays, and a female announces her arrival by tapping him with her antennae. Using his own antennae as feelers to find his way up, the male mounts the female. Leaning well forward on her head, he now 'plugs' her antennae into special sockets in his own and begins to sway them from side to side. Now he also strikes up a steady 'treading' with his middle legs, each up-step concluding with a mighty abdominal spasm. Leg-lifts cease as antennation commences and he finally finishes off with front-leg-stamping before he at last tries to couple his genitalia. This routine differs in a number of important details from that of a 'sibling' species. With their blind males, differences in the minutiae of courtship are probably of special importance as species isolating mechanisms.

Some elements of courtship in the Torymidae are similar. Antennal rubbing, bobbing movements and wing-flicking are typical ingredients in *Monodontomerus montivagus* and *M. clementi* (Goodpasture, 1975). The procedure in *M. saltuosus* is particularly interesting in containing a 'head-swivelling' session. The male, mounted atop the female, rotates his head rapidly from side to side, lashing his antennae vigorously against those of the female.

Fighting in fig wasps

Wingless male torymid fig wasps are notorious for the savageness of their fighting. In the New World genus *Idarnes*, up to 50 per cent of combatants may be fatally wounded (Hamilton, 1979). The males are often amazingly dimorphic. At the one extreme are relative giants with big heads, amply furnished with huge sickle-like jaws capable of rending an opponent in two with one bite. At the other extreme are tiny runts which look superficially like a completely different species. In between these two extremes are middle-sized males with 'normal' jaws. The larger males are born warriors, instantly taking on any rival and fighting to the death. Only under certain conditions, such as an unusual abundance of virgin females, do these committed pugilists ignore one another. If there are plenty of females to go around, then there is no point in risking injury or death fighting over them. The smallest males always try to avoid fights and resort instead to 'sneak' tactics. Their small stature then becomes an advantage, enabling them to insinuate their way between the fig flowers in order to reach females ensconced in the galls within. These would be inaccessible to the large heavyweight gladiators. Levels of mortality up to and beyond 50 per cent are rare in any animal, usually because there is some ritualistic method of settling who wins before matters turn really nasty. The most surprising aspect of lethal fig-wasp battles is that, even when most of the wasps are closely related – and that usually means brothers – there is no indication of any tendency to fight less often or less brutally (Murray, 1989). This is awkwardly contrary to evolutionary theory, which predicts that close relatives should be 'nicer' to one another and not 'go for the neck' as a first resort. However, fig-wasp males cannot be constrained by such niceties, as they have little or no chance of ever mating outside the microcosm of their own fig. If winning the reproductive race means killing the competition – and the competition consists exclusively of brothers – then there is not much point in reining in the savagery .

Some Old World fig wasps, such as *Philotrypesis*, are just as violent, yet share their fig receptacles with several other wasps in which animosity between males is notably lacking (Murray, 1987). In Malaysia, *P. pilosa* males are typical fighting machines, sturdily built with heavy-duty mandibles on armoured heads and tiny wings. They are adept at identifying females, which are still enclosed in their galls. With their can-opener jaws, these formidable males do not wait around for the occupant to emerge but set about chewing their way down to reach her. Once an entry passage has been breached, the male sticks his rear end down inside and copulates with the helpless resident. Once he has mated, the male begins an intermittent stint of non-contact-guarding of the female. He wanders off to reconnoitre for further galls which might conceal a female but regularly returns to his original mate to repulse any intruding males and mate with her again. He often manages to tot up as many as seven copulations, spread over a period of about 20 minutes. After this, he returns no more and concentrates instead on new explorations. Fights between males can be quite vicious, particularly if there is a nearby female at stake. Heads are sometimes guillotined in champing jaws but usually it is just a leg or two which is left to litter the battleground.

Competition in ichneumonids

The large American ichneumonids *Megarhyssa atrata, M. macrurus* and *M. greenei* all parasitize larvae of the horntail *Tremex columba* (Siricidae) within tree trunks. The males hatch first and gather in mixed-species aggregations (Crankshaw & Matthews, 1981). Sexual behaviour is dominated by the inescapable fact that the females only mate once. This results in a single over-riding imperative, which motivates the daily activities of every single male – to be first in the queue for the newly emerging females.

Scramble competition rather than female defence is the only viable option for rather a peculiar reason – the males are unable to determine the identity of a wasp which is just about to emerge from beneath the bark. It could be a virgin female of their own species but it could just as easily turn out to be a male, or even a male or female of one of the other two species. Given this 'blind spot', it is not worth fighting to monopolize an emerging wasp when it could prove to be not only the wrong sex but the wrong species as well.

Males locate emerging adults by stroking the wood with their antennae, listening for the faintest sounds of chewing. The finder settles down and waits but quickly becomes a focus of attention for his rivals, so that soon quite a party builds up, often composed of males belonging to all three species. Preparing for imminent action, many of the onlookers 'thread' their abdomens forwards through their legs; the genitalia now rest near the edge of the prospective exit hole. As the intensifying sound of chewing announces an incipient break-out, the waiting males begin to 'elbow' one another in an effort to be first to make contact. When the moment of breakthrough arrives, it probably releases a waft of species-specific pheromone. Males who receive a 'wrong-species' message now begin to melt quickly away. However, at this stage it is still too early for details of the emerger's sex to be definite, which means that an emerging male still retains a considerable retinue. It takes some 5–20 minutes more for the wasp to struggle free of the hole. During this time, several bystanders will try to slide their abdomens down inside, insinuating themselves close beside the occupant in an attempt to steal a march on their rivals by copulating before the final emergence.

Up to eight males – sometimes belonging to all three species – may simultaneously achieve partial abdominal insertion. Eventually, one male may manage to insert himself completely up to his thorax and mate with the occupant (if it is a female). As he withdraws, he will be replaced by a succession of others. The first male often prevents this by remaining fully inserted for a very long period, monopolizing the female for as long as possible. Following the first successful prolonged insertion, the queue begins to decline, probably because the tail-enders realize when the game is lost; it then pays them to cut their losses and resume searching as soon as possible. When she finally walks free, a female will still be subjected to numerous mating attempts. She usually deals with these by kicking unruly males off her back, but now and again a retreating female allows a male to copulate. This probably indicates that the 'inserters' do not always carry out successful insemination within the confines of the hole. Such 'mistakes' by apparent winners therefore make it worthwhile for stragglers to hang around until the last moment.

Stridulation in mutillids

In the velvet-ants (Mutillidae), there are substantial differences between the sexes. The males are usually fully winged and usually much larger than the tiny wingless females, who are swept off their feet by the males during mating. Mutillids use stridulation during a pre-mounting courtship, operated by a file-and-scraper mechanism present on the edges of two opposing abdominal tergites; when these are pumped in and out, sound is produced. The precise role of stridulation has been much disputed, but in-depth studies of the American *Dasymutilla foxi* (Spangler & Manley, 1978) have helped enormously in elucidating its function. Both sexes can stridulate. The female produces a 'chirp' of varying quality, while the male, in addition to 'chirping' can also give vent to a 'honk' by vibrating the thorax and 'parked' wings.

The male drops on to a female, usually announcing his arrival with a series of 'honks'. She reacts with a burst of stridulation as they struggle together in the dust.

▲ As in several American *Megarhyssa* sp. ichneumonids, these male *Rhysella approximator* from Europe can detect the imminent emergence of 'something' from the host tree. However, as in the American species, they can determine neither the sex nor the species of the emerger. Although both males here are preparing to mate with the owner of the head sticking out of the wood, this does in fact belong to their host – the alder wood wasp, *Xiphydria camelus*. This indicates an inability to determine even family until the insect has fully appeared.

► Mutillid females are wingless and much smaller than the males, who literally sweep them off their feet, carrying them away in a nuptial flight. This is a *Trispilotilla* species from Kenya.

As soon as he succeeds in coupling his genitalia, he stops 'honking' and starts to stridulate, striking up a duet with the female for the duration of coupling. Unreceptive females change the tone of their stridulation, probably as a signal of their unwillingness to mate. Male 'honking' seems to be essential for success, as non-honkers seldom manage to couple their genitalia.

Nuptial feeding in thynnine wasps

The males are much the larger sex in the thynnine wasps (Tiphiidae), many of which exhibit conjugal habits which are unique among the aculeate Hymenoptera – the males make an investment by feeding the female during copulation. The wingless females spend most of their lives tunnelling beneath the ground in search of scarabaeid beetle larvae, their usual hosts. The winged males divide their time between feeding on flowers and flying around looking for females. They also visit aggregations of homopteran bugs and lap up the exuding honeydew; this, along with any nectar, is stored in the crop.

The Australian *Hemithynnus hyalinatus* gathers nectar at 'provisioning' trees and transports it up to 800 m to areas where females are likely to be present (Ridsdill Smith, 1970). The female broadcasts a pheromonal 'request' for a mate by 'calling' from just beneath the surface. So definite is her need that, if no males show up within a day or two, she increases her 'calling' range by climbing up a plant stem, allowing the pheromone to disperse more widely. She senses a male's imminent arrival and hurries out to meet him,; he mounts her and she looks upwards so that he can nibble gently at her mouthparts. The male is usually in a hurry to carry her away before some of his rivals also answer her call. The female now swings beneath the male and adopts a tightly curled posture beneath the rear of his abdomen. Later she stretches forward again, so that their two mouths almost touch, and starts to caress her mate's abdomen with her rear legs. This induces him to regurgitate a quantity of food from his crop. A copious droplet oozes on to his mouthparts; the female buries her mandibles in it and drains it dry. Only a single such meal is on offer and the female usually takes 2–3 minutes to finish it. Each female mates several times during her lifetime, calling up males as and when needed between strenuous bouts of tunnelling. It therefore seems that one mating may be needed per egg laid, which would explain the female's ability to attract a male 'on cue'. It also seems that the male's nuptial gifts constitute a vital fuel for the earth-bound female's labours. If a female is separated from her mate after copulating – but before feeding on the gift – she quickly resumes her 'calling' to summon another gift-bearing male (Alcock, 1981).

▼ Various feeding attitudes of female thynnine wasps during copulation (the female is the smaller of the two). (a) The female feeds herself on honeydew or flowers; the male facilitates this by positioning himself so as to give her maximum access to the food source. (b) The male regurgitates a droplet of food and the female feeds on it directly while held to one side. (c) As in the last drawing, but the female is held directly beneath the male. (d) The female curls up and feeds directly on the male's exuded anal droplet. (e) The male regurgitates a droplet on to a leaf and the female then feeds on it. (After Given, 1953)

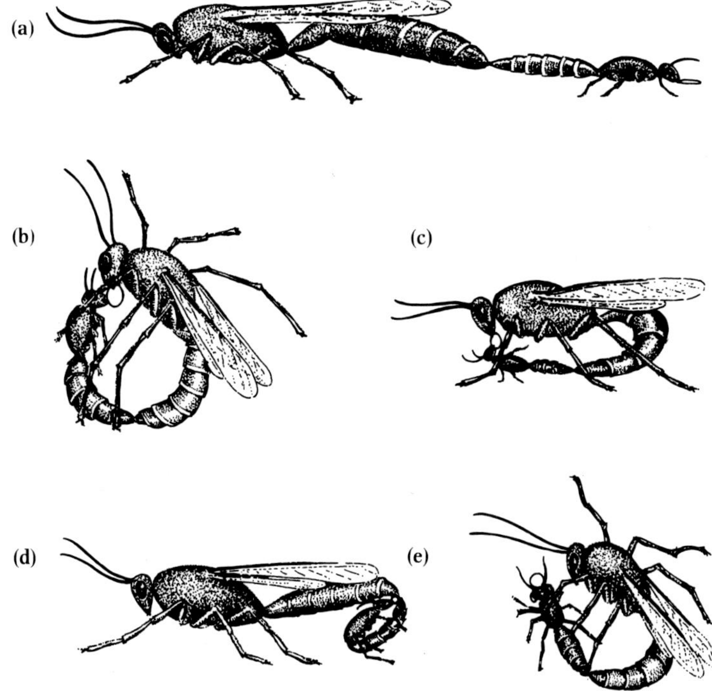

H. hyalinus is one of a minority of Australian species (also including some *Tachynomyia* spp.) in which the male brings the food to the female at the breeding site, rather than plucking her off the ground and ferrying her to a feeding area in so-called 'phoretic copulation'. *Tachynomyia* males carry the liquid food (honeydew or nectar) for distances of up to 1600 m from provisioning trees to the site of mating. The remarkably large droplet, which may be a great deal bigger than the head itself, is carried in a cavity on its underside, from which it bulges down noticeably; it is kept from falling by a fringe of long, recurved hairs (Given, 1953). In some species, the females disconnect their genitalia while feeding, and may even just cling on with their mandibles while in flight. In most of the South American species, the male ferries the female to suitable flowers and she feeds herself while *in copula*. Self-service feeding is the norm in some Australian thynnines, including certain *Rhagigaster* (Alcock, 1981) which transport the females to eucalyptus flowers. In *Thynnoides*, the coiled female sups directly on a nuptial meal consisting of a droplet of syrupy liquid which oozes from the male's rear end. In a number of other species, the male regurgitates a gob of fluid from his crop on to a leaf. He then moves so that the female can uncoil herself, reach forward and feast on the offering.

The males devote an inordinate proportion of their lives to feeding on flowers, building up the reserves which they will later donate during copulation. As they invest substantial resources in their mates, it would be surprising if they were not 'choosy'. In return, the females should be less fussy than normal about accepting a male. In practice, this does indeed seem to be the case, with males rejecting 'calling' females more frequently than the reverse (Alcock, 1981). As small females will give a less satisfactory return on investment than large females, males may exercise size-related discrimination. This would not be difficult to operate, given that the male has to lift the female, which provides him with a quick and easy method for accurately estimating her size and weight. If she is too small and light, he dumps her.

Elements of courtship in solitary and social wasps

Post-mounting courtship (where it exists) is usually simple and stereotyped. The male plays his antennae across those of the female, perhaps lassooing them and pulling them upwards; often he rubs his abdomen down the sides of the female's and may also 'tread' with his feet.

In *Parancistrocerus pensylvanicus* (Eumenidae), there is an interesting post-copulatory ritual in which the male 'bounces' vigorously up and down on the female for 12–13 minutes before disengaging and dismounting (Cowan, 1986). Similar behaviour is seen in the Australian eumenid *Abispa ephippium* and certain parasitic wasps, as well as the bee *Centris pallida*. This 'bouncing' is thought to induce a state of unreceptiveness in the female; in many hymenopterans this is brought about by the physical reception of sperm from the first male to mate with a virgin. However, in the highly competitive mating systems of these 'bouncers', it is always on the cards that a male will be forcibly displaced by a superior rival, even after some sperm has been transferred. It may therefore be good practice for a female to remain receptive until after her first mate has 'proved' his fitness. He must do this by staying in place long enough to perform his full 'bouncing' display, which lasts considerably longer than the brief (less than a minute in *P. pensylvanicus*) copulation.

Mating strategies in wasps and bees

Male wasps, and bees, can adopt one or more of the following strategies; these are largely dictated by two salient facts. Firstly, it is the males who usually emerge first and await the emergence of the females; secondly, in many species, the females only mate once, putting the males under great pressure to stake sole claim to available virgins.

(1) The male can wait by the natal nest and mate with his sisters as they emerge (harem-defence polygyny). He can then take up patrolling on flowers in the hope of coming across females from other nests who have been missed by their own brothers (scramble competition). As only one male can dominate the area near the nest, his own brothers will have to take up patrolling right from the start.

(2) He can hang around near a large nesting area and compete with other males to mate with any females who emerge (scramble competition). In large nesting aggregations, these would not normally be sibling matings. There is intense competition, which often gives large males a considerable advantage. In highly social species, such as the honey bee, mating takes place in swarms.

▲ Typical post-mounting courtship in a wasp; this male *Ropalidia* from Madagascar is lassooing the female's antennae with his own while wiping his abdomen back and forth across hers. Despite more than an hour of this she still refused to mate, probably because females in this species are monandrous and she had already mated.

(3) He can patrol on and near flowers or other resources, such as water or mud, and mate with females as they arrive (scramble competition). Depending on the species, this can be a primary or alternative strategy. If the latter, it suffers from the disadvantage that females may already have mated with a male by making use of the other strategies.

(4) He can establish a territory and mate with females who pass by or who are specifically attracted to features of the territory. A territory may be established near a nesting site (harem-defence polygyny) or beside resources, such as flowers, water, mud, resin-sources or hibernation sites (resource-defence polygyny). A territory may alternatively lack any resources useful to females and be purely landmark-based. Females recognize certain features of the territory and visit it purely to find a mate. When several males fight to establish closely adjacent landmark-based territories they form a lek, enabling visiting females to 'choose' between males (lek polygyny).

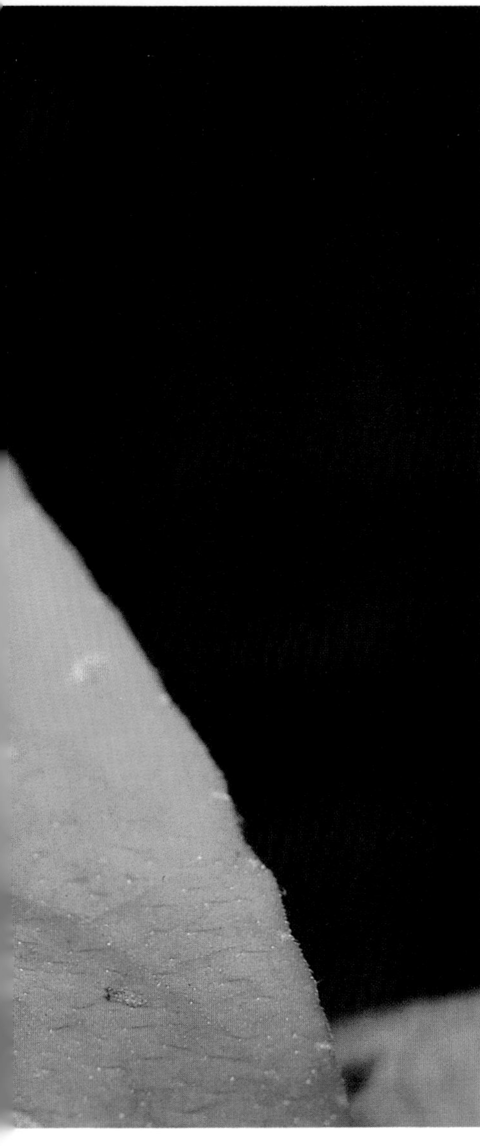

(5) He can move in with a female who is provisioning a nest and exchange helpful services as a nest-guard for frequent copulations.

Sibling matings

This strategy, which involves certain genetic risks inherent to in-breeding, is well known in some of the very small parasitic wasps. For example, males of the minute egg-parasite *Trichogramma dendrolimi* (Trichogrammatidae) achieve 100 per cent fertilization of their sisters within the confines of the butterfly egg in which they have developed. This seems to be the preferred method, as no attempts are made to mate with 'strange' females once the males leave the egg. In-breeding may be necessitated by the poor ability of the males to disperse from their natal area; many of them die inside the host egg after mating, so never have any chance of outbreeding (Suzuki & Hiehata, 1985).

In contrast to the microcosmic nature of this incestuous situation, large vigorous aculeate wasps have every chance to visit 'foreign' nests and inseminate non-kin, so the strategy of deliberate in-breeding is rare. Males of the American *Euodynerus foraminatus* (Vespidae) are the first to hatch from the outermost nest cells inside beetle borings in tree trunks (Cowan, 1979). Their sisters occupy the inner cells, an arrangement which is expedited by the species' haplodiploid genetic system. This produces diploid females from eggs fertilized by sperm, whereas unfertilized eggs produce nothing but males. The female can control the exit of sperm stored in her spermathecae at the time of oviposition, so she can control the sex of her offspring with great accuracy. This enables her to allot certain cells either to males or females (Cowan, 1981).

The males in the outer cells hatch first. A series of short fights quickly decides which of them is going to gain sole access to their sisters. Once the winner's rivals have been expelled, he remains on the nest mainly in the mornings, when his sisters are likely to hatch. In the afternoons he joins his evicted brothers and any other local males in patrolling flowers, on the lookout for females from other nests. He mates briefly with each sister as she makes her exit, but cannot possibly cope with the rapid egress of the entire brood, so a proportion pass him by while he is busy copulating. Even so, he still manages to inseminate around 40 per cent, at least in isolated nests. Any virgins who escape untouched are quickly mated by the numerous patrollers; these are now joined permanently by the former nest-guard, giving him the chance to mate with non-siblings from a different nest. By adopting a mixed strategy, he maximizes his reproductive success, monopolizing a good percentage of his sisters, while still maintaining a significant option on mating with non-kin through daily patrolling. Similar sibling matings are known in the American *Ancistrocerus adiabatus* and the Indian vespid *Paraleptomenes miniatus*. In the latter species, a brother and sister spend about 2 hours together, during which time numerous 'quickie' copulations are repeated at intervals of 15–20 minutes (Jayakar, 1966).

Scramble competition at nest sites

This tactic is common in wasps and bees which form large nesting aggregations in the ground. The males emerge first and await the first females. In the sphecid tribe Bembecini, these vigilant and highly 'keyed-up' males adopt a typical mode of behaviour called the 'sun dance'. They rapidly patrol back and forth in a sinuous pattern over particular sectors of the nesting area; but only if the weather is sunny. This sun dance is regularly performed by *Bembix* males, such as those of *B. rostrata*, which nest in large aggregations in sand-dunes. As more males emerge, the 'sun dance' patrols become virtual swarms; increasingly the males land and run over the sand in search of clues to the impending breakout of the females from their subterranean cells (Schöne & Tengö, 1981).

A promising spot soon lures a crowd of up to 50 males, all frantically pushing and shoving in order to muscle their way through to the central spot, where the digging is at its most frenzied. They cannot wait for the female to make her own way to the surface but vie with each other to be first to unearth her and copulate. Similar 'digging-out' behaviour is seen in many bees and wasps which nest in dense aggregations in the ground. However, at this pre-emergence stage, the diggers (as in *Megarhyssa* described on p. 116) are unable to distinguish the sex of the buried wasp. So their frantic efforts may merely hasten the exit of another male competitor whom they could well do without.

The appearance of a female sparks off a free for all as the males pile in on top of one another. This scrummage rapidly takes shape as a 'mating ball' of scuffling bodies – with the female buried somewhere at its centre. Even when a male does manage to couple, he still has to resist the efforts by the huddle of rivals to lever him off his mount. To evade this he often carries his prize away in a nuptial escape flight, but usually succeeds only in trailing behind him a long streamer of tenacious pursuers, who crowd into a new 'mating ball' as soon as the eloping pair comes to rest; they may even be precipitated into making a forced landing by sheer weight of numbers. The consort only needs some 30–60 seconds without disturbance, copulation being brief, but even this may be difficult to snatch. The females only mate once and then reject further attempts by curling the abdomen downwards and emitting a penetrating buzz, which appears to deter the male.

Similar 'scramble' tactics occur in the American sphecid *Bembecinus quinquespinosus*, but with a subtle difference (O'Neill & Evans, 1983). The males occur in two forms: a large, bright-yellow form which is unusual in being as big as the females, and a small dark form. The large males swarm over the emergence areas, dig out females and form mating balls in which it is usually the heftiest members who win the females. On occasion the sheer ferocity of the infighting leads to a Pyrrhic victory for the eventual 'winner'; he is left grasping a headless and partially legless female corpse, sad evidence of what a merciless pummelling from dozens of powerfully built males can do. It may be to escape such damaging or even fatal harrassment that the females shift their nest site *en masse* each year, to a nearby but different location (Evans *et al.*, 1986). In the bedlam surrounding an emerging female, the males sometimes become confused and accidentally form a 'mating ball' around each other. This allows the female to sneak out unnoticed and fly off and this happens often enough for about 10 per cent of females to be claimed later by non-digging males. These will include some of the small dark males who are unable to hold their own in the scrummage of a mating ball. However, they can succeed in picking up some of the sexual crumbs by patrolling near the nest site and scooping up virgins who have given the diggers the slip.

Landmark territories and lekking

Territorial behaviour is common in solitary wasps. Sometimes territories may be clumped, usually rather loosely, so as to resemble leks. In Oregon, *Eucerceris flavocincta* males perch on conspicuous but small 'mini-landmarks' such as stones, dead plant stems and low bushes (Steiner, 1978). While 'on duty', which only lasts a few hours in any one spot, the male adopts a characteristic posture: his sombre wings and antennae are raised and splayed apart, while his head is held high to display the prominent 'facial mask'. In common with most territorial males of bees and wasps, he flies up to challenge the majority of intruding insects, but only if the trespasser is a conspecific male does a real aerial skirmish develop.

Landmark-based lek territoriality is seen in the tarantula hawk, *Hemipepsis ustulata*, in Arizona (Alcock, 1981a). Males set up and vigorously defend territories on *palo verde* trees on ridgetops. There is a regular change in ownership, with individual tenure lasting an average of 8 days, and a general tendency for large males to hold the best territories. The smaller males are often excluded from territory-holding and are forced to resort to the less rewarding alternative ploy of patrolling along the ridgetop. The masarid wasp *Pseudomasaris maculifrons* has similar hilltopping habits, perching on rocks on ridgetops in the Arizona desert. An 'owner' can be very faithful to a particular rock, returning regularly for up to 29 days. As happens in a number of landmark-based species, the same areas of the habitat – and even the self-same rocks – prove to be favoured year after year. The males are only mildly territorial, probably because rocks are not a scarce commodity in the area, and they may rely on speed to decide who claims a visiting female (Alcock, 1985). Hilltopping is probably the only practicable mating strategy in this wasp; the nests are very rare and scattered and the females, who gather nectar, do so from a broad variety of widely dispersed flowers. As the males have no viable way of encountering a female, it is left to the latter to seek out a mate; and the best way to do this is for prospective partners to hang around in an area whose visual 'trademark' can be innately recognized through generation after generation. In *P. vespoides*, the males are faced with quite another set of circumstances: their females forage almost exclusively from *Penstemon* spp. As these tend to grow in large clumps, there is a good chance that a male who inspects the plants regularly will come across a female before too long.

The repeated application of male pheromones to stems and twigs within a territory, aimed at increasing its attractiveness to nubile females, is a widespread practice in both wasps and bees. The method of applying the scent is fairly standard, as exemplified by *Eucerceris arenaria* from the USA. The male makes regular little 'marking' walks along a stem, dragging the tip of his abdomen along it (Alcock, 1975) and depositing a fragrance which is clearly detectable by the human nose. Similar behaviour is found in *Philanthus multimaculatus* and numerous other sand wasps, and there is evidence that females alight directly on a 'smelly' perch while the owner is in flight nearby. *P. multimaculatus* males defend their territories fiercely, engaging intruders in an escalating aerial contest which commences with a whirlwind swirling flight (Alcock, 1975a). This is followed by an upward zoom and then perhaps a drop-strike – one of the contestants plummets on to the other's back, possibly sending him crashing to the ground. Despite the severity of such contests, occupancy of a perch is quite ephemeral, with any one male often staying only for an hour or so before leaving and being replaced by another. Yet, in the European species *Philanthus triangulum*, a male will defend a territory for several days around a large object, such as a pine tree, attempting to reserve sole access to visiting females for as long as possible. In *P. basilaris* from Colorado, territories tend to be clumped into groups of from 5 to 50, thus forming a lek (O'Neill & Evans, 1981).

Territorial patrols in stenogastrine wasps

Scent-marking of perches is also typical of some male vespids. *Liostenogaster flavolineata* from Malaysia flits rapidly around, stopping regularly to drag the tip of his abdomen across a perch (Turillazzi & Francescato, 1990). *Metischnogaster drewsoni* from the same region acts rather like a syrphid fly, hovering in some forest glade and 'displaying' by stretching out his abdomen to flaunt its whitish banding. Similar behaviour is found in several other species, in which some form of patrolling behaviour which attracts or intercepts females seems to be the main mating strategy. Moreover, males of *Parischnogaster nigricans*, *P. mellyi* and a few others bear a cluster of glands on the abdomen, which may disseminate a female-attracting pheromone as the male patrols. In *P. nigricans* and *P. mellyi*, the 'white-stripe' abdomen-stretching display is often given to rival males. It probably functions as a form of harmless visual jousting, establishing dominance around forest landmarks, such as prominent clumps of leaves, without the need for the vigorous aerial clashes seen in many other territorial wasps and bees (Turillazzi, 1983).

SOCIAL WASPS

Mating strategies in the social wasps

Mating habits in the social wasps vary greatly, although there is no evidence that males ever try to mate with their sisters on their natal nest. Strategies may be dictated by whether the species occurs in the temperate or the tropical regions. In temperate zones, hibernation sites, such as cracks in rocks, holes in trees or the eaves of houses, are a valuable resource to which females can reliably be expected to come as autumn draws near. Males can therefore either set up territories around such a site (resource-defence polygyny) or gather in leks and let the arriving females do the choosing.

In contrast to this, nesting in the tropics may continue throughout the year, or at worst be interrupted by a dry season during which the occupants cluster inactively on the nest or on leaves nearby. Under circumstances of constant nesting, patrolling around flowers may be a more reliable strategy for locating females. However, in some areas, a dual strategy comes into play, depending on the point reached in the nest's development when the males start to emerge.

Alternative tactics – patrols or leks

In Arizona, males of *Mischocyttarus flavitarsis*, which emerge in summer, assume a patrolling tactic, usually along streamsides where females gather to forage (Litte, 1979); males also perch on leaves near the water and await the arrival of females. The patrollers settle regularly to scent-mark twigs and leaves, actively defending such marked sites. Males which emerge in autumn head directly for hibernation sites where they form leks. In-coming males are attracted by these, so that eventually each resident establishes a discrete micro-territory extending for 10 cm around himself, delineated by frequent sessions of abdomen dragging. An arriving female is not harried by any of the lek's members but is given free rein to 'choose' a male from the line-up by landing near him. Although she might well mate with him there and then, she could just as easily reject him and fly off; why females 'change their minds' at this last moment and 'decide' that all the males are, after all, not up to scratch is unknown. Other species of *Mischocyttarus* are not territorial; some, such as *M. drewseni* from tropical Brazil do not even scent-mark while conducting their regular patrols along likely routes (Jeanne & Bermúdez, 1980)

Strategies in *Polistes*

Polistes is a large worldwide genus with a broad diversity of mating habits. *Polistes jadwigae* from temperate Japan adopts a resource-defence system based on hibernacula, but does not form leks (Kasuya, 1981). Instead, each male tries to secure sole ownership of the site – but only for a single day. He spends his temporary tenancy with his abdomen bent upwards in typical 'calling' stance, during which pheromones are almost certainly released. Virtually any female who arrives proceeds quickly to mate with him and he may even enjoy as many as seven copulations with different females within an hour. Given that such a 'caller' has to fight hard to defend his territory, such a high rate of sexual return certainly repays the effort and might explain why a day's tenure may be all that most males can physically manage; only a few 'super-studs' seem able to hold out for a maximum of 3–4 days.

P. fuscatus males in the USA defend territories in female nesting and hibernation sites, marking them by abdomen-dragging. Such sites are mainly monopolized by larger males; smaller males have to fall back on the less rewarding strategy of patrolling on flowers. They have a hard time 'persuading' females to mate, presumably because the females 'know' that patrollers are inferior to territory-owners (Post & Jeanne, 1983). *P. erythrocephalus* males in tropical Columbia hang around near newly founded nest sites and try to pounce on passing females. In two other North American species, *P. exclamans* and *P. annularis*, males head for potential hibernation sites in autumn where they await the arrival of the females; *P. commanchus* is a 'hilltopper' in the deserts of the southwest USA.

Guarding in exchange for sex in *Trypoxylon*

Perhaps the most fascinating sexual strategy of all is evinced by *Trypoxylon* (Sphecidae). In the subgenus *Trypargilum*, the males render assistance in guarding the female's nest in return for regular, guaranteed copulation. Although such a 'help-for-sex' swap seems quite reasonable, it is not quite as logical as it appears. The nest-owner will not only be laying eggs fertilized by her sentry; she will also be laying approximately equal numbers of unfertilized male eggs. It is therefore only his daughters who are of direct importance to the sentry as a return on his investment of time. It would therefore be in his interests to try and coerce the female into increasing the number of 'female' eggs she lays at the expense of the 'male' variety.

Experiments on the American pipe-organ mud-dauber, *T. politum*, have established that the male is not able to influence directly the female's 'decision-making' on the sex allocated to any particular egg being laid (Brockmann & Grafen, 1989). The only possible remaining mechanism which might increase the relative number of daughters could be an increase in foraging efficiency. This could come about if the female is able to leave the nest under some kind of secure stewardship – and this is exactly what the male provides. As daughters require more food than sons, this increased efficiency could perhaps result in the production of more daughters. However, little is known about how the females 'decide' to allocate resources, so any influence on the percentage of male versus female offspring must remain speculative.

Whatever the wisdom of his actions, the *T. politum* male makes a studious and attentive sentry. He will have singled out a nest (a vertical structure built of mud against a wall or cliff) by flying from one to another. If he bumps into a resident sentry, he tries elsewhere; if not, he moves in. Some males may be quite itinerant, staying for a few minutes or an hour, and then moving on to another nest or even a succession of nests. Yet, despite the considerable competition to become sentries, there still remains a proportion of nests (usually a little over 10 per cent) in which the females have to go it alone. A sentry often 'helps around the house', assisting the female to shove spiders into a cell, although he may also hinder her by trying to copulate. However, during provisioning, she is only likely to acquiesce if a really pushy male climbs on her back. In his role as sentry, the male is very effective, lunging with gaping jaws at parasitic flies and wasps while emitting an intimidating buzz. One of his most invaluable services is to fend off scout ants, putting them off the scent before they can discover the valuable contents of the nest and rush back to recruit reinforcements. Other males are received with equal hostility; sometimes the intruder makes an attempted take-over, clamping the defender's slender 'waist' in his jaws and trying to drag him from the nest. The victor announces his triumph with a high-pitched buzzing lasting a minute or two. These battles can be both nasty and protracted, lasting up to nearly 30 minutes, although there is no special bias towards victory by the resident.

When the female is ready to lay an egg on the completed store of paralysed spiders, the sentry engages in a particularly long period of copulation. Even when she tries to turn and lay her egg within the cell, he drags her out and insists on mating once again. He may enforce his wishes in this way up to nine times before he finally allows her to get on with her task. At no time does the female seem to object to this rough treatment by parrying the male's approaches. Once she has laid her egg, she is permitted to leave and gather mud for closing the cell; the male meanwhile resumes his guarding. Once the cell is safely sealed, he may quit the nest and try his luck elsewhere; alternatively, he may stay on and guard the next cell.

The behaviour of some other species is broadly similar, though with a few idiosyncrasies. In *T. tenoctitlan* from Costa Rica, it seems that the females have come to rely completely on their male security guards, not even attempting to make provisioning sorties unless the nests are left in the secure custody of an 'adopted' male (Coville & Coville, 1980). The females' demonstrate their reliance on having a sentry on duty, as a pre-condition for beginning to forage, by actively soliciting males when the time is right. When the nest reaches a stage at which a sentry is required, such as just prior to stocking a newly built cell, the owner attracts a male by making short 'soliciting' flights near the nest. During the final closure of a cell, the sentry, his job done, often becomes foot-loose and starts to wander off in search of a new nest. If he cannot find an unguarded nest, he will often forcibly evict the resident male before taking over his job. In many cases, the resident ends up moving in with his assailant's former mate. This 'wife-swapping' behaviour seems to be a typical activity within the group, presumably because it benefits each male to spread his genes around as many females as possible. In *T. rubrocinctum*, female and sentry greet one another after even brief separations with a special ritual – also seen in similar guise in other species (Paetzel, 1973). The male takes an active part in housekeeping, helping to clean out and stock the cells, using his head to ram the spiders home in the early stages of provisioning and then turning to stuff them in with his feet or the tip of his abdomen when the cell begins to fill up. He may even assist in closing the cell with mud. He usually leaves the vital business of copulation until the last moment before the female turns to lay an egg. Some males help more than others; really dedicated 'husbands' even return with the female to inspect the closed nest.

Help-for-sex swaps in other wasps

Similar home-help behaviour has been noted in *Pison strandi, Dynatus nigripes* (Kimsey, 1978) and *Trachypus denticollis*. In *Oxybelus subulatus*, a North American sphecid which nests in the ground, the male sits on watch close to the nest entrance, driving off all intruders save the female nest-owner (Peckham, 1977). When she arrives back from a provisioning trip, she will be carrying her prey spiked on the tip of her sting. This would normally allow her to slip smoothly into her open nest entrance without pause. However, instead of doing this, she alights a short way off and approaches the nest on foot. This allows the male to dart out and mate with her, during which the prey is normally pushed off the sting and on to the ground. Copulation only lasts from 5 to 10 seconds, after which the female turns around to re-impale the prey on her sting. Males are not guaranteed to be on hand to volunteer their services as sentries, so a female can only rely on having assistance for perhaps 75 per cent of the time. The presence of a male does seem to reduce the rate of parasitism from various miltogrammine flies.

Males of *O. sericeus* behave similarly and may even act as 'decoys' for larvipositing *Senotainia* flies. These tend to be fooled into following the male rather than the female after mating has taken place outside the burrow, permitting the female to finish her provisioning in security (Hook & Matthews, 1980). The males are highly territorial around the burrows and some become completely dominant, making successful takeovers of other males' territories day after day. The rewards for such bellicose behaviour can be enormous – up to 20 copulations per day. A female will not usually submit willingly to such repeated copulation attempts, especially when made by up to four 'subordinate' males between her moment of landing and her eventual dash into her burrow. This is time-wasting for her, particularly as it is only her own dominant 'nest-guard' who will actually fertilize many of her eggs, due to last-male sperm precedence. However, reduction in parasitism is undoubtedly an adequate compensation for time wasted in avoiding subordinate males.

Males of *Trigonopsis cameronii* in Colombia also chase ants away from nests, which they 'adopt' each day with the objective of being on hand when a female virgin emerges (Eberhard, 1974). She expedites her exit from her mud cell by regurgitating a chemical fluid on to its hard clay cap, softening it up. The sex-specific fluid is detected by the waiting male who settles on to the back of the emerging female. However, he will bite and harry an emerging male, presumably to try and make life as difficult as possible for a potential sexual competitor.

BEES

Mating strategies

The mating strategies of male bees and wasps are broadly similar and the use of more than one strategy, depending on circumstances, is quite common. For example, when nests are densely clumped, the males may remain close by and mate with emerging females. When nests are dispersed, the males resort to patrolling at flowers instead. Much male behaviour is devoted to being the winner who secures the one and only copulation which the females of most species will permit. After this single mating, the female stores enough sperm to fertilize all the eggs she can lay. It is not in her interests to interrupt vital foraging trips to mate again with males whose own driving force is to inseminate as many females as possible. She stands to lose a considerable proportion of her hard-won pollen load during the ensuing struggle, this being a particularly likely hazard for many halictines, such as *Dialictus*. Fortunately, most

◀ **Scramble competition at its most intense – a 'mating ball' of the bee *Colletes cunicularius celticus* in the British Isles. After clustering frenziedly around the emerging female for several minutes, the males quickly disperse once she has acquired a mate from among their number. This suggests that the female may emit a scent informing the throng that she is no longer a virgin, thereby minimizing unwanted interference during copulation.**

female bees have ways of avoiding, or at least minimizing, the degree of disturbance. Many can snuff out their sexual allure simply by switching off the emission of an attractive pheromone, this being redundant once they are no longer virgins; males then simply lose interest. For example, a *Centris pallida* male will only briefly touch a pollen-loaded female – or perhaps just fly near her – before breaking off the engagement. He obviously 'feels' it is worth checking but soon loses interest due to the absence of a stimulating 'virgin' odour. A female's unwillingness to mate can also be communicated visually. At the first signs that she is about to be pestered, the female sticks her rear legs up in a kind of two-legged 'get-lost' gesture, while also bending the tip of her abdomen downwards.

In bees, as in wasps, virgin females rapidly become an 'out-of-stock' commodity. The same kinds of pressures to be first to mate as females hatch or start to forage are also present in bees and here too have given rise to 'mob-rule' scramble tactics around emergence sites. The males emerge a week or so before the females and spend this period familiarizing themselves with the layout of the nesting site. Then, as the first females start to nose their way towards the surface, the eager male hordes are ready to greet them.

Scramble competition around nests

Scramble competition around emergence sites is found in many ground-nesting bees from different families. 'Mating balls' are commonly formed by species such as the Eurasian *Colletes cunicularius* and *C. succinctus* (Colletidae), and *Tetralonia malvae* (Anthophoridae). In *C. cunicularius*, several males will frantically dig down to meet an emerging female, hauling her bodily to the surface. The huddle of scuffling males which rapidly gathers around such a virgin rapidly melts away once she has established genital contact with one of their number. This sudden eclipse of her former charm spreads so quickly through the throng of former 'admirers' that it probably indicates the emission of a 'deflowered' scent by the newly mated female. Once this spreads around, the pair can peacefully copulate in full view of dozens of feverishly searching males without being interrupted.

Small 'mating balls' containing only a handful of males are also found in the American *Centris pallida* (Anthophoridae). The males are able to detect the buried females at some depth, and a single dedicated tunneller may toil away for up to 20 minutes in order to unearth a virgin (Alcock *et al.*, 1976). As soon his quarry crawls into the hole he has made, the tunneller clamps his jaws behind her head, clasps her tightly with his legs and hoists her into the air. His haste is motivated by the need to make a quick getaway before any other males notice what has happened. Even so, their efforts at elopement may be thwarted by a 'tail' of pursuing males, whose best chance of appropriating the female for themselves is to ram the couple, sending them plunging to the ground or even splitting them up. *C. pallida* is one of relatively few bees known to have a comparatively complex post-mounting courtship, during which the male caresses the female with his legs and antennae and drums against her with his abdomen.

There is an enormous variation in the size of *C. pallida* males. Large ones specialize in patrolling the nest site and enjoy a definite overall advantage in the race to win females, partly because they are able to make successful take-overs of excavations begun by smaller rivals. Conversely, the rivals will not be in a position to make trouble for a large male who is busy digging away. In practice, it is only the burliest and most resolute of males who will manage to unearth a female and copulate, such is the stamina needed to parry an average of four aggressive intrusions and yet still carry on with the strenuous task of digging. Being hopelessly outclassed at the nesting areas, most of the smaller males resort to an alternative strategy. They hover near plants away from the emergence areas, where there is a chance of picking up females who escape the clutches of the big patroller-diggers (Alcock *et al.*, 1977). While such female runaways do undoubtedly come the way of the hoverers, their mating success is greatly inferior to that of the large patroller – diggers, who secure the vast bulk of the matings.

Some of the large ground-nesting *Proxylocopa* spp. carpenter bees (Anthophoridae) also patrol nest sites and dig females out (O'Toole & Raw, 1991).

'Mating balls' in a bumble-bee

Bumble-bees adopt a number of different mating strategies, including 'circuits and trails' and landmark territoriality. In several species, such as the European *Bombus ruderarius* and *B. subterraneus*, groups of males wait near nest entrances and compete to mate with emerging virgin females (gynes). Although this had always seemed to be a straightforward 'scramble' system, studies on the American *B. confusus* indicate that, at least in that species, something far more sophisticated may be in operation (Lloyd, 1980a). It appears that the relationships between the males gathered around the nest entrance, and between them and the females inside, may in fact be highly complex and probably involves the power to discriminate down to the level of the individual.

Several males (5–20) usually adopt a waiting brief outside the nest hole. As well as jostling one another for the best positions, they also engage in frequent grooming activites. Those nearest the entrance also spend much of their time fanning their wings, apparently directing a draught of air down into the nest. On occasion, a serious fight erupts, the combatants wrestling for several minutes and trying to wreck one another's future sexual prospects by inflicting permanent damage upon the wings. This is accomplished partly by worrying away at the wings with the jaws and partly by employing the following highly original technique: one bee sits on his opponent's back, grips a wing base in his jaws and performs several revolutions designed to wrench the wing away from the body, or at least damage it so severely that the owner is too handicapped to constitute an effective rival. A further male activity is to 'dive-bomb' one another, apparently in an effort to make staying right beside the entrance too difficult to be endured for long. When a virgin emerges, she is instantly engulfed by a ball of males, all struggling for supremacy until one finally rides off on her back. If she is somehow missed by the waiting males, she will repeatedly return until she acquires a consort, illustrating the 'desire' of the virgins to mate at this juncture. Once she lands, the female repeatedly turns somersaults with the male on her back.

Lloyd suggests that the males' behaviour is designed to enable each female to choose a mate before she leaves the nest. The males' wing-fanning wafts into the nest a 'personalized' pheromonal message whose total duration reflects the 'fitness' of the sender; only a top-notch 'alpha' male can retain the best wing-fanning spot at the entrance. Weaker males adopt 'cheat' tactics by dive-bombing the 'alpha' males, obliging them to canvass on behalf of the 'cheats' by transmitting a blend of pheromones. Only when the fanner next grooms himself can he remove the cheat's scent; this explains the regular grooming. When a female emerges she 'chooses' the male whose scent has held sway for the longest period, fighting off the rest until he is in place. If she still ends up with the wrong mate, she somersaults, reducing his capacity to copulate and maybe unseating him. She might even return later to the nest to try and pick up her chosen partner.

Circuits, trails and patrolling at flowers

Many male bees establish a regular aerial beat or circuit, often following a naturally defined trail, such as the line of a hedge, ditch, gulley or ridgetop. The trails of many different males may overlap at certain points to form the equivalent of multi-laned highways, in which all the males move in the same direction like well-ordered traffic. Circuits and trails usually contain a number of flowers, which may be visited by females, but the males do not attempt to establish sole occupancy around such floral resources, which always occur at low densities not conducive to territorial behaviour.

In some oligolectic bees, the males do, however, restrict their patrolling to areas in and around the female's specific provisioning sources, interacting little if at all with one another but pouncing on any females present. *Dufourea novaeangliae* (Halictidae: Dufoureinae) males from North America patrol mainly near clumps of pickerel weed, *Pontederia cordata*, an aquatic flower on which the females are strictly oligolectic (Kukuk *et al.*, 1985). Even so, males are not tied to this strategy and if the host plant is too thin on the ground to be easily patrolled, and the nest site is warm and sunlit, they will patrol there instead. This gives them the option of intercepting emerging females as well as patrolling near nectar sources.

Flexi-time in *Nomadopsis puellae*

Females of *Nomadopsis puellae* are also oligolectic, on *Malacothrix* dandelions in the deserts of the southwest USA (Rutowski & Alcock, 1980). The tractable yet finely-tuned behaviour exhibited by the males, both in mate location and actual copulation, is a perfect example of the innate flexibility found in the mating behaviour of many species of bees. When confronted by massed nests, the males patrol the 'home front' and intercept incoming females. When nests are scattered, and the chances of being at the right 'front door' just as a female arrives are too remote, the males turn to patrolling the *Malacothrix* flowers, starting at around 09.00 hours. Subtlety is not required; all the male has to do is pounce. The female rarely resists and usually allows the male to copulate once he has climbed securely into place on her back. Each female submits to several such intrusions each morning, often being mated each time she leaves the nest on a provisioning trip. Such promiscuity is rare in bees but the females probably have little choice, given their restriction to just a single host plant, where the males can easily find them. Under such 'sitting-duck' conditions, mating is probably less damaging and time-consuming than resistance.

As the morning wears on three things happen. Firstly, a mating pair will be interrupted more and more by ram-raiding intruders, who try to knock the male off his mate. The consort responds with an effective and equally heavy-handed countermeasure, tossing his body abruptly backwards and cannoning the assailant away. The second development is for more and more males to take up loitering around the nest site, trying to steal females arriving *in copula*, while mating with any flying in alone. This interference can become so intense that mating pairs are forced to go into a 'holding pattern' and fly around the area until they can land without being molested. This sudden increase in the arrival of pairs *in copula* is due to the third development of the advancing morning – each male now stays with his mate. This replaces the promiscuous behaviour earlier in the morning, when the guiding principle was to copulate briefly and then hurry on to fresh conquests. During the first part of the morning copulation may last as little as 10–15 seconds. By noon, just before the females enter the nest after their final foraging trip, the copulation period may extend to 40 minutes and the male accompanies the female right back into the nest. The reason for this increase in male fidelity seems to relate directly to the time of the morning and this, in turn, relates to how far the female has progressed with her provisioning. A female with a full pollen load is likely to be making her last flight back to the nest, where the probability that she will lay an egg is very high. By staying with such a heavily laden mate the male can ensure his paternity of that egg.

The pros and cons of the situation earlier in the morning are quite different. If he opts to stay with a female carrying a small pollen load, i.e. one who is only just starting her day's work, he would squander his chances to mate with several females. Even worse, he might also lose his one and only mate at the last minute to a hijacker, ending up with nothing. The male is well equipped for making a 'to-stay-or-not-to-stay' decision, being able to estimate his mate's current foraging status by performing a unique act just before breaking off copulation. He shuffles forward on the female's back, pushes her wings wide open and uses his middle legs to stroke the sides of her abdomen and the bulging pollen loads on her hind legs. His 'leave-or-stay' decision seems to be based on information so gleaned, a remarkable example of the ability of a male insect to adjust his behaviour hour by hour so as to maximize his chances of reproductive success under the prevailing circumstances.

Ambushing tactics on flowers

Some solitary bees are less flexible, only mating with females as they forage on flowers and never waiting around near nest sites. Males of *Perdita opuntiae* from the deserts of Colorado do not patrol but sit quietly inside flowers of a prickly-pear cactus (Bennett & Breed, 1985), either *Opuntia polyacantha* or *O. compressa*. As these are the sole plants from which the oligolectic females will gather pollen, it is a safe bet that sooner rather than later a female is going to turn up. When she does, the male jumps on to her, needing to hang on long enough for her struggles to subside, when he will be accepted. This is by no means guaranteed and females often resist strongly and break free, probably indicating that they mate only once. It is important that superfluous extra matings be avoided, as during copulation females remain stationary on the flower, losing precious foraging time. By contrast, *Perdita texana* females in Texas just keep on foraging, despite being loaded down with a copulating male; males either actively patrol among flowers (usually the cactus, *Opuntia phaeacantha major*) or else wait inside them to stage a sexual ambush (Barrows *et al.*, 1976). Unlike in the previous species, the females seldom reject a suitor, indicating that multiple mating is probably the norm. She can presumably 'afford' such a liberal policy because of her ability to continue collecting pollen at a rapid rate, relatively unencumbered by the male. Other oligolectic bees which mate on flowers probably include all the North American species of *Melitoma* (Anthophoridae), and certainly *M. taurea*; all these are restricted to species of *Ipomoea* (morning glories).

The European andrenids, *Panurgus banksianus* and *P. calcaratus*, are oligolectic on *Hypochoeris radicata* flowers (Compositae). The males bed down inside the closed-up flowers, which only open from about 09.00 until 12.00 hours (Tengö *et al.*, 1988). Once the flowers are fully out, the males take up regular patrols along specific routes, looking for females. All encounters are likely to be productive, as the females will mate repeatedly throughout the flight season (polyandry), a common feature in panurgine bees. In both species, mating probably place takes place exclusively on *H. radicata* flowers, which the males are well able to distinguish from other similar-looking yellow-flowered 'dandelions' (a skill which in human beings is limited to reasonably experienced botanists). The female is able to continue gathering pollen during copulation, although she may be temporarily delayed when buried under 3–4 males, such is the intensity of male competition. It is evident that, in oligolectic bees generally, males adopt a resource-based strategy. This is likely to pay the best dividends in view of the willingness of most females to mate more than once, coupled with their regular visits to the same species of flower over a long period of nest-provisioning. The males are thus faced with a relatively predictable encounter rate. This is especially so compared with species whose females visit a wide variety of flowers, making it impossible for any kind of effective patrolling to be mounted. However, with the flexibility typical of many bees, some individual males of *P. banksianus* elect to remain in a single flower for perhaps several days until a female drops by. Also, as in many other ground-nesting solitary bees (and wasps), whenever the nesting aggregations are densely packed, the males capitalize on this by staying nearby; they then drop the flower patrols.

Mating in primitively social bees

Males of primitively social species encounter certain problems not faced by purely solitary bees. These mainly centre around the more constant availability of females and the different 'classes' of females on offer. For example, in *Halictus ligatus* from Florida, the continuous nesting cycle means that young gynes (potential queens) and workers (non-reproducing but potentially fertile females) are available. This poses a dilemma: how to choose the most rewarding mate (Packer, 1986). The males use the typical halictine strategy of patrolling around patches of plants on which the females prefer to forage (in this case *Bidens pilosa*), barging into them and catapulting them into the surrounding herbage, where they mate. However, not all virgin females are treated thus, many being ignored by males who thus seem to be manifesting a remarkable degree of choosiness. Large virgins who are destined to be queens are preferred, while particularly small virgins, who will probably spend their lives as workers, are often passed over. Even so, some of these 'second-raters' may eventually end up being mated, probably because there is always the possibility that they will be 'promoted' to queen status if the original foundress dies.

Patrolling around circuits

Many European *Andrena* males use a very different kind of patrolling. As their females visit a wide variety of flowers, it would not pay to adopt resource-based patrolling. Instead, the males establish circuits following certain natural features, such as ditches, hedgerows or even up and down tree trunks, depending on the species. Special constituents of the habitat, such as flowers or low bushes, are odour-marked by the bee using the mandibular glands. This helps to separate species, which look rather alike and fly at the same time, but spatial separation is also achieved by patrolling at different heights: thus *A. helvola* flies from ground level up to 0.5 m; *A. haemorrhoa* just above this at 0.5–1.5 m; and *A. scotia* above that at 1–3 m; all overtopped by *A. tibialis* at 2–6 m (Tengö, 1979).

Territories at nest sites. Some bees fight over the right to monopolize a favourable perch from which females emerging from a nest can most easily be intercepted. Such battles over plum perch sites are typical of the European megachilid, *Osmia rufa* (Raw, 1976). The contestants begin by facing up to one another in a mutual appraisal, then mount an aerial charge which sends them tumbling to the ground, biting and kicking until the loser disenganges and flies off. Not all face-offs result in fights; sometimes one male recognizes that he cannot win and withdraws without coming to blows. In this way, the males gradually establish a dominance hierarchy around the nest site. A concentrated site – such as an old fallen log riddled with beetle borings – might reward the dominant male with a considerable number of matings. However, when nests are scattered, the males turn to searching among flowers, or perhaps adopt a dual strategy.

Males of the primitively social American bee *Lasioglossum rohweri* (Halictidae) also take up territoriality as part of a dual strategy (Barrows, 1976). When overwintered gynes (females capable of becoming queen reproductives) start to emerge in mid-summer, several males set up so-called 'micro-territories' around each nest entrance. Sometimes as many as five males may neatly arrange themselves around a hole, all facing inwards with antennae touching and 'concentrating' on one another. This 'staring' behaviour is thought to reduce aggression and leave more time and energy for dealing with emerging females. Frequent biting matches probably serve to establish a dominance hierarchy and low-ranking males (usually small ones) are forced to the back; here the best they can hope for is a sneak mating with a female missed by the dominant males while they were too busy fighting among themselves. However, not all males take up micro-territories; some patrol flowers nearby and probably mate with gynes missed by the territorial males. On flowers, there is also the chance of coming across the substitute queens which are produced by some nests in late spring and early summer.

▼ When nests are clumped, males of the European solitary bee *Osmia rufa* establish territories close to the nest entrances, where they will be in a good position to intercept emerging virgin females. When nests are scattered, they turn to searching for females on flowers.

Male bumble-bees employ a wide variety of mate-locating strategies, depending on the species. Some are territorial on non-resource-based landmarks (see page 118), while others patrol a circuit which 'calls in' at a number of nests liable to produce virgin females. Still others battle it out for ownership of perches close to a nest from which virgins are emerging. The most typical tactic, utilized by the majority of bumblebees is the circuit, in which the males repeatedly mark certain spots, such as twigs, with a species-specific scent produced from the labial gland. A virgin queen orients to the circuit where she is arrested by the scent-marks, mating with the first male who happens to come along while she is waiting. This could be one of several males, as circuits are usually shared, but due to their great length, there is seldom much interaction between the individual males. As in many hover flies, flowers are 'neutral territory', reserved excusively for feeding.

Resource defence

Many bees practice resource defence. The defended resource usually consists of flowers but it can be something else of potential value to a female. In the European *Osmia aurulenta*, it is a snail shell lying on the ground; its value lies in its possible use as a nest; this species is one of several Eurasian bees which make their nests exclusively in abandoned snail shells. The male probably excercises a certain amount of discrimination in his choice of shells, popping inside to look around a fair selection before finally deciding on one. A female on the lookout for a promising shell will probably join him in a close inspection of his accommodation. If she likes his choice, they will mate at the entrance (O'Toole & Raw, 1991).

Resource defence on flowers

In a number of bees, the males establish territories based around flowers, this being especially common in species with oligolectic females. For example, the megachilid *Hoplitis anthocopoides* is oligolectic on viper's bugloss, *Echium vulgare*, and the males' territories always include at least some stands of this plant, as well as some rocks or bare soil to act as perch sites (Eickwort, 1977).

The endemic Madagascan anthophorid *Pachymelus limbatus* establishes territories around the densest flowering stands of a number of plants in the central plateau (Nilsson & Rabakonandrianina, 1988). The females' preferred nectaring flowers are species of *Plectanthrus* (Labiatae), so a male is in a practicable position to annex a reliable rendezvous site by defending the prime patches of these plants. He encircles a patch of flowers in an 'odour-fence' up to 1200 m long, dashing around the periphery on a 'whistle-stop' tour, dabbing his scent-marks on to twigs. He has to repeat this exercise every 30–80 minutes to reinforce the scent-points; during each brief touch-down he marks the twig by wiping his legs against it. The last scent-marks in each 'round' are deposited near the favourite perch site. These are then monitored by the male, presumably to assess when the next round of marking will fall due. He chases or pounces on other insects visiting 'his' flowers. The inclusion of butterflies and wasps on his hit-list hints that this behaviour is probably a deliberate effort to reserve a 'private' nectar source for use by visiting females; conspecific males are fiercely expelled. Any females who do enter the territory tend to play remarkably 'hard to get', so they probably only mate once (monandry), in contrast to the multiple matings allowed by *Anthidium* females under similar circumstances. After one or two fruitless attempts at copulation, he leaves a female free to plunder his resources for as long as she likes. This singular bee is the only one so far known to use odour-marked, resource-based territoriality designed to attract monandrous females.

The males of *Protoxaea gloriosa* from Arizona also guard a variety of plants which females visit (Cazier & Linsley, 1963). The 'owner' hovers with dangling legs, ferociously attacking intruders and violently knocking even tough scarab beetles to the ground; the only large insects which seem immune are *Xylocopa* carpenter bees, which are merely 'buzzed' and then left alone. The owner begins his hovering early in order to snap up virgins during their first foraging flights. He is subsequently capable of identifying non-virgins and allows them to forage unmolested within his territory. These are probably his erstwhile mates, who regularly 'shop' for nectar and pollen in his protected store. The females are therefore on to an exceptionally good thing, exchanging just the one initial mating for prosperous future foraging.

The most aggressively territorial members of the Megachilidae are the Anthidini, in which the males are usually larger than the females, who are noted for their willingness to submit to repeated matings; this is so in *Dianthidium ulkei* from the western USA. The males pugnaciously defend territories encompassing either nesting sites or resin-runs on trees, where females gather supplies of resin for use in nest-building (Frohlich & Parker, 1985). Territorial defence involves aerial charges and head-butting and an 'owner' returning after a temporary absence to find a 'sneak' snatching a quick mating will bulldoze him off his ill-gotten mount with considerable violence. The females are rather unusual in actively soliciting copulation, making several visits to a favourite perch site and even gently butting the owner until he takes notice and mates. During copulation, the male does not actually grip the female's legs or wings, leaving her free to kick him off if she is unwilling. Conversely, a female may keep a recalcitrant suitor in place by reaching backwards and fastening on to his appendages. An unreceptive female can easily dissuade a male with a brief charge or head-butt, in contrast to the lengthy and possibly damaging struggles seen in many other bees.

Males of the North American *Anthidium maculosum* set up territories around flowering clumps of the mint *Monarda pectinata* (Alcock *et al.*, 1977). The owner spends most of each day flying 'combat patrols' around his patch, chasing off intruders and ramming any which do not flee at once; even large carpenter bees are soon expelled. Intruding *A. maculosum* males either retreat immediately after a brief aerial stand-off or take the offensive and grapple with the owner in a serious bid at a take-over. In high-density populations, holding a territory probably means forcibly countering an incursion every 3–4 minutes, leading to a high turn-over. A male who is not able (perhaps only temporarily) to establish his own territory can try one of two strategies: he can either become a 'satellite' within one sector of a particular territory, keeping his head down and trying to avoid the 'owner'; or he can be a 'floater', following a circuit which takes in a number of territories, in each of which he can surreptitiously attempt to 'sneak' copulations. 'Satellites' and 'floaters' tend to be smaller than 'owners', who are therefore easily able to maintain their dominance. The advantage in owning a territory is clear; owners secure an average of 4.5 times as many copulations as non-owners. Even when a 'sneak' does procure an illicit copulation it may be too brief to be worthwhile. The returning owner either 'panics' the sneak by smashing into him, forcing him to uncouple

▲ The female carder bee *Anthidium manicatum* has little choice but to mate with the male who 'owns' the patch of high-quality flowers on which she forages. In return for repeated matings, she does, however, acquire regular access to a resource, whose plunder by competing insects has been reduced to a minimum by the formidable ability of the male to reserve it for his mates' 'private' use.

prematurely, or else he uses his superior size and strength to drag the 'sneak' off the female and then mates with her himself. The females behave in a somewhat ambiguous way while visiting a territory. Some dally for long periods, docilely accepting the 'owner's repeated insistence on mating. Other females keep their stay short and try to evade the 'owner'. Several females cover a regular circuit through a number of territories, whose 'owners' are therefore able to mate repeatedly with a limited 'pool' of females. Immediately after copulation, the 'owner' usually hovers near the female as she forages among 'his' flowers. This is apparently a form of non-contact-guarding, excluding the possibility of a 'sneak' waylaying the female before she leaves the territory.

The European *Anthidium manicatum* is of special interest due to the incredibly high level of violence vented by territory-owners against trespassers (Severinghaus *et al.*, 1981). The males are the 'hit men' of the bee world, armed with a row of spines situated at the tip of the abdomen. After landing on an intruder's back, the male curls his abdomen upwards, forcing his spines into the other male's tissues, which often crumple under the stress. His dedication to the cause of keeping his territory free of trespassers may be evident from the dead and dying insects scattered on the soil beneath 'his' plants. The females are subjected to several matings each day but this does not stop them making repeated visits to the sexual 'free-fire zone' within the males' territories. They probably have little choice, given that the best plants are normally within a male's private domain, with copulation being the inevitable 'fee' for entry. This may in fact be a fair swap, as the female does gain access to high-quality resources whose depletion by competing insects, such as fast-working honey bees, has been reduced to a minimum by the territory-owner's policy of reserving his floral resources for the females. The value of these is evident from the fact that males who guard high-quality flower patches secure more matings than those with poorer-quality patches. 'Sneaks' do occur in this species but they are quickly chased away, usually turning tail and fleeing with little fuss. Actual physical contact is therefore rather rare, possibly because the kind of lengthy grappling matches often seen in other bees could prove fatal in these heavily-spined assassins.

The male Australian banksia bee, *Hylaeus alcyoneus*, sits on top of a large banksia flower and disseminates a female-attracting pheromone by whirring his wings (Alcock & Houston, 1987); he also 'marks' nearby stems. Larger males tend to win the numerous ferocious battles for territories, intense rivalry being inevitable because, although banksia flowers are often abundant, certain individual flower spikes are preferred, normally those with at least some flowers still left unopened. Smaller males who cannot win a territory take up patrolling among the flowers, always on the lookout for a vacant territory, even though it might only be temporarily free.

Landmark-based territories

In bees, it seems that scent as well as the topography of the site plays a role in attracting females to the trysting grounds; the males often 'mark' certain points with mandibular gland extracts. Males of *Centris dirrhoda* and *C. decolorata* in Jamaica scent-mark the boundaries of their territories in this way (Raw, 1975). *C. decolorata* establishes its territory within large nesting aggregations where females are guaranteed to occur. The boundaries of each territory touch one another to form an interconnecting mosaic which serves to spread the males evenly over the site. This eliminates the frenzied scramble competition seen in species such as *Colletes cunicularius*. This neat spacing is possible because the males establish a dominance hierarchy by fighting.

Males of *Eulaema* and *Euglossa* (Apidae) have strange and rather enigmatic habits which have earned them the name of 'orchid' bees. They collect naturally occurring aromatic products (often from specific orchids) and modify them before applying them as territorial markers. *Eulaema meriana* and *Euglossa imperialis* set up territories in tree-fall areas in the rainforest (Kimsey, 1980). As such gaps in the canopy are always rather scarce, each one usually contains several males. This accidental clumping of males may have given rise to the apparently erroneous belief that the main purpose of the scents gathered by the males is to attract other males to form leks.

The tree-fall areas themselves contain no resources of interest to the females, who forage on flowers high in the canopy. However, the special value of a tree fall can be judged from the way in which certain tree trunks may provide territorial perches for a succession of males in different years. It seems likely that, during his period of tenure, each 'owner' impregnates the bark with some durable scent. This is presumably derived from the aromatic substances which the males collect so assiduously, taking days off in order to gather fresh supplies. (*Euglossa* species usually collect orchid fragrances; *Eulaema meriana* and other members of the genus are commonly seen scraping aromatic material off rotting logs.) The territorial behaviour of the two species is similar and incorporates two elements. Firstly, males 'display' by perching head-up on a favourite trunk in a specific 'display' stance with the rear end tilted upwards and the closed mandibles resting against the trunk. *Eulaema meriana* males also produce regular 'buzzes' by vibrating the flight muscles, synchronizing each 'buzz' with an opening and closing of the wings. This serves to show off the smartly banded cream and black basal segments of the abdomen contrasting with its showy orange tip. The function of the buzz is uncertain, but it may serve to send a blast of pheromones into the surrounding forest. Bouts of displaying (or 'calling'?) are interspersed with patrolling flights out into the territory.

Trespassing males are not treated as roughly as in some other bees; grappling is absent. Instead intruder and 'owner' merely fly at one another and engage in

some aerial jostling and nudging. Females probably find the territory by scent and by recognizing its qualities as a mating site, to which they fly specifically for the purpose of mating. This is indicated by a female's rapid assumption of a 'mating posture' on the male's perch site; but only after she has made a brief circling 'inspection', during which she possibly assesses the male's qualities as a potential mate. Several other euglossines probably exhibit similar territorial behaviour, including *Eulaema cingulata*, *Eulaema polychroma* and *Euglossa ignita*.

Territoriality in carpenter bees

Most *Xylocopa* carpenter-bee males (Anthophoridae) are territorial. In many African species, including *X. torrida* and *X. flavorufa*, a spray of flowers comprises the centrepiece of the territory and the male cruises endlessly back and forth around it. Mating in *X. torrida* takes place on the wing and is particularly interesting because of the male's method of holding the female while in flight (Anzenberger, 1977). As he rides on the female's back, he jams a special

◄ This African *Xylocopa* sp. carpenter bee male will spend several hours a day constantly patrolling within a small territory around a sprig of flowers, waiting for a receptive female to arrive.

femoral spike down behind her wing bases, locking the wings in a horizontal position. He places his exceptionally hairy pretarsi across the female's eyes as 'blinkers', while his elongated metatarsi curl up snugly beneath her head and keep him firmly in place. It seems that the idea is to 'blind' the female so that she is discouraged from trying to use her wings, leaving all the flying to the male; if she does try to fly, she unbalances them both and they rapidly dip earthwards.

As in many other bees, multiple mating strategies are often called into play. *X. virginica virginica* from the USA uses as many as five possible tactics, the preferred one being to establish a territory around a nesting area. Failing this, a territory centred on a group of flowers or a prominent landmark – such as large rocks or a house – will suffice (Barrows, 1983). Males also patrol flowers, pouncing on females (scramble competition), while small males who have little prospect of competing realistically, either for territories or in 'scramble' contests, adopt typical 'sneak' tactics. Territorial males are typically pugnacious; an 'owner' hovering within his domain rushes out to inspect and drive away any insects coming within a radius of around 20 m. He also shows an interest in such diverse and more distant objects as passing birds and high-level jet-planes. Such confusion, due to an apparent inability to judge distance and scale, is common in *Xylocopa* and other territorial bees. Intruding males are greeted with vigorous chasing and the occasional brief grapple, yet an 'owner', hovering constantly in an adjacent territory, is ignored, presumably because he is recognized as a 'rightful' neighbour. This avoids repeated antagonistic encounters between neighbours, which would be wasteful in both time and energy, yet gain little.

However, maintaining a hovering patrol is hard work. If there are no flowers within the territory, the owner will have to leave now and again to visit the nearest floral refuelling station. This is the opportunity for which a smaller 'sneak' has been waiting. He will loiter inconspicuously near the edge of the territory until the owner vacates it for a while; then the sneak dodges in and mates with any females who arrive. Sneaks are finely attuned to the territory-owner's movements, rushing in as soon as he leaves; they also pursue and mate with incoming females, with far more urgency and assertiveness than the rightful owner. This perhaps indicates the pressure on a 'sneak' to squeeze in as many matings as possible while the going is good. His reign will only endure for the period of the owner's absence, and that might only be for a minute or so.

Sisters as male refuelling stations

Many male carpenter bees devote a part of each day to an energetic hovering defence of landmark territories around trees or rocks in which nectaring sources are always wanting. Deserting such a territory, even temporarily, to go and feed would invite a take-over by another male, a visit by a 'sneak' or even a missed opportunity to mate. Any device which can prolong 'on-station' hovering time would be invaluable, as it would increase overall mating success. In *Xylocopa nigrocincta* from Brazil, territorial males are able to procure distinct improvements in their energy budgets with a little help from their relatives (Wittmann & Scholtz, 1989). The territorial males return to their natal nest each day, where they are fed nectar by their mother and sisters. The males then go and sit at the nest entrance and dehydrate this free hand-out by regurgitating a droplet between the mouthparts (galeae) so that evaporation will concentrate the sugar content. There is nothing unusual in this process in itself – females of both solitary and social bees 'ripen' honey in this way before storing it. However, *X. nigrocincta* males capitalize on the ability of the process to reduce the weight of the copious fuel supplies necessary for extended hovering flight (going from a starting weight-to-energy ratio of 1:4.9 to a 'high-octane' ratio of 1:8.6), thus permitting a longer sojourn in their territory. A male can thus leave the nest each evening and spend up to 3 hours in constant circling flight 2–3 m above the ground, usually around the sterile foliage of trees or shrubs. The importance which the female nest members attach to helping the males to maximize their reproductive success is clear. During the mating period, the males are not only assigned the biggest share-out of nectar within the nest – they also receive their share first.

Due to this matronly pampering, *X. nigrocincta* males never have to bother feeding themselves at flowers, unlike most other species of carpenter bees. Moreover, any young sisters who have yet to leave the nest actually assist in the task of nectar-dehydration, revictualling a returning male in a highly efficient 'pit stop' so that he can race straight back to his territory without losing much time. Three other species, *X. frontalis*, *X. augusti* and *X. varians* are sympatric with *X. nigrocincta* in southern Brazil. In none of these do the males profit by receiving assisted fuelling by nest mates. Instead they are kicked out of their natal nests early in their lives and subsequently form all-male 'hostels' in deserted nests. They have to call in at flowers to collect their own nectar but are able to concentrate it themselves before embarking on mating flights, as in *X. nigrocincta*; this emphasises that this useful behaviour is not dependent on services rendered by female nest mates.

Dispersed leks in *Xylocopa varipuncta*

The arid territories occupied by males of *X. varipuncta* in Arizona are also short on nectar sources. Males hover for up to 2 hours, dispersing a sweet-smelling pheromone in territories established around creosote bushes on ridgetops, or around large ironwood trees down below in nearby dry washes (Alcock & Smith, 1986). As both ridgetops and washes follow an approximately linear alignment, it is likely that they funnel females into areas likely to contain territorial males. Several territories usually occur closely adjacent to one another and it has been suggested that this might constitute a 'dispersed lek'. This would enable arriving females to 'choose' between a number of potential mates by visiting each territory in turn. In fact, females do show a high degree of 'fussiness', usually taking the time to carry out an inspection when two or more males are 'on show' around a single creosote bush. If satisfied, a female might choose one of their number, or perhaps even reject all of them and move on to inspect another selection farther along the ridge. This could explain why a single bush may be able to accommodate several males living in apparent amicability – if the female chooses males on some characteristic intrinsic to each one individually, then the extra cost of maintaining personal territories would be pointless. This is backed up by a male's response to an arriving female; he mounts a 'land-and-walk' display, parading around on a sprig of leaves and rubbing his legs against them. He may repeat this several times, interspersing it with hovering flights near the female. This whole 'strutting-the-catwalk' drama is designed to spotlight the individual male's qualities. The female – who is hovering nearby and closely scrutinizing the action – is being invited to land and copulate. Females who lose interest (as most do) and fly away are not followed, presumably because the males 'realize' the futility of doing so. The females thus appear to exert complete control over choice within a system (the dispersed lek) which seems the only practical means of bringing the sexes together under conditions which are able to demonstrate in some way a male's 'fitness'. No alternative strategy could really be viable. Nests are widely distributed and difficult to locate, thus effectively precluding territorial behaviour around a nest site, while the female's main foraging plant, the creosote bush, *Larrea tridentata*, is the most abundant flowering plant in their habitat, which rules it out for either scramble competition or resource defence (Marshall & Alcock, 1981). A similar situation applies to *X. pubescens* in Israel.

Landmark territoriality in bumble-bees

A number of species exhibit landmark-based territoriality. Males of *Bombus rufofasciatus* in Kashmir spend the day on perches, defending them with such ferocity against intruding males as to inflict actual injury. Both this bee and the highly territorial American species, *B. nevadensis*, have enlarged eyes to assist in scanning their surroundings for invaders. Males of *B. nevadensis auricomus* and *B. griseicollis* scent-mark a number of spots around their territorial perches in Arizona (Alcock & Alcock, 1983). There is a considerable turn-over of territories in *B. griseicollis*, with individual males only remaining for a short time before moving on. Females put in only the occasional appearance and mating takes place briefly while the couple are airborne.

Macrocephalic guards

There is a certain parallel between the nest-guarding behaviour exhibited by the males of *Trypoxylon* wasps and the habits of certain flightless 'macrocephalic' male solitary bees which have grotesquely enlarged mandibles. 'Macrocephalics' are found in a number of solitary-bee species, including *Halictus latisignatus*, *Perdita portalis*, *Lasioglossum dimorphum* and *Lasioglossum erythrurum*. However, the precise *raison d'être* for these oddities is uncertain, especially as normal males are also usually produced by the same nest. However, it is only the giants who remain at home. In *L. erythrurum*, the macrocephalic residents fight fiercely, often to the death, for the sole right to stay behind and mate with the females in the nest. Only a proportion of these is likely to be the winner's sisters, given the communal nature of *Lasioglossum* nests (Kukuk & Schwarz, 1988). Once in sole command, a macrocephalic male has no need to leave the nest as he is well fed by the females. The return on this investment seems to be the macrocephalics' superior capabilities as nest-guards, a job which they can carry out with formidable efficiency, given their meat-cleaving mandibles. The normal males, who leave the nests soon after their emergence, probably mate with females from other nests, as in most bees.

Mating swarms in highly social bees

The sexual arrangements of the most highly social bees, such as honey bees and stingless bees, are unique in that large numbers of virgin queens are released simultaneously from several nests and mate with males who have been released some time earlier.

Trigona and *Melipona* stingless-bee queens often do not have far to go in order to find a mate, as hordes of males gather in swarms near the nests. Honey-bee queens are released from the nest about 1 hour after the males, who by then will have had sufficient time to foregather in special trysting arenas known as congregation areas. These areas are remarkable for being the assembly points for thousands of drones from many different nests (often from several kilometres distant). Somehow all these males have headed unerringly for the same target, measuring just 50–200 m across. An amazing feature of the congregation areas is their regular use, year after year, by bees who can have had no previous knowledge of their existence and location. A queen heading for such an aggregation trails behind her a pheromone-stream up to 30 m long, which lures as many as 100 males to compete for the chance to mate with her. Copulation for the winner occupies only 2–4 seconds but this will be his sole sexual conquest. Unlike in other bees, males of both honey bees and stingless bees mate only once. It could not be otherwise, as, during uncoupling, the male's genitalia are virtually blown free of his body and left affixed to the tip of the queen's abdomen. As the queen herself mates several times during her nuptial flight, a number of these disembodied genitals – perhaps as many as ten – can be seen attached to her abdomen. The workers pull them off after the queen's return to the nest and they will also remove any excess sperm exuding from the female's genitalia, the result of an orgy of matings. Monogamy in the male honey bee makes sense because his chances of ever claiming a second female are effectively nil, given the huge excess of males over females in the swarm and the brief duration of its existence.

ANTS

There are no solitary ants and mating strategies are broadly similar. Certain weather conditions act as a trigger for the simultaneous release of winged males and females (alates) from many different nests. These mate on the wing, on trees and bushes or on the ground, after which the mated females cast off their wings and try to establish a new nest.

Swarming in *Atta* leaf-cutting ants is fairly typical. In late spring, the first winged adults of *Atta texana* from the USA start to mass in upper cavities in the underground nests; cavities with rich fungus gardens are preferred (Moser, 1967). They spend 1–2 months here before finally flying from the nest, although the reason for such a long wait is unknown. Release is always prefaced by sufficient rain (at least 7 mm) to wet the ground above the nest over a reasonably long period (more than an hour – a short shower will not do). The relative humidity must be between 98 and 100 per cent and there should be no wind. Shortly after the downpour, and if conditions are still dull, major workers flood out on to the nest. In this pre-release phase they are in an extremely jittery mood and rush around with gaping mandibles, freely attacking anything which moves, including leaves and insects which have picked the wrong night for a stroll; even large intruders, such as lizards or frogs, are simply torn apart. With their security force of hyperactive guards milling around up above, the alates can now start to emerge. They do this apparently without any overt management or coercion by the workers (unlike many other ants in which the workers exert considerable control over the movements of the alates), congregating on the surface of the nest in the early hours of the morning. They eventually make a mass take-off just before dawn. Only around half the colony's alates leave in this exodus; the rest remain inside the nest and make their own flights over several more nights. When many of the female alates fly the nest, they act as involuntary carriers for female *Attaphila fungicola* cockroaches, an obligate myrmecophile which reaches new nests by hitching a free ride in this way; several species of mites do likewise.

Male investment in bees. In most bees, the only investment made by the male is to contribute a small amount of sperm; however, there are one or two exceptions. Males of some (perhaps all) *Nomada* parasite-bees are able to synthesise (in their cephalic glands) several compounds which are chemically close to the secretions from the Dufour's-gland of their own specific *Andrena* solitary-bee host (Tengö & Bergström, 1977). Each *Nomada* species male produces the correct chemical 'fit' for his own host, yet the *Nomada* females cannot produce these complex compounds at all. During copulation, the male sprays the chemical on to the top of the female's thorax, where it is remarkably durable. It probably serves the important function of lending her the 'nest odour' of her specific host. In this way she is enabled to enter the *Andrena* nest and lay her eggs unmolested and unrecognized by the rightful owner.

◄ Snuggling together on a flower, a group of male *Eucera* bees (an undescribed species from Israel) spend the night in a temporary truce, taking a night's break from the fierce sexual rivalry which dominates each day. Male bees often bed down peacefully together in this way, regardless of how ruthless the battle may have been during the previous few hours.

► *Camponotus fellah* alate ants are shepherded from the nest by their worker chaperones in the Israeli desert.

Just after alighting on the ground, each female de-wings herself, forcing the wings out, one at a time at right angles with the back legs, so that they snap off at fracture points. She then starts to dig a new nest. The procedure in most other ants is broadly similar. However, in tropical species, such as *A. cephalotes*, swarms are released during the rainy season, rather than in springtime. Direct rain does not always seem necessary for swarming – *A. sexdens* in subtropical southern Brazil can detect imminent rain and releases the alates accordingly. The nocturnal release of *A. texana* is also somewhat unusual; most ants swarm during daylight.

Assembly stations in harvester ants

The mating system of *Pogonomyrmex* harvester ants from the southwest USA is unique among ants for its lek-like mating aggregations, which often occur in the same places year after year. This provides an interesting parallel with the leks established by some birds which use traditional arenas, and with the congregation areas of honey bees (Hölldobler, 1976). However, the male harvester ants are not strictly territorial within the 'lek' but engage instead in the kind of 'mating-ball' scramble competition also seen in some bees and wasps. It is perhaps inappropriate therefore to use the term 'lek in its strict sense.

In Arizona, several *Pogonomyrmex* species often occupy the same habitat but the timing of their flights helps to keep them reproductively separate. Thus, the release time for *P. maricopa* is 10.00–11.30 hours; for *P. desertorum* 11.00–13.00 hours; for *P. barbatus* 15.30–17.00 hours; and for *P. rugosus* 16.30–18.00 hours. Not all nests even within a single species release their alates on a given day; most nests make several releases during a season. The workers exert considerable control over the alates, shepherding them out on to the surface and then dragging them back inside if they show signs of imminent flight before 'lift-off' is officially due. The workers are also exceptionally aggressive at this time, shouldering their heavy responsibility for the successful launch of the vital alates by descending on intruders in savagely biting hordes. In general, the males are the first to take wing, heading for trysting sites, which vary according to species: the tops of hills, rooftops and chimneys in *P. occidentalis*; the crowns of tall trees for *P. californicus* and *P. comanche*; areas of flat ground for the alates of *P. rugosus* who fly upwind and zero in with no obvious visual landmarks whatever. This is perplexing, as millions of ants from many different nests still manage to end up there. The males arrive first and dash frantically around on the ground, soon to be joined by the females, each of whom is instantly buried in a cluster of jostling males as soon as she alights. Every male is desperately trying to secure a firm grip on the female's thorax with his mandibles and front legs. This will anchor him in place against the pressure from the melée until he can obtain further attachment with his genitalia. However, retaining any kind of grip is difficult and males are frequently dragged off their mount, despite having especially strong mandibles designed to clamp firmly to the female's thorax. A female will sometimes lose her abdomen (gaster) when it is accidentally snipped off as her rider applies his jaws rather too tightly around her slim waist.

A female does not accept the first male who manages to climb on board but for some time resists all efforts to couple, letting her many suitors fight it out until the strongest or more persistent finally secures her with an immovable grip; this is almost always the largest male (Davidson, 1982). The first male to connect successfully usually attempts to retain his position for as long as possible, presumably because several rivals will be stacked up on top of him, queuing up to be the next to mate once the female falls vacant. A female will indeed mate several times but with each succeeding occasion she reduces the time spent *in copula* before she forcibly ejects her consort. She communicates her wishes by turning around to nibble at his gaster. This is supposed to be a harmless tactile 'time-is-up' message but it can become serious enough to cause wounds, presumably when the male fails to take the hint. A male who has managed to be first to mate can ensure his mate's future chastity in a rather drastic way. If he grossly outstays his welcome, despite the female's best efforts at gentle persuasion, she may eventually terminate the copulation so violently that her mate leaves behind his genitalia during his ejection. This effectively blocks up the entrance to the female's genital opening. As a chastity belt, this rather resembles the sphragis of certain butterflies, although in a far more once-and-for-all manner. The habit of hanging on until forcibly ejected is much more common in first-in-line males,

indicating that it may be a deliberate strategy based on the principle that one good mating is better than several poor ones, if indeed these ever transpire. Males who are further down the queue have less to lose and do not act in this way, opting instead for trying to achieve as many (individually less effective) matings as possible.

Once the female has concurred in several copulations she becomes increasingly intolerant of the males around her. She communicates her unwillingness to mate again by stridulating in a kind of 'end-of-game' signal which avoids further wasted time and effort both for herself and for the competing males. Overnight, the males cluster together like bees in nocturnal harmony, only to take up the competitive cudgel next day. The same mating grounds are apparently used year after year both in this species and in *P. barbatus*, whose arenas are similarly deficient in obvious landmarks. *P. maricopa* and *P. desertorum* aggregate in trees, and again it is the same specimens which are utilized in more than one year.

How the first males to arrive at these sites manage to locate them so accurately year after year is not known, especially as the topography is so unremarkable, there being no obvious landmarks to act as guidance. However, once the 'pathfinder' males have unerringly fetched up once again at the correct spot, they expedite the recruitment of more males and females by discharging their mandibular-gland secretions, releasing a clearly discernible sweet-smelling odour. This forms a downwind plume which can be followed by the new arrivals, all of which can clearly be seen to be flying upwind towards the olfactory signpost.

Females who 'call' – the role of pheromones

In a number of myrmecine ants, nuptial relations are far more subdued. Females of the European slave-making ant *Harpagoxenus sublaevis* and the American *H. canadensis* climb plant stems near the nest and adopt a characteristic 'calling' stance with their abdomens held erect (Buschinger, 1983, Buschinger & Alloway, 1979). From the tip of the extruded sting, a small droplet of pheromone is exuded, which serves to attract winged males and elicits sexual behaviour. Within the myrmecine ants, the pheromone is produced by either the poison gland or the Dufour's-gland. Several members of the Leptothoracini, including *Doronomyrmex pacis*, *Leptothorax kutteri* and *Formicoxenus nitidulus*, all of which are social parasites, as well as the non-parasitic *Leptothorax muscorum* and *L. gredleri*, exhibit similar behaviour.

'Calling' behaviour has also been recorded in *Rhytidoponera metallica*, an Australian representative of the primitive subfamily Ponerinae (Hölldobler & Haskins, 1977). Many species of *Rhytidoponera* are unusual in that winged reproductive females are not produced. The role of 'queens' is delegated instead to certain workers or 'ergatoids' which develop functional reproductive organs. As these ergatoid workers are wingless and cannot swarm in a nuptial flight, they have to utilize some other means of arranging a liaison with the winged males. They do this by crawling out of the nest and assembling in small 'calling' parties. Each 'caller' raises her abdomen off the ground and extends the rear segments to release a stream of pheromone from the tergal glands. Males are greatly stimulated by this pheromone and quickly home in on the 'calling' females.

The degenerate slave-maker *Epimyrma kraussei* has an unconventional reproductive system. Brothers and sisters mate while still in the natal nest and dispersal to new nests is on foot (Winter & Buschinger, 1983).

Egg-laying Behaviour

It is usual, though not inevitable, for a female animal to go to some lengths to choose a suitable site in or on which to lay her eggs. In doing so, she may be providing them with protection from dehydration, climatic factors, predators or parasites. The actual act of oviposition may be straightforward or may involve some fairly complex behavioural activity in order to ensure success.

◀ **Wind spiders, along with scorpions and most mygalomorph spiders, dig a burrow in which to lay their eggs. Here a female *Solpuga* species from South Africa is digging her burrow at night in savannah. She uses her front appendages like a bulldozer to remove loose soil as she digs.**

CHILOPODA Centipedes

The female of the American geophilomorph centipede *Geophilus rubens* often lays her eggs beneath bark. In order to oviposit, she first coils her body so that the genital segments are close to the antennae. As each egg appears, it is held briefly in her poison claws before being placed in a heap, around which her body is coiled. Following the deposition of about a dozen eggs, all subsequent layings take place directly on to the egg pile without manipulation by the poison claws. The female European *Necrophloeophagus longicornis* lays her eggs beneath the soil. Immediately before egg-laying she coils her body into a loop, bringing together the body segments at the closure of the loop to form a platform on which she then deposits the eggs. The apical abdominal segments are held off the surface as the sticky egg is forced out. The end of the abdomen is then swung over and the egg is placed on the platform. The scutigeromorph centipede *Scutigera coleopatra* does not produce an egg batch and then brood it, although she does sometimes produce a loose nest containing between seven and ten eggs. Instead, as each egg is laid, it is alternately covered with fluid from her rear end and then wiped on the soil surface until it is covered in a layer of soil particles. Finally it is dropped into a crack in the soil and the female then tramples the area with her hind legs; sometimes eggs are just left on the surface.

ARACHNIDA Spiders, Scorpions, mites, etc.

Whereas the scorpions and pseudoscorpions retain their eggs and produce live young, solifuges lay their eggs in burrows.

Most opiliones lay their eggs in cracks and crevices, but one notable exception is the Neotropical harvestman *Zygopachylus albomarginis*, whose males build a nest and then guard the eggs which the females lay in it. Males construct their nests on logs and tree bases, using materials collected from the trunk nearby, mainly flakes of bark prised off in the chelicerae and mud scooped out of crevices. This is fashioned into a rather lumpy ball about 1 mm across and ferried to the nest site in the chelicerae. Saliva is probably added to the material during the subsequent building process to act as a glue. A single-layer disc-shaped floor around 3 cm across, forms the foundations, to which is added a vertical boundary wall just under 1 cm high so that the finished product resembles a small crater. The whole process takes 1–4 days and the males subsequently carry out repairs as necessary, damage usually being caused by weather and ants. Another essential maintenance task performed by the male is to graze down the hyphae of fungi which attempt to colonize the nest materials. Eggs are normally rationed to just one or two per female. The male shows the female the vacant plot by tapping on her body and the nest floor with his front legs, eventually inching beneath her from the rear as he taps away on a particular spot on the floor. This causes her to stand up high on all her legs, extrude her ovipositor and then briefly squat low as the egg appears. Then, to the accompaniment of a fusillade of leg-tapping by the male, she uses her wiry front legs to firm down the soil so that the egg becomes half-buried. She might then be allowed to lay another egg, or she could just bow out and disappear.

Females sometimes try to 'dump' eggs in a nest without 'paying' for the privilege by mating. The males are far too canny to be duped by this and invariably make a meal of such 'cuckoo' eggs, while never touching their 'own' fertilized eggs. The males are rather fussy over which females they accept and reject many supplicants looking for an egg-laying site. Females sometimes 'adopt' a particular nest and remain nearby for several weeks, aggressively driving off any other females who show up. Being a nest-owner is essential if a male is to attract a mate, and some males try to take a short cut to sexual riches by stealing nests rather than building their own. This results in regular take-over fights. A successful invader always prepares the nest afresh for his own reproductive efforts by making it his first job to eat the existing eggs fertilized by the deposed owner.

It is only in the spiders that we find any degree of sophistication in egg-laying behaviour involving the production of a silk-covered egg sac. Oviposition in spiders is inextricably linked with the production of the egg sac, which varies from a few strands of silk loosely tying the eggs together, in primitives such as the Pholcidae, to a full-blown, multi-layered sac in most of the higher families. Descriptions of the actual construction behaviour are somewhat limited, since the process occurs either during the relatively safe hours of darkness or in recesses beneath bark, under stones and in other secure sites.

One species for which it has been observed is the Neotropical araneid *Nephila clavipes*, at the northern limit of its range in Louisiana in the USA (Christenson & Wenzl, 1980). The egg sac is usually constructed beneath leaves that can be shaped to form a protective canopy and, in this part of its range, two to three egg sacs are produced per female per season. Having found a suitable leaf, the female spider hangs below it and applies her spinnerets to it many times before she eventually covers about half the leaf surface with a mat of silk roughly 3 mm thick. She then lays her eggs on to this mat, pushing them up against the force of gravity with her abdomen so that they adhere to the silk. Once egg-laying is completed, she swathes the whole mass in silk. She now moves from the egg sac to the twig which holds the leaf, pulling a line of silk out behind her, flips over the twig and returns to the egg sac. By repeating this movement a large number of times, she produces a loop over the twig which holds the egg sac in place. This presumably prevents it from falling to the ground should the leaf dehisce before the spiderlings are ready to hatch.

The egg sacs carried around by females of the lycosid genus *Pardosa* are somewhat more complex than those of *Nephila* and their manufacture involves more complex behaviour. Egg-sac construction in the North American *P. lapidicina* has been observed in the laboratory (Eason, 1969). The female's first task is to construct a raised structure of silk strands spanning two suitable objects. Once this is wide enough she spins a circular mat, spinning in first a clockwise and then an anti-clockwise direction. While the mat is still <0.025 mm thick, she fashions a slight rim all the way round it. Once the mat is complete she then lays a mass of fluid-covered eggs on to it. She now positions herself over the egg mass and begins to cover it with silk, attaching the silk first to one side of mat. She then raises her abdomen high over the eggs and touches them with the spinnerets once or twice before finally touching the silk on to the opposite side of the mat. By alternately moving clockwise and anti-clockwise, a thin canopy of silk is woven over the eggs. Next, by moving her abdomen from side to side, starting at the circumference of the mat and moving to the centre of the canopy, she covers a sector of the egg mass in silk. By repeating this four or five times she completely covers the surface of the egg mass. She now reverses direction and repeats the entire procedure several times. Once the canopy is roughly the same thickness as the original mat, she raises the strands supporting the silk mat with her palps and cuts them with her chelicerae.

As soon as the as-yet-incomplete egg sac is free of its ties she lifts it and holds it in her third pair of legs. Now, using these legs to turn it, and her palps to guide it, she rotates the whole structure and turns over the seam between the mat and the canopy so that the sac loses its fried-egg shape and assumes its familiar lenticular form, the whole becoming more taut in the process. Still holding the sac with her third pair of legs and her palps, she spins a tiny silk disc near to the centre of one side of it, leaving it attached to the spinnerets by a short bundle of threads. After a break of a few minutes she detaches the egg sac from the spinnerets and then, rotating it first around the seam and then around the flattened sides, she spreads a colourless fluid, which oozes from her mouth, all over the silk covering. She finally re-attaches it to her spinnerets and her task is complete.

MANTODEA Mantids

The variously-shaped egg masses produced by female mantids may be found attached to vegetation, on or under rocks or in any place thought to be suitable by a particular species. Having found an appropriate site on which to oviposit, the female first secretes a liquid, produced by abdominal glands, from her genital opening, which on contact with the air becomes a foam. She then shapes the foam mass, using the end of her abdomen, so that a series of receptacles are formed down the centre, into which she deposits her eggs. Each egg ends up in its own separate chamber with a valve at the top which allows the young an easy means of escape.

ORTHOPTERA Crickets, locusts, crickets and grasshoppers

Female katydids (Tettigoniidae), crickets (Gryllidae) and mole crickets (Gryllotalpidae) have long ovipositors with which they lay their eggs into the soil or plant tissues, or on to the surface of plants. Female grasshoppers (Acrididae) lack such a structure and lay their eggs in soil (by extending the abdomen and digging, using special valves borne by it), amongst grass or other

▲ Many, but by no means all, grasshoppers lay their eggs in the soil, protected from desiccation by a layer of foam. In some species, the male may remain in attendance on the female while she lays, although this *Loboscéliana* sp. grasshopper, laying her eggs in a sandy path in the Shimba Hills in Kenya, is on her own.

▼ Female mantids secrete a liquid from their abdominal glands which foams on contact with the air. This is then shaped to form a series of receptacles to receive the eggs, the foam gradually hardening and forming a protective envelope. Illustrated is a female *Sphodromantis viridis* from Israel who is in the process of constructing her ootheca.

herbage or, occasionally, in plant tissues. During laying, acridid eggs are covered in a layer of foam which hardens as it dries, giving a degree of protection from desiccation, predators and parasites.

The grasshopper *Zonocerus variegatus* is an important pest of the food-crop cassava in Nigeria. Mass oviposition takes place diurnally under the shade of trees and bushes adjacent to the cassava fields (McCaffery & Page, 1982). Gravid females approach these sites from downwind, apparently drawn by the pheromones produced by the males who are already present. Mating takes place and the females then begin probing the ground as close to the base of a suitable bush or tree as possible. The females oviposit in the ground, usually in the late morning, the peaks of oviposition taking place after heavy rainfall.

The oviposition behaviour of grasshoppers, which lay their eggs inside plant tissues (i.e. endophytically), seems to be very similar in those species for which it has been described in any detail, that is, the North American species *Leptysma marginicollis marginicollis* and *Stenacris vitreipennis* and the Costa Rican tropical-forest species *Microtylopteryx hebardi* (Braker, 1989). While the American species lay their eggs in culms of the giant bulrush, *Scirpus californicus*, the Costa Rican grasshopper lays its eggs in the leaf rachis and sometimes in the stems of up to four of 15 usual food plants. Oviposition was observed in caged individuals and takes place as follows. The female grasshopper climbs on to the underside of a leaf and makes her way to its rachis where she walks up and down, apparently testing it with her antennae. Having found a suitable site, she then bites a hole into the rachis tissue and enlarges it until it is roughly the diameter of the end of her abdomen. She then moves to face the tip of the leaf and inserts the tip of her abdomen into the hole she has cut. She then bores her way down the rachis, using the serrated valves of her ovipositor, in a manner much the same as that employed by locusts and other grasshoppers when laying in the ground. She lays her eggs in a pink foam which she also uses to cap the hole in the rachis once laying is completed. Finally, she bites off small pieces of rachis and pushes these into the foam plug, effectivly camouflaging it. No such attempt to camouflage the oviposition site has been reported for the two North American species.

PHASMATODEA Stick insects or walkingsticks

Stick insects and leaf insects show three basic behaviour patterns in relation to oviposition. The more primitive species produce elongated eggs which are laid in the ground. Eggs of less primitive species are elongated to a more or less spherical shape and are glued to some substrate, such as the surface of plants. The most advanced species lay spherical eggs which are either dropped to the ground or are thrown, with an active flick of the abdomen, up to several metres away from the female. The advantage of the latter behaviour, which is found in the well-known laboratory stick insect from Australia, *Extatosoma tiaratum*, is that it separates the eggs from the faeces, which merely fall to the ground beneath the food plant. This means that egg-predators, or parasites which might home in on faecal odour, will not also find the eggs (Carlberg, 1989).

ODONATA Damselflies, dragonflies

The oviposition behaviour of the Odonata is often separated into a number of categories, although quite a number of species show variations both within and between them. The basic oviposition behaviours are as follows:

(1) The female submerges and lays her eggs in plants beneath the water.

(2) The female sits and oviposits into mud or amongst mosses, dead wood etc. at the water's edge.

(3) The female sits and oviposits directly into the water.

(4) The female oviposits into mud or sand while in flight.

(5) The female oviposits into water while in flight.

(6) Non-contact oviposition occurs while in flight.

Most of these encompass no particularly complex behaviours, other than the choosing of a suitable oviposition site. What follows are some examples of more involved behaviours.

The North American damselfly *Enallagma hageni* is one of a number of species whose females submerge themselves completely before inserting their eggs into plant tissues (Fincke, 1986). The male remains in tandem with the female while she is still above the water surface and, during this time, she may make a few experimental probes into emergent stems, occasionally laying a few eggs. Finally, however, all females submerge in the same manner. First they back down the plant stem until the thorax contacts the water, whereupon they curve round the stem and walk head-first down into the water; at this point the males let go of them. Eggs are laid into the stems of sedges and pondweeds, usually at depths of between 5 and 22 cm, the female moving under water to another stem or plant once the first is full of eggs. The eggs are laid in batches of between 10 and 30, partially encircling the stems. Once laying is complete, the female, buoyant with air trapped beneath her wings and in her air sacs, lets go, and rises to and breaks through the water surface, from which she can take flight. Females of this species seldom walk up the plant stem and out of the water.

► **A female migrant hawker dragonfly *Aeshna mixta* from Europe laying her eggs in rotten wood at the side of a pond.**

There is some evidence to indicate that the oviposition behaviour may alter with changes in the nature of the plant into which the eggs are being laid. For example the African tropical-forest damselfly *Phaon iridipennis* most commonly inserts one egg per cut into the leaf-like bracts of the sedge *Cyperus involucratus*. If, however, as it less often does, oviposition occurs in the stem just below the water surface, the eggs are laid in a single row, up to 90 at a time, through a single long slit (Miller & Miller, 1988). These two distinct behaviours possibly result from the fact that the force required to penetrate the *Cyperus* stems is 2.38 times that required to penetrate the bracts.

The widespread libellulid *Tholymis tillarga*, which is found throughout much of Africa, the Far East and Australia, oviposits on to the surface of water plants while in flight. This is one of the dragonflies which may be guarded by a male during egg-laying (Miller & Miller, 1985). The female has a structure on the underside of the ninth abdominal segment, the egg basket, in which the egg batch is probably prepared prior to oviposition. The female swoops down above a horizontal floating leaf, strikes the water surface with the tip of her abdomen and then deposits the egg batch on the leaf surface. When guarded by a male, she usually oviposits on the same leaf but, if unguarded, she may deposit her eggs on several different leaves.

Although most libellulids oviposit by dipping the tip of the abdomen into the water and washing the eggs off, the Mexican species *Dythemis cannacrioides* is an interesting exception (Gonzalez Soriano, 1986). The eggs are laid on to the roots of lianas and other plants which hang down into the water of rivers and streams. After copulation, the female of this dragonfly perches for a short time before ovipositing. During this short period of time, she prepares a large egg mass, which bears some resemblance to a bunch of grapes. Once the egg mass is ready, she flies down and deposits it, in flight, on the roots where they enter the water. The filaments which hold the egg mass together are sticky and this helps it to stay entangled in the roots.

The female of the African libellulid *Zygonyx natalensis* shows some plasticity in her oviposition behaviour; this is an adaptation to the limitations and unpredictability of usable oviposition sites and the high level of intraspecific competition between males (Martens, 1991). This species frequents areas of rivers where there are waterfalls and rapids, and consequently, it relies upon current-washed mats of plant roots as oviposition sites. Males and females remain in tandem after copulating while a suitable oviposition site is sought. The female usually then separates from the male, settles on plant stalks or roots and lays her eggs, which are washed away in the current. The gelatinous secretion which covers the eggs swells up very rapidly on contact with the water and the eggs then stick to any soil or roots that they contact. Alternatively, the female *Z. natalensis* may either assume a sitting position and deposit her sticky eggs directly on to the substrate or may lay them straight into the main water current while dipping in flight.

TRICHOPTERA Caddisflies

Female caddisflies normally lay their sticky masses or strings of eggs on objects close to or in the water in which the larvae will develop. In a North American sericostomatid caddisfly, oviposition, of a more sophisticated nature has been observed to take place at dusk (Wood *et al.*, 1982). The female first extrudes her egg mass and carries it at the tip of the abdomen. She then flies over an appropriate stretch of water, dips the end of her abdomen on to the surface and releases the egg mass.

LEPIDOPTERA Butterflies, moths

The choosing of a suitable oviposition site by a female lepidopteran takes place in two stages. In flight, she must search for the correct food plant, or plants, or other food source on which to land, and then, having alighted, she must check that it is perfectly suitable for the development of her larvae. In making her in-flight choice of a suitable plant she may use both visual and olfactory cues, while, after landing, its suitability as a food may be tested using visual, olfactory, gustatory and tactile cues. Since suitable food plants may be few and far between, the female butterfly may have to adopt a strict foraging pattern in order to ensure that she is as efficient as possible at finding them. In other instances, food plants may be more abundant but of more than one kind and she may then have to make a choice as to which one she uses as an oviposition site. Clearly, precise details of butterfly host-plant selection and oviposition behaviour in the field require many hours of careful observation and, even then, interpretations of what has been seen can contain errors. Despite these problems, some very useful work has been carried out in the field to give us an insight into the way in which a number of different species of butterfly seek out plants on which to lay their eggs. If a female butterfly does choose one plant in preference to another then she has exhibited selection and, under these circumstances, it would be expected that a particular plant would be used as an oviposition site more often than dictated by its relative abundance.

BUTTERFLIES

In a series of laboratory experiments to increase our knowledge of how butterflies choose oviposition sites, females of *Pieris brassicae*, a cluster-layer, were given simultaneous choices of two different cues (Rothschild & Schoonhoven 1977). In one experiment, for example, they were offered a choice of two similar clean cabbage leaves: one with eggs already on it, one without. In another experiment they were offered two further similar leaves: one with feeding caterpillars, one without. These experiments revealed that *P. brassicae* females lay fewer eggs on:

(1) Leaves which already bear batches of their own eggs.

(2) Leaves bearing batches of model eggs.

(3) Leaves wiped with crushed egg contents.

(4) Leaves bearing fluids from the tips of female abdomens.

(5) Leaves bearing feeding caterpillars.

(6) Leaves which have once borne batches of eggs, since removed.

► It is usual for butterflies to lay their eggs either singly or in batches, on the leaves or stems of the caterpillar food plant, having first carefully examined it for its suitability. In the process of laying a batch of eggs on a vine in Trinidad is a female king cracker butterfly, *Hamadryas amphinome* (Nymphalidae).

It was shown by their response to the yellow-painted model eggs, that these butterflies respond to visual cues and that, if these eggs were painted green, they had a less inhibitory effect on the numbers of eggs laid. The importance of air-borne cues was indicated by the way in which recently damaged leaves, or leaves smeared with crushed leaf extract, also inhibited oviposition. The reaction of the butterflies to various extracts placed on the leaf, and to the leaves from which eggs had been removed, all of which were explored, reveals that chemoreception also plays a part in oviposition-site choice. This inhibition, caused by leaves from which eggs had been removed, indicates that the response may be due to the presence of pheromones purposefully laid down with the eggs to reduce further oviposition, a behaviour found in other insect families. Experiments with two species which lay single eggs, the small white, *Pieris rapae*, and the monarch, *Danaus plexippus*, revealed that they respond to much the same cues as *P. brassicae*. One important difference between the three of them is the way in which they test the host-leaf surface: *P. brassicae* uses her tarsi, *P. rapae* the tip of her abdomen and *D. plexippus* her antennae. Assuming all butterflies put all or some of these host-appraisal mechanisms into practice, how does this relate to choosing a suitable larval food plant in practice?

Like *Pieris brassicae* and *P. rapae*, the green-veined white, *P. napi*, which is distributed across the temperate regions of the northern hemisphere, uses various Cruciferae as larval food plants. Oviposition behaviour of *P. napi macdunnoughii*, and also *P. occidentalis*, was studied at two sites in the Rocky Mountains in North America (Chew, 1977) with the following results. Females of the two *Pieris* species oviposit only on members of the Cruciferae but some species are ignored, in particular, though not always, those whose leaves are lethal to the butterfly caterpillars. The distribution of the eggs on the various plant species tended to reflect their suitability as caterpillar food plants. There is a tendency on consecutive egg-laying sessions for females to choose plants of the same species, though over the whole of their life they probably oviposit on all of the species of crucifer used by the population as a whole. Choice of oviposition sites does not appear to be perfect for, at one of the sites, *P. occidentalis* laid more than half its eggs on *Thlaspi arvense*, which is lethal to its caterpillars. There is, however, a plausible explanation for this. *T. arvense* is probably not native to the area and the butterflies have come into contact with it only in relatively recent times. The tendency, however, of a number of females to lay a substantial proportion of their eggs on plants with rosettes too small to feed the caterpillars is less easily accounted for.

Research on another pierid, the orange-tip butterfly, *Anthocharis cardamines*, seems to indicate that female butterflies are not always perfectly adapted in their choice of a larval host plant. The eggs of this species have been recorded from as many as 33 different species of crucifer. Eggs are laid on the flowering head of the host plant and the larvae feed there upon the developing fruits. Females find the plants visually by the presence of the flowers, from which they and the males feed. At a number of riverside sites in northern England, it was found (Courtney, 1981) that, despite the ample presence of the butterfly's two most ideal food plants, in terms of larval development, a proportionately greater number of eggs were laid on the host plants which yielded the lowest larval survival.

In the USA, the pierid sulphur butterfly, *Colias philodice eriphyle*, can be a pest on the legume alfalfa, though in the wild it will oviposit on a number of other leguminous plants. In a study carried out in Colorado (Stanton, 1982), where this butterfly is bivoltine, several legumes are available as oviposition sites. Within the study area, which is several hectares in extent, grow *Vicia americana*, *Trifolium hybridum*, *T. longipes*, *Lathyrus leucanthus*, *Astragalus decumbens* and several *Lupinus* spp. It was found that, within the area, females of *C. p. eriphyle* are most likely to lay eggs after landing on three of the above legumes: *V. americana*, *T. hybridum* and *T. longipes*. Even on these plants, however, oviposition takes place on only 50 per cent of the visits made. The investigation revealed that these plants are not chosen entirely at random but, instead, the females have particular flight paths along which the plants are over-represented when compared with the area as a whole.

Thus it would appear that host-plant finding may be subdivided into an initial search for a locality in which it grows followed by a search for the plant within the locality. *C. p. eriphyle* females consistently fly upwind in their search for oviposition sites, and it may well be that they are orienting to volatile substances emanating from the legumes and being carried some distance. Once a female finds an area rich in potential oviposition sites, the length of her flights becomes shorter, which of course increases her exposure to the legumes within that area but decreases the likelihood of her finding new patches of preferred food plants. Female *C. p. eriphyle* are apparently not perfect in their recognition of these preferred food plants. In the study area, they tended to confuse their preferred *V. americana* with the non-preferred *L. leucanthus* and *A. decumbens* in that they often landed on these species but seldom oviposited on them. Early in the season it is not easy to distinguish the three species since they all have similar leaves and, if the female butterflies use leaf shape as a cue for landing on the plant, this may well explain the errors they make. In 1978, first-brood females landed as readily upon young plants of *L. leucanthus* as they did upon those of *V. americana*. Late in 1978, as the leaves of *V. americana* dried up, so the females landed but did not oviposit upon the leaves of *A. decumbens*. It may be concluded, therefore, that females of *C. p. eriphyle* land selectively upon leguminous plants but the final identification as to whether they are suitable for oviposition is made post-landing.

In another bivoltine North American butterfly, the pipevine swallowtail, *Battus philenor*, host abundance plays a part in the choice of oviposition sites but so also does juvenile survival. In east Texas, where a study on its oviposition habits was carried out (Rausher, 1980), it uses two host plants on which to lay its eggs, both members of the Aristolochiaceae. They can be distinguished visually in that *Aristolochia reticulata* has broad, ovate leaves whereas *A. serpentaria* has long, narrow leaves. One very important aspect of the life cycle of this butterfly is that a single plant of either species seldom bears enough edible leaves to support the growth and development of even a single caterpillar. Once a caterpillar has eaten the available foliage from one plant, it will then seek another of either species to continue feeding, irrespective of the species on which it hatched. It may in fact have to feed on as many as 25 different plants in order to obtain sufficient food to complete its development. Mortality tends to be high while the larvae are seeking new hosts, with the smaller caterpillars suffering the most.

From this, it seems obvious that the female *B. philenor* not only has to choose the right host plant but also has to judge its abundance and, in the east Texas study, this proved to be the case. When females of the spring brood are on the wing, *A. reticulata*, which is by far the most common species of host plant, is the most common oviposition site so that a broad-leaved visual cue seems to satisfy their needs. Interestingly, however, when they do land by accident or design on a plant of *A. serpentaria*, they are more likely to oviposit on it than on the other species. The behaviour of the second summer-brood females is the opposite to that of the first brood. In these second-brood individuals, the search image is that of a narrow leaf and *A. serpentaria* is the preferred oviposition site. The reason for this relates to the state of the plants in summer. The caterpillars of *B. philenor* can only feed on the young leaves of *A. reticulata*, which are few and far between in the summer; thus, although this species is even more abundant than *A. serpentaria* at this time, it is no longer the preferred food plant. Caterpillars can feed on *A. serpentaria* leaves at all times of the year, which would account for this plant being the second-brood females' preferred oviposition site. This change of preference brings another advantage, for although survivability of the spring caterpillars is roughly the same on each host-plant species, in the summer, caterpillars on *A. serpentaria* have higher survival rates than those on *A. reticulata*.

When this study was later extended (Papaj & Rausher, 1987), it was discovered that *B. philenor* are able to discriminate between host plants differing in quality both before and after they have landed. Host plants on which females land and oviposit are smaller, have longer buds and a higher proportion of high-quality leaves than those plants on which the females land but do not oviposit. Additionally, such host plants are not themselves a random sample of the plants available in the habitat; they have more good-quality leaves, fewer low-quality leaves and longer buds than adjacent plants which are not visited. In a similar study (Damman & Feeny, 1988), the female zebra swallowtail butterfly, *Eurytides marcellus*, was found in the field to be able to distinguish the characteristics of its food plant and its suitability as larval food either before or after landing on it. She lays preferentially on pawpaws (*Asimina* spp.) with young leaves and rejects those with older leaves and also non-host plants.

In a study of sympatric troidine swallowtails in Brazil (Brown *et al.*, 1980), differences in oviposition behaviour were noted between females of both *Parides* spp. and *Battus polydamas*. Female *Parides* lay single eggs on the undersides of *Aristolochia* leaves while *B. polydamas* lays small batches of eggs on the apices of the *Aristolochia* vines. Each species of butterfly prefers its own particular range of the *Aristolochia* species available and this, coupled with a choice of a drier or damper, open or enclosed habitat and high- or low-growing plants, helps to reduce competition between the species to a minimum. Before laying their eggs on a particular leaf, these butterflies appear to inspect the leaf both visually and by palpating it with their front tarsi. Two Asian pierid butterflies, *Colotis amatus* and *C. vestalis*, are not only sympatric but use the same larval food plants, *Salvadora persica* and *S. oleoides*, in New Delhi. Competition between them is minimal as a result of their differing oviposition behaviours (Larsen, 1988). Females of *C. amatus* lay about 30 eggs at a time, in batches, on the upper surface of leaves at the outer extremities of the host bush or tree. *C. vestalis*, on the other hand, lays her eggs singly, on twigs or branches deep inside the plant, often some distance from the nearest leaves. Caterpillars of *C. amatus* feed only on young leaves while those of *C. vestalis* feed only on mature leaves and the researcher noted that he had never observed both species consuming the same leaf.

Female butterflies may use cues other than those produced by the food plant in order to decide the exact site for oviposition. The American tiger swallowtail butterfly, *Papilio glaucus*, for example, can make use of more than 20 different tree species from 13 different families as host plants on which its caterpillars can feed. Whichever host it choses, it oviposits on the tips of branches which are exposed to sunlight, almost exclusively at heights <3 m. In a study in the North American state of New Jersey, 50 per cent of observed eggs were laid on the west-facing sides of trees, with 25.4 per cent on the east-facing and a mere 7.6 per cent on the north-facing sides (Grossmueller & Lederhouse, 1985). By laying eggs on the sunny sides of the tree, the caterpillar development rate was increased by 15–35 per cent over those in shaded positions. This particular oviposition strategy allows the species to produce two broods in the northern part of its range, where the number of daily hours at a temperature suitable to maintain growth are often not sufficient for the development of the second brood.

Ants as oviposition stimulators

Since as many as one-third of all lycaenid butterflies have relationships with ants, one of the cues that the female uses in choosing an oviposition site is the presence of these associates. It has been shown, for example, that females of the Australian lycaenid *Jalmenus evagoras* are more likely to choose a food plant bearing the ants of the genus *Iridomyrmex* than one without ants (Pierce & Elgar, 1985). Egg-batch sizes, however, are the same whether or not the ants are present. Before settling on a potential food source, gravid females respond to the presence or absence of the ants but, once they have landed, they lay eggs whether the latter are present or not. They respond even more strongly to plants on which the ants are not only present but also are tending homopterans. Indeed the presence of homopterans as well as ants may induce *J. evagoras* to oviposit on a plant which is not the normal food plant. One of the apparent consequences of this behaviour is that lycaenids that apparently use ants as cues for oviposition sites choose a far greater range of larval food plants than non-myrmecophilous lycaenids. *J. evagoras* females, for example, were seen to choose several new species of *Acacia* as oviposition sites during one season of study but, in each case, they chose only those plants which already played host to ants and membracids. The caterpillars apparently had no difficulty in accepting the alternative food sources. Females of a second Australian lycaenid, *Ogyris amaryllis*, will also only oviposit on their mistletoe host in the presence of *Iridomyrmex* ants. Their choice of a suitable oviposition site is, however, complicated by the fact that their major host plants, *Amyema* spp., often grow adjacent to a second mistletoe of the genus *Lysiana*, whose leaves are toxic to the butterflies' caterpillars, but also host *Iridomyrmex* ants. During a field investigation (Atsatt, 1981) near Alice Springs in the Northern Territory, it was discovered that, in a very high percentage of cases, female *O. amaryllis* were able to distinguish between the two mistletoes and detect the presence of the ants, so that the majority of the eggs laid were on *A. maidenii* populated by *Iridomyrmex* ants tending homopteran bugs. The act of oviposition in this butterfly normally takes place following tactile stimulation by the ants.

The caterpillars of the southern African lycaenid *Erikssonia acraeina* shelter in the nests of the ant *Acantholepis* during the

hours of daylight and then emerge at night to feed on their host plant, *Gnidia kraussiana*. In common with most butterflies, the female flies around looking for host plants, landing on those which appear suitable and investigating them with her antennae. If she lands on a host plant, she uses her antennae to detect traces of the trail pheromones of her host ant. If these are present, she then descends to the ground beneath the plant and begins to oviposit between soil particles or beneath leaf debris (Henning, 1984). A similar oviposition pattern had earlier (Henning, 1983) been shown to be true for the southern African lycaenid *Aloeides dentatis*. Females of this species seek out their food plant, *Hermania depressa*, and if it is close to a nest of their host ant, *Acantholepis capensis*, they will oviposit on its leaves. In a laboratory test conducted alongside the field observations, captive lycaenid females failed to lay eggs on host plants unless ants were present. The North American lycaenid *Satyrium edwardsii* again has caterpillars that are ant-protected. In this case, females lay eggs in old wounds in the bark or beneath loose bark of their host plant, *Quercus velutina* (Webster & Nielsen, 1984). Females of the Sulawesi lycaenid *Allotinus major* deposit their eggs either on the brooding females or on the food plant of the membracid bug whose nymphs form the prey of its caterpillars (Kitching, 1987).

MOTHS

The life cycle of the pyralid moth *Cryptoses choloepi*, the Panamanian sloth moth, is inextricably linked to those of two-toed and three-toed sloths. Both male and female moths spend their lives living amongst the fur of their host. When the sloth descends to the ground to defaecate at the base of a tree, the female moths leave it and fly down on to the fresh dung where they lay their eggs. The caterpillars are coprophagic and development takes several weeks, after which newly hatched adults fly into the canopy to find sloth hosts and begin the life cycle once again (Waage & Montgomery, 1976).

The saturniid moth *Hylesia lineata* is found along the Pacific side of Central America and its caterpillars have been

◀ It is not uncommon for moths to oviposit in rings around twigs and stems on the plant on which their caterpillars will feed. This female *Thyretes negus* moth is in the process of laying a large batch of eggs around a grass stem in Kenya (Arctiidae: Ctenuchiinae).

found feeding on at least 46 plant species from 17 different families. Individual females, however, prefer a smaller number of species on which to oviposit (Janzen, 1984). Females construct egg nests and, in the dry season, they always choose a dry twig 1–4 mm in diameter as an oviposition site, whereas in the wet season 2–4 m diameter leaf petioles may also be utilized. Nest sites tend to be chosen in the more open parts of the forest along streamsides, roadsides, sites of tree falls etc., these presumably being the most favourable areas for growth of the host plants. Having chosen a twig or petiole on which to oviposit, the female moth hangs from it and curls her abdomen upwards until she contacts its surface. She then lays all her eggs at one go, stuck together in a single mass, at the same time twisting her abdomen round both them and the substrate so that long, loose abdominal hairs break free and form a dense, felt-like egg covering. Simultaneously, short urticating hairs are mixed in with the long hairs. As nest-building nears completion, the female may let go of the substrate and twist herself around in the air, suspended from a thin string of her own hairs; this apparently has the affect of tightening the felt covering around the egg mass. Occasionally, she may be unable to break free from her connection with the egg nest and she may thus die *in situ*; this is no great loss, however, for all females die within 4 days of becoming fully gravid, whether or not they have laid their eggs.

The life cycles of the yucca moths are inextricably linked with the life cycles of their host plants, the yuccas, in North America and Mexico. The prodoxid moth *Tegeticula maculata* oviposits on *Yucca whipplei* (Agavaceae) in southern California and the Baja peninsula and a careful study has been made of the relationship between the two organisms (Aker & Udovic, 1981). Adult moths emerge from their overwintering sites throughout the the time when host flowers are available. Following mating, the females undergo a pre-oviposition rest period inside a flower, during which they resist any further attempts at mating.

Collection of pollen by the female moth can take place both before and after mating but, whenever it takes place, once she has her pollen mass, she disperses away from the plant on which it was collected. These dispersal flights involve the moths flying high above the surrounding vegetation, usually downwind, ignoring flowering yuccas adjacent to the one they have just left. The females use their highly modified maxillary tentacles to remove pollinia directly from the anthers. Once sufficient numbers of pollinia have been accumulated, the female packs them into a ball which she then carries around held against the thorax with the mouthparts.

The female moths are very choosy about their oviposition sites and they may crawl in and out of a number of different flowers before they find one that is suitable. The main choice seems to be freshly opened flowers whose anthers have not yet dehisced. Normally, they palpate the ovary wall with their antennae before inserting the ovipositor into it. Often these insertions are very brief and they are not followed by pollination. It seems therefore that the females are simply making a further check on the suitability of the ovary for oviposition. At a chosen egg-laying site, the ovipositor is inserted for much longer, roughly 34 minutes on average. The fact that there is rarely more than one caterpillar in a single seed row in the ovary indicates that only a single egg is laid. Not long before she withdraws the ovipositor, the female moth begins to move her maxillary tentacles over the pollen ball. Then, immediately after withdrawing the ovipositor, she either moves to the top of the pistil and draws her tentacles back and forth over the stigma to effect pollination or she leaves without pollinating. Since a single yucca-flower ovary is made up of six locules, each containing a row of seeds which can support a single moth caterpillar, most moths will attempt to oviposit more than once in a flower. Following a second oviposition in a flower, the moth is very much less likely to pollinate it again.

During the final 2 weeks of the yucca's flowering season, the females of *T. maculata* show a change in oviposition behaviour in that they may be found ovipositing on the developing pods of inflorescences on which most of the flowers have already begun to wilt and are thus unattractive to the moths. The presence of at least some open flowers is necessary to attract the moths to the pods, for none are seen on inflorescences whose flowers have already died. The females still carry pollen but do not attempt to pollinate the wilting stigmas. This behaviour seemingly relates to a recognition by the female moths that the very last flowers on the upper part of the inflorescence have a very much lower chance of forming pods, even when pollinated. It is therefore more advantageous for them to oviposit in an already developing pod lower down the inflorescence.

COLEOPTERA Beetles

Females of many carabid beetles oviposit into the ground, with some species burying themselves completely before commencing to lay their eggs. *Abax ater* and some other species of the genus *Abax* oviposit into a specially constructed soil sheath. The female uses the styli on the end of her extendable apical abdominal segments to scrape off particles of moist clay. This is then used to form a thin covering over the end of the abdomen. She then presses the abdomen against the ground and deposits a single egg into this earthen sheath. As she withdraws her abdomen, the dorsal part of the sheath closes down on to its lower half; she then fills any remaining gap with horizontal side-to-side movements of her rear end. The female of the closely related carabid *Percus navaricus*, on the other hand, uses her mandibles to mould a bowl-shaped mass of soil which she then raises to her abdomen with her legs for oviposition to take place. This structure is then either left in position or attached to a piece of earth before being provided with an earthen lid. *Agonum dorsale* attaches its earth-covered eggs to the underside of leaves. In the North American species *Tecnophilus croceicollis*, the female produces an adhesive substance from accessory glands which sticks together particles of soil loosened by the tip of the abdomen. Once enough particles have adhered to the abdominal apex she climbs on to, for example, a twig, presses her abdomen against it and then releases a drop of fluid. She now pulls her abdomen away from the twig, drawing out a thread of the fluid which then hardens to form a silk-like strand. Finally, she deposits an egg into the ball of soil particles which is then released and left dangling from its silk thread.

The females of most lampyrids oviposit into the soil in sites which are favourable for the development of their predacious larvae. The larva of the Japanese lampyrid *Luciola cruciata* is also a crawling predator but is an inhabitant of flowing water. Oviposition in this species occurs nocturnally and, most unusually, gregariously (Yuma & Hori, 1981). Three criteria have to be met for an oviposition site to be suitable: firstly it must be on the underside of an overhanging substrate, secondly it must be above water, and thirdly it must be covered in mosses. Gregariousness results from the initial discovery of such a site by a single female who then glows as she oviposits. This glowing attracts further females who also glow and so on.

Females of many species of cerambycid beetles choose oviposition sites that need little or no preparation, for example, cracks or crevices in bark or injured areas of the host plant. In some instances, the females may make a minimal preparation of the oviposition site, for example those of *Saperda inornata* gnaw a shield- or horseshoe-shaped egg niche in the outer bark of *Populus tremuloides* in which usually a single egg is laid. Females of the subfamily Lamiinae, however, go to a lot more trouble to prepare the host plant to receive their eggs.

From southern Texas and northeast Mexico comes *Oncideres pustulatus* which uses various species of leguminous trees as hosts (Rice, 1989). Mated female beetles girdle suitable branches of host trees by using their strong mandibles to chew through the bark and down to the heartwood. This has the effect of killing the branch although the initial bite is not deep enough to sever it. The female then prepares oviposition sites by inserting her mandibles deeply into the bark of the upper regions of the branch and, by working them back and forth, removing a chip of wood at each site. She then turns round, inserts her ovipositor into the slit and lays a single egg beneath the bark. Once one female has started to girdle a branch, other females may oviposit upon it without assisting in the girdling process. The larger the branch prepared, and the greater the number of lateral offshoots, the greater the number of eggs deposited on it.

The American aspen beetle, *Gonioctena americana*, is a viviparous chrysomelid whose larvae feed on leaves of the trembling aspen, *Populus tremuloides*. A gravid female positions herself with the end of her abdomen towards the tip of the lower surface of the leaf, holding on to the leaf edges with her tarsi. She then places the tip of her abdomen against the leaf surface and ejects the larva tail-first, raising her abdomen as she does so with the result that the larva stands temporarily perpendicular to the leaf and attached to it by its rear end (Mason & Lawson, 1982).

▲ **Like many other weevils, which perform the same task, this giraffe-necked weevil, *Trachelophorus giraffa*, from the rainforest of Madagascar is rolling up a leaf in which she will later oviposit.**

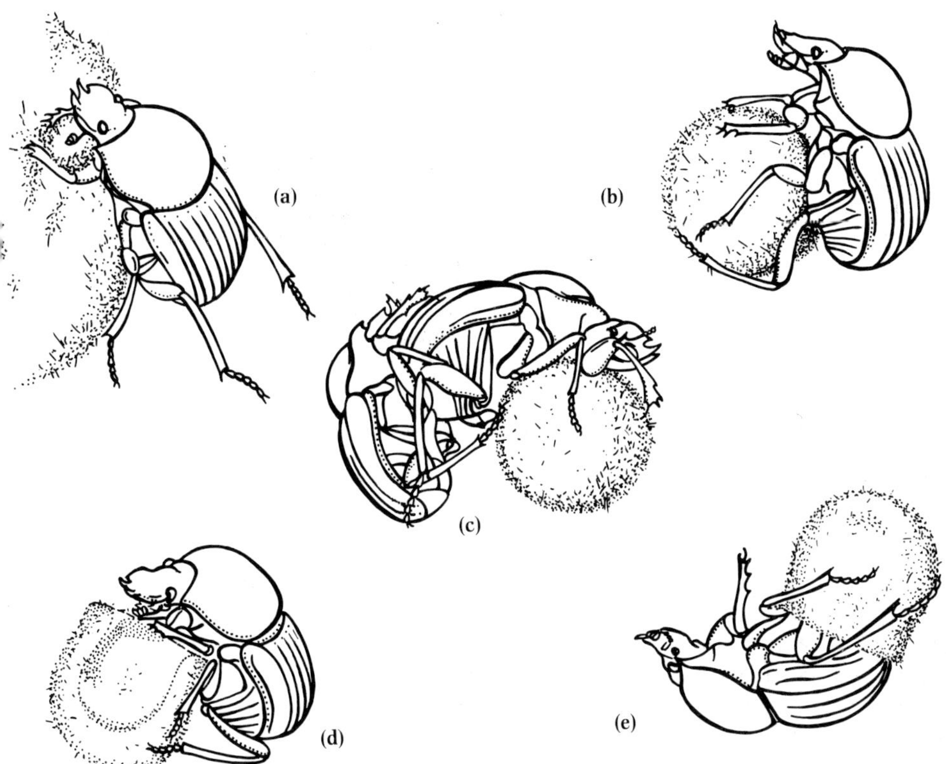

► **The scarabaeid *Cephalodesmius armiger* does not utilize real dung to feed her offspring but instead she employs synthetic 'dung' made from processed plant material. Initially (a) she tears a portion off the prepared brood mass and, assuming a sitting position (b), she forms it into a smooth ball. Whilst she is carrying out this process the male approaches and mates with her (c). Using her front legs, she now fashions the ball of brood material into a cup (d) then, lying on her back, she inserts her abdomen into it (e) and lays a single egg. Finally she closes off the opening into the cup and re-forms the material into a sphere.** (After Monteith & Storey, 1981).

The females of a number of weevils of the subfamily Rhynchitinae oviposit in leaf rolls of their own manufacture, which provide both protection and a source of food for their larvae. *Byctiscus populi* in Europe, for example, rolls leaves of its main host plant, the aspen *Populus tremula*, and occasionally those of other species (Daanje, 1975). A female who is ready to lay her eggs first searches out a suitable young aspen leaf. She reaches the leaf blade via the petiole, makes her way up one edge of the leaf and, about half-way, bites a lump out of the upper side and eats it. If the leaf tastes wrong, i.e. it is the wrong food plant or the leaf is too old, she leaves it straight away; otherwise she begins to roll the leaf.

First she makes her way back to the petiole, walking on the leaf blade in a manner which causes the tarsal claws to be driven into the leaf surface and often biting small holes in the outer leaf cells with her mandibles: so-called 'perforating' behaviour. Back at the petiole she straddles it and, clinging firmly to it with her legs, bores a hole into it, leaving only a thin layer of cells linking the leaf blade to the portion of the petiole connected to the stem. She now moves back on to the leaf blade and, starting close to one edge, begins perforating again. Perforation can take place on both under and upper surfaces or on one surface alone; she also makes cuts in the secondary

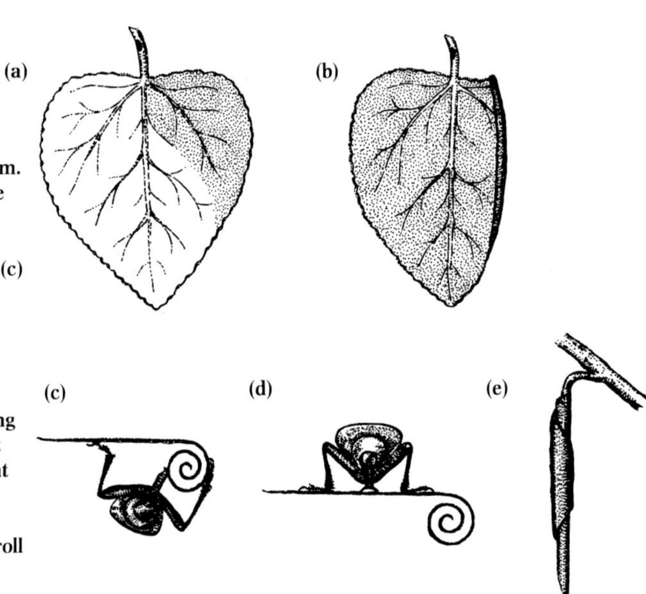

▶ **Leaf-rolling by the weevil *Byctiscus populi*. (a) A leaf showing the hole bitten in the petiole and an area on the right-hand side which has been punctured by the weevil's rostrum. Both processes serve to make the leaf more flaccid and therefore more easy to roll. (b) A fully punctured, partly rolled leaf. (c) The stance used by the female weevil while rolling the leaf. She anchors her left legs on the leaf surface, pulling her tarsal claws hard into it, and pulls the leaf across with her right legs, tucking the leaf in with her rostrum as it rolls. (d) A leaf which is reluctant to roll may require further puncturing with the female's rostrum. (e) The completed leaf roll into which the weevil will have oviposited.** (After Daanje, 1975).

veins. The overall effect of this behaviour is to reduce the turgidity of the cells in the perforated area, making the leaf more flexible and thus easier to roll.

Following perforation, the female now moves to one edge of the leaf and, from the underside, begins to roll it parallel to the main vein. The rolling action is performed by holding on to the undersurface of the leaf with the legs on one side, reaching over with the legs on the other side, hooking the tarsal claws in and pulling the edge of the leaf towards the body, tucking it in with the snout as the leaf begins to curl inwards. If rolling proves impossible at first, the leaf may have to be perforated further or the hole in the petiole may have to be enlarged, both actions further reducing the leaf's turgidity and resistance to rolling. Having rolled the leaf a little, she now has to reposition herself, which causes the leaf to uncurl somewhat, so further perforation is carried out on the upper surface, parallel and close to the rolled section, before further rolling commences. By a continuation of all these processes, she eventually rolls a portion of the leaf into a fairly tight cylinder.

She now crawls a short distance into the roll, positions herself longitudinally within it and bites a groove in the leaf coil, pushing her snout into the cut to enlarge it and make sufficient space for an egg. The *B. populi* female then turns through 180°, pushes her abdomen into the slit and lays her first egg. Usually the egg is positioned about half-way along the rolled section of leaf. She now rolls the leaf some more before laying another egg and, if the leaf is large enough, three eggs may be laid in it, each in its separate hole.

In all species of weevils from the subfamily *Attelabinae*, the females lay their eggs in leaf rolls. The technique employed by the Japanese species *Apoderus balteatus* on leaves of its host plant, *Stachyurus praecox* (Flacourtiaceae), is somewhat different from that of *B. populi* (Sakurai, 1986). For example, the leaf is first folded in half along the mid-rib and only then is it rolled up, but at 90° to the mid-rib, starting at the leaf tip. As with *B. populi*, eggs are laid at intervals as the leaf is rolled and, the bigger the leaf, the greater the number of eggs that are laid. Most cradles contain one or two eggs, rarely three. Unlike *B. populi*, *A. balteatus* produces two types of cradle, one of which remains suspended by the leaf petiole while the other falls to the floor below the plant.

Weevils which do not roll leaves use their rostrum to cut a hole into structures such as twigs, buds, fruits and seeds and then oviposit into them. The female of the Neotropical curculionid bottle-brush weevil, *Rhinotomus barbirostris*, for example, oviposits into fallen palm trunks into which the larvae bore and feed (Eberhard, 1983). The female searches a log for a suitable oviposition site, usually a natural depression in the bark, by walking slowly along and palpating the surface with her antennae. As she does so, she swings her head from side to side, occasionally stopping to examine a particular area more carefully. At an appropriate oviposition site, she stops and touches the area for drilling with her mouthparts, while she repeatedly touches the adjacent area with her antennae. Now, dragging her legs laterally, so that her tarsal claws grip the bark, she brings her rostrum into position and begins to drill. Drilling takes about 5 minutes, after which the female brings her rostrum out of the hole a number of times with a sharp flicking movement, which apparently cleans debris out of the hole. With drilling complete, she now turns round, inserts the tip of her abdomen into the hole and oviposits. She then fills the hole with a creamy-white fluid which soon becomes brittle on hardening, thus forming a protective cap.

Females of the North American scolytid beetle *Ips pini*, the pine engraver, construct an egg gallery prior to oviposition and the behaviour involved has been observed under laboratory conditions (Schmitz, 1972). Construction of the gallery commences

▲ Of the more than 300,000 known species of beetles (Coleoptera) only a handful is known to be viviparous. This *Eugonycha* sp. chrysomelid beetle from Brazil is depositing live young onto a *Solanum* leaf. In this species, the evolution of viviparity could be tied in with its lack of warning coloration in any of its stages. Most insects which associate with plants of the family Solanaceae are warningly coloured, reflecting their ability to handle the plant's defensive toxins and to sequester them for their own use, an ability that this *Eugonycha* apparently lacks. It probably, therefore, needs to exploit its foodplant rapidly at the beginning of the wet season, before toxins in the leaves rise to unacceptably high levels. With no time-wasting gap between egg-laying and hatching of the larvae, viviparity would give the beetle a head start in the race against the plant's defences.

◀ In weevils (Curculionidae) the female uses teeth at the tip of her rostrum to gnaw a hole in the food plant; she then turns and lays an egg in the cavity she has made. This *Pissodes pini* pine-weevil female from Europe is just starting to bite a hole in a fallen pine tree. She is accompanied by her mate, who guards her against intruding males until she is finished. Not all females are thus chaperoned, as oviposition often takes place alone.

immediately after mating has taken place in a specially constructed nuptial chamber. Starting from the chamber, the female bites off lumps of phloem and pushes these under her body and behind her with her legs. The beetles walk in the gallery on the distal ends of the tibiae, with the tarsi of the front legs folded back against the anterior tibial surface and those of the middle and hind legs against their posterior tibial surfaces. They walk in the gallery on a spine located on the distal end of each tibia and these, along with three smaller, adjacent, laterally situated spines are used as rakes to clear the gallery as boring proceeds. Once sufficient debris has accumulated, the female backs it down the gallery, employing her elytral declivity as a scoop, and the male, in a similar way, moves it down into the nuptial chamber and out of the entrance hole to the exterior. At intervals, the female cuts a niche into the side of the gallery, keeping the detritus from this for later use. She then returns to the nuptial chamber, turns round, backs up the gallery and lays an egg in the niche. She returns to the nuptial chamber once more, turns round again and moves head-first back to the egg niche. Here she uses her mandibles to pack the retained phloem fragments around the egg. Eggs are laid on alternate sides of the gallery, which is defended by the occupant if another female attempts to take over.

The colydiid beetle *Lasconotus subcostulatus* lives in association with another North American *Ips* species, *I. paraconfusus*. The female colydiid begins to search for oviposition sites on the second day after mating (Hackwell, 1973). She uses her mandibles to pull frass from the sides of the *Ips'* galleries, and from the egg niches, then turns round and probes the space with her ovipositor. Although the act of laying was not observed, eggs were eventually found carefully concealed in the *Ips'* egg niche. No eggs were found in the open galleries.

HYMENOPTERA Sawflies, wasps, ants, bees

Female wood wasps (Siricidae) have long, stiff ovipositors which they use to drill holes through the bark and into the wood of trees in order to lay their eggs. At the same time, female *Sirex* and *Urocerus* wood wasps inject arthrospores of a symbiotic fungus into the tree, as well as a mucopolysaccharide which conditions living trees to the fungal growth. The wasp larvae feed on

▲ ***Rhysella approximator*** **is a parasitic wasp whose larvae develop in the larvae of the alder wood wasp,** ***Xiphydria camelus*****. The female** ***Rhysella*** **is able to detect the presence of the wood-wasp larva in its tunnel in an alder trunk and she then drills down to it with her long ovipositor. She is able to assess the wood-wasp larva's degree of development with sense organs on the tip of the ovipositor and, if it is sufficiently well grown, she will lay an egg on it.**

wood in the tunnels that they excavate, as well as on the introduced fungus which grows there. The arthrospores are kept in a pair of sacs, called mycangia, situated at the base of the ovipositor, and the mucopolysaccharide is produced by a pair of mucus glands and stored in a large, centrally placed reservoir. For each insertion of the ovipositor through the bark, female *Sirex* sp. make one to four drills, the more drills per group the greater the number of eggs laid per drill. Some drills in a group contain no eggs, in which case they always contain more of the fungus than those containing eggs. Drills of *Urocerus* sp. are longer than those of *Sirex* sp. and are only single, each containing several eggs with fungus deposited between each of them (Spradbery, 1977).

Ichneumonids of the genus *Rhyssa* are parasites of siricid wood wasps. *Rhyssa persuasoria* is a widespread parasite of *Sirex* sp. in Europe and its oviposition behaviour has been studied in detail (Spradbery, 1970). The first thing that the female ichneumonid does on a suitable tree is to carry out a preliminary investigation in which she walks over the bark, moving her head from side to side and tapping it with the tips of her antennae. In a particular area of interest, she may stop and rapidly tap her antennal tips on the bark. Indications are that she is searching for cracks in the bark and wood in which to commence drilling. In order to do so, she raises her ovipositor until it is roughly vertical. She may then make short probes of < 3 mm into the tree before embarking on a longer session of drilling. The holes that she drills may be anywhere between 75° and 90° to the horizontal and they are not always straight. The drill does not, therefore, necessarily start vertically above the host larva or pupa, although, in this survey, of 11 drills which resulted in host detection only two were immediately above the host. Further experimentation has shown that the female *Rhyssa* is responding to the presence of siricid larval frass in her choice of a drilling site. The cynipoid wasp *Ibalia leucospoides* lays its eggs in the eggs of *Sirex* wood wasps, using the oviposition hole drilled in the wood by the host species. *I. leucospoides* locates the hole by detecting the presence of the fungus which the wood wasp introduces into it.

Rhysella approximator is a parasite of the alder wood wasp, *Xiphydria camelus*, and her life style is similar to that of *Rhyssa*. She assesses the stage of development of the wood-wasp larva with sense organs at the tip of the ovipositor and, if it is fully grown, or nearly so, she then lays an egg on it. This, however, is not the end of the story, for her egg-laying activity may be observed by a second species of ichneumon, *Pseudorhyssa alpestris*. The latter is a kleptoparasite and its larva kills that of *R. approximator* and then itself proceeds to feed on the wood-wasp larva. *P. alpestris* is unable to drill into wood herself so she has to force her ovipositor down the hole drilled by *Rhysella* in order to lay her egg. Although

Oviposition in a braconid wasp. *Diolcogaster facetosa* is a parasite of the green cloverworm moth in the eastern USA and the females exhibit an unusual form of oviposition behaviour (Yeargan & Braman, 1986). Female wasps antennate the leaves of the cloverworm's host plant in what appears to be a random search for the latter's larvae. When, however, the wasp comes across feeding damage caused by the larvae, then the antennation and searching behaviour intensifies. Occasionally a larva may be parasitized on the plant but, more often than not, it escapes the wasp's immediate attentions by dropping down on a silken thread. The wasp's response to this escape behaviour is to locate the thread and then make her way down it, head-first, until she is within a few millimetres of the larva. She then thrusts her ovipositor forwards and attempts to introduce an egg into the larva. Sometimes both wasp and larva fall to the ground, where a violent tussle may take place. If oviposition is successful, the wasp returns to the plant to continue her searches or to groom; if she loses contact with the larva on falling, she may spend some time on the ground looking for it. Both parasitized and unparasitized larvae return to their host

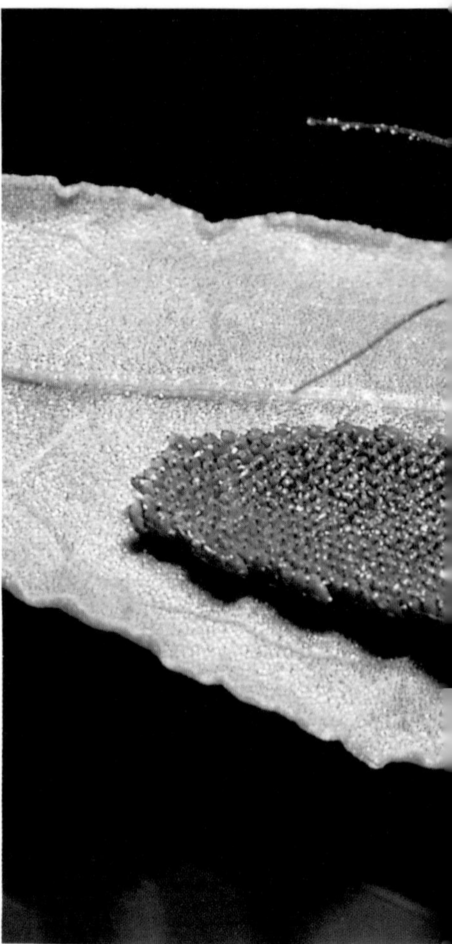

plant and continue feeding. Cloverworm larvae will suspend themselves on a thread during breaks in feeding and also as a result of threats from other predators. In this case, the female wasp is able to detect the thread and make her way down it to attempt oviposition.

▶ While some parasitic wasps oviposit into the eggs of their hosts, others do so into their larvae. The braconid wasp *Apanteles glomeratus*, here laying her eggs in the newly hatched caterpillars of the pierid large white butterfly, *Pieris brassicae*, is an important agent in controlling the numbers of this pest.

▼ A female alderfly *Sialis lutaria* (Neuroptera) laying her eggs on a leaf beside a river in England. This species often engages in clumped oviposition with several females arriving at different times to add their contribution to the existing mass of egg batches. In this way the stems of some waterside plants may become covered in a coat of *S. lutaria* eggs. Also in attendance is a minute female of the parasitic wasp *Trichogramma semblidis*, who is ovipositing into the newly laid alderfly eggs. Such egg-parasitic wasps are known to mark the host eggs in one way or another to reduce superparasitism.

she often finds the hole by observing *Rhysella* ovipositing, *Pseudorhyssa* is also able to find the hole by detecting secretions left behind by *Rhysella*. *Xiphydriophagus meyerinckii* is also a parasite of the alder wood wasp. It is a small wasp and, in order to reach the host's larvae, it crawls along old wood-wasp emergence tunnels deep into the alder wood. It somehow detects when it is adjacent to a new, occupied tunnel, bites its way through the intervening wood to the wood-wasp larva and then oviposits. The evanioid wasp *Aulacus striatus* is yet another alder-wood-wasp parasite; in this case, however, the female lays her egg into the host's egg by using the hole drilled through the wood by the host female.

In a family of their own, the Agaonidae, are the fig wasps whose life styles are inextricably intertwined with the flowering and fruiting of *Ficus* sp., in the plant family Moraceae. Around 900 species of figs are presently known and each species has its own special pollination mechanism and is pollinated by its own species of fig wasp. With such a diversity of species and pollination mechanisms, the number of them which has been investigated in any detail is quite small.

The pollination of two species of monoecious figs, *F. citrifolia* and *F. aurea*, by the wasps *Pegoscapus assuetus* and *P. jimenezi* respectively has been studied both in the field and in the laboratory (Frank, 1984). The familiar fig 'fruit' is derived from a syconium, which is in fact a globular inflorescence containing hundreds of pistillate flowers and a smaller number of staminate flowers, lining a sealed cavity. Entrance to this cavity is afforded via the ostiole, a series of closely packed, overlapping scales. Once the syconia are ready to be pollinated, it is fairly certain that they emit a species-specific chemical attractant and this is supported by the fact that the wasp inevitably enters the correct species of fig. It is the female fig wasp who is winged and she will have left the fruit in which she underwent her development after first mating with a wingless male inside the fruit. She now has to find a developing syconium of the correct species, enter it through the ostiole and then lay her eggs before pollinating the female flowers in order to ensure that there are hosts for future generations of the wasps.

The ostioles of the two fig species investigated differ somewhat from each other but the behaviours evolved by the two wasp species to enter them are very similar. On arriving at a syconium, the wasp assumes a characteristic stance and, arching the antennae so that their distal ends contact it perpendicularly, she antennates its surface. During this assessment period, the female wasp walks seemingly haphazardly over the syconium for anything between a few seconds and 5 minutes. She may then either fly off, walk along the same twig in search of other syconia or, having decided that the syconium is acceptable, she may commence the entry phase. The ostiole forms a lip on the smooth syconial surface and the female searches for this by pressing her distal five or six antennal segments flat against it. On finding the ostiole, she then uses the sharp, tough horns on the third antennal segments to loosen the upper scale. If the scale does not come free with this treatment, she then approaches at an angle, forces a horn under it and then prises it up, using her forelegs and head.

Having opened the entrance, she now applies the ventral surface of her body to the surface of the syconium and spreads her legs laterally. At the same time she compresses her abdomen dorso-ventrally, places her antennae forward and flat against the surface of the syconium, and raises her wings into a vertical position. She now pushes first her antennae, then her head, under the upper ostiolar scale and, as she moves forwards, she arches her body to an angle of about 120°. As her thorax passes under the scale, so her raised wings become detached; ostioles through which a number of females have passed may bear a tuft of wings. Having pushed under the upper scale, she then contacts the lip of the second scale and pushes her head beneath this. In this way, she makes her way beneath all of the scales until she reaches the cavity of the syconium; as well as losing her wings, the female also loses the distal to the fourth or fifth antennal segments as she enters the ostiole.

Once inside the syconium, the wasp manoeuvres her ovipositor into a perpendicular position and grooms it with her back legs. She now probes the mass of intertwining stigmata, which form a mat lining the inside of the syconium, with the tip of the ovipositor, searching for a suitable ovary into which to lay. The pistillate florets are polymorphic, some with short styles so that the wasp's ovipositor can reach the ovary and some with longer styles so that it cannot. Eggs are deposited in the short-styled flowers where they hatch and the larvae feed on the developing seeds; long-styled flowers are eventually pollinated and produce seeds. The wasps usually pollinate these flowers following each oviposition.

Towards the end of laying a single egg, the female performs a series of complex movements with the result that the fore-coxae loosen a few pollen grains and move them to the forward edge of the thoracic pockets. These contain pollen picked up while she was in the syconium from which she hatched. This loose pollen is then picked up by the arolia, opening into membranous scoops, of the fore-tarsi. She now reaches down with her front legs and wipes the pollen upwards along the anterior edge of the hypopygium, a movable sclerotized plate which lies against the ventral surface of the abdomen. Finally, she drags the pollen-covered hypopygium over the surface of the stigmas, effecting pollination.

Females spend 1 or 2 days laying eggs before they die within the syconium. By the time the wasps have completed their larval and pupal phases and are ready to hatch, the syconium has reached its so-called male phase and the anthers are ripe. The wingless males exit from the ovaries and then go in search of pistillate florets, in which female wasps are lying quietly awaiting their arrival. Having found a floret, the male chews a hole in it, inserts his telescopic abdomen and mates with the female inside.

The final role of the male or males is to chew an exit hole through the wall of the syconium to the exterior. The female now widens the hole in the ovary wall, which was cut by the male, and makes her way into the syconial cavity. Following a cleaning-up period of grooming she then goes in search of an anther which still contains pollen. She opens the anther along its dehiscence line with an antennal horn and then, placing her head into it, loosens the pollen grains with her mandibles. She now uses the arolium of each foreleg to scoop up pollen and place it into the coxal pocket of the opposite foreleg. In a series of complex movements, similar to those used when emptying them, she now fills the mesothoracic pockets with the pollen from the coxal pockets. If there is insufficient pollen on a single anther, she will move to another to complete the filling process. Finally, she makes her way to the outside through the tunnel cut by the males and flies off to begin the cycle over again.

Wasps of the ichneumonoid family Aphidiidae and the chalcidoid family Aphelinidae

are parasitoids of aphids. The female wasp deposits a single egg into the host abdomen and the resultant larva feeds internally on the host. Eventually the wasp larva kills the aphid, at which point its exoskeleton hardens and changes colour to form a so-called 'mummy'. The larvae of these primary parasitoids are in their turn devoured by hyperparasitoids from five families within the Chalcidoidea, Ceraphronoidea and Cynipodea. Within the hyperparasitoid genera involved, there are two distinct categories of ovipositional behaviour. The female wasp of the endophagous species lays her egg inside the primary parasitoid larva while it is still feeding on the live aphid, before the latter becomes mummified. The egg only hatches after the mummy has formed and the larva then proceeds to feed internally on its host. In ectophagous species, the female wasp lays an egg on the outside of the primary parasitoid larva after the aphid is dead and mummified. Inside the mummy, the hyperparasitoid larva then feeds externally on its host. *Alloxysta, Phaenoglyphis, Lytoxysta* (Cynipidae) and *Tetrastichus* (Eulophidae) are endophagous hyperparasitoid genera, while *Asaphes, Pachyneuron, Coruna* (Pteromalidae) and *Dendrocerus* (Megaspilidae) are ectophagous hyperparasitoid genera. *Aphidencyrtus* (Encyrtidae) has an essentially endophagous larva but the female wasp can indulge in either type of ovipositional behaviour.

Although the majority of cynipoid wasps are phytophagous gall-formers, a number of them are parasites. The oviposition behaviour of one of these, *Callaspidia defonscolombei*, a parasite of aphidophagous syrphid larvae, has been studied in South Wales (Rotheray, 1979). Laboratory studies have revealed that the female (this species

◄ A female *Alloxysta* sp. braconid wasp ovipositing as a hyperparasite into the larva of the primary parasite inside the aphid *Macrosiphoniella artemisiae*. Such behaviour involves certain risks not faced during oviposition by the primary parasite. The latter can use a 'stand-off' technique by threading the highly telescopic abdomen through the legs so that only the abdominal tip, with its ovipositor, comes into minimal contact with the aphid. The *Alloxysta*, by contrast, is faced with the problem of locating the primary parasite's larva within the aphid. This can only be accomplished by climbing on to the host's back, thereby greatly increasing the chance of contacting the aphid's waxy defensive cornicle secretions. These harden upon contact, cementing the hyperparasite in place with such finality that it usually struggles vainly until it dies.

reproduces by parthenogenesis) responds to the presence of the aphid rather than to the fly larvae. In the field, female wasps have been observed amongst colonies of a range of aphid species. In their turn, aphidophagous syrphids are not host-specific.

Since *C. defonscolombei* utilizes a number of different syrphids, a response to any aphid colony is likely to result in the discovery of one or more suitable hosts. The female wasp detects the host larva using her antennae and then climbs on to it; the larva becomes motionless and retracts its head end. The wasp now inserts her ovipositor laterally and apparently injects venom, for the larva now becomes comatose. She removes her ovipositor and inspects the larva with her antennae before re-inserting her ovipositor mid-dorsally and probing the larva. Eventually she removes the ovipositor, moves to the head end of the larva, inserts the ovipositor into the head, probes, becomes still, with antennae held out to the side, and deposits an egg into the host's cerebral ganglia. Finally, she removes the ovipositor, raises her antennae, moves off and grooms.

Dissection of host larvae indicates that a single egg is always laid in the cerebral ganglia. The advantage of this over haemocoel deposition, which on the other hand takes less time, is that the parasite is not attacked by the defence cells present in the body fluid. In a later series of investigations (Rotheray, 1981), another cynipoid wasp belonging to the family Figitidae, *Melanips opacus*, was described as having oviposition behaviour similar to that of *C. defonscolombei*, with the egg also being placed in the cerebral ganglia.

Oviposition behaviour in the diplazontine ichneumonids studied is much less complex, the eggs being laid rapidly and directly into the syrphid larva's haemocoel. The exception to this is the female of *Enizeum ornatum*, which, on mounting a larva, stabs with the ovipositor and bites the host's integument. This ill-treatment results in the larva raising its front end and emitting secretions from the mouth. At this point, the wasp inserts her ovipositor just behind the host's mouthparts. During the same study, the importance of scent in finding aphid colonies, and thus host larvae, was demonstrated for a number of species of ichneumonid parasitoids of syrphids. Field observations revealed that the ichneumonids *Diplazon laetatorius*, *Enizeum ornatum* and *Homotropus pictus*, as well as *M. opacus*, could be found on a wide range of aphid species in the field. Within an aphid colony, the females moved slowly around, probing with their antennae, presumably in a search for syrphid larvae.

The Australian tiphiid wasp *Hemithynnus hyalinatus* is a parasite of larvae of a number of different species of scarabaeid beetles. The female wasp is wingless and searches for her subterranean hosts by digging through the soil with her mandibles. She moves beneath the soil in a small chamber, clearing soil away in front of her and then pushing it away behind her, travelling at an estimated 30 cm per day (Ridsdill Smith, 1970). In the laboratory, it took females between 0 and 12 days, with a mean time of 3.2 ± 2.3 days, to find a host larva. Having found one, the female wasp then holds it behind the head, using her legs and jaws, and thrusts her sting into the first or second thoracic segment between the legs, in the region of the ganglia of the ventral nerve cord. She then builds a chamber round the larva and malaxates it before she lays her eggs upon it. This she does by moving back and forth along the host, chewing at the underside and also stinging it. Some haemolymph exudes from the bites, on which it is assumed the wasp feeds. The laying of a single egg normally occurs within a few hours of paralysing the host larva.

The North American tiphiid *Pterombrus rufiventris hyalinatus* specializes in parasitizing tiger-beetle larvae (Cicindelidae). In searching for prospective hosts, the female wasp runs rapidly around on the ground in an apparently random way, antennating all holes or cracks until an occupied larval burrow is found (Knisley *et al.*, 1989). The wasp enters the burrow head-first and moves down until her head contacts that of the retreating beetle larva, whereupon she curves her abdomen forwards and stings it beneath the head or thorax. The wasp then waits a few minutes before again touching the larva's head with her antennae; any movement by the larva may elicit a further sting. The wasp now raises the larva slightly in its burrow, by grasping it with her mandibles and hauling it up, or by pushing it up either from beneath or by positioning herself next to it and grasping its front legs in her mandibles. The larva may then be stung one more time on the underside.

Once the larva has been re-positioned, and sometimes following oviposition, the wasp uses the tip of her abdomen to pack down the bottom of the burrow. In order to oviposit, she places her ventral surface in contact with that of the host and moves the tip of her abdomen around until she finds the correct position to lay her egg, i.e. the larva's second abdominal sternum on the mid-line. The wasp may sting the host or chew it around the oviposition site. Having laid her egg, she now moves above the larva and constructs a primary and secondary plug in the burrow. The primary plug is placed 2–4 cm above the larva. The wasp curls herself around the burrow, pulling down soil with her mandibles and collecting it against her abdomen; this soil is then compacted with the hind legs and pressed into place with the abdomen until a circular shelf of soil encircles the burrow. She then fills the hole in the centre of the shelf by breaking off pieces of soil from near the top of the burrow. The completed primary plug is 15–30 mm thick and takes between 7 and 15 minutes to construct. Now the wasp leaves the burrow and searches around for suitable materials with which to make the secondary plug; these include bits of soil, small stones, small twigs etc. These she picks up in her jaws and just drops into the burrow to form a loose plug. In some sites, the wasp simply pushes loose sand into the burrow with her hind legs before capping it with larger pieces of material.

Members of the family Drynidae are parasitic exclusively on the homopteran suborder Auchenorrhyncha. The females of most of the drynid genera are wingless and they possess chelate front legs with which they catch and hold their prey during feeding and oviposition. Studies of five European drynid species from five different genera revealed many similarities but some small differences in foraging and oviposition behaviour (Waloff, 1974). *Gonatopus sepsoides* feed on their hosts, which include several different species of cicadellid, both as adults and as larvae. In searching for a leafhopper, the female appears to walk around randomly, frequently palpating the substrate with her antennae. She usually becomes aware of it from a distance of about 1.5 cm and she then either runs and catches it or else, but less often, jumps on it; before jumping she momentarily remains motionless with her chelate front legs held forwards. The wasp holds the prey at right angles with her chelate legs grasping the host's hind legs and her other legs spread out in support. Between 5 and 50 seconds after catching the leafhopper, she bends her abdomen through her legs and stings it on the underside. The first leafhopper captured

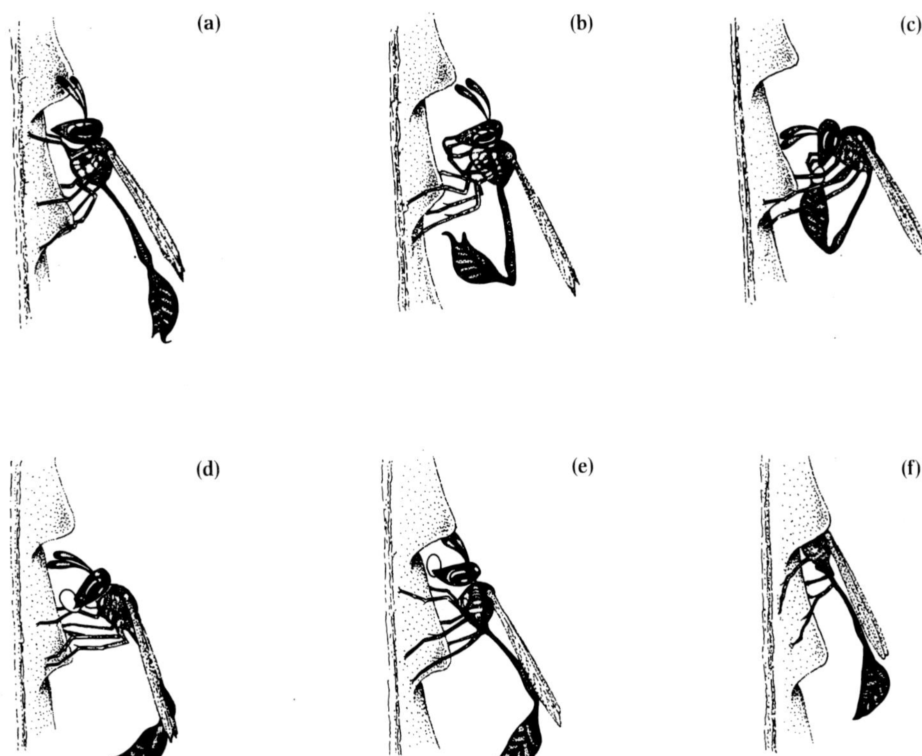

◀ **Steps involved in egg deposition in the wasp *Parischnogaster nigricans* (Vespoidea: Stenogastrinae) (Turillazzi & Pardi, 1982). With the cell completed, and prior to oviposition, the female wasp produces an abdominal secretion, which she places in the cell. She then readies herself for egg-laying (a), extending her abdomen and extruding her sting dorsally. She now bends her abdomen beneath her, simultaneously vibrating it (b) and, as its tip approaches her jaws, she emits a single egg (c). Sometimes a drop of clear fluid is visible in the jaws before they gently grasp the egg (d); the wasp moves towards the cell (e) and places the egg inside (f). A further portion of the gelatinous abdominal secretion is then placed in the cell.** (After Turillazzi & Pardi, 1982).

is usually consumed by the wasp. Further hosts caught may be perfunctorily fed upon before oviposition, though these seldom survive the effects of the feeding activities, or used just for oviposition. An egg is laid between the intersegmental membranes of the host abdomen. A second drynid species, *Pseudogonatopus distinctus*, found and attacked its delphacid host in much the same way as the previous species but dragged it around for up to 30 seconds before ovipositing between the second and third abdominal segments or, occasionally, between the thorax and abdomen. *P. distinctus* killed a lower percentage of hosts than *G. sepsioides*. Unlike the previous two species, the delphacid parasitizing *Dicondylus bicolor* did not oviposit on any hosts on which it had first fed. *Anteon lucidum* parasitizes cicadellids and, having subdued one she sometimes cuts a small hole into a thoracic intersegmental membrane; this hole may be used as an oviposition site.

Larvae of the scelionid wasp *Rielia manticida*, as the name implies, are parasites of mantid eggs. The female wasp seeks out a female mantid. She then clings with her jaws to the mantid's abdomen or wing bases and then discards her wings. She now has to wait until the female mantid is ready to lay her eggs before she herself can oviposit. During this period, she maintains herself by feeding on the mantid's body fluids. The mantid lays her eggs in a mass of froth, which hardens with time, and the wasp lays her eggs on the mantid eggs while the froth is still soft.

The wasps of the family Pompilidae use a single spider as a food source for their developing larva. In many species, this is placed in a burrow but, in just a few, the egg is laid on the spider and the larva develops as a parasite on the active host rather than as a predator on the comatose one. Observations on how the parasitic forms deposit their egg on the host spider are few and far between but at least one instance is recorded (Grout & Brothers, 1982) from a garden in Zambia. The female wasp, assumed to be the widely distributed African species *Paracyphononyx africanus*, was observed tracking a large lycosid spider. She overtook the spider from behind and faced it; the spider reared up aggressively and the wasp immediately attacked, seemingly stinging the spider several times both on the prosoma and opisthosoma. The stings reduced the spider to a comatose state whereupon the wasp mounted it and laid a single egg on the front dorsal surface of the opisthosoma. The wasp departed within seconds without returning; the spider recovered within 15 minutes, apparently unaware of the fact that it now carried the wasp's egg.

Stelis montana, an anthidiine kleptoparasite on other megachilid bees, is widely distributed across western North America. Its variable oviposition behaviour in relation to its particular *Osmia* host has been studied in detail (Torchio, 1989). *S. montana* females are able to analyse the internal arrangement of the food stores of the host and vary their oviposition accordingly. In the nest of *Osmia lignaria propinqua*, for example, she taps the leading edge of the food store with the tips of her antennae and assesses its angle to the horizontal. The surface of a nest in the early stages of provisioning is horizontal; as the process proceeds, so the angle increases. If the *S. montana* female detects that the angle is sufficiently great, i.e. the nest is at least half-provisioned, then she uses her mandibles and front legs to move some of the pollen–nectar mixture to form a pyramid on the medio-posterior surface of the main mass of provisions. She then re-positions herself and lays an egg immediately behind the pyramid.

Why she goes through this complex series of activities becomes clear when the host returns with more supplies. First she chews along the surface of those provisions already present, depositing nectar as she does so. On meeting the pyramidal barrier constructed by *S. montana*, she continues to chew along its outline, thus missing the egg which the latter has laid behind it. The host then crawls out of the nest, turns round and backs into the nest before emptying her abdominal scopa on to the now-moist provisions that are already there. The second species used by *S. montana*, *Osmia californica*, constructs a cell threshold and, in this instance, the former uses her antennae to measure the distance between it and the

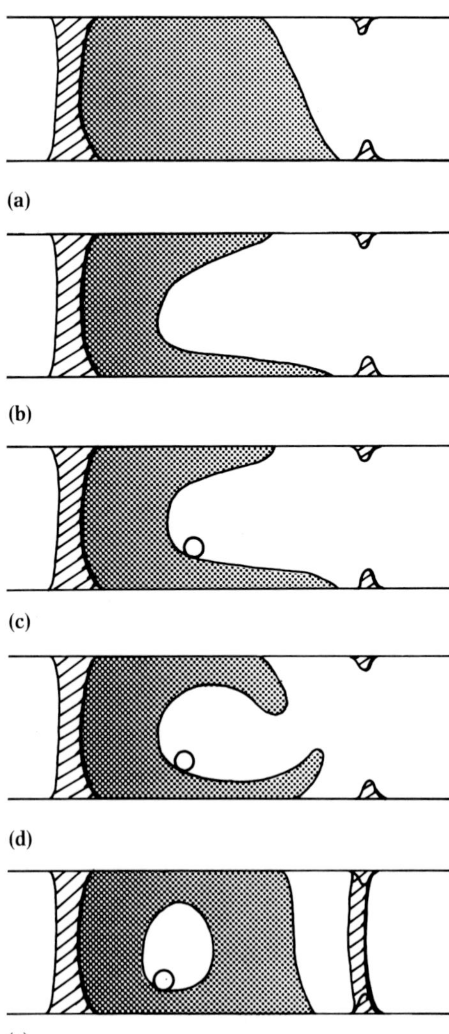

▲ Egg-laying in *Osmia californica* (Megachilidae). Having completed the process of provisioning her cell (a), the female bee now excavates a hole in it (b) into which, when completed, she lays a single egg (c). She then uses the tip of her abdomen to bend the edges of the provisions inwards (d) until they meet and the egg is then completely encapsulated (e). Finally she caps the cell. (After Torchio, 1989).

edge of the provisions. If the distance is greater than 2 mm, she leaves the nest; if it is less than 2 mm, then provisioning is two-thirds to three-quarters completed and she remains in the nest. She then carves a chamber into the nectar–pollen mix, turns round and then lays an egg in it. Next she picks up a large lump of moist pollen in her jaws, fills up the egg chamber with it and then tamps it down with the front of her head. The way in which *S. montana* deals with the nest of *Osmia montana montana* has elements of both of the previous examples. The state of provisioning is detected by the angle of the pollen–nectar mixture, as for *O. l. propinqua*, but egg-laying is in an excavated cavity, as in the nest of *O. californica*. Of interest is the fact that *S. montana* actually mimics the oviposition behaviour of its host, i.e. *O. l. propinqua* lays on the surface of the provisioning while the other two species oviposit in excavated chambers.

Looking more like wasps than bees is the Holarctic nomadine genus *Nomada*, cuckoo bees that are kleptoparasitic on soil-nesting Andrenidae, Halictidae, Melittidae and Anthophoridae. Part of the oviposition behaviour involves a search for a suitable host and, in this, they are assisted by the males. Male *Nomada* spp., whose females oviposit in the nests of *Andrena* spp., produce from their cephalic glands a secretion which has the same odour as, and therefore mimics, the secretion produced by the Dufour's-gland of the host female (Tengö & Bergström, 1977). In the host–parasite pairs of *Andrena haemorrhoa/Nomada bifida* and *A. carantonica/N. marshamella*, the chemical *all-trans* farnesyl hexanoate is the dominant component in the glands of the respective species. In the pairs *A. helvola/N. panzeri* and *A. clarkella/N. leucophthalma*, the main component is the substance geranyl octanoate. The male cuckoo bees fly around the nest aggregations of the host species and, presumably by secreting a similar scent, make it easier for the *Nomada* females to home-in on the area. During copulation, the female cuckoo bee also picks up some of this odour which, it is believed, may make it easier for her to enter the host's nest.

The host-seeking behaviour of *N. articulata*, a kleptoparasite of the communal-nester *Agapostemon virescens* in the USA, is perhaps typical of nomadines as a group. The female cuckoo bees normally visit the *Agapostemon* nesting area once or twice during the time of day that they are active, staying for between 15 and 45 minutes and making repeated attempts to enter host burrows (Eickwort & Abrams, 1980). They fly around the burrow area a few centimetres above the ground and alight on a suitable perch a short distance from a nest entrance. For anything between 5 seconds and 10 minutes, they sit facing the burrow entrance, with their antennae directed towards it, before they make any attempt to enter. Of attempts to enter the nest burrow, 50 per cent are made after the host has left on a foraging trip; otherwise they are made just after the host has entered the burrow or when there is no visible activity at the nest entrance. At no time do *N. articulata* females attempt to enter when the host is at the nest entrance. Having decided to investigate a burrow, the female cuckoo bee advances towards it with her wings vibrating. At the burrow entrance, she stops and then inserts her antennae into it for 1 to 2 seconds, her wings still vibrating. The wing-vibrating then ceases and she walks into the burrow to lay her eggs.

Nomadine bees of the Holarctic genus *Epeolus* are kleptoparasitic upon bees of the genus *Colletes*. Though a large genus, little of their biology is known but one species, *E. compactus* from the southwestern USA, has been investigated in depth (Torchio & Burdick, 1988). It is associated with the bee *Colletes kincaidii*, which constructs linear rows of polyester-lined cells which it provisions with food for its larvae. Having entered the nest of its host, the female *E. compactus* uses one of two oviposition sites. If the cell is fully provisioned and capped, the egg is laid through the cap but, if it is only partly provisioned, then the egg is laid through the upper side-wall. The latter method is only possible when the host is utilizing old burrows made by previous generations and there is thus sufficient space between the new cell wall and the burrow wall for an egg to fit. In order to lay her egg, *E. compactus* first exserts metasomal sternum VI, which bears a pair of spined projections, through the apex of the abdomen. She then uses these to cut a U-shaped hole through the polyester cell wall into which she then deposits her egg. The egg is inserted through the hole until its flat top is roughly flush with the inner cell surface. A close examination of the egg *in situ* reveals that, as it is laid, a secretion, probably from Dufour's-gland, is applied to its neck region. This dissolves the cut edges of the cell wall and the liquid thus produced fills the space between it and the egg. The liquid then solidifies, firmly gluing the egg into the host-cell lining.

DIPTERA True flies

Mosquitoes lay their eggs either one at a time or in rafts, and the oviposition behaviour of *Culiseta ornata* is an example of the latter (Pappas & Pappas, 1982). Prior to ovipositing, the female mosquito lands on the water surface and then crosses her hind legs and moves them towards the top of the abdomen to form a kind of cradle in which to deposit her eggs. She then lays the first egg in the apex of the V formed by the crossed legs. The next row of two eggs is laid above this egg, starting at the left and

▲ **The larvae of the European bee fly *Bombylius major* are ectoparasites of the developmental stages of solitary bees. Here, a hovering female is dropping eggs into a solitary-bee colony. The eggs will hatch into active first-instar planidium larvae which will make their way into the bees' burrows to find their hosts.**

moving to the right. The third row, of three or four eggs, is laid in a similar manner to the second, i.e. starting on the left and moving across to the right. From the fourth row on, four to six eggs are deposited but the pattern is altered. She starts laying on the left but, when she reaches the centre, she moves across to the right and lays from there to the centre, repeating this sequence for the remainder of the raft. During the whole process, the abdomen remains in the same position and it is the cradle which is moved by the action of the legs. Also, between the laying of each egg, the legs move the developing raft away from the body and then back into position to receive the next egg. Females of *Toxorhynchites* mosquitoes, which breed in rot-holes in trees, flick their eggs into the water contained in these holes while in flight.

Rhagionid flies of the genus *Atherix*, e.g. *A. variegata* in North America and *A. ibis* in Europe, indulge in communal oviposition. Female flies oviposit on leaves overhanging streams and each female then remains hanging over her eggs while others settle on her and lay their eggs. The eventual mass of dead and dying females and their eggs has the appearance and shape of a small swarm of bees.

Female Asilidae oviposit in a number of different situations: in the soil, on vegetation and in wood to name but a few. The behaviour of two species which oviposit in soil, *Proctacanthella leucopogon* from Mexico and *Efferia frewingi* from the USA, is very similar (Lavigne & Dennis, 1975 & 1980). Females of both species initially probe the soil with their ovipositor as they seek a suitable place to lay. *P. leucopogon* then inserts her ovipositor, followed by her whole of the abdomen, into the soil and bounces slightly in an apparent tamping action as she lays her egg batch. She then withdraws the ovipositor and uses it briefly to sweep the soil before flying off. The female *E. frewingi* pushes her ovipositor into the soil with sideways sweeping movements but inserts the abdomen only to the fourth or fifth segment. While she is laying her batch of three to seven eggs she keeps her wings folded over her back. Having laid a batch of eggs, she then withdraws the ovipositor with rapid sideways and/or rotational movements and then uses it to sweep the soil around the oviposition site, effectively hiding its location.

Whereas *E. frewingi* oviposits in the soil, in another North American species, *E. varipes*, eggs are laid on vegetation. Females of the latter have long, flattened ovipositors with which they deposit their eggs between the leaf and stem of grass plants. Females searching for egg-laying sites land head-up on vegetation, curl their abdomens under at an 80° to 90° angle at segment four or five and begin to probe with their ovipositors. They crawl up and around the plant as they probe and, on finding a suitable oviposition site, they straighten their abdomens to a slightly-bent state. They then reverse their ovipositors into the fissure between leaf and stem and begin to lay between 10 and 40 eggs in a row.

Females of the asilid subfamily Laphriinae lay their eggs in dead wood. Observation of *Laphria fernaldi* in North America (Lavigne & Bullington, 1984) revealed that, of the oviposition sites chosen 24 per cent were the holes of carpenter bees, 20 per cent were cracks in the surface of the dead wood and 14 per cent were under the bark. The female searches for these sites by flying along logs about 50 mm above the surface, hovering just above any dark area such as a bee hole or a crack. If the site turns out to be acceptable, she lands next to it, reverses back to the opening and then extrudes her ovipositor into it. While in this position, she intermittently extrudes and then retracts her ovipositor. If she is disturbed and put to flight for any reason, such as the emergence of a bee from a hole, then she will return to the hole as soon as the coast is clear.

Asilus gilvipes has been studied also, in southeast Wyoming (Schreiber & Lavigne, 1986), where it is associated with the presence of the burrows of mammals such as ground squirrels and badgers. These apparently act as a focus for calliphorid flies, which form the main part of the asilid's diet. Oviposition is quite straightforward: on finding a suitable burrow, the female sits at the entrance and drops either single eggs or packets of three down into it. Although ground-squirrel burrows were utilized, the asilids indicated a preference for the somewhat larger burrows of the badger. Sympatric with *A. gilvipes* for part of the summer is the smaller *A. formosus*, whose females opt for smaller burrows as oviposition sites.

The oviposition behaviour of the South African asilid *Neolophonotus dichaetus* is somewhat unusual for the family as a whole (Londt & Harris, 1987). The female oviposits on pinnate-leaved members of the Leguminosae and she employs her legs to hold two leaflets together while she deposits her eggs. About 12 narrow strips of froth are laid down between the leaves without withdrawing the ovipositor and eggs are laid along these strips, except for the two outer ones which are not usually used. Females have been observed to produce several egg batches, often in quick succession.

Most bombyliid larvae are ectoparasites of other insects, such as the larvae of aculeate Hymenoptera; some are predacious on locust eggs and others are endoparasites in the pupae of tsetse flies. Often as not, the female bombyliid drops her eggs in flight or lays them on the ground in the vicinity of the hosts' nest and the planidium larva which hatches makes its way into it.

Anthrax gideon from Costa Rica and *A. analis* from the USA have larvae which parasitize the larval stages of tiger beetles (Cicindelidae). In both species, the females take some care in searching out the host's burrows (Palmer, 1982). Where these are at low density, they fly rapidly and erratically, appearing to sample specific areas at random but, where the burrows are at high density, they fly in a methodical way, seeming to cover every square metre. Upon finding a burrow, the female fly hovers in front or over it and may land briefly at the edge, as long as no beetle larva is at the opening. She then hovers 30–50 mm from the opening, curls the tip of her abdomen towards it, makes a short dart towards it and then releases an egg into the burrow. This behaviour pattern is repeated if further eggs are to be deposited into the same opening.

The larvae of many species of phorid flies live in some form of association with social bees, wasps, ants and termites. Females of *Metapina pachycondylae*, for example, oviposit on the larvae of the ant *Pachycondyla harpax*. The fly larva then encircles the ant larva behind the head and feeds alongside it when the worker ants bring it food.

The Conopidae are normally parasites of bees and wasps, the female flies following and pouncing on their hosts in flight and then ovipositing on them. Flies of the genus *Stylogaster*, uncomfortably included in the Conopidae, follow columns of the tropical doryline army ants. The females are interested not in the ants themselves but in the other insects which the ants flush out as they rampage across the forest. They swoop down upon cockroaches and various flies and stab them with their substitute ovipositor, laying an egg, which has a barbed point, into the wound.

Three kleptoparasitic, larvipositing, sarcophagid flies associated with the solitary hunting wasp *Tachysphex terminatus* have been observed in the northeastern USA. The wasp nests in aggregations and always closes it burrow temporarily when it departs to hunt for acridid prey. On its return, it drops the grasshopper near the burrow, which it opens and reverses into, pulling the prey behind it. This behaviour may be repeated several times until sufficient prey for the wasp's larva is present in the burrow. The provisioning behaviour of the wasp therefore leaves a number of openings for fly larviposition and each is exploited by a different sarcophagid (Spofford *et al.*, 1986).

The female of *Phrosinella aurifacies* takes advantage of the wasp's hunting forays to find an unoccupied burrow. She flies around just above the sandy ground in a zigzag fashion, landing on occasion to investigate disturbed areas. In searching for a burrow entrance, she circles around on the sand and probes with her front legs and antennae. On finding a burrow, she digs a hole 1–2 mm deep in the sand, closing the entrance, and larviposits into it; she then departs leaving the maggot to make its way into the wasp's burrow. The same fly employs the same larviposition behaviour when kleptoparasitizing wasps of the genus *Oxybelus* (Peckham, 1977).

Females of the fly *Senotainia trilineata* sit on suitable perches in the area around the wasp burrows awaiting the arrival of the host carrying prey. The flies then take off and follow close behind the wasp, on occasions contacting the prey in the air and apparently forcing the wasp to land. Then, as the wasp re-adjusts the prey before continuing its flight to the burrow, the flies dart in and larviposit. More often, however, the flies stalk the wasps after they have landed

Oviposition in tephritids. These flies have a noticeable ovipositor with which they introduce their eggs into developing fruits or into the individual flowers on composite capitula. The Mediteranean fruit fly, *Ceratitis capitata*, for example, lays its eggs in a crack in the rind of ripening citrus fruits and the larvae then burrow into and feed on the flesh. *Rhagoletis basiola* and *R. pomonella* are tephritids which oviposit in rose-hips and apples respectively, although, in the American state of New England, *R. pomonella* makes use of rose-hips as well. Having laid its egg into a rose-hip, *R. basiola* drags its ovipositor over the surface of the fruit and lays down a trail of pheromone. It has been found that this deters other females from laying their eggs in that particular fruit (Averill & Prokopy, 1981). Interestingly, though *R. basiola* was unable to recognize the marking pheromone used by *R. pomonella*, this species showed a partial recognition, when using rose-hips, of the *R. basiola* pheromone.

◀ Flies of the family Tephritidae, such as this female *Terellia serratulae* on a woolly thistle, *Cirsium eriophorum*, extrude a noticeable ovipositor with which they introduce their eggs into developing fruits or into the individual flowers on composite capitula. The larvae feed on the developing seeds.

▲ The larvae of the East African muscid fly *Stomoxys ochrosoma* are scavengers in the bivouacs of the army ant *Dorylus nigricans*. This fly has evolved oviposition tactics which keep the female well clear of the jaws of these ferocious ants (pers. obs.). The female fly is seen here hovering above a column of these ants and seeking out a worker not carrying anything in its jaws. She will then release her batch of about 20 eggs so that they fall accurately within the range of this worker's antennae and jaws. The ant will then carry them to the bivouac.

with their prey, with up to four flies in attendance on a single host at any one time. *S. trilineata* always deposits larvae when the wasp is present with the prey and it shows no interest in unattended grasshoppers. *S. trilineata* is also a kleptoparasite of the hunting wasp *Oxybelus subulatus* and employs the same larviposition behaviour as that associated with *T. terminatus* (Peckham, 1977).

The female of the third fly species, *Senotainia vigilans*, perches in the vicinity of digging wasps and, once the burrow is complete, waits for the host to return with its prey. She then either larviposits on the prey, left unattended while the wasp is opening the burrow, or follows the wasp into the burrow and presumably larviposits on the prey underground. Finally, she may deposit larvae in the burrow entrance, either before or after the wasp has taken in the prey.

These three flies do not get things all their own way for the wasp, in some instances, is apparently able to detect whether larviposition has occurred. If it has, she will clean the grasshopper thoroughly, searching for maggots with her antennae and removing them with her jaws.

At least one species of tachinid and one sarcophagid have been demonstrated to have females which are attracted to their larval host by sound (Fowler, 1987). Gravid females of the sarcophagid fly *Colcondamyia auditrix* are attracted to singing males of the North American cicada *Okanagana rimosa*, on which they larviposit. The presence of the parasite in the abdomen of the host seems to prevent further sound production by the male cicada, an apparent adaptation which reduces the likelihood of multiple parasitism. In Brazil, the females of the tachinid fly *Euphasiopteryx depleta* are likewise attracted to the singing males of *Scapteriscus* mole crickets.

Dermatobia hominis, the New World human warble fly, has a very unusual method for ensuring that its eggs reach their target on the skin of a large mammal such as a cow. Rather than laying them there herself, the female lays her egg batch on the abdomen of other Diptera, while they are either on or in the vicinity of the host animals. These, for one reason or another, then visit the host where the warmth of its body causes the warble-fly larvae to hatch. Vectors for the warble's eggs include mosquitoes and, in Central America, *Sarcopromusca* spp. and *Fannia* spp. (Mourier & Banegas, 1970). The female warble fly probes and manipulates the vector to assess its size before laying and, in general, the bigger the vector, the larger the egg mass – up to a maximum of about 35 eggs per mass.

Stomorina lunata is a calliphorid fly whose larvae feed in locust egg cases. The gravid female fly sits in vegetation adjacent to ovipositing locusts, descending from time to time to examine any depressions in the soil which might mark the site of a locust egg pod. On finding one she lays a single egg above it and then covers it with soil. As the locusts move on to new oviposition sites so the flies follow them.

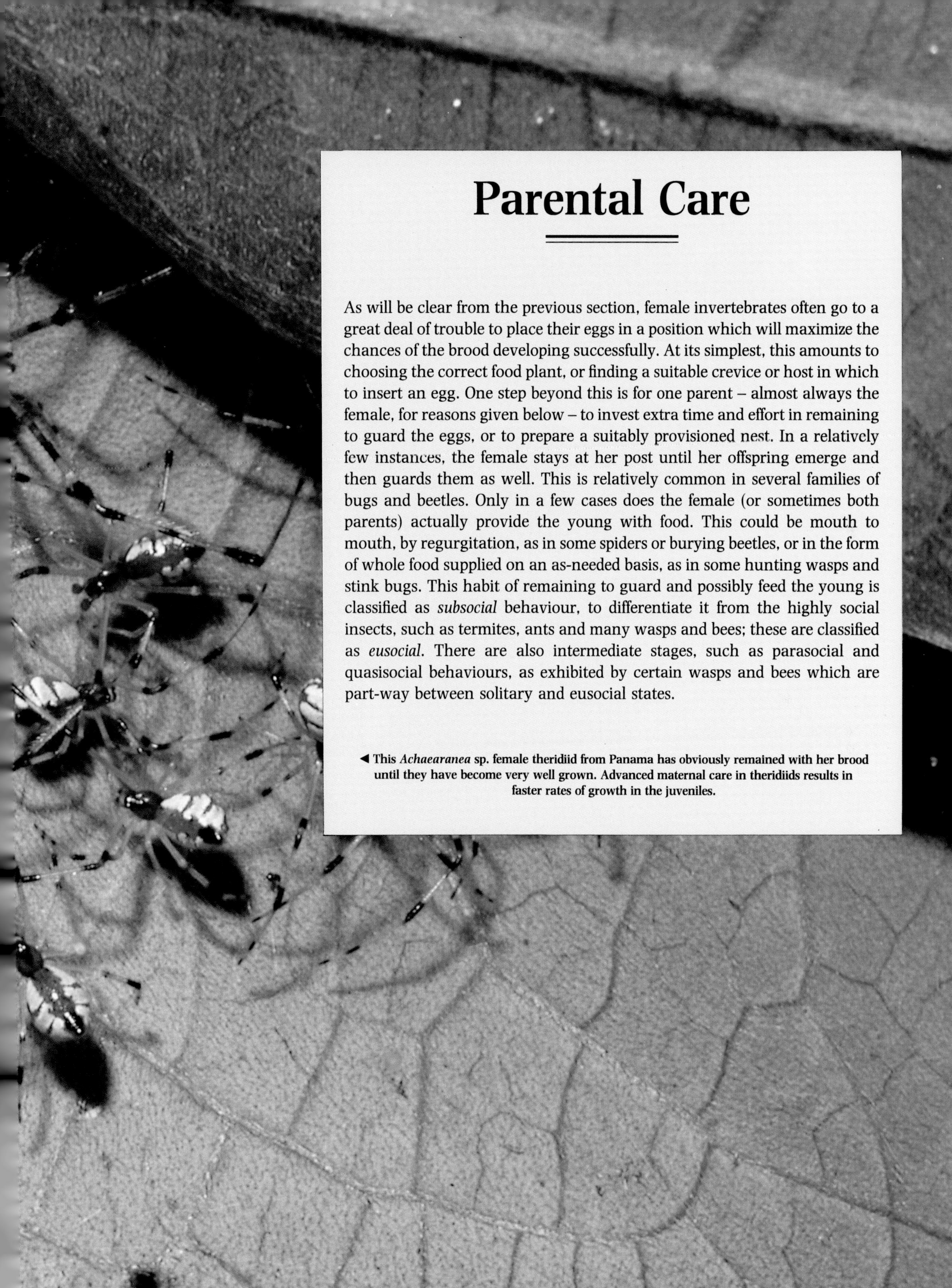

Parental Care

As will be clear from the previous section, female invertebrates often go to a great deal of trouble to place their eggs in a position which will maximize the chances of the brood developing successfully. At its simplest, this amounts to choosing the correct food plant, or finding a suitable crevice or host in which to insert an egg. One step beyond this is for one parent – almost always the female, for reasons given below – to invest extra time and effort in remaining to guard the eggs, or to prepare a suitably provisioned nest. In a relatively few instances, the female stays at her post until her offspring emerge and then guards them as well. This is relatively common in several families of bugs and beetles. Only in a few cases does the female (or sometimes both parents) actually provide the young with food. This could be mouth to mouth, by regurgitation, as in some spiders or burying beetles, or in the form of whole food supplied on an as-needed basis, as in some hunting wasps and stink bugs. This habit of remaining to guard and possibly feed the young is classified as *subsocial* behaviour, to differentiate it from the highly social insects, such as termites, ants and many wasps and bees; these are classified as *eusocial*. There are also intermediate stages, such as parasocial and quasisocial behaviours, as exhibited by certain wasps and bees which are part-way between solitary and eusocial states.

◀ This *Achaearanea* sp. female theridiid from Panama has obviously remained with her brood until they have become very well grown. Advanced maternal care in theridiids results in faster rates of growth in the juveniles.

Subsocial behaviour is the most basic form of interaction between parents and their progeny, and has evolved independently in terrestrial crustaceans, myriapods, several groups of arachnids and at least 13 orders of insects. The prime benefits of the behaviour varies. In certain spiders, direct regurgitative feeding of the babies accelerates their growth. In ground-nesting insects, the mother's main role may be to inhibit the growth of moulds which flourish in such enclosed habitats. In many membracid bugs and at least one fungus beetle, the mother assists the young in their feeding by preparing or finding food sources. In stink bugs and leaf beetles, it is the mother's abilities as a guard, able to repel numerous predators, which is the main benefit conferred on her offspring. In certain cockroaches, the mother herself carries her babies around and not only acts as a defensive shield but also ensures that they all remain in the best micro-environment – a habit found in other animals, such as scorpions. In the dipluran *Dipljapyx humberti*, the female lays her eggs at the tip of a leaf stalk, then keeps them clean until they hatch, when she continues to guard the larvae (Pages, 1967).

It has been hypothesized (Wilson, 1971) that parental care is likely to evolve under certain environmental conditions. These are:

(1) When the immature stages are particularly vulnerable to elevated levels of predation or parasitism. Included in this category are the numerous subsocial bugs and the handful of subsocial beetles which live exposed on leaves.

(2) When animals, by developing subsocial behaviour, are able to take advantage of exceptionally harsh environments, which might otherwise be impossible to colonize permanently. Several intertidal beetles come into this category.

(3) When animals exploit a patchily distributed and limited, but very rich, resource for which levels of competition are very high. Many subsocial insects come into this category. The classic examples of such a resource are dung and small animal corpses, thus the widespread evolution of subsocial behaviour in dung beetles and burying beetles is not surprising. These have to compete with other exploiters of dung and carcasses, such as certain flies, which have evolved their own answer to the problem – 'internal' maternal care in the form of viviparity and the dropping of voracious living larvae.

If we exclude the pre-ovipositional investment made by males in their offspring in the form of nuptial gifts of various kinds, purely *paternal* care is rare, although male–female co-operation in brood-rearing is more common, especially in the labour-intensive activities of dung and carrion beetles. The reason for the rarity of male guarding behaviours is related to the costs involved compared with the alternative strategy of attempting to inseminate as many females as possible and leaving egg-laying and care (if any) to them. The most essential prerequisite for paternal guarding is that the male should be able to assure paternity of the guarded eggs, otherwise he may be wasting his time caring for a rival's offspring. In male guarders, this paternity is assured by staying with the female as she lays her eggs. However, becoming a guarder may also actually increase a male's sexual success by attracting extra females in a kind of role-reversed sexual system whereby females come to males, rather then vice versa. Male guarding and brooding services seem to be greatly in demand in the harvestman *Zygopachylus albomarginis* and in reduviid and belostomatid bugs. All these species are predatory and male guarding behaviour has the advantage that females can continue active foraging. This enables them to produce more eggs, which can then be guarded by the males, with no loss of fecundity by the female, as would happen if she were to stop feeding in order to become a guarder herself. Male guarders are therefore probably assured not only a regular supply of willing females, but also mates who are highly fecund.

The greatest advance in brood care has been the evolution of eusocial behaviour. Its development in termites has been linked to the need for successive generations to pass on to one another certain essential internal symbionts. However, in the order Hymenoptera, the evolution of social behaviour has been connected with the haplodiploid method of sex determination and the resulting increased likelihood of kin selection (Hamilton, 1964). Male hymenopterans develop from unfertilized eggs and females from fertilized eggs. As a result, hymenopteran sisters share 75 per cent of their genes compared to the 50 per cent shared by sisters in most other animals. Because of this degree of relatedness, a female, rather than trying to go it alone and rear her own progeny, may find it a better bet to stay with a sister and help rear her offspring. These, therefore, will be nieces and will share three-eighths of the aunt's genes – only fractionally less than the half which would be shared with her own daughters. However, for an aunt to end up passing on more genes in this way, and to make her social behaviour worthwhile, she has to be confident of producing at least one-third more nieces than she would offspring of her own, had she opted for going it alone. Indeed, levels of pay-off, from reasonably good to quite substantial, do seem to occur in the primitively eusocial wasp *Polistes exclamans*, in which the advantages of group living have been clearly demonstrated in terms of 'offspring equivalents' (Strassmann, 1981). In this species, a female who founds and runs a nest by herself manages to rear about three offspring, while females who join a sister and then help to raise nieces produce from 3.75 to 10.38 'offspring equivalents', depending on the size of their nest and their position in its egg-laying hierarchy.

Haplodiploidy is not restricted to the Hymenoptera – it is the exclusive method of sex determination in the Thysanoptera (Kiester & Strates, 1984). Maternal care in thrips is indeed quite common but what could be called a kind of parasocial behaviour is so far known in only one species. The lack of evolution of truly eusocial behaviour in thrips may be due to the paucity of opportunities for sisters to help one another in a mutually beneficial form of brood-rearing. Baby thrips can feed themselves, while, in marked contrast, the larvae of social hymenopterans are confined in a cell, quite helpless, unable to fend for or defend themselves and dependent for all essential services on the teamwork of a group of sisters.

CRUSTACEA Woodlice, crabs

Parental care, with parents and offspring living together in family units employing a sophisticated form of social behaviour, is found in two genera of terrestrial isopods, *Hemilepistus* and *Porcellio*. The sole example in the former genus is *H. reaumuri*, a woodlouse which inhabits the parched deserts of North Africa and Asia Minor. This is the driest habitat conquered by any species of crustacean and it is probably the exigencies of surviving in such a harsh environment – particularly the need to dig and then maintain possession of a burrow – which have predisposed the evolution of social behaviour in this highly successful animal (Linsenmair, 1984). In so doing, it provides one of the relatively few examples of paternal care among invertebrates.

Burrows are critical to survival, leading to constant take-over attempts which can only be effectively countered by the kind of joint defensive system available to a socially cohesive family unit. An individual living alone is faced with two stark alternatives: it can either stay at home and ultimately starve, or leave to forage and return to find its home appropriated by a stranger. Only by staying with his mate and taking care of his offspring can a male ensure any degree of reproductive success in such a competitive and hostile environment. Parents and offspring are able to recognize one another and interact via a refined system of communication based on genetically determined messenger compounds. Thus an individual who has gone out to forage is recognized on its return by the guard at the burrow entrance, who antennates the incomer and grants admission only after detecting and assessing the correct chemical 'family password' (Linsenmair, 1972).

A burrow is usually initiated by a female and, although she initially excludes all intruders, she eventually yields positively to the protracted 'admission ritual' performed by a single male. As the female is committing herself to this male for life (or at least until one of them dies), it is not surprising that his entry plea is required to be lengthy, presumably so that his potential qualities as a life-long mate can be gauged. Once ensconced, the male takes his share of the work involved in extending the burrow and eventually helps care for his offspring. The young woodlice – which may number up to 100 – spend their first 2–3 weeks in the nursery burrow, where they are provided with food by their parents. After that, they go out to forage, being re-admitted by virtue of the family chemical badge. However, there is also the problem of recognizing their own family residence among thousands. They do this by antennating an embankment of smooth rectangular faeces deposited around the entrance, forming a chemically-impregnated wall.

Parental care in decapods, involving the formation of a family group consisting of mother and offspring, is restricted (as far as is known) to a single species, *Metopaulias depressus*, from Jamaica (Diesel, 1989). This crab also inhabits rather a harsh environment, the forest-covered limestone hills of Cockpit County, Jamaica. It also requires a specific kind of home, namely terrestrial or epiphytic specimens of the bromeliads *Aechmea paniculigera* and *Hohenbergia* spp., for which there is considerable competition. Although these plants are abundant in the area, there is a catch – they are only of use if they can trap a water body large enough to provide a reliable and stable breeding environment for the female and her offspring. As this size of water body is greater than the average size of bromeliads in the population, there is considerable competition for the natural tanks available, especially because, in this rather hot and arid environment, these vegetable tanks are the sole enduring source of water.

Even when she finds a suitable tank, the female crab still has to prepare it for the reception of her offspring, as the conditions within may be unsuitable for their development. She prepares an aquatic nursery chamber by evacuating a leaf axil of accumulated, decaying organic detritus, such as rotting leaves. This action probably raises the levels of dissolved oxygen, as well as reducing the level of acidity to somewhere near neutral; the latter process is further aided by the female's habit of leaving *in situ* any empty snail shells (which contain lime), these being strictly excluded from her otherwise thorough cleaning job. She stoutly defends her developing brood against aquatic predators, such as damselfly nymphs, and brings them food consisting of small invertebrates gathered from the rosette. The baby crabs remain with their mother for about 8 weeks, gaining weight more rapidly than orphaned broods and enjoying a correspondingly reduced rate of mortality. The babies are slow to desert their mother to set up home for themselves, which results in the formation of family units comprising the young from several successive broods.

CHILOPODA Centipedes

Maternal care of the eggs and young is found widely in the centipedes and has been known for many years in both the Geophilomorpha and Scolopendromorpha. The American *Geophilus rubens* (Geophilomorpha) lives and breeds in crevices beneath bark. She tends her eggs by stroking them regularly with her antennae and mouthparts, also turning them over with her poison claws. Such behaviour is commonly seen in brooding mothers and probably serves to inhibit the development of mould on the eggs. It is likely that glands situated on the female's head produce a fungicidal secretion which is applied to the eggs during her tending operations. Female geophilomorphs do not feed while brooding their eggs, which means fasting for up to 50 days or more. A female of the European *Necrophloeophagus longicornis* will vigorously defend her eggs from a predatory incursion by another female, maintaining a secure rampart around her eggs by curling the rear two-thirds of her body around them, leaving the front third free to engage in the struggle. Females stay with their developing offspring through several moults.

Maternal care is common in the spiders, and can be very highly developed, but is rarer or more basic in the other subclasses. Female scorpions give birth to live young, which usually emerge into a 'brood basket', normally formed by the mother's front two pairs of legs. The newborn babies clamber up and across either their mother's chelicerae or her rear legs (depending on species) and on to her back, usually arranging themselves in a random fashion, save in *Vaejovis*, in which they form up into neat rows. First instars do not appear to receive food hand-outs from their mother, but may share in her prey when they reach their second instar, after which they soon disperse. However, it is possible that first instars receive some nutriment from their mother via lipids absorbed directly from her exterior surface. Whether or not this is so, the main functions of the early close association between mother and larvae may lie in defence against predators and the mother's ability to remain in the optimal microclimate for larval survival.

Some female Solifugae also exhibit maternal care. *Galeodes vorax* guards her offspring in her burrow, driving away insects which intrude but fully able to recognize any of her own young which have been experimentally removed and then replaced. Similar habits are found in other members of the genus, as well as in *Solpuga caffra* from South Africa. Many female pseudoscorpions lay and guard their eggs inside a closed igloo-like chamber which is constructed from silk produced by their chelicerae. This is combined with small particles of detritus which are manipulated with the pedipalps and chelicerae.

OPILIONES Harvestmen

Remarkable paternal care

Perhaps the most remarkable example of parental care seen in the Arachnida occurs within the Opiliones. It is the males of the Neotropical harvestman *Zygopachylus albomarginis* who build a nest and guard the eggs (Rodriguez & Stella Guerrero, 1976; Mora, 1990). This is the sole known example of paternal care in arachnids and one of only a handful of examples in all terrestrial invertebrates. Paternal care in this harvestman is based around the unusual mating system, in which it is the females who seek out the males and their nests, and the males who exhibit choosiness over which suitors they accept. As the males subsequently guard any eggs, they are assured of paternity, thus making a continued investment in time and energy worthwhile. When the offspring hatch they are guarded by their father for a few days, but he does not feed them; shortly afterwards they disperse.

Females wander freely around on the tree trunks, sometimes visiting several nests in turn and trying to court a succession of owners. The female initiates proceedings by tapping the male gently, first with her front and then with her middle legs, stimulating him to reply likewise. Eventually, she grasps him by his cephalothorax and draws him up close. However, many females do not succeed in getting this far, as the males discriminate actively between possible partners, often refusing to respond to their leg-tapping advances and even driving them fiercely off the property. The male might even harry the departing female by nipping her more tender appendages, such as the leg joints, with his chelicerae. Females never attempt to reply with equal hostility, and withdraw from the nest if its owner persists in meting out a hot reception. However, a successful female suitor is allowed to stay and mate, although she is still under the command of the male, who exerts his authority over all goings-on inside his property by dictating to the female where she should deposit her eggs. If she trys to overstay her welcome, the male makes his intolerance of her presence quite clear by biting and harassing her until she makes a rapid exit. The owner makes an effective guard against certain common predators, such as ant scouts, which are deftly flicked off the trunk before they can detect anything of value, thereby preventing the recruitment of reinforcements and total loss of the nest. There is no defence against a raiding swarm of ants and the occasional insidious attentions of marauding flatworms. However, the main threat comes from other harvestmen of the same species, both male and female, who will raid the nest and eat the eggs if given the chance. The male's role as a guardian is vital to the survival of his offspring, as it has been demonstrated that unguarded nests suffer a high rate of predation compared with guarded ones.

There are also examples of maternal care in harvestmen, such as the cave-dwelling *Hoplobunus boneti* from Mexico (Mitchell, 1971). The females stand protectively over clutches of eggs, which are attached firmly beneath the roofs of caves, and, later, over clusters of youngsters. Maternal care has also been seen in harvestmen from New Zealand, Uruguay, Chile and Brazil.

ACARI Mites, ticks

Parental behaviour

There are few examples of parental care in the Acari. The argasid tick *Ornithoderus moubata* remains beside its eggs until some of the young have hatched, while the prostigmatid mite *Cheyletus eruditus* guards its eggs and defends them against intruders. The female of the ixodid tick *Ixodes kopsteini* from Indonesia 'defends' her developing brood by retaining them within her own body (Anastos *et al.*, 1973). The female's dead body thus acts as a 'brood sac', protecting her offspring until they hatch and become mobile.

Two species of mite appear to have attained some degree of subsociality. A colonizing female of the moth ear mite, *Myrmonyssus phalaenodectes*, modifies the moth-host's tympanic organs by biting through the tympanic and countertympanic membranes (Treat, 1956). This provides an entry to the tympanic air sac for her newly hatched brood and, at the same time, deafens the moth in that ear. The mother remains with her offspring until they reach maturity, often resisting attempts by other females to enter the brood chamber, but otherwise not exhibiting any further

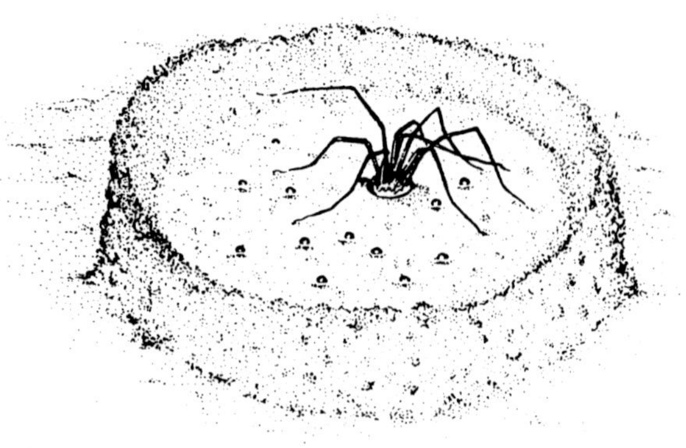

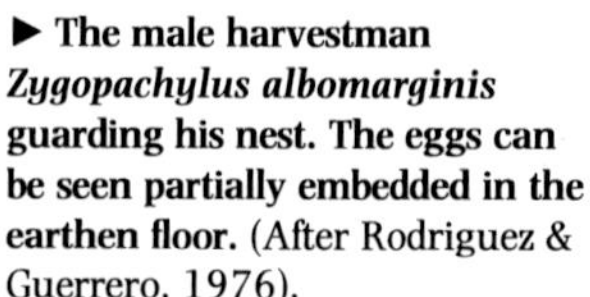

▶ **The male harvestman *Zygopachylus albomarginis* guarding his nest. The eggs can be seen partially embedded in the earthen floor.** (After Rodriguez & Guerrero, 1976).

maternal solicitude. The mites divide up the moth's three-chambered ear into specific areas devoted to egg-laying, feeding, copulation, defaecation and the disposal of moulted skins. Several females and their broods may eventually colonize a single chamber, yet despite the inevitable overcrowding, the second vacant chamber is never invaded, thus ensuring that the host is still able to hear approaching bats and take avoiding action. As this is essential for preserving the further existence of the colonizing mites, their habit of sticking to one ear only is obviously adaptive, but just how they 'decide' among themselves to be so conservative is uncertain.

The spider mite *Schitzotetranychus celarius* forms subsocial 'family nests' on bamboo stems in the Far East (Saito, 1986). Both the female and, most interestingly, the male take part in defending their family against enemies, in particular the predatory larvae of the mite *Typhlodromus bambusae*. Guarding males will mount a ferocious attack against invading larvae of this mite, often being able to kill them and provide total protection for their offspring.

ARANEAE Spiders

Maternal care

Maternal care is very widespread in the spiders (Araneae). The habit of remaining on guard near or upon the egg sac, or carrying it around attached to the jaws or spinnerets, is common to the females of many different families. In many instances, the mother dies before the eggs have hatched, thus she never sees her offspring. In other cases, the female overlaps with her brood but does not interact with them directly, while, in a select few examples, she actually provides food for her developing progeny. However, considering the abundant nature of maternal care in spiders, it is surprising how little is known about the degree of protection provided by the mother, although she may also provide services which are unrelated to defence, such as thermoregulation.

Most female salticids construct a silken nest between leaves, twigs or stones, and stand guard over their egg sac. *Lyssomanes jemineus* from Colombia provides active defence for her eggs, lunging at predators such as lacewing larvae and eating them (Eberhard, 1974). Ants are treated to a different reception, being flicked neatly off the leaf with the spider's front legs. The female remains with her brood as they hatch, enveloping them in more silk as they emerge (for reasons unknown), but she deserts them as they begin to stray away soon after their moult. *Portia fimbriata* from Australia spins a nest on a suspended leaf, remaining to defend her eggs against insect enemies and other females of her own species, who will often try to evict a guarding female (Jackson, 1982). A successful evictor will quickly eat the former tenant's eggs and then lay her own in new silk deposited atop the ruins of the old eggs. Females of a *Euryattus* species from Queensland lay their eggs in an aerial nest formed by gradually hoisting a leaf into position on a network of silken lines in a most complicated exercise (Jackson, 1985).

▲ This female *Hersilia bicornuta* from South Africa is standing guard above two egg sacs on a tree trunk. She has carefully camouflaged the sacs with pieces of bark prised off with her chelicerae.

Bark-inhabiting spiders, such as most hersiliids, and sparassids, such as the Australasian lichen spider, *Pandercetes gracilis*, place their flattened egg sacs in full view on the bark. They are usually then camouflaged with flakes of bark and lichen prised off in the chelicerae and carefully incorporated into the surface of the sacs. The highly cryptic mother sits on guard head-downwards above the sacs, where her prime function is probably to nip in the bud any possible ant invasion by dissuading their lone scouts from coming too close.

Carrying the eggs around

The main function of maternal care in many of the wolf spiders (Lycosidae and Pisauridae) may be to ensure the best possible micro-climate for the optimal development of the eggs. This is especially important in species from temperate countries, where long periods of inclement weather would slow things down, as well as increase the possibility of moulds invading and destroying an egg sac which is constantly damp. Pisaurid females carry their large, globular egg sacs speared on the points of their fangs, which prevents them from feeding while they are thus encumbered. After a spell of cool wet weather, the first warm day draws out females in some numbers, sitting around in exposed positions and airing their egg sacs by tilting the body backwards so as to expose the maximum surface area of the sac to the sun's rays.

Egg-sac-toting female pisaurids and lycosids sense when the young are about to hatch, and tease apart some of the tough enveloping threads with their chelicerae. If the female is prevented from rupturing the egg sac, the young are unable to emerge and so die within their silken prison. The female *Pisaura mirabilis* constructs a silken nursery tent in which she deposits the sac just before it hatches. She straddles the top of the tent until the spiderlings have moulted and dipersed, although she neither catches food for them nor interacts in any other way. This life-style is highly successful, for up to 14 of these conspicuous nurseries may be present within a square metre of low vegetation. However, despite carrying her sac in her chelicerae for an extended period, the female seems incapable of preventing *Trychosis* parasitic wasps from creeping up unseen and ovipositing within it. She may thus spend her final days carefully nurturing a nestful of developing wasps, the chances of this happening increase as the season advances and the wasps become active. This effectively means that there is a race on between the spiders and the wasps each spring. This places a premium on a spider's ability as a predator, as the more rapidly it can feed up early in the year, the quicker it will grow and be able to produce an egg sac before parasite levels increase. The nuptial gift provided in this species by the male during mating may serve to augment the female's rate of growth, thus contributing towards boosting her eventual reproductive success (Austad & Thornhill, 1986).

◀ In spiders of the primitive family Pholcidae, the female holds her egg batch in her chelicerae until they hatch, as seen here. Unlike in pisaurids, which fast while carrying their egg sacs, the female pholcid will temporarily hang the egg cluster in her web in order to feed.

▶ Lycosid babies cluster on their mother's back immediately after hatching, and are carried around in this way for several days. This is *Pardosa nigriceps* from Europe.

▼ After a spell of cool weather, the female nursery-web spider, *Pisaura mirabilis*, from Europe will 'air' her egg sac by sitting in a warm spot. She carries the sac impaled on the tips of her fangs and fasts while so doing.

The habit of carrying eggs around in the chelicerae is also found in the primitive web-building Pholcidae, although the wrapping is more insubstantial so that the eggs are clearly visible, being held together only by a few scrappy threads. Unlike in the pisaurids, a pholcid female may briefly relinquish her egg sac in order to feed, or even to mate, the web providing temporary secure stowage for her precious cargo. The spiderlings begin to emerge while the eggs are still held in the chelicerae, so that the female appears to sport a large, more-or-less spherical 'moustache' composed of numerous long-limbed hatchlings. These soon disperse among the silk of their mother's web, where they remain for a while without feeding until they undergo their first external moult (spiders normally moult from a deutovium, or first stage, while still *inside* the egg sac). The equally primitive *Scytodes* has similar habits, except that the female slings her egg sac beneath her, held between her palps and a line drawn from the spinnerets. She thus has sufficient freedom of movement to be able to catch prey and feed, during which time she hangs up the egg sac nearby. In the Eurasian *S. thoracica*, the female has to tease apart the silken case on the egg sac to allow the young to emerge, although in some other species the eggs are virtually naked and such assistance is presumably unnecessary.

Female wolf spiders, such as *Pardosa* and *Lycosa*, carry their egg sacs attached to their spinnerets and can therefore continue to feed while waiting for the young to hatch. This event is probably accelerated by the activities of the females, who select the correct micro-climate, such as a sunspot on a log or leaf, where they spend most of their time basking. Despite the egg sac's apparently inaccessible position at her rear end, the female can sense the imminent emergence of her offspring and facilitates their exit by tearing the sac around its seam with her chelicerae. As they pour forth from the split in the envelope, the spiderlings crawl on to their mother's back and assemble in a neat mound. The stimulus for this highly ordered activity seems to be the peculiar knobbed hairs which cover the relevant part of the female's abdomen. These not only indicate to the spiderlings where to assemble, they also help them to cling on. (In the related Pisauridae, which do not carry their babies on their backs, these hairs are absent.) The female scampers freely around for several days unhindered by her load of babies, which are able to re-board via their mother's legs if they are dislodged during her wanderings, finding their way back by following the silken lines which they always trail behind them. The babies are also said to descend from their mount in order to drink, re-boarding up the

▲ **This female green lynx spider, *Peucetia viridans* (Oxyopidae), in Costa Rica will have stood on guard over her egg sac and then assisted her offspring to emerge by tilting the egg sac and tearing an opening in it; the babies soon disperse without being fed by their mother.**

legs when they have finished. The spiderlings do not feed during this jockeying period but survive on the vestiges of their yolk before finally moulting and deserting their mobile home.

Benefits of care in the green lynx spider

The adaptiveness of maternal care is often taken for granted, it being assumed that investing time and energy in staying with a brood must be the best bet. However, there is an alternative option, which is to leave the eggs to take care of themselves and invest the resulting savings in time and energy in producing a second clutch. In temperate spiders, this choice is usually unavailable, as there is only enough time to produce a single egg sac anyway, so the logic of taking good care of the sole investment in the future is self evident. However, tropical and subtropical spiders may be able to produce two or more egg sacs in a lifetime, and then the wisdom of investing everything in the first brood is not so clear-cut.

The manifest advantages accruing to guarding behaviour by the green lynx spider, *Peucetia viridans* (Oxyopidae), in a subtropical habitat in Florida, have however been firmly established (Fink, 1986 & 1987). The female attaches her rather lumpy whitish egg sac to a leaf by a number of silken guy-lines. She stands guard above it, apparently oblivious to the approach of a tiny mantispid larva, which bores into the sac and eventually destroys it while she is still on duty. Yet she is alert to, and well able to attack and kill, the most common threat in any warm country, the omnipresent ants. However, if several ants mount a persistent incursion which threatens to overwhelm her defences, she modifies her response. She gives up trying to rebuff the enemy and instead cuts the sac free from most of its guy-lines so that it hangs from just one or two. Then she clambers on to the sac, thus making it less likely that the questing ant scouts will be able to re-find either the egg sac or the owner. If a scout does succeed in making its way along one of the remaining silken ties, it can easily be tackled and removed from the fray. An alternative strategy adopted in the face of repeated invasions by ants is to remove the sac altogether from the danger zone and relocate it at some distance. The female does this, not by carrying the sac in her chelicerae, as a pisaurid would do, but by a cutting-and-attaching sequence, whereby new lines are attached and old ones cut, thus gradually drawing the sac along a predetermined path. By this rather clumsy-sounding method, distances of a metre or more can be covered.

The guardian may also have to defend her sac against attacks from other spiders, such as salticids, as well as other females of her own species who will eat her eggs if they can, after killing her. Just before her offspring hatch, the female spins a silken tent over the sac and assists in their emergence by biting at the silken envelope, pushing her chelicerae in beneath the lid and tugging the bowl-shaped egg container downwards to leave a gap. She also tilts the sac downwards after opening it in order to expedite the egress of the spiderlings from the tear in its lower side (Randall, 1977). The babies exit much more slowly from an untilted sac, which results in a higher mortality from cannibalism among the babies stuck inside. Their mother's habit of tilting the sac therefore leads to a higher rate of emergence for her offspring. She executes her sac-opening operations when she receives some cue from within, probably after detecting movements by the spiderlings. This is unlike the situation in *Trochosa* (Lycosidae), in which the stimulus to open a sac arises from within the female, and is probably regulated by an internal mechanism linked to the same external processes which govern the development of the spiderlings themselves.

If the *P. viridans* mother fails to open the sac, the babies can still bite their own way out (unlike in the pisaurids and lycosids, in which the sac becomes a tomb from which the babies cannot escape without their mother's help). The spiderlings may take their share of insects caught by their mother, but she does not deliberately

Advanced care in theridiids and eresids. In a number of spiders, the female not only remains with her progeny within her web but catches food for them as well; two of the best-known examples are *Theridion pictum* and *Achaearanea riparia*. When a female of the latter species has caught and subdued an insect, she strokes the web as a signal to her offspring to come and feed. However, if danger threatens, she vigorously plucks the silk in quite a different message, sending her babies rushing from their meal and into their silken retreat.

In a small and very select band of spiders, the female actually provides her babies with food on a direct mouth-to-mouth basis by regurgitating a special liquid. This habit has so far only been noted in a few species, but remarkably these belong to two different and unrelated families, the Theridiidae and Eresidae. It has long been known in the very common European *Theridion sisyphium*. The first meal for the newly hatched spiderlings comes from their mother's mouth. The food comprises a blend of pre-digested food derived by the mother from her insect prey and her own intestinal cells. During the first few days out of the egg case, the tiny spiders exist almost entirely on their mother's offerings, clustering around her regularly to sup at the droplet as it oozes from her mouth. If the food is not offered regularly enough they may caress her legs, which apparently induces her to regurgitate on demand. However, during these first few days, and increasingly thereafter, their mother also catches insects which blunder into the web. She prepares these for consumption by chewing the tough exterior with her mouthparts, thus allowing the contents to leak out over a wide area of the integument. By doing this, her tiny babies, with their inadequate mouthparts, can deal with prey which would be much too large for them to tackle alone. As they grow larger, the babies begin to take their own part in subduing an insect, throwing over it masses of their own slender silk as it struggles to free itself. It is common to see the mother and a cluster of babies all feeding harmoniously together on a hover fly or beetle, and even distasteful species such as the cantharid *Rhagonycha fulva* are apparently acceptable fare to both mother and brood. By providing a combination of regurgitated liquid food and pre-prepared insect food, the female *T. sisyphium* considerably accelerates the development of her progeny, who gain weight so fast that they are ready to moult after only a week.

Similar habits are found in *Theridion impressum*, although in this species the mother's final contribution to her brood's welfare is her own corpse. This is not unusual as, in at least ten spider species from the genera *Amaurobius, Coelotes, Theridion, Eresus* and *Stegodyphus*, the mother bequeaths to her offspring a posthumous funeral banquet comprising her own defunct body. Female *Stegodyphus* (Eresidae) actually digest their own tissues at the termination of their brood-care period, turning their inner selves into a kind of convenient liquid soup, stored inside the still-intact outer husk. *Stegodyphus* also provide regurgitated food to their offspring, containing liquefied intestinal tissue as in *Theridion*, so the provision of a pre-digested whole corpse at the end of the female's useful life is a logical extension of the feeding process. Babies which have enjoyed the benefits of regurgitative feeding have subsequently been observed eating their deceased mother in the solitary or occasionally social species *Stegodyphus lineatus* and *S. pacificus*, but the habit is also found in the social species *S. mimosarum* and *S. sarasinorum*. It has perhaps been longest known in the agelenid *Coelotes terrestris* which is common in woods in Europe. The spiderlings solicit regurgitation by stroking their mother's chelicerae with their palps, finally prospering at her demise from the banquet provided by her corpse.

▼ For the first few days of their lives the European spider *Theridion sisyphium* feeds her babies mouth to mouth with regurgitated fluids (top). When they are hungry and need a meal, the babies 'gang up' on their mother and tug vigorously at her legs (bottom). If she is not prepared to feed them she will shake them off, sometimes even kicking out and flicking one or two of the tiny spiderlings across the web, usually without harm. If incoming prey is scarce, the females rapidly become shrunken, indicating the heavy toll taken of their bodily tissues by this investment in their offspring's future.

provide them with such food, as do some other spiders, and in most cases the babies probably do not feed at all before dispersing. The female thus invests a total of 6–8 weeks in her brood, but the benefits from reduced predation by ants and the reduced destruction of unguarded sacs by bad weather outweigh the costs, at least in Florida. The female's lifetime fitness is thus increased by guarding a first egg sac, rather than abandoning it and producing a second, as would be theoretically possible.

This is a relatively short guarding period compared with the New Zealand species *Desis marina* (Desidae), in which the female remains with her brood for about 5 months (McLay & Hayward, 1987). This lengthy investment in time is due to this spider's strange habitat, for the nests are affixed to seaweeds on the shoreline. The female's help is essential in releasing the hatchlings from the egg sac, after which the babies remain in their home for about 2 months. This extended developmental period within the nursery (much longer than in any other spiders, save *Desis kenyonae* from Tasmanian waters) is governed by the low temperatures in the nest, which is submerged beneath chill coastal waters for much of the time.

Social spiders

Some form of sociality is found in about 33 species of spiders in 11 families (Dictynidae, Amaurobiidae, Oecobiidae, Uloboridae, Pholcidae, Eresidae, Araneidae, Theridiidae, Agelenidae, Dipluridae, Lycosidae (Burgess, 1978)). The existence of brood cannibalism of the mother by her offspring is something of a surprise in *Stegodyphus mimosarum* and *S. pacificus*. This is because one of the basic criteria for the establishment of harmonious group living was thought to be the elimination of the cannibalistic tendencies which are common among non-social spiders. However, it seems that this is not necessarily the case.

The second-instar spiderlings of the highly social *S. sarisinorum* from India have been seen feeding on their mother while she was still alive (Jacson & Joseph, 1973). *S. mimosarum* from Africa is one of the most highly social of any spiders, yet in the silk nests which lie within the communal webs, clusters of juveniles may be found sucking the juices of their still-living yet moribund mothers (Seibt & Wickler, 1987). There is even considerable evidence that they are not very fussy about whose mother they tuck into, and they dine indiscriminately upon any corpse lying handy. However, neither juveniles nor adults of this species can be induced to feed on healthy members of their own species, nor even upon other co-inhabiting species of spiders. This seems to indicate that females who are approaching the ends of their active lives actually emit some kind of 'eat-me' signal. This switches off the normal aversion to cannibalizing healthy individuals, which is necessary for the maintenance of social harmony. Senile mothers seem to expedite the cannabilistic process by forming groups or 'body pools' within the chambers, enabling juveniles from different mothers to feed indiscriminately on the available provender. Seibt and Wickler have coined the term 'gerontophagy' for this process, which among social animals is so far restricted to *Stegodyphus*.

The New World theridiid *Anelosimus eximius* exhibits a particularly well-developed and sophisticated form of social organization. Individual females are not restricted to caring just for their own broods, but feed all young unselectively via regurgitation (Christenson, 1984). The huge webs may contain thousands of individuals from many different generations, each member undertaking various tasks necessary for the good maintenance of the colony. Thus, there are close similarities with many social wasps, even to the extent that only some of the females in the spider colony get to mate and reproduce.

Repair of the web is carried out daily by adult females, and by juveniles at or beyond their third instar. The adults preponderate in this mundane work, but if sudden catastrophe causes major damage to a large part of the web's structure – such as a branch crashing down – it is the juveniles who are recruited in their hundreds to perform the work of repairing the damage. Web sanitation is also undertaken by females and juveniles, who clear away bits of debris, old egg sacs and the indigestible bits of prey by tossing them on to the ground. Prey capture is carried out on a co-operative basis and up to 100 spiders may simultaneously crowd around an exceptionally large meal, although a female who helps in subduing prey may not necessarily feed on it. The propensity for rapid recruitment of helpers is also useful in nest defence, when trespassing *Trypoxylon* wasps intent on capturing an individual spider are often driven away by the rapid onrush of a number of adult females. However, despite the generally concordant nature of the interactions between such a large number of predatory animals, the harmony is not perfect and there are occasional instances of cannibalism, often during fights over the possession of prey.

Egg sacs may be cared for by a succession of females, and these nursemaids also feed the spiderlings indiscriminately by regurgitation. As in most communal spiders, the males do not participate in either web repair or egg-sac guarding (an exception being the African *Agelena consociata* in which the males take part in communal activities). There are considerable advantages to the social way of life for *A. eximus*, not least of which is the principle that a female can be certain that her offspring will be nursed by other females if she dies. Set against this is the fairly high risk (10–20 per cent) of losing an egg sac through cannibalism, although the reasons for this are uncertain. In fact, it could even be possible that egg sacs serve as 'emergency food stores' when prey is scarce, a situation in which having more mouths to feed would make matters even worse.

Females who strike out on their own in an attempt to establish new colonies meet with little success. Single webs just do not seem to be viable and their occupants quickly succumb to a fatal combination of weather and predators. Expansion of existing colonies is the normal method of growth, although it is a mystery why several females do not leave and set up home together, with all the start-up advantages of communal care, as happens in *Achaearanea wau*. By contrast, any deliberate mechanism for the establishment of new colonies in *Agelena consociata* (other than accidental splitting of a colony by an animal blundering through a web) seems to be strangely absent (Roeloffs & Riechert, 1988). This is inferred from the fact that inhabitants of colonies situated only a few metres apart are genetically no more closely related to each other than they are to members in colonies many kilometres distant. This has led to a high incidence of relatedness within each colony and an associated highly developed social life. Egg cases are deposited in a communal chamber, where they are protected not only by adults but also by older juveniles (females of the tiny Mexican *Oecobius civitas* (Oecobiidae) go one step further and construct joint egg masses). Small spiderlings cluster in large crèches, where they are brought food indiscriminately by a variety of females, as well as receiving meals via regurgitation. Even capture of prey insects is undertaken co-operatively, with the advantage that larger prey can be tackled than would be possible for one spider operating alone.

Regurgitative feeding had been thought essential for the evolution of sociality in spiders, due to the intimate contact and the high degree of tolerance between individuals, which are required for its efficient functioning. However, the highly social *Achaearanea wau* from New Guinea has been shown not to employ this habit, and the newly emerged spiderlings are capable right from the start of feeding on prey captured by the females (Lubin, 1982). Similar habits probably apply to *A. mundula* and *A. kaindi* from the same area, as well as to *A. disparata* from West Africa.

BLATTODEA Cockroaches

Viviparity occurs within the Blattodea, but is so far known only in a single species, *Diploptera punctata*. In this fascinating cockroach, the eggs are exceptionally small and do not contain sufficient yolk for the complete development of the embryo. To compensate for this deficiency the female has developed complex 'milk glands' which provide the necessary nourishment.

A temporary association between mother and offspring is common in ovoviviparous species, in which the ootheca is carried around inside the female's body until the eggs hatch. The pale, soft, newly hatched nymphs usually cluster around their mother for a few hours until their cuticle hardens and darkens, when they disperse. In the Neotropical *Homalopteryx laminata*, the female has to stilt high on her legs in order to accommodate the mass of white nymphs clinging beneath her for the few hours needed for them to harden off and leave; similar behaviour is seen in *Nauphoeta cinerea*.

The mother–offspring relationship in a few other species is far more enduring, although but little studied as yet in some of the more interesting of these. The nymphs of the Cuban burrowing cockroach, *Byrsotria fumigata*, cluster beneath their mother for several days. They recognize her position and orient to her through an aggregation pheromone which she releases. However, their maternal ties fade after their moult into the second instar, after which they will happily cluster beneath a strange female. Their mother does nothing save make her body available as a shelter, but even this is probably sufficient to increase her brood's chances of survival beyond that of an untended brood.

Observations on several other species of ovoviviparous cockroaches suggest a much closer and more enduring bond between mother and offspring. The Sri Lankan *Phlebonotus pallens* acts as a kind of 'armoured personnel-carrier' for her brood, which are tucked away in a gap between the top of her abdomen and the underside of her rather-domed wing cases (Roth, 1981). A dozen or more nymphs can be neatly stowed away in this mobile chamber, invisible to the outside world and at least temporarily protected by their mother's ability to avoid danger and seek out the most favourable micro-climate.

In a number of other species, the nymphs are carried externally, clinging beneath their mother's body which is usually rather heavily armoured. *Pseudophoraspis nebulosa* from Borneo trundles her nymphs around in this way, as do at least two species of *Perisphaerus* from the Far East. *Perisphaerus semilunatus* from Thailand resembles a pill millipede and, like a millipede, she can roll up tightly into a protective ball. It is surprising that, in so doing, the female cockroach is able to fit in a cargo of at least nine babies, which cling to her underside, without nipping any of them between the armour plates as they close together. Once fully rolled up, the shiny-backed female and her brood are relatively impervious to the attacks of the most pernicious of invertebrate enemies, the ants. However, the relationship between mother and brood may amount to much more than just transport and protection. The nymphs have strange elongate proboscis-like mouthparts, which appear to 'plug in' to corresponding cavities (intercoxal-gland orifices) on the mother's underside. Although no glands have been found to open into these cavities, it is still possible that the nymphs are able to batten on to some maternal secretion while they are being trundled around. The fact that the first two instars are also blind and probably quite helpless lends weight to the theory that the mother's body might function as a mobile vending machine but this has still to be elucidated. As there are at least 17 species in the genus, at least some of which are known to co-exist on plants as mothers with broods, it is likely that all are subsocial in some respect. In *Leucophaea maderae*, the nymphs also accompany their mother, not clinging beneath her but foraging on their own account, like chicks with a hen.

The primitive brown-hooded cockroach, *Cryptocercus punctulatus*, forms family groups in rotten logs. Wood comprises the sole diet for this species, which shows many striking similarities to the termites (Isoptera). The hind gut of both *C. punctulatus* and the primitive termites is home to cellulose-digesting protozoans. It had long been suggested that the need to pass these essential symbionts from one colony member to another at the time of moulting (when an individual termite loses its crop of symbionts) was the stimulus for the evolution of a full social life in the termites. A newly moulted termite re-establishes its symbionts by lapping up the faecal fluids of a nest mate, thus making group life obligatory. It had been suggested that newly hatched nymphs of *C. punctulatus* acquired their protozoans in the same way, from their elder relatives. However, recent observations (Seelinger & Seelinger, 1983; Nalepa, 1984) point to a different method of transfer in the cockroach, and so no significant clues about the evolution of social life in the termites can be provided.

The brown-hooded roach lives in galleries in rotten logs, each family having its own gallery, although, in a heavily colonized log, partitions between galleries, erected by the roaches themselves, may be almost paper-thin. Young larvae spend a considerable amount of time grooming one another and the two adults. The adults patrol the galleries, usually feeding some way off from their brood, but at certain times one of the adults (often the female) becomes a focus of excited attention for up to 20 nymphs who cluster around, jostling for position in order to line up close to the adult's rear end. This activity seems to be linked to the secretion of gut contents from the adult's anus. This is licked up by the larvae and they are thus provided both with pre-digested food and a full complement of their vital symbiotic protozoans. Sometimes several nymphs are able to push their heads deep inside the anal chamber of the adult. The adults offer this food at regular intervals, although how its imminent appearance is communicated to the nymphs is unknown. The nymphs take their first faecal meal when only a few hours old and still feed occasionally in this way after nearly a year.

ISOPTERA Termites

The behaviour of the brown woodroach is rather similar to that of a number of termites (Isoptera), especially the primitive Darwin termite, *Mastotermes darwiniensis*, from Australia, which is both morphologically and behaviourally similar. Termites are by far the oldest social animals, having evolved a complex social organization long before the ants, bees or wasps. All termites are social, with many generations living together in the same nest, each succeeding generation helping to raise the next – the hallmark of truly social animals. There is usually a complex nest which was originally founded, after a brief nuptial flight, by a single male and female pair, the 'king' and 'queen'. Colony members (usually blind in most species) are divided into separate castes as in ants: workers for foraging, nest repair and brood-raising; soldiers for defence; and the king and queen for procreation. However, unlike hymenopterans, in which the sterile worker caste is normally composed solely of females, in termites both males and females are involved.

The diminutive king usually remains by the queen's side in the royal chamber, occasionally mating with her in order to fertilize the prodigious output of eggs, which may reach 30 000 per day, the female's bloated body having become an egg-producing factory. This founding pair are the primary king and queen and they have come a long way since their nuptial flight. They would have nurtured their first brood on secretions from their own bodies and, even after a full year, they would probably only have reared a handful of workers. Once the nest is established and growing, the primary queen will continue producing eggs for many years but, when she eventually dies, or perishes prematurely, her place may be taken by a secondary queen. She is drawn from one of the brood of nymphs which were being reared for eventual release as winged reproductives. A special food supply is given to several selected nymphs, who develop sexual organs and become secondary queens, able to continue the job of supplying eggs for the colony, although at a reduced rate compared with the primary queen. Should no nymphs be available, then immatures originally destined to be workers can be specially reared to become a tertiary king and queen but, due to her small size and limited egg output, such a queen is unable to cope with natural wastage of nest members and the colony eventually declines and perishes.

MANTODEA Mantids

Most female mantids abandon their ootheca as soon as it is finished. With its tough exterior finish, it is well suited to resist the elements, but it may be vulnerable to attack by parasitic wasps. In a number of species of *Oxyophthalmellus*, *Galepsus* and *Tarachodula*, however, mainly from Africa, the female remains behind to guard her ootheca. In all these, the ootheca is long and narrow, being designed to fit beneath the mother's elongate body. She closely straddles the ootheca and is presumably effective in preventing parasitic wasps from gaining access to those eggs directly beneath her, although peripheral eggs are likely to be at risk. It seems likely that the East African *Tarachodula pantherina* may 'fly a warning flag' on her ootheca, as her black-spotted legs and orange-banded abdomen constitute a typical 'warning' uniform. Also, her habit of

▲ **This female *Oxyophthalmellus somalicus* mantis from Kenya is standing guard over her newly hatched brood. As the egg case was not found nearby, she must have moved with her brood to the safety of this isolated dead-end stem. She is in the perfect position to repel scout ants or marauding assassin bugs. In order for such behaviour to be effective, her natural predatory instincts must be suppressed and she must be able to recognize her own young.**

cementing her ootheca to an exposed twig, rather than hiding it away, seems to back this up. By contrast, the stick-like *Oxyophthalmellus* and *Galepsus* seek out a concealed spot in which to lay their eggs. *Galepsus* females are remarkably adept at selecting a background (e.g. dead bleached doum-palm leaves), which match perfectly both their own colour and that of the ootheca.

Little is known about how long the African species stay with their eggs, in particular whether they remain until they hatch. This is important, as if the mother co-exists alongside her progeny, then there must be some mechanism for suppressing her normally powerful predatory instincts. In non-guarding species, the adults often cannibalize small nymphs, which make particularly easy prey. In this context there is at least one observation from Kenya of a female *Oxyophthalmellus somalicus* apparently standing on sentry duty next to a brood of quite newly hatched nymphs (pers. obs.). These were grouped at the tip of a twig – the perfect 'peninsular' refuge for minimizing the chances of being located by searching ants, the main enemies of mantis nymphs – while the female was stationed between, presumably her brood and the only access route, back along the twig. She resisted efforts to nudge her along the twig away from the nymphs and seemed most reluctant to move far away from them. When pushed into the group of nymphs, she made no effort to attack them. As the

◄ **A number of African mantids stand guard over their egg cases. This Kenyan *Tarachodula pantherina* looks to be warningly coloured, in which case she could provide possible protection against visually hunting birds; the more usual enemy would be tiny parasitic wasps, which the guardian is capable of driving away or even killing.**

ootheca from which the brood had emerged was not visible nearby, it seems that the nymphs must have migrated some considerable distance from this position, where they were guarded for the first few vulnerable hours (or maybe even days) by their mother.

▲ Certain behaviour patterns may evolve in one area of a species' range and not in another. There is no evidence of egg guarding by female *Acanthops falcata* mantids in Panama, yet this female in Peru refused to leave her ootheca, which is obviously not newly constructed. The guardian is also noticeably emaciated, indicating that her sojourn may have been protracted.

DERMAPTERA Earwigs

The maternal solicitude of earwigs has been known for many years (since at least 1773), and all the species studied so far (only about a dozen out of 1000 or so) exhibit parental care (Lamb, 1976). However, the extent of this care varies greatly from species to species. In the ovoviviparous *Hemimerus talpoides*, a member of a strange group (Hemimerina) which lives on African pouched rats, the degree of assistance may merely extend to the female's efforts to free her emerging offspring from their 'birth membranes'.

The best-known species, *Forficula auricularia*, from Europe, exhibits complex maternal care, extended over a long period and repeated several times through successive broods. The female digs a nest, usually beneath a stone and normally with two chambers at the lower end (some species dig a single blind tunnel, others a complicated labyrinth with several exits). By choosing a site beneath a stone, the female is able to regulate the temperature of her eggs by carrying them up and down the tunnel. In effect, she exploits the rapidly warming stone as an incubator by placing her eggs against it when it starts to warm up, then taking them down below when the stone becomes too hot later in the day. When tunnelling, she prises off soil particles with her jaws and carries them to the surface. Sometimes a male helps with the digging, and is allowed to remain with the female until she is ready to lay her eggs. At this point, her tolerance evaporates and she ejects him forcibly from the nest, spurring him on with her mandibles and tail forceps; females of various species may actually kill (and sometimes eat) the male at this time.

The female stands guard over her eggs in the nest, which by this time she has sealed up, so that she is unable to feed during the rest of her vigil. She is attentive to her maternal duties. If the eggs become scattered, she gathers them up in her mouthparts or rakes them in beneath her body with her legs. She is unable to recognize her own eggs and will 'steal' those of another female if given the chance, going on to tend a double clutch as if all the eggs were her own. Throughout her time on guard, the female regularly turns the eggs over with her mouthparts, 'licking' them as she does so. This seems to be essential for the welfare of the eggs, as when they are left untended they are soon destroyed by mould. Such 'disinfecting' behaviour is also seen in

Euborellia annulipes, E. cincticollis, Anisolabis maritima, Labidura riparia and *Forficula tomis*, and it seems likely that an active fungicide is applied from the female's mouthparts. However, in some species, the eggs hatch perfectly well without such treatment. The guardian is steadfast in the defence of her eggs, and will even parry the advance of a predatory ground beetle, driving it from the nest. She is able to detect the imminent hatching of the eggs and responds by spreading them out in a single layer on the soil, making it easier for the tiny nymphs to emerge.

Once the nymphs have all hatched, the female drapes herself over them as they form a pile beneath her, rounding up any strays which wander away and carrying them back in her mouthparts like a cat with a kitten. She now supplies her brood with provisions, bringing food down into the burrow from above or offering it in liquid form by regurgitation. The nymphs seem to solicit regurgitative feeding by nuzzling her mouthparts with their own. However, the nymphs of *F. auricularia* are quite independent from an early stage, leaving the nest to forage by themselves, unaccompanied by their mother who goes out by herself. Once the babies have finally left their nest, their mother fails to recognize them and, in predacious species, they may fall victim to her. Similar behaviour patterns, involving licking of the eggs and young nymphs, feeding of nymphs in the nest and fasting by the mother, have been observed in a number of other species in several genera. Earwigs are therefore among the most interesting of subsocial insects and it remains to be seen whether more complex social care has evolved in any of the numerous, so far unstudied, tropical species.

ORTHOPTERA Crickets, locusts, grasshoppers

Subsocial care in this large group is apparently restricted to burrowing species, in which the female is in a position to guard her eggs or brood in the privacy and security of a nest. Female *Gryllotalpa* mole crickets (Gryllotalpidae) lay their eggs in subterranean brood chambers, which may be over 30 cm below the ground. The female deposits her eggs (usually 100–300) over a period of a week or two and leaves them in a mound on the floor of the chamber. She remains on guard, regularly licking the eggs, much as earwigs do and probably with the same anti-fungal function. As in many earwigs, if taken away from the female, the eggs are usually rapidly destroyed by mould. After hatching, the tiny nymphs stay with their mother for a few weeks and are able to feed themselves on humus and small rootlets protruding into the tunnel walls.

Females of the large brown cricket, *Brachytrupes achatinus* (Gryllidae), from India, guard their eggs at the end of a subterranean gallery from which the nymphs disperse a few days after they hatch. However, advanced parental care involving provision of food to the young by their mother has so far only been found in the American cricket *Anurogryllus arboreus* (=*muticus*). After mating, the female excavates a nest chamber in the earth and stocks it with a food store of berries and grass (West & Alexander, 1963). She spends most of the time beside her 20 or so eggs, often picking them up and turning them over in her mouth, presumably as an anti-fungal measure. She also lays a stockpile of miniature eggs which seem to fulfil the same role as the trophic eggs of ants, being supplied to her brood as baby-food. The nymphs cluster around these eggs as they are ejected from their mother's body, vigorously disputing ownership and the right to take the first bite. In marked contrast to this feverish gourmandizing behaviour with trophic eggs, the nymphs make no attempt to eat normal eggs which remain in the chamber. The female performs the essential chore of nest sanitation by carrying her faecal pellets in her mouth and dumping them in a 'lavatory chamber' in the lowest part of the nest. She continues to stock the burrow with grass and fruits, but eventually dies 'in harness', after which she is probably eaten by her offspring, as in certain spiders, such as *Coelotes terrestris*. She thus makes a massive 'all-or-nothing' investment in a relatively small number of offspring (compare her 20 or so eggs with the 100–300 normally produced by a similar-sized, non-guarding cricket), whose survival and subsequent growth must receive a major boost from this assisted start in life.

EMBIOPTERA Web-spinners

All web-spinners are regarded as subsocial, and it is exclusively the females who take care of the eggs and nymphs. Certain species have established a form of communal

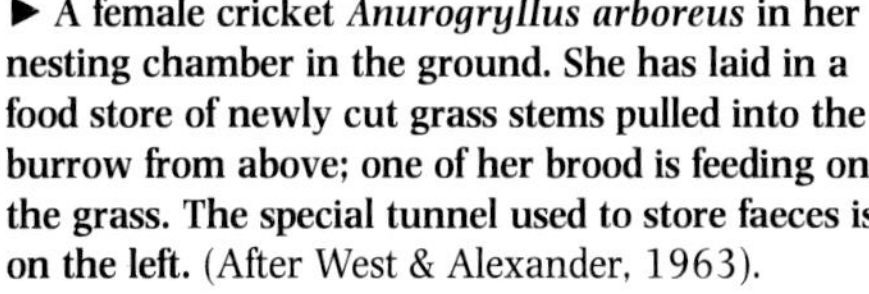
▶ A female cricket *Anurogryllus arboreus* in her nesting chamber in the ground. She has laid in a food store of newly cut grass stems pulled into the burrow from above; one of her brood is feeding on the grass. The special tunnel used to store faeces is on the left. (After West & Alexander, 1963).

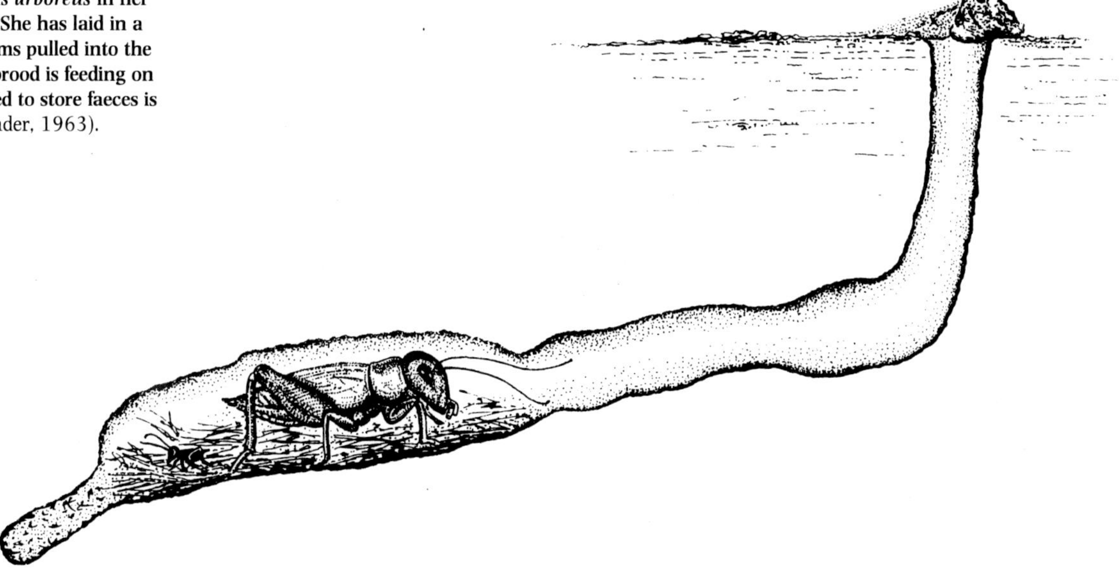

life-style, involving the co-habiting of several females, each caring for her own brood within a network of silken galleries, which owe their construction to a joint effort. The silk is derived from enlarged glands in the forelegs and can be produced by juveniles, as well as by adults of both sexes. The technique is to apply multiple strands of silk across a tree trunk by sweeping the forelegs back and forth across its surface. The thick mat of silk, which covers the brood chambers with their precious eggs and nymphs, seems to be the females' main contribution to the survival of their broods, protecting them from the inroads of ants and parasitic wasps, which are unable to penetrate the dense covering. This certainly seems to be the case in *Antipaluria urichi* from Trinidad and may apply to other species as well (Edgerly, 1987 & 1988).

However, the females of this species also safeguard their eggs in other ways. While ovipositing, the female picks each egg up in her mouthparts and then devotes about 20 minutes to adding a thick 'muffler' to it, by coating it with a mixture of chewed-up food, flakes of wood and silk. The final pile of eggs is then thoroughly camouflaged and protected with a layer of chewed-up lichen topped by yet another layer of silk. This forms an effective barrier to enemies, but has to be removed by the female in order to allow the nymphs to emerge. Once she has completed the laborious task of constructing her 'nest', the female spends the bulk of the next 6 weeks in close attendance, not feeding and seldom moving far away. She vigorously drives away intruding parasitic wasps by lunging at them and vibrating her body. However, she does not even pretend to defend her brood against invading ants, which are simply left in peace to devour the eggs. Any attempt at countering such an invasion would probably be fruitless anyway, given that ants often kill foraging adult web-spinners. The female guardians also fail to prevent the occasional attacks of a parasitic wasp, *Probethylus callani*, which enters the galleries and lays its eggs on the nymphs, after first stinging them to induce an ephemeral paralysis. Nest sanitation is somewhat unusual in that the faecal pellets are not thrown away outside the nest but are incorporated into the silken walls of the galleries. The pellets are moved to the relevant building site in the female's mouth and often finish up being fashioned into a silk-wrapped pile, which does useful service as a support column for the attachment of further silk.

The females of some other species enjoy a much more intimate relationship with their offspring by providing them with food, without which they weaken and die. *Embia ramburi* from Europe furnishes her young brood with a pre-chewed form of convenience-food. The first-instar nymphs remain close to their egg mass and depend on this maternal food delivery for their early survival, although they are able to forage for themselves once they are larger (LeDoux in Edgerly, 1988).

Oligotoma ceylonica is remarkable for building its galleries within the complex webs of the social spider *Stegodyphus sarasinorum*, which itself exhibits complex and advanced subsocial care (Bradoo, 1967). The web-spinners construct a subset of tunnels which follow the spiders' own communication galleries, deep inside the latter's large communal nest. Short side-tunnels lead out, via small exit holes, on to the surface of the nest, where the spiders' silk becomes impregnated with a rich growth of green algae. This constitutes a source of food for the web-spinner nymphs. The mothers camouflage their eggs, much as described above, and their brood follows them around through the galleries like a family of ducklings. Interestingly enough, it is the nymphs in their later instars who are responsible for breaking out from the home nest and establishing migrant colonies. They do this by extending a system of silken tunnels outwards from the original spiders' nest, until it connects with a neighbouring nest.

THYSANOPTERA Thrips

A number of thrips exhibit subsocial behaviour. Lone females have been observed guarding eggs and young in at least five species, including *Actinothrips trichaetus*; while both males and females of *Bactridothrips brevitubus* are known to behave in this way.

Elaphrothrips tuberculatus from the eastern USA is a species which is oviparous early in the season and then becomes viviparous as the year advances. An oviparous female lays her eggs in some sheltered spot on a leaf (Crespi, 1990). She remains on guard over the eggs, reacting to danger by arching the tip of her abdomen up over her back and directing it towards any intruder. If she is attacked, the guardian exudes a droplet of brownish liquid from the tip of her abdomen, where it is held in place by a tuft of short hairs. This liquid, which contains juglone, a toxic chemical, is dabbed on to an assailant with a deft flick of the abdomen. This mode of defensive behaviour is common in this group of thrips (Tubulifera). It seems to be especially effective against jumping spiders and, in the thrip *Bagnaniella yuccae*, against predatory ants of the genus *Monomorium*. Yet strangely enough, it is not deployed against conspecific females intent on egg-cannibalism, nor even against parasitic wasps heading with deadly intent for the first-instar larvae. Intrusions by such wasps are greeted with a different and less sophisticated method of defence – the guardian charges at the intruder and strives to grab it with her front legs.

The female remains on guard only until the eggs have all hatched, after which she may remain nearby but mounts no kind of defence of her brood. The chances of any kind of intruder mounting a successful attack against a clutch of eggs are considerably increased by the fact that, unlike most other egg-guarding insects, the female thrip does not fast while baby-sitting. Instead, she regularly deserts her eggs in order to wander off in search of food, usually when the leaves are wet and covered in edible fungal spores. While she is busy feeding, her undefended eggs often fall prey to non-brooding, viviparous females, who are ever alert for just such an opening, which allows them to guzzle the eggs of another female without fear of reprisals. If the guardian does return to find her brood being cannibalized, she will attack the perpetrator, usually driving her away. If the intruder is much larger than the guardian, then any attempt at effective defence is probably doomed to failure and all the eggs will be lost. The viviparous females have much to gain through their cannibalistic plunder. Eggs are a highly nutritious diet, enabling the consumers to achieve a significant increase in their own fecundity. Such regular depredations by viviparous females have probably been the main stimulus for the evolution of maternal guarding of eggs by the oviparous females. This is now so essential that, in the absence of a protectress, few eggs survive to hatch.

A remarkable example of apparently advanced parasocial behaviour is evidenced by *Anactinothrips gustaviae* from Panama (Kiester & Strates, 1984). This thrips forms dense colonies, usually composed of both adults and nymphs, on the trunks of smooth-barked trees. Each colony may be separated into several individual clusters or 'bivouacs' to which the inhabitants return each day after foraging. This is conducted on highly organized lines, with bands of thrips marching in tight formation up the trunk in order to graze on micro-fungi growing on lichens. The band may move from one lichen to another, following a trail pheromone, which is capable of lasting for 2 to 3 days. These thrips are thus behaving very much like the highly social ants, which follow similar pheromone trails. Within the bivouac, egg-laying is also a highly organized activity. Many different females add their own contribution to a communal cluster containing as many as 300 eggs, which are subsequently guarded by the females. The evolution of such complex behaviour in thrips is only to be expected, given the haplodiploid method of sex determination within the whole order. This should theoretically promote the evolution of kin-related social behaviour, as seen in the similarly haplodiploid bees, wasps and ants.

HEMIPTERA True bugs

Subsocial behaviour is widespread in both suborders. Within the Homoptera, maternal care is common in the Membracidae and related families. The Heteroptera gives us many of the major examples of paternal care in insects, distributed around no fewer than three families, while examples of maternal care are legion.

HOMOPTERA

Maternal care in membracids

Female membracids perched rather incongruously on top of an outsize pile of eggs are a common site in the tropics and subtropics, becoming less so in temperate regions. As in most subsocial bugs, the female's behaviour undergoes a marked change once she has laid her eggs. Once firmly enthroned above her clutch, she sits tight, refusing to budge when confronted by threats which would previously have caused her to seek safety in flight. However, the degree of stolidness in the face of danger varies from species to species. Some are probably reluctant to desert their egg masses or nymphs, as they are unable to re-locate them, or to identify which is their own. Others, such as *Guayaquila compressa*, appear able to do so, and so there is less reason for them to be steadfast in confronting an enemy (Wood, 1978). Females of the latter species are unusual in herding their nymphs from one leaf to another on the food plant, presumably to maximize the quality of forage.

The female's role has been well studied in the 'thorn bug', *Umbonia crassicornis*, which is widely distributed from the southern USA down into South America (where it sometimes lives in mixed groups with *U. spinosa*). Females lay their eggs in a group on a twig, remaining on guard atop the egg cluster for up to 30 days (Wood, 1974 & 1976a). If eggs are left unattended, they are soon destroyed by a mould and attacked by predators, such as anthocorid bugs. When the offspring begin to emerge, their mother paves the way for them to take their first meal by cutting a series of spiral slits in the bark below the egg mass. The nymphs soon congregate here to feed, apparently being unable to pierce the bark by themselves. The female sits below them, facing upwards, ready to thwart any tendency for any of her brood to break away and descend the stem. If this happens, she gently strokes any errant nymphs on the back with her front legs, which seems to convey a persuasive message to return back to the group. Predators such as ladybirds (Coccinellidae) can soon devastate a group of unattended nymphs but will be sent packing by the mother if she is present. In fact, few predators seem willing even to hazard an attack on the nymphs while the female keeps her vigil, but any group of orphaned nymphs will be wiped out within days.

The female's response to an approaching threat, such as a ladybird, is to fan her wings, often producing an audible buzz. She will move out to confront the intruder if it still menaces her brood. Wing-fanning is the main method of defence against predators in both homopterans and heteropterans which practise egg-guarding. In some species, the fanning action also serves to blast a wave of repellent (and even toxic) chemicals into the enemy's face. An intruder may also be physically dislodged from the twig by the impetus of the female's wing-buzz. Ladybirds are easily repelled by the defensive tactics of *Umbonia crassicornis* and do not attempt to press home an attack, while hover-fly larvae (Syrphidae) and assassin bugs (Reduviidae) are also effectively rebuffed. Females respond visually to the approach of predators by turning to confront them but are also alerted to an assault against their brood by an alarm pheromone, which is released by the nymphs when they are injured.

Similar alarm pheromones have also been reported in other membracids, such as the gregarious *Vanduzeea arquata* and the subsocial *Publilia concava* and *Entylia bactriana* (Nault *et al.*, 1974), and are probably widespread in the family, whenever guarding of nymphs is undertaken. However, in the three above-mentioned species, the pheromone serves to initiate a scattering reaction in the clusters of nymphs. In marked contrast, *U. crassicornis* nymphs sit tight and rely on their mother to respond to the alarm and come rushing to their successful defence.

The general behaviour in *Platycotis vittata* from the USA is very similar. The female remains on the eggs for a considerable period and prepares the way for her nymphs' first meal by cutting slits in the bark (Wood, 1976). She accomplishes this by forcing her ovipositor into the bark with a see-sawing action and then drawing it forwards to plough a furrow. Once the nymphs have moved off the egg mass and clustered on the twig, their mother keeps them together by gently stroking them with her front legs, which also serves to coax a stray back to re-join its fellows. As in other bugs, she wing-fans when assailed by predators, even heading off assassin bugs in this manner. In common with *Umbonia crassicornis*, few eggs or nymphs survive if the mother is absent, underlining the essential role of maternal care in *P. vittata*. In both these species, the females stick it out to full term, until their offspring reach adulthood, thereby making a considerable investment in time and energy.

In *Entylia bactriana*, the females reduce the duration of their investment by only remaining with their nymphs up to the second instar. They can afford to cut short their parental obligations in this way, because they effectively hand over the job of baby-sitting to mercenary caretakers in the guise of ants (Wood, 1977). The presence of ants is essential to the survival of the nymphs right from the start, even while the female is present with the first two instars. With this being so vital, the female will desert her egg mass if no ants show up within a certain time, moving over to a bigger group of treehoppers, which is more likely to attract the necessary protectors. Although some species are more effective than others, the ants generally constitute a formidable and efficient standing army which can be amazingly sensitive to the needs of their charges. For example, *Formica subobscura* is able to detect the aphid alarm pheromone trans-β-farnesene and rushes across to tackle the predators which have caused its release (Nault *et al.*, 1976).

The association between ants and membracids is a characteristic feature, particularly of the tropics. Its significance for the parental role of the female has also been studied in *Publilia reticulata* from the USA (Bristow, 1983). Nymphs and adults alike secrete honeydew, which draws in a constant supply of ant attendants belonging to at least seven species. The female membracids normally stick with their eggs until they hatch, about a fortnight after they are laid. However, the subsequent hold which the nymphs exert over their mother is related to the presence of ants on the plant. The females seem capable of determining whether ants are coming around often enough to make them a safe bet as substitute guardians. On plants which are well attended by ants the females abandon their eggs to the proxy protection of these ferociously efficient bodyguards. This frees the female to lay a second batch of eggs while in no way reducing the reproductive success of her first batch. In fact, the ants prove to be better at the job than the females themselves, being well able to detect and evict a wide range of potential enemies, such as spiders, ladybirds and lacewing larvae, all of which rarely dare to trespass on to ant-defended territory. Nymphs guarded by ants alone therefore survive in greater numbers than those guarded by their mother, while untended broods fare much the worst. Being free to lay a second brood is particularly significant for the females as, by the time this comes about, the weather is warmer and more favourable for nymphal development. Ant numbers too will have risen, so the chances of securing an armed guard for the second brood are virtually guaranteed. In both cases, the mother's presence initially seems to be a useful bait to attract the ants and get them hooked. Otherwise, groups of tiny nymphs, not yet capable of producing rewarding quantities of honeydew, could easily go unnoticed.

Membracid-ant mutualism is widespread, but relations between these two insects are not always beneficial to the bugs. Some species, presumably because they are unable to produce honeydew 'bribes' to keep the ants sweet, regard ants as enemies to be chased away by a guarding female. An unusual kind of mutualistic relationship seems to occur between individual females of the membracid *Polyglypta dispar* from Costa Rica (Eberhard, 1986). Although female guardians are reasonably effective at putting to flight predators such as ladybirds, they are almost powerless when up against agile and well-armed opponents such as assassin bugs and jumping spiders, both of which are capable of killing the guardian herself. However, the main enemy, against which the guardian has little defence, is a tiny *Gonatocerus* parasitic wasp which is able to roam almost at will over large parts of the egg mass, this being simply too large to be adequately covered by the parent. Rates of parasitism may therefore average nearly 30 per cent, with the peripheral and rearmost eggs receiving the highest attack rates.

Female guardians always desert their nymphs before they are fully grown, but they also tend to desert egg masses. These are then often taken into care by other females who guard them for varying amounts of time. In some cases, the adoptive female might well have a vested interest in the egg mass, owing to the habit, of females generally, of adding eggs to another female's pile. This happens to such an extent that a conspicuously large mass may be the product of eggs from several different females. A guardian usually makes a bid to repel such an imposition by a stranger, but eventually she always gives way and allows the eggs to be added to her own. The fact that the guardian always capitulates in the end might indicate that this is not a genuine attempt at resistance, but rather a 'statement' that she is not about to quit her duties, making it worthwhile for the stranger to add some more eggs.

Additional eggs placed around the vulnerable periphery will tend to act as an outer bulwark to protect the guardian's own eggs against parasitism, so it is probably not in her best interests to repel an insistent egg-layer. In many instances there is no objection anyway to a fresh layer of eggs being tacked on around the edges of the mass. Females often switch roles, guarding their own eggs for a while, then switching to life as a roaming contributor to other females' stocks, and finally perhaps taking over the guarding role from another female. As these bugs form dense colonies, it is likely that a high degree of relatedness exists between all these females, which would help to explain the tendency to adopt another female's egg mass. The whole scenario is rather similar to the 'egg-dumping' described on page 180 for the tingid bug *Gargaphia solani*, but with a greater degree of mutual benefit being likely in the case of the membracids.

Subsocial behaviour in the Aetalionidae. There are several examples of subsocial behaviour in the closely related family Aetalionidae. *Aetalion reticulatum* occurs in a variety of habitats from Costa Rica to Brazil, feeding on a wide selection of bushes and small trees. The females lay clutches of about 100 eggs, which are covered in a viscous liquid secreted by the female. These egg masses are often much larger than their creators who perch on top, periodically sweeping their hind legs down the sides of the mass, which thereby assumes a rather polished appearance (Brown, 1976). *Pterygogramma* and *Gonatocerus* parasitic wasps are often dislodged from the egg mass by this frequent sweeping action, which may therefore represent a kind of 'automatic' discouragement to parasite attack. However, not all guarding females engage in leg-sweeps. Even so, the periphery of the egg mass is often heavily parasitized, especially the rear section, which is probably exposed when the female leans forward to suck the plant stem.

A. reticulatum forms sizeable colonies, comprising guarding females with their eggs, along with large clusters of nymphs belonging to various instars. Such aggregations are normally attended throughout the day by several species of ants. These probably combine with the females to reduce attacks by predators, while doing little or nothing to counteract the malevolent attentions of parasitic wasps. In Costa Rica, the ant *Camponotus sericeiventris* is exceptionally pugnacious in its defence of *A. reticulatum* nymphs, while showing no such inclination to put up a fight when lapping up the sweet exudations from extrafloral nectaries nearby. *Trigona* and *Melipona* stingless bees also visit the bug colonies for honeydew. In the absence of their ant protectors, the nymphs quickly succumb to predators, whose marauding intrusions after dark are rebuffed by a 'night watch' mounted by a second species of *Camponotus* which makes the rounds during darkness.

► As in many membracids, female aetalionids stand guard over large egg masses which are afforded extra protection by the presence of ants. This is *Aetalion reticulatum* in Brazil.

HETEROPTERA

Subsocial behaviour

Within the Heteroptera, subsocial behaviour is found in most families of the Pentatomoidea, the 'stink bugs' or 'shield bugs', as well as in the Belostomatidae, Coreidae, Gerridae, Reduviidae and Tingidae. There are probably many more examples awaiting discovery, particularly among the tropical Pentatomoidea.

The habits of the 'parent bug'

The maternal devotion exhibited by the very common European shield bug *Elasmucha grisea* has been known since 1764, and is found in at least three other members of the genus. The female *E. grisea* lays her clutch of 40–50 eggs in a compact diamond-shaped batch on the underside of a birch leaf, straddling the eggs until they hatch. She is not a passive guardian, but tilts sharply over towards any perceived threat, using her body as a mobile shield between an aggressor and her brood. She will also react to a variety of menaces, such as a human being coming too close, by 'standing on her head' and fanning her wings with a loud buzz. This is an effective deterrent against larger insect predators, while having sufficient 'punch' to flick tiny parasitic wasps clear off the leaf. The female's responses are both finely tuned and deftly executed, making her a highly effective guardian of her eggs, so that rates of parasitism are usually very low, in contrast to some tropical species (see page 182). After hatching, the tiny nymphs remain clustered below their mother for a few days and then move away to feed, closely escorted by the mother who will keep her vigil beside them until their penultimate instar.

As they move around, mother and brood stay in contact via a trail pheromone, which is laid down by the larvae as they walk over the leaf (Maschwitz & Gutmann, 1979). The nymphs also emit a defensive chemical which acts as an alarm pheromone, inducing the female to rush towards the source of trouble and buzz her wings while the nymphs scatter temporarily. The female usually dies at her post just before the nymphs moult into the final instar. With this safely accomplished, they aggregate on a leaf and are probably no longer in need of any maternal protection. They are amply capable of deploying a powerful chemical counterattack against even a large aggressor, producing an irritant smog able to invoke quite severe nasal and bronchial distress in human beings from a range of several centimetres (pers. obs.).

▲ Maternal care by the European stink bug *Elasmucha grisea* pays handsome dividends. Eggs are seldom successfully attacked by parasitic wasps and survival rates of nymphs are very high. These nymphs will shortly moult into their final instar, when their mother will finally leave them, probably to die soon afterwards.

Maternal care in *Antiteuchus tripterus*

A broad, diamond-shaped egg mass is the norm in most stink bugs, which stand guard over their eggs on leaves. The shape and size of the egg mass has several implications for the efficiency of guarding behaviour, which is beautifully illustrated in the Neotropical *Antiteuchus tripterus limativentris* (Eberhard, 1975). From this species, we can derive a fascinating insight into the likely evolutionary arms race which rages between a bug, its predators and its egg-parasites. Unlike *Elasmucha grisea*, the female *A. tripterus* is unable to prevent a high rate of parasitism by two species of wasps, *Trissolcus bodkini* and *Phanuropsis semiflaviventris*. The female is compelled to remain on guard because, if she leaves, the egg cluster will be completely destroyed by a medley of predators within just a few hours. Yet her essential presence actually *increases* the toll from parasites because her own body acts as a target beacon, pointing out to the wasps, in the most helpfully conspicuous way, just where to find her eggs. Take away the large, brightly coloured and rather strong-smelling female and the wasps may easily walk straight past the eggs without seeing them. This is particularly true for *P. semiflaviventris*, whose parasitic attacks drop off markedly if the guardian is removed.

The female's defensive behaviour itself is quite inadequate for counteracting the wasps, whose incursions are deadly. This tends to be especially so around the periphery of the egg mass, whose very breadth makes it impossible for the guardian to mount an effective defence of any save the innermost eggs. This ineffectuality is mainly due to the wasps themselves having evolved a suite of tactics which easily checkmate the guardian's measures at contravention. The latter mainly consist of scraping movements with the forelegs and kicking actions with the middle and hind legs. The wasps have, at least for the moment, gained a lead in the arms race. The bugs do not deal with the wasp problem by using the most devastating weapon in their armoury – a shower of defensive chemicals squirted from ducts near the leg bases – probably because this gas attack is so potent that it can kill the bugs themselves, and their eggs, if they are confined in an enclosed space with the chemical vapour.

The wasps exploit the guardian's inability to mount a proper guard on the outer eggs in a number of ways. *T. bodkini* females loiter just out of range of the bug's antennae and then dart in quickly, make a quick about-turn and ram an egg of their own into one of the outermost eggs. The guardian is usually sufficiently 'on the ball' to detect such a lightning attack, and reacts by scraping and kicking with her legs and flailing her antennae, sometimes successfully booting the invader clear of the egg mass. However, attacks from the rear are likely always to be successful, as the bug cannot bend her back legs beneath her abdomen, leaving a 'dead zone' in which the wasp can oviposit in reasonable security. The rearmost eggs are therefore sacrificed for the sake of those at the front, which always remain protected because of the female's refusal to turn round to tackle an intruder operating at the rear.

The other parasite, *P. flaviventris*, adopts completely different tactics, relying on the extreme elasticity of her abdomen to telescope her eggs into the mass, using the kind of 'stand-off' oviposition strategy also seen in aphid-parasites. The slim tip of the wasp's abdomen is therefore the only part exposed to the scraping actions of the guardian's legs, which have little hope of connecting well enough to do much damage. Several wasps may also 'gang up' on a single guardian, overloading her capacity to respond effectively. The wasps even profit by the bug's natural maternal reactions as, when they hatch from their host eggs, the brooding female conveniently lifts her abdomen, much as she would on detecting her own newly hatching babies crawling around beneath her. This allows the fragile parasites freedom of movement to emerge, mate and disperse in comfort.

The female bug remains with her nymphs after they hatch and is highly effective in defending them against a variety of predators. However, when it comes to defence of the eggs, this species seems to have run into an evolutionary 'Catch 22' situation. Egg masses which are left unguarded are normally completely destroyed. This is presumably because the evolution of guarding behaviour itself has led to the loss of a tough shell on the eggs. This protects the eggs of non-guarding species against generalist predators but is less efficacious against egg-parasites. However, from an initially favourable situation whereby guarding increased the level of protection against both predators and parasites, the point has now been reached when the female is forced to stay on guard, despite acting as a target beacon for the wasps. Meanwhile, the wasps themselves have evolved suitable countermeasures, giving them the advantage in the current state of matters.

Parental care is also seen in some other members of the genus, including *A. piceus*, *A. mixtus*, *A. variolus* and *A. macraspis*. However, in all of these, the degree of care is restricted to just the physical protection of eggs and nymphs. In a major advance on this, the Brazilian *Phloeophana longirostris* apparently nurtures her brood of tiny babies as they cling beneath her body. In this position, they are said to feed either on secretions from her anus or on fluids regurgitated from her rostrum. An additional rather more bizarre suggestion is that they tap directly into her digestive tract by piercing the intersegmental membranes of her abdomen, her 'soft underbelly', with their mouthparts. This rather extreme level of maternal benefaction is said to be necessary because the puny rostrum possessed by the small nymphs is incapable of piercing the tough bark of the host tree, *Terminalia catappa*.

Most 'guarding' stink bugs lay down broad diamond-shaped egg masses similar to those of *A. tripterus*, although there are exceptions. The Australasian *Tectocoris diophthalmus* attaches a a large cluster of eggs (usually over 100) in the form of a broad collar encircling a twig. This is a sizeable clutch for a bug which guards in the open, exposed on the food plant, but more common in species which nest in a chamber in the soil (see below) The guardian sits on top of her eggs, but is obviously unable to maintain any kind of visual surveillance of those on the opposite side of the twig. However, her long flexible antennae are capable of extending this far, while her legs can reach all areas of the egg mass. When confronted by a parasitic wasp, she acts in a very agitated manner, vibrating her body and twitching her legs and antennae, making it difficult for the enemy to find an opportune moment to dodge in and lay an egg. Nevertheless, her defences are probably easily overwhelmed by a multiple attack, as would happen under certain combinations of weather, food-plant geography and the relative abundance of eggs and parasites. These variables are such that average rates of parasitism run from as low as 0.3 per cent to punishingly high levels of 91 per cent. The highest figures may only occur early in the season when egg masses are scarce and wasp numbers high. The wasps have the Achilles' heel of being severely affected by exceptionally hot weather, a factor which does not adversely affect their host (Ballard & Holdaway, 1926).

There is at least one exception to the general rule that stink bugs exhibiting maternal care produce diamond-shaped egg masses which protrude beyond the edges of the body. An unidentified green pentatomid from Costa Rica was seen to stand guard over just two single lines of eggs (pers. obs.). Such a small number could probably be afforded total protection by the mother's body, with no eggs to form a vulnerable periphery. Presumably, by reducing wastage to a minimum by not committing resources to extra eggs, which would be doomed to failure through unavoidable parasitism, the female is able to produce several successive, highly successful batches of eggs.

Soil-nesting stink bugs

One strategy for reducing parasitism may be to forsake the exposed leaves of the food plant and place the eggs in a more easily guarded nest in the soil below. The European pied shield bug, *Sehirus bicolor*, lays her eggs in a rather loosely formed ball in a small cavity, which she excavates in dry soil. She sits over them for nearly 3 weeks, regularly turning them over with her rostrum. When the babies hatch they follow their mother to the food plant, *Lamium album*, at which point she leaves them. *S. luctuosus* has similar habits, as does *S. cinctus cinctus*, the only member of the Sehirinae from the USA. This species is special in another way as well, as the female has been seen to furnish her tiny first-instar babies with seeds of the food plant, *L. purpureum*, while they are still in the nest in the soil (Sites & McPherson, 1982). The first-instar nymphs stay with their mother, often crawling around over her, and do not leave the nest until the second instar. The habit of providing food for the babies in a ground-based nest is also found in the spectacular Asian *Parastrachia japonensis* (Tachikawa & Schaefer, 1985). When leaving the host tree, *Schoepfia jasminodora* (Oleaceae), to dig her nest in the soil, the female carries on the tip of her rostrum a drupe (fruit) from the tree, which will serve as provisions for her newly emerged brood. She guards her 50–100 eggs below a protective covering of leaf litter, on to which she and her second-instar nymphs later emerge to refresh themselves on fallen drupes beneath the host trees. A lone nymph seems unable to cope with the tough drupes, while aggregations seem to manage very well, possibly indicating that the mass injection of enzymes during communal feeding helps digest the fruit.

Brachypelta aterrima is possibly unique among bugs (see also *Phloeophana longirostris*, above) in providing her first-instar nymphs with a special 'baby-food' consisting of her own anal secretions. These come complete with the necessary dose of digestive bacterial symbionts – vital organisms which other bugs derive from their own egg shells, which are smeared with symbionts by the female during oviposition (Schorr, 1957).

Maternal care in coreids

Maternal care in the large family Coreidae (squash bugs) seems to be rare. The best known example is the large *Physomerus grossipes* from Thailand (Hemmingsen, 1947). As in stink bugs, the female sits on the eggs and is apparently regularly beleaguered by gangs of parasitic wasps intent on ruining her eggs. The guardian also remains with the nymphs and, as there is often a male loitering nearby, it is possible that baby-sitting may be a joint effort between the sexes. When threatened by larger predators, the adults respond by squirting a stream of strong-smelling fluid from the anal orifice. This is a common defensive reaction in large coreids and is easily induced in other tropical examples, such as several large *Pachylis* species from Mexico.

Maternal care and egg-dumping in lace bugs

Maternal care is known in a number of different lace bugs (Tingidae). Its operation has been most closely studied in the American egg-plant lace bug, *Gargaphia solani*, a species whose females utilize Solanaceae (potato family) as the sole food plants, and guard their nymphs right through to maturity. Despite her small size, the female lace bug is capable of driving off predators, such as ladybirds, which are of greatly superior size, as well as anthocorid bugs and ants. She accomplishes this rather astounding 'David-and-Goliath' feat mainly by wing-fanning, as seen in most other subsocial bugs. She backs this up by rushing full tilt at the enemy and using a daring head-butt to compel its retreat (Kearns & Yamamoto, 1981).

Nymphs usually scatter if some of their number are killed, probably impelled by the release of an alarm pheromone. When the nymphs scatter and head for a new and 'safer' leaf, the female may guide them along a 'preferred' route by blocking alternative pathways with her body, and also often by wing-fanning. Her chaperonage is most effective, and nymphs who do not enjoy its benefits suffer a far higher rate of loss than those escorted by their mother – in fact only a miserable 3 per cent of unguarded nymphs survive (Tallamy & Denno, 1981). The advantages to the mother of her attentiveness are therefore considerable. Her vigil results in the successful raising of around four times as many progeny during her whole reproductive lifetime as she could manage by the alternative strategy of investing all her efforts into the production of a much larger number of eggs, but abandoning them immediately and leaving them unguarded to the mercies of every predator in the vicinity.

Some *G. solani* females manage to enjoy the best of both worlds by practising the art of 'egg-dumping'. In effect they become cuckoos, foisting their offspring and their care on to another female of their own species, who is already committed to guarding her own young (Tallamy, 1985). Whenever it is practicable, a female will choose to add her eggs to a cluster already being guarded, thus increasing her lifetime output of eggs without losing the benefits which guarding bestows on her nymphs. Just why the female at the receiving end of the dumping sits passively by and allows herself to be imposed upon is uncertain. It may be to the advantage of her own nymphs to be mixed in with the largest possible herd, thus reducing their chances of being the ones to be picked off by a predator. As looking after an increased number of nymphs probably does not incur any extra risks, or investment in time or energy by their guardian, it would make sense for her to accept the extra eggs, once she has committed herself to being a non-dumper. However, as a mega-crowd of nymphs may well attract extra predators, as well as taxing their leaf to the point of exhaustion, the wisdom of accepting dumped eggs is still far from clear. What is not certain is how it is decided who should be dumpers and who the dumped-upon. Some females always dump, while others always guard, even when given a chance to dump. Dumpers tend to be younger females with a long reproductive life ahead of them.

▼ The victim of egg-dumping? This female egg-plant lace bug, *Gargaphia solani* (Tingidae) (top right), in Mexico seems to be taking care of a particularly large family, some of which may have been foisted upon her as eggs by other females.

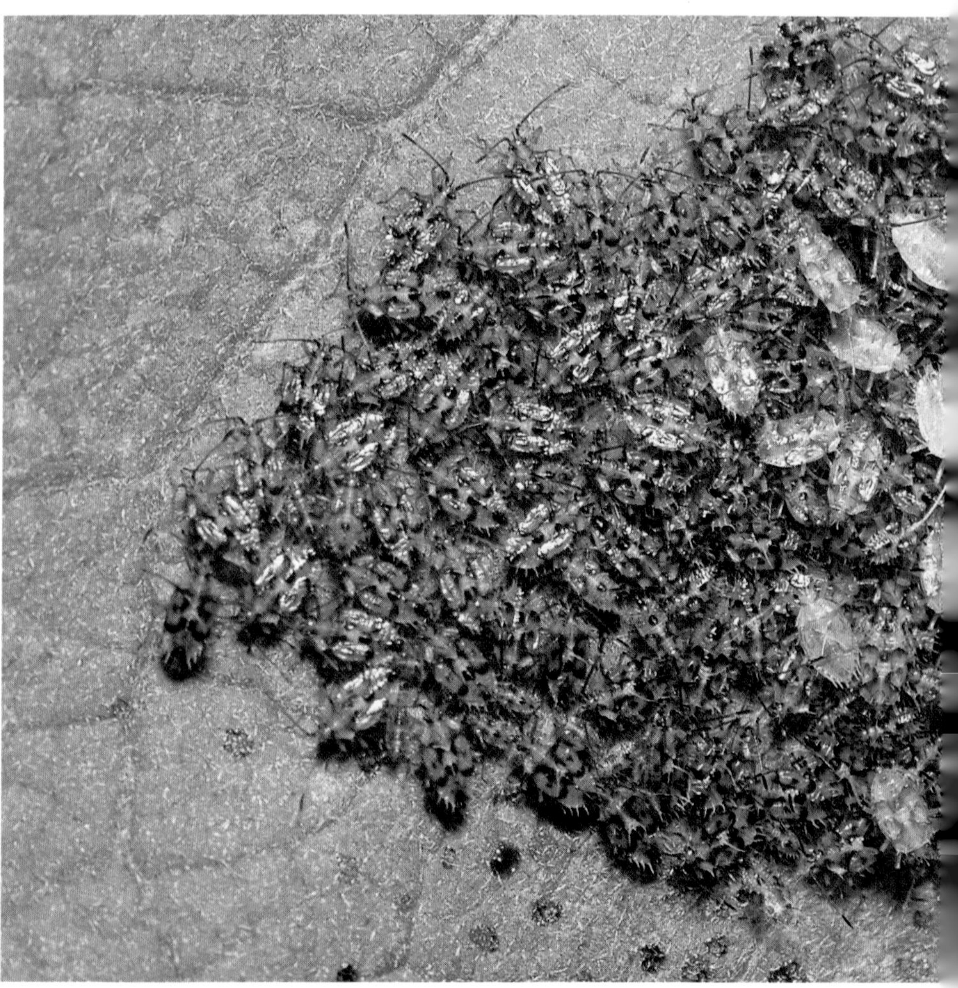

For these, avoiding extra risks by not becoming a guarder at an early age makes reproductive sense (Tallamy, 1986).

The females of another American lace bug, *Corythuca hewitti*, also guard their brood, but not in the aggressive way seen in *G. solani* (Faeth, 1989). Instead, the female merely guides her brood to a safer part of the plant, leading the way by regularly stopping to vibrate her abdomen in the air.

Maternal care in assassin bugs

Maternal care among predatory bugs seems to be rare. *Ghilianella* (Reduviidae) are slender stilt-like bugs in which the female has been seen carrying her nymphs around on her back, their long slender legs wrapped around her underside. The brood size is very small – just one or two offspring – but these are cared for in a way which ensures a very high chance of survival. Small broods are typical in such instances in all animals. Moreover, the female is probably able to feed her tiny riders with prey which she herself catches. The females of two African species, *Pisilus tipuliformis* and *Rhinocoris carmelita*, stand guard on or near their eggs (Parker, 1965; Edwards, 1962).

▲ Paternal care is rare in invertebrates. When it does occur, the male has to be sure of his paternity of the eggs in his charge. This male *Rhinocoris tristis* assassin bug (Reduviidae) from Kenya is ever alert to possible enemies, particularly the tiny parasitic wasp seen on his right. He will chase these off and attempt to spear them on his rostrum.

Paternal care

Paternal care is rare among insects for a variety of reasons, not least of which is the absolute requirement for the male to be sure of being the father of any eggs or nymphs which are guarded. This is a problem which females obviously do not encounter. Any female can always be certain that her offspring are her own (although egg-dumping complicates matters), but males can easily be cuckolded in a number of ways. The common requirement to be last to mate in order to achieve paternity through sperm precedence has already been emphasized in the section on Sexual Behaviour (page 9). It has major implications for a male's behaviour before and during guarding operations.

In the African assassin bug *Rhinocoris albopilosus*, the male has taken on the role of guardian in a way which has a dual pay-off. He increases his progeny's chances of survival, while also picking up a substantial bonus in the guise of additional nubile females, for whom a male already occupied in guarding an egg batch acts as a considerable attraction (Odhiambo, 1959). He will allow them to add their eggs to his batch, but there is a price to pay – he must be allowed to fertilize them first. Before first setting himself up as guardian, he mates with a female several times. He then either remains mounted while she lays her egg batch or else stands nearby and keeps a watching brief, quickly moving in to take up his position over the eggs as soon as the female has done her part and left. He is a committed parent, withstanding the full force of the torrential rains and buffeting winds which are typical of the wet season. Nor does hunger weaken his constancy and he will not desert the eggs in order to feed, although he will take the chance to prey on any potential victim which strays too close.

During his vigil atop the eggs, the male is visited regularly by his original mate, who again copulates with him and adds some more eggs to the original batch. Other females may do likewise. Once the nymphs hatch it is obvious that the father's predatory instincts have been 'switched off' as far as they are concerned, an essential prerequisite for the guarding of nymphs by any predatory insect. In fact, he ignores them, even allowing them to clamber up on to his back, but he does not attempt to catch prey for them and eventually abandons the whole clutch once all the nymphs have safely emerged. While on guard, the males seem very alert to the dangers posed by parasitic wasps. One observer watched a male *R. albopilosus* chase away a pteromalid wasp which was attempting to gain access to his eggs. The male of *R. tristis* is very agile in defence of his eggs and seems able to deal quite capably with a simultaneous assault by as many as three trichogrammatid wasps. He will seek to outmanoeuvre them, as they approach from different angles, and will even turn and try to spear one on his long probing rostrum (pers. obs.). This slim stiletto makes a far more rapidly manipulated and deadly weapon than anything which has to be faced by parasites attacking plump, squat, plant-eating stink bugs.

Similar guarding behaviour of multiple egg masses is also seen in a *Zelus* species from Colombia (Ralston, 1977). However, the degree of consideration for the welfare of the offpring has advanced a stage further than in the African species. *Zelus* males catch insects for their brood and hold them at the tip of the rostrum while the nymphs cluster around and take their fill. The advantages of guarding are clear, for 55 per cent of unguarded egg masses were found to be parasitized, while the rate for guarded eggs was only 21 per cent. The bark bug *Neuroctenus pseudonymus* (Aradidae) probably exhibits paternal care, as the female seems to leave immediately after laying her eggs. These are then guarded by a second individual, presumed to be the male with whom she mated prior to oviposition.

The best-known and most complex examples of paternal care are found in the giant water bugs (Belastomatidae). In these large bugs, the males do not sit and guard the eggs – they carry them around cemented in dense rafts to their backs. This habit is also found in *Phyllomorpha* (Coreidae), with the major difference that the male has to cope with only two eggs affixed to his spiky back. Male belastomatids, by contrast, may be burdened with as many as 150 or more eggs which form a close-fitting mat almost covering the elytra. In *Abedus herberti*, the male ensures paternity of the eggs by insisting on mating with the female repeatedly over a very long period of oviposition, lasting up to 48 hours (Smith, 1979). In fact, the male insists on one copulation for every 1.4. eggs laid by the female. The male controls the whole process, indicating when oviposition should cease and another session of copulation commence and even stroking his own back regularly to detect vacant slots left by the female. He then patiently shows her where these spaces are, so that she can fill them in with more eggs. It is in his interests to do this, as the most intensive possible use of his back area is desirable if he is to derive maximum reproductive benefit.

▼ Paternal care in the giant water bug, *Abedus herberti*. (a) The male sits at the water's surface and 'airs' the eggs on his back. (b) He strokes them regularly with a back leg.

(a)

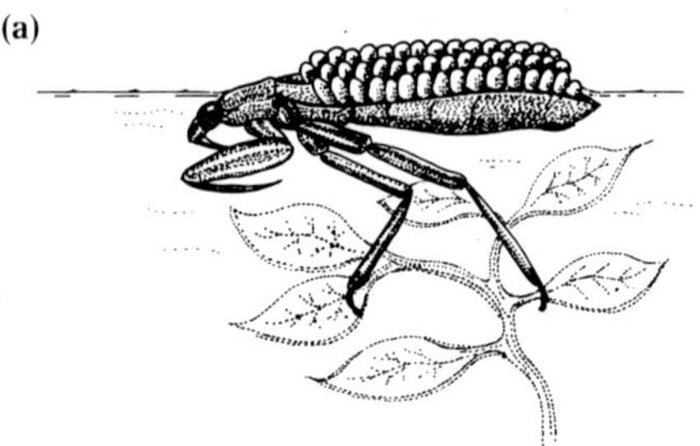

(b)

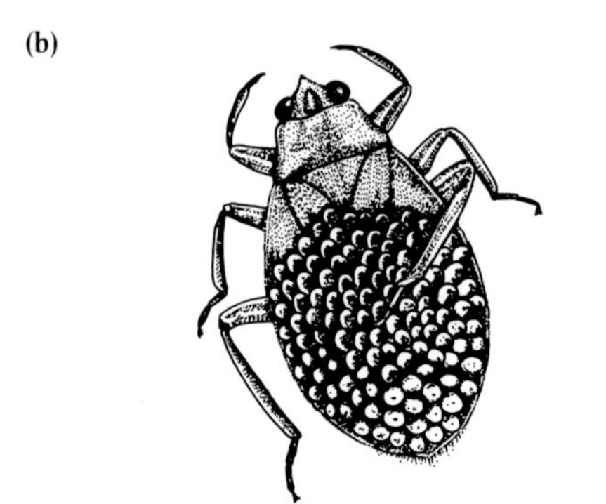

The male demonstrates his readiness to accept the eggs with a series of accelerated 'pumping' exercises (see-sawing up and down), which function as a courtship procedure. Later, in a less vigorous form, he uses pumping to ensure successful hatching of the eggs by increasing the amount of oxgyenated water flowing over them. Eggs which are deprived of paternal care do not hatch, indicating the vital nature of the male's role (Smith, 1976). Similar behaviour has been seen in most other members of the Belostomatidae.

COLEOPTERA Beetles

Despite the vast number of species within the Coleoptera, examples of subsocial behaviour are relatively scarce and restricted to ten families: the Staphylinidae, Silphidae, Hydrophilidae, Scarabaeidae, Passalidae, Tenebrionidae, Chrysomelidae, Erotylidae, Scolytidae and Platypodidae. In the majority of these, the brood is reared in a nest constructed in the ground or in galleries within timber. Only in a few cases does the female carry out her maternal duties in an exposed situation on a leaf or log.

Maternal care in tortoise beetles

The best-known examples of maternal care without a nest are in the tortoise beetles (Chrysomelidae: Cassidinae). Out of more than 3000 species of tortoise beetles, distributed worldwide, only four species, namely *Acromis* (=*Selenis*) *sparsa*, *A. spinifex*, *Omaspides pallidipennis* and *Pseudomesomphalia thallasina*, are definitely known to exhibit brood care. Two further species of *Acromis*, namely *A. venosa* and *A. nebulosa*, may do so. All of these hail from the Neotropical region.

The life history of *Acromis sparsa* is probably typical for all four of the brooding species. The female lays her eggs in a 'rope' attached to the underside of a young apical leaf of the vine *Merremia umbellata* (Windsor, 1987). Before she deposits the first egg, the female chews at the mid-rib of the leaf, close to the site where she proposes to oviposit. As a result, the leaf droops somewhat during the 12 days of her vigil upon the eggs, possibly providing her with extra protection from the torrential tropical downpours and helping to conceal her from enemies. This 'umbrella-constructing' behaviour is similar to that of the tent-making bats and for similar purposes.

The egg rope juts out beneath the leaf at an angle of about 45°, providing a kind of 'high-chair' for the female. The attachment of the eggs, at a single point on the leaf, makes it less likely that they will be discovered by a marauding scout ant. If this does happen, the mother has the relatively easy task of protecting a 'peninsular' structure with just a single attached end from which a small ant – the most abundant potential predator – can approach. However, parasitic wasps intent on attacking the eggs are adept at circumventing the guardian's defences and even ride around on her back. Up to 85 per cent of eggs in a batch may be parasitized, although the average rate is only 26 per cent. By contrast, the egg masses of a non-guarding *Stolas* tortoise beetle from Brazil suffered a 100 per cent rate of loss through parasitism in 86 per cent of its clutches (Carroll, 1978). Egg-guarding thus seems to ensure that everything will never be lost through allowing a parasite to enjoy uncontrolled access to an egg mass. When this happens, the wasps have all the time they need to check out and parasitize every last egg within the clutch. The effectiveness of an *A. sparsa* female's defensive efforts against the ever-present scourge represented by myrmecine ants is far higher than her success against the parasites. Unprotected clutches suffer complete annihilation within a few days, while clutches which are guarded remain intact over the same period.

The newly hatched larvae remain in a compact herd, which first grazes the leaf's lower surface. As they grow larger they go

▲ **Few species of leaf beetles (Chrysomelidae) are known to exhibit maternal defence of eggs and young. This female tortoise beetle *Acromis spinifex* from Central America is perched on her eggs beneath a leaf.**

on to destroy the whole leaf except the tough mid-rib. Their mother remains close by but does not act as a shepherd in any way. When not feeding, the larvae congregate in a close-packed mass, enabling their mother to climb on top and form a shield with her body. She leaves them temporarily to make a pre-emptive strike against an intruding enemy such as an ant, an essential course of action if a 'scout' is not to discover the location of the larvae and 'report back' to recruit reinforcements. The larvae are, however, more vulnerable when they move down the narrow stem to another leaf, leaving their guardian to trail in the rear. From this position, she is unable to prevent (or perhaps is unaware of) lightning egg-laying attacks by dive-bombing chalcid wasps. Predatory shield bugs also seem to exploit this temporary vulnerability by persistently attacking a migrating group so that it eventually splits into two. As the guardian cannot be in two places at once, the bug can now safely attack the 'orphaned' group without risk of being charged and rammed by the mother.

The mother's solicitude extends throughout the whole of her brood's development and she even sits on guard atop the pupae, which are clustered together around a stem near ground level. She does not finally forsake her charges until they have emerged as adults from the pupae. Her job done, she finally leaves her post and eventually lays another clutch, producing several in succession during her lifetime. Parental care in this species undoubtedly pays considerable dividends in survival of eggs and larvae, although there is inevitably a slow but continual rate of loss, mainly to egg-parasites but also to a few predators.

Maternal care in a leaf beetle

It is only recently (Choe, 1989) that a definite example of maternal care in a leaf beetle of the subfamily Chrysomelinae has come to light, in rainforest in Costa Rica. *Labidomera suturella* does not seem to go in for the large families typical of the tortoise beetles, but produces instead a small and manageable group of only 1–4 larvae whose security can be adequately assured. When she detects a predator, the female

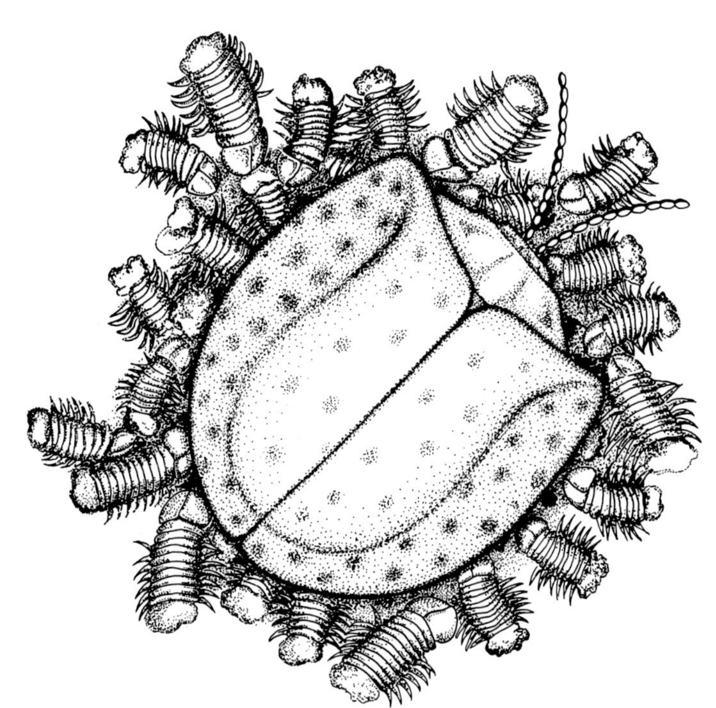

▼ **A female *Acromis sparsa* tortoise beetle stands guard directly over her brood.**

climbs on to the larvae and hangs on tightly by gripping the leaf beneath with her legs. Protection is therefore purely physical, with the female making no attempt to bulldoze predators from the vicinity. Rather she reacts to overt aggression by clamping down ever more firmly over her brood. Such passive armoured protection does seem to work wonders against the giant ponerine ants which roam the rainforests. Larvae are quickly spirited away by these ants when the mother is removed, while any larvae which are not swiftly snapped up by predators seem to become disoriented without their mother, roaming around aimlessly before dropping to the ground, presumably to linger and die.

Shepherding behaviour in a fungus beetle

The ephemeral nature of the larval food may have been the main stimulus behind the evolution of maternal behaviour in the Neotropical fungus beetle *Pselaphicus giganteus*. The following account is based on the author's observations of beetles on a log in rainforest in Trinidad. A female beetle was first encountered sitting on top of a mass of about 200 larvae, which from their small size were probably in the first instar. Shortly afterwards, the female scuttled off across the log and spent several minutes in a comprehensive search of its surface. She paid particular attention to some tiny buds of a bracket fungus, which were just in the early stages of erupting from the damp bark.

During their mother's absence, the larvae were very restless, almost boiling with activity so that the compact mound gradually began to lose its cohesion, as more and more larvae began to stray from their companions. When the female returned, she immediately resumed her position on top of the mound and proceeded to gather in the errant larvae, forming them once again into a compact mass beneath her. Several more times she set off to inspect the log, returning after a few minutes to gather in her brood. However, after what proved to be the final return, she nudged her brood with her mouthparts and turned to face the direction from which she had come. She must have given her larvae some kind of 'go-now' signal, for they began to stream across the log in the direction indicated. The female was in the vanguard, with half the brood scurrying along beneath her body, while the rest trailed along behind. Now and again she would turn round, run back to the rear and round up any stragglers, like a sheep-dog dealing with strays, keeping the whole party together in a broad and compact band.

▲ In a rainforest in Trinidad a female fungus beetle, *Pselaphicus giganteus* (Erotylidae), gathers her young larvae into a neat pile before going in search of a fungus in suitable condition for them to eat.

With such careful maternal guidance, the larvae swiftly arrived at a tiny fungus bud, which they fell upon and devoured within a few minutes. Their mother then again gathered them together and led them across to a second bud. Over the next few days, the larvae grew at a rate commensurate with the rapid expansion of the fungi themselves, which, within a day or two, developed from tiny buds to flat-topped brackets 5–8 cm across. The female remained with her brood but it was not seen whether she still guided them to their food each time, although it is likely that she did. The larvae stuffed themselves with fungus on such a continual basis that a whitish faecal fluid oozed more or less constantly from the anus. Within a few days, they were larger than their mother and within a week all had disappeared, presumably to pupate. By this time, any uneaten specimens of the fungus, which had provided the sole source of food, were already beginning to decay and disintegrate.

The sequence of actions performed by the female thus needs to be highly synchronized with the state of the fungal food source, its rapid development and equally swift decay. She must be ready to lay her eggs (although viviparity must be considered a strong likelihood here) just as the fungi start to develop. Because a single bud of the immature fungus only provides a quick meal for the larvae at their tiniest and least mobile stage, their mother's ability to move around rapidly and locate a continuous succession of suitable fungi, and then lead her brood straight to the spot, is essential. Conversely, she cannot delay the start of reproduction until a single fungus is large enough to nourish a whole brood of larvae in one go because, by the time the larvae had reached their second instar, their short-lived food would have collapsed and rotted around them. Maternal care in this species thus seems to be directed towards exploiting a species of fungus with a short 'use-by' date, something which the larvae left to themselves could not do.

▲ Once she has shepherded them to a small bud of a bracket fungus, she sits on top as they gather round to feed.

▼ Through continuous feeding on fungi located by their mother, the larvae grow incredibly fast under her close supervision.

Ground-beetle nests

All other known examples of parental care in beetles involve the construction of some kind of brood chamber. At least 14 species of ground beetles (Carabidae) in the subtribe Pterostichini exhibit maternal care (Thiele, 1977), all employing a similar procedure. The female excavates a brood chamber in the ground, usually under a stone or log, or else in soft rotten wood. She stays with the eggs for about 3 weeks, during which she fasts. Apparently her main duty is to protect the eggs from moulds, as no defensive reactions to intruders have been noted.

In a few species, the scope of brood care has increased, with a corresponding decrease in the number of eggs laid. In the European *Carterus calydonius* (Harpalini), the female constructs a labyrinth of cells beneath the ground, laying an egg in each one. She stands guard over the single entrance and feeds, as do her developing larvae, on a stock of seeds which she has laid down both in the larval cells and in her own living quarters.

Egg-brooding in staphylinids

One of the most hostile environments colonized by any beetle – the intertidal zone around European shores – provides board and lodging for *Bledius spectabilis* (Wyatt, 1986). The female constructs a burrow, shaped like a narrow wine-bottle with a mouth only some 2 mm across. This small diameter makes use of certain desirable properties of surface tension, which prevents the incoming tide from flooding quickly into the burrow as soon as the waters spill over its entrance. This buys precious time for the female, who requires about 4 minutes to insert a stopper formed from mud. She rears her brood within the burrow, going out at low tide to forage for algae, which she feeds to her offspring for the first week after they hatch, after which they leave home to dig their own burrows. A similar style of burrow construction combined with brood care is also seen in other saltmarsh beetles, such as *Bembidion pallidipenne* (Carabidae) and *Heterocerus maritimus* and *H. fossor* (Heteroceridae). This concentration of species exhibiting such a generally rare trait as maternal care, suggests that subsocial behaviour is an important adjunct for survival in the harsh and changeable environment of the saltmarsh.

The first example of subsocial behaviour in the Staphylinidae was noted in *Platystethus arenarius*, a common European species breeding in cow dung (Hinton, 1944). The female excavates a spherical brood chamber within the cow pat itself. She attacks and drives away or even kills any intruders, and prevents the growth of fungi within the humid conditions of the chamber by 'mowing' any incipient hyphae with her mouthparts. After hatching, the larvae stay with their mother until just before their first moult, when they leave and construct their own feeding cells within the dung.

Brood-rearing in dung beetles

Dung is normally both a scattered and an ephemeral resource, for which there is often considerable competition between different kinds of insects, such as beetles and flies. The desirability of being first in line to claim a share of this short-lived commodity, and then keep it out of the hands of competitors, has probably fuelled the evolution of parental care in many dung beetles. In order to deny rivals the chance to cash in on what is always only a windfall supply, it is essential that any early arrivals at the source should appropriate their share as quickly as possible. Large quantities of dung can be rapidly stashed away to serve as larval (or adult) food using two main methods:

(1) The dung can be buried in a chamber beneath or close to the original source.

(2) A portion of fresh dung can be moulded into a ball which is easily rolled well away from the source of any interference or competition, after which it is buried.

The first method is practised by many

members of the Coprini and Geotrupinae but suffers from the disadvantage that conditions below the pat may become rather crowded, so that developing larvae break through into one another's living-quarters, with the possibility of fatalities caused by fighting. This is minimized in many Scarabaeinae by the larva's ability to effect rapid repairs to any damage by applying a rendering of liquid faeces to the breach. This rapidly hardens to form a resilient wall and some species can repair remarkably large gaps in a short time. In species without parental care, the adults often mould the dung into large sausage-shaped masses within the subterranean burrow. However, in many Coprini and Scarabaeini, the mass of dung is formed into free-standing brood balls, which have sufficient free space around them to allow the presence of a guarding female. She is able to move with enough freedom to check on her brood's state of health and take any necessary remedial action.

The high risk of intruders taking over a nest in the crowded conditions beneath a pat may have led to the evolution of horns in males nesting in this situation, most notably in the Geotrupinae. The *Typhoeus typhoeus* male co-operates with his 'mate' to build a branched brood burrow in sandy ground (Palmer, 1978). The female packs the lower levels of the burrow with several brood sausages, using dung collected and delivered to her by the male. He also guards the burrow in its upper portion, turning sideways to block the passage with his broad back if a gatecrashing male tries to displace him as consort. Only if the assailant is of superior size can he gain purchase with his correspondingly longer horns, hook them beneath the resident's elytra and lever him triumphantly aside. Males also meet head-on and engage in trial-of-strength contests to decide who shall remain and who shall be expelled. Meanwhile the female busies herself down below with the preparations for her brood, ready to mate with whichever male is the victor.

The process of nidification in the common European *Copris lunaris* is normally a co-operative venture between male and female. This is a feature common to many dung-beetle nesting operations, although females of *C. lunaris* are not alone in being able to complete the whole job perfectly well by themselves if necessary. The couple dig a chamber 100–200 mm below a cow pat and fill it with dung, which they drag down from beneath the pat without disturbing its hard outer crust (Klemperer, 1982). Once a mass of dung is safely in place, the labourers enlarge the chamber around it and expel the spoil into the cavity which they have just created beneath the dung's crust. During digging and provisioning operations, the female plays a dominant role, shoving the soil up to the male above her and 'ordering' him either to shove it right out or to go back up for another load of dung by drumming on him with her legs. Once a large enough mass of dung is in place, the female finally expels the male from the brood chamber itself, although he often takes up residence in his own cell higher up.

The female now seals up the brood chamber, and there is a pause in activity for 3–7 days before she begins to fashion the rough-hewn cache of dung into some 4–7, neat, ovoid brood balls. She accomplishes this by tearing off small pieces of dung and gradually building them into a large ball. In the upper pole of each ball, she excavates a small egg chamber, lining its lower part with semi-liquid dung. After laying her egg within this chamber, the female seals it up, using a semi-porous plug. She forms this by using raking movements of her front legs and mouthparts, which probably serve to extract much of the liquid and finer matter, leaving a relatively dry, fibrous plug. This acts as a kind of 'lung' for the ball, the larva within its walls always keeping open an air-conduit between its feeding chamber and this porous area. The female remains with the brood balls, regularly licking their surface. This is a common habit among brood-caring dung beetles, and probably hinders the development of moulds. The guardian remains in place until at least one of her progeny has emerged as an adult, and then she finally deserts the chamber.

The procedure in related species is broadly similar, although a number of them coat the brood balls with a thick layer of soil (e.g. *Synapsis tumulus*, *Copris boucardi* and several *Heliocopris*). In the small Mexican *Copris laeviceps*, the presence of the female seems to be necessary to prevent the invasion of the brood balls by the larvae of mycetophilid flies. She also kills any 'foreign' larvae which infiltrate the chamber, an ever-present possibility in the crowded conditions beneath a pat. This can lead to larvae of the European *Aphodius rufipes* becoming kleptoparasitic if they penetrate from their normal feeding area near the pats down into the chambers beneath them made by other species (Klemperer, 1986).

The Asian species *Oniticellus cinctus* is one of a number of species of Oniticellini and Onitini which have also developed the habit of brood-ball construction. *O. cinctus* exhibits an impressive advance on the behaviour of most other known dung beetles, in that the female practises a kind of progressive provisioning by expanding the size of the brood balls as the larvae develop within (Klemperer, 1983). This is possible because the state of the balls (usually around 20 in number) can be monitored within a brood chamber constructed inside the hollowed-out shell of the dung pat itself. The female expands each ball by tearing off dung from the inside of the chamber and plastering it on to the ball. When the larva is fully grown, it bites an aperture in its ball and seals it with its liquid faeces. By repeating this process many times, the dung, which formerly constituted the walls of the ball, is gradually replaced with a thin shell consisting of dried faeces. This is probably less porous than the dung and may serve to maintain the humidity within the ball, an important factor given the above-ground situation of the nest and the now very desiccated state of its protective cow pat.

The female remains with her brood until the first adults hatch. She wields an impressive range of custodial behaviours, attacking and often killing intruding beetles, repairing the brood balls and mending holes in the chamber's outer walls. As the broods of an African species of *Oniticellus* are liable to be invaded by various fly and wasp parasites, as well as by histerid beetles, the adaptive nature of such concentrated maternal care is clear.

Dung-rolling beetles

Dung-rollers adopt the type 2 strategy, i.e., that of avoiding the cut-throat scene around and beneath the dung by trundling away portions of it to a more private nesting site. The males of many species manufacture a dung ball which attracts a female. This is the so-called 'nuptial ball', which precedes mating, and the subsequent fashioning of a second ball, the brood ball, is a co-operative venture between male and female. Beetles of both sexes also construct food balls, which serve exclusively for their own consumption. In many species, the female works alone to form a single ball, which she buries and then abandons after laying upon it her single egg. She then makes a further ball, repeating the whole process several times.

In a number of species, either the female

▲ Dung is a valuable yet ephemeral resource for which there is considerable competition between diverse groups of insects. The race to win a share of the dung has probably stimulated the evolution of various complex parental behaviours in the dung beetles (Scarabaeidae). Most of the beetles at work carving out small dung balls from this buffalo dung in South Africa (top) are *Gymnopleurus aenescens*. The best way of safely appropriating a share of the dung is to roll it quickly away from the scene. In most dung-rolling beetles, the male pushes the ball with his back legs, with the female riding on the ball, as in these *Gymnopleurus aenescens* (bottom) from South Africa.

alone, or a male and female acting in consort, make a single large ball which is later sectioned up into several smaller ones. These serve as food for several larvae whose development is watched over by the female. She therefore invests a considerable amount of time in nurturing just a few offpring. This form of subsocial behaviour has recently been studied in a number of African dung-rollers. In *Kheper platynotus* from East Africa, it is the female who inaugurates the process of moulding the first ball from the raw dung, which usually originates from elephants (Sato & Imamori, 1986 & 1987). In other species it may be the male who starts things off, e.g. in *K. aegyptiorum* and the American *Canthon pilularius*. It is while she is working on the dung that the female *K.platynotus* is joined by a male, who will usually begin their relationship by trying to copulate. If he is not given a hostile reception, he joins the female in her labours.

The task of pushing the finished ball, with the female riding on top of it, now falls to the male – the 'standard' procedure in dung-rollers. However, *Kheper platynotus* is special in making probably the largest ball of any dung-rolling scarab – up to 9 cm in diameter. As such an outsize load is often too heavy for a small male to handle, his place is unprotestingly taken by the female, illustrating the high degree of flexibility in the behaviour of this beetle. This flexibility also extends to other situations in which the male runs into difficulties – such as when running the jumbo-sized ball slap up against an insurmountable obstacle. When this happens, the female riding the ball will dismount and give a hand to set the ball rolling again, before resuming her position. On occasion, the pair may pause deliberately to roll the ball 'on the spot' in a patch of loose sandy soil, thereby giving it a good coating of soil. This may serve to retard the onset of desiccation, which is always a problem in the hot savannahs where the species lives; similar behaviour has also been noted in *Canthon pilularius* from the USA (Matthews, 1963).

It is now also the male who buckles down and buries the giant ball, still with the female clinging on top as the male's exertions cause it gradually to subside into the earth. The male remains in the nest for a further day or two before departing. In this species, copulation often takes place at any time between the first meeting of the sexes and the completion of the burial (although pairs probably always copulate at this last stage, regardless of whether or not they have mated earlier on). In other species, copulation is restricted to the period after burial, as only then can the male be sure that his work will be rewarded by the certainty of fertilizing the eggs laid on the brood balls.

During the first one or two days after burial, the female *K. platynotus* occupies herself with applying a 2–3 mm thick coat of soil to the ball, finishing it off by scraping the surface smooth with her curved, toothed, front legs. Then for 10 days, she daubs the surface with a similar thickness of her own semi-liquid faeces, which probably acts as an envelope, allowing the anaerobic fermentation of the dung to proceed within. Similar methods of ball-coating are seen in other dung-rollers. Once sufficient fermentation has taken place, after some 10 days, she splits a piece off the ball, converts this into a brood ball, coats it with her faeces and, finally, makes a deep dish-shape in its top in which she lays an egg. In sealing this up she converts the ball into a finished brood pear; she then repeats the process with what remains of the initial ball, finishing up with 2–4 completed brood pears, depending upon the initial size of the ball of dung collected. Her final act of preparation is to add a topping of soil which she compacts with her front legs. She stays with the balls, regularly licking their surfaces and scraping off larval faeces, which are ejected onto the upper 'ceiling' of the pear. This scraping action reduces the thickness of the wall in the upper area. It probably increases the rate of air-flow into the ball, while at the same time reducing the risk of moulds or bacteria thriving unhealthily in the larva's heaped-up excrement.

The female *K. platynotus* thus devotes about 2 months to the rearing of just 2–4 offspring. In *K. aegyptiorum*, a similar period may be required to rear 1–2 larvae. In *K. nigroaeneus* from South Africa, however, the female makes an incredibly high investment (possibly the highest of any insect) in just a single larva. However, she does not strictly place all her eggs in one basket, as she holds a second fertilized egg in reserve within her body, as a back-up in case the first should fail to hatch or the larva perish in its early stages (Edwards & Aschenborn, 1989). Once the egg has hatched, she daubs the ball with a covering of soil mixed with her own faeces. This probably helps prevent infection of the dung by bacteria and fungi, while reducing the speed of its drying out. She does not coat the whole surface, but leaves free a kidney-shaped patch which permits gas exchange.

The female's investment in time is apparently rewarded by a greatly increased rate of survival for her offspring, compared with those in unguarded nests. These suffer the depredations of moulds, ants, termites and dung-feeding flies and beetles. The male's investment is also high, for he often constructs, rolls and buries the bulky ball on his own, only finally bringing a female's attention to the interred ball by 'calling' and emitting a pheromone (Edwards & Aschenborn, 1988). Even when assisted by a female from early on, he still performs the majority of the work. As he is providing food not only for his single offspring, but also for the female during her internment (she uses 60 per cent of the ball as a reserve for her own consumption), he is in effect presenting a substantial nuptial gift.

In the much smaller *Canthon cyanellus* from Mexico, the male and female co-operate in moulding and rolling the ball and in excavating a shallow nest chamber (Halffter *et al.*, 1983). It is always the male who originates ball construction on the dung, where he is joined by a female. The process of making a ball is fairly rapid and good use is made of the spatula-shaped front legs and the spade-like front to the head. These are used in combination as the beetle turns full circle, thus forming a circular groove. Still turning in a circle, the labourer then rakes in a mass of dung with its forelegs, so that it soon forms a roughly globular shape beneath it. When the mass reaches a certain size, the beetle carves away the dung at the base, thus leaving a free-standing ball. This is further shaped into a smoother free-running ball by repeated patting with the front legs, while the beetle is either sitting astride its handiwork, or lying to one side and using the legs to revolve and work the ball at the same time.

There is often severe competition wherever dung is on offer and there are always males who are ready to cut down on their own workload and accelerate the process of attracting a female by trying to steal complete or partly made balls from their rivals. Fights frequently break out, with the ball's owner having the advantage of the 'high ground', standing on top of the contested property and jabbing his front legs under his opponent's underside so that he can be flipped over on his back. Attempts at take-overs are rarely successful unless the assailant is considerably larger than the ball-owner.

The female *Canthon cyanellus* prepares two brood pears from several balls which are rolled in by the male. If he is laggard in returning with a fresh supply – possibly because he has been tied up in skirmishes with other males – the female leaves the nest and goes out to assist him in rolling in a new ball. She covers each resulting pear with a coat of soil, often adding it to an already-generous coating picked up by the ball's sticky exterior on the journey to the nest. The male usually remains with the female, guarding the developing brood until they reach the third instar or even beyond. The magnitude of the male's investment in his offspring thus considerably exceeds anything so far known in African dung-rollers. This even seems to extend to the chemical protection of his progeny's food supply, which he impregnates with a substance capable of repelling the potentially invasive larvae of dung-feeding flies. The process takes several minutes, which he spends 'standing on his head' as he rubs his hind

▶ Nest of the dung-roller *Canthon cyanellus cyanellus* with five brood balls. The male and female both remain in attendance on the nest. The spherical knobs on the brood balls are larval excrement voided to the surface. If the female is removed, the larvae die, despite the continued presence of the male. (After Halffter *et al.*, 1983).

▲ Rather than carving out a dung ball from a mass of dung, some dung beetles utilize a mammal dropping which is delivered to them ready made. This *Helictropleurus quadripunctatus* from Madagascar will only collect the full-size droppings of the sifaka lemur, *Propithecus verreauxi*, ignoring the much larger splodgier droppings of the ring-tailed lemur, *Lemur catta*. However, because sifaka droppings are not spherical, the beetle has to drag them to the nest backwards, an unusual habit in dung-rollers.

legs down across a gland on the underside of his abdomen. Once his hind legs are thoroughly impregnated with the secretion from this gland, he transfers the repellent chemical to the ball, either during its manufacture or while rolling it and fashioning the brood pear (Bellés & Favila, 1983). The chemical impregnation of the ball may also serve the dual role of trail-marker, enabling the male accurately to re-locate the nest chamber during his visits back and forth to the dung for extra supplies.

Despite the generous extent of his input, the male alone cannot successfully raise a brood. In the absence of the female, all the larvae soon die, exactly as they do if both parents are absent. When the female alone is present, or both parents are in harness, the brood develops as normal. The parents' main role is probably to exclude klepto-parasitic insects and keep the development of moulds in check. Unlike in *Kheper*, the guardians do not remove larval faeces from the surface of the brood balls, which soon assume a knobbly appearance as a result.

Artificial manure in Cephalodesmius

Not all so-called dung beetles actually exploit dung for making their brood balls. Some utilize plant material from which they synthesise a kind of artifical 'dung'. Although a number of species in various genera are known to provision their nests in this way, the only ones to exhibit advanced subsocial behaviour as well are several Australian *Cephalodesmius*. These beetles are remarkable not only for their use of leaves and fruits as fare for their brood, but for their habit of progressively provisioning the growing larvae (Monteith & Storey, 1981). This is only possible because of the unusually high degree of involvement by the male, who conducts all the foraging expeditions. The male and female initially team up to excavate a brood chamber in the ground. Once the early

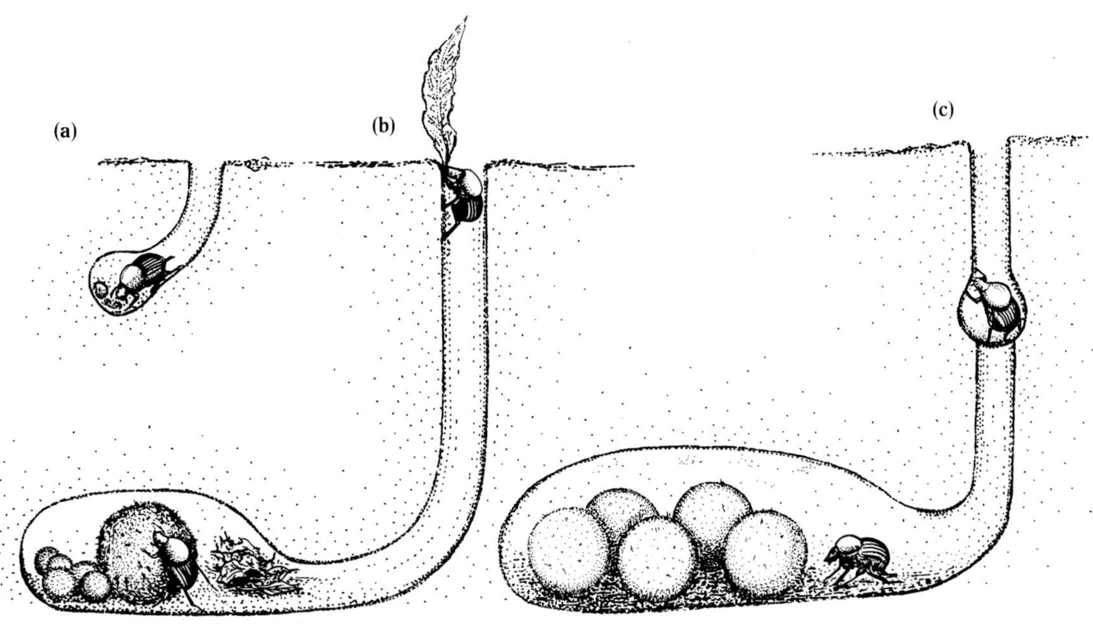

► Different stages in the nesting behaviour of the dung beetle *Cephalodesmius armiger* from Australia. (a) Small feeding burrow made by adult males or females in late summer. (b) A nest burrow excavated by a monogamous pair of adults. The male is pulling a leaf down into the burrow, while the female adds material to the brood mass which she has manufactured from the leaves. On the left are several brood balls containing larvae. (c) The male has now been expelled from the nesting chamber and the female alone tends the brood balls; these are now much larger as she has been adding extra material during progressive-provisioning activities. (After Monteith & Storey, 1981)

stages of nest establishment are over, the female remains inside while the male makes regular trips to the surface to collect fallen leaves and fruits. These he drags to the nest backwards, with a rather shuffling motion. The female uses her front legs to crush and process this raw material into a dung-like substance, which she adds to a central brood mass. This does service as a food store, both for the adults' own consumption and for the production of brood balls.

The female constantly manipulates the brood mass with her front legs and kneads it with her mouthparts, while both she and her partner donate their faeces to the growing mass. Eventually this reserve grows larger than its maker, by which time fungal activity has becomes rife. This plays a vital part in breaking down the rather rough material into a spongy, easily worked medium, ripe for forming into brood balls. The female performs this highly skilled task, first constructing a rather small ball in which she hollows out a large cavity, so that the 'ball' now resembles an urn. She lies on her back and holds the 'urn' above her, this being the only practicable method of inserting an egg into such a basically fragile structure, notably lacking in the robustness of a true dung ball. The male takes a close interest in all this, waiting for the appropriate moment to approach and mate, thereby being certain of his paternity of each egg laid. Once the female has constructed about five brood balls (at a rate of one per day) she stops work for a day or two, waiting for the first egg to hatch.

The initial store of food supplied to each ball is rapidly exhausted by the larva feeding within, but the female monitors events closely and regularly draws supplies from the central store to add to the balls. She treats each one individually, adding material as and when required. So precise is her judgement of the larva's needs that the wall of the ball is maintained at a constant thickness of about 2 mm. The male, meanwhile, continues his foraging so that the food store remains adequately stocked. The female's close supervision of her invisible offspring may be facilitated by their ability to inform her of their needs by stridulating. A larva does this by a method believed to be unique: it doubles its body over and 'scratches its chin with its tail'. This may enable the female to assess the thickness of the ball's wall. Alternatively it is the larva, which does this mechanically and then communicates the fact to its mother in a 'more-food-needed' message.

The female's last service to her brood, just before pupation, is to plaster down large blobs of semi-liquid faeces which are ejected on to the outer surface of the balls, to form a smooth surface. This faecal wall later sprouts a growth of fungal fruiting bodies, which probably provide the emerging adult with its first meal. It seems likely that this is the mechanism for providing the adult with a crop of essential symbiotic fungi. These will be needed later to 'digest' the plant material in the brood mass, after it has been inoculated with fungal spores from the adults' faeces. At least two other *Cephalodesmius* species, *C. laticollis* and *C. quadridens*, exhibit broadly similar behaviour, which is undoubtedly the most complex in any scarab. This particularly applies to the impressive range of manipulative talents demonstrated by the females.

Advanced care in burying beetles

The acme of parental care is probably found in the burying beetles (Silphidae), which evince a spectrum of behaviours surprisingly similar to those of some birds. As in dung beetles, the evolution of complex parental care is probably linked to the ephemeral nature of the food source – small animal corpses – and the high level of competition for it. The battles over ownership of a corpse and subsequent mating behaviour have been fully described in the previous section (page 71). Once the corpse has been buried, it remains in the sole charge of either a female working alone or a female and her mate, depending on species. During the burial operations, the beetles use their mouthparts and legs to remove any hair or feathers, and then work the suppurating flesh into a more or less globular shape, which can be easily pushed into a chamber dug beneath it. Once the chamber or 'crypt' is sealed, the female scatters a number of eggs around on the floor a short distance from the brood mass.

Upon hatching, the larvae are left to make their own way to the odorous ball, in which the female chews out a bowl-shaped depression (Pukowski, 1933). The larvae pour into this ready-made 'trough' and wait there like chicks in the nest. They even 'beg' for food in a most bird-like way, rearing up in consort and thrusting their heads, one at a time, into the female's gaping jaws, from which each receives a droplet of regurgitated food. This intimate feeding association between mother and offspring continues throughout their development. However, it does gradually diminish in regularity and importance, because the larvae are able to browse directly from the brood mass when only 5–6 hours old. The tendency of this vital and perilously perishable food store to disappear prematurely through putrefaction and disintegration is thought to be inhibited by antibiotic substances applied by the adults in their saliva and anal secretions. The parents are also highly efficient at repelling possible enemies, such as predacious beetles or fly larvae. These are crushed by the guardians' powerful jaws and contribute to the general food reserves by being eaten.

In the European *Necrophorus vespillo*, it is the female alone who cares for her brood. In *N. vespilloides*, however, the male's role is on a par with his mate's, for he too takes his share of feeding the brood by regurgitation (Bartlett, 1988). In fact, his capabilities and competence are so substantial that, in the absence of the female, he is fully able to rear the whole brood by himself.

One question which has vexed many researchers is why the male should closet himself away and invest so much time and energy in caring for a single brood. He could just as easily stay outside and attempt to mate with as many females as possible. The reasons for his loyalty to mate and brood have recently become evident. It seems that, despite the crypt being buried some way beneath the surface, it is not immune to detection and invasion by rival males (and females). Take-over raids are quite frequent in the American *N. orbicollis*, in which the male's role is similar to that of *N. vespilloides* (Trumbo, 1990), although females also often nest alone. It is very much in a resident male's best interests to repel an invader, in view of what will quickly take place if he loses the battle for ownership of the crypt. His successor's first action will be to start eliminating the resident larva, which will be killed and eaten over the next few days. This may allow the newcomer to father a brood of his own, assuming that the corpse is still large enough to support it. It is significant in this context that larger corpses seem to incite more take-over attempts than small ones, presumably with the prospect of a replacement breeding effort. A lone female will also try to commandeer a crypt, and commit infanticide as a first act, thereby appropriating the remaining food mass for the development of her own brood. If a male is present, she will mate with him; if not she will use her supply of stored sperm.

One vital factor is the reduced number of

take-over attempts against pairs compared with single females. This emphasises the importance of the male's staying to guard and feed his offspring, a commitment which will always be necessary in view of the general shortage of suitable carcases and the resulting high levels of competition for them. However, now and again, the new owner does not slaughter all the resident larvae, and he may then care for a mixed brood consisting of his own offspring along with those sired by the previous resident. Similar mixed-brood guarding is also occasionally seen when two *N. vespilloides* females share a crypt after an inconclusive fight over the ownership of a corpse. In such a case, the dominant female may care for a few larvae belonging to her subordinate companion (Müller *et al.*, 1990).

Infanticidal tendencies are not restricted to invaders – sometimes it is a brood's own parents which may be responsible for their slaughter (Bartlett, 1987). This seems to be brought about by the female's inability to relate the number of eggs she lays to the size of the corpse. If all the eggs hatch, the brood mass is soon overwhelmed with larvae, whose numbers plainly outweigh the capacity of their rations to support them. The end result would be a brood of very undersized adults, doomed always to be losers in the competitive whirl of burying-beetle economics, due to the inability of small adults to secure a corpse or a mate. In order to avoid the reproductively pointless production of such a load of non-starters, the *N. verpilloides* parents take remedial action at an early stage to 'thin out' their brood, by killing and eating some of the excess. Just how they judge which larvae should get the chop and how many should remain is not presently known. Avoiding such superfluous production of larvae in the first place, by laying fewer eggs, would seem to be more sensible. However, it does appear that the policy of laying extra eggs is an 'insurance' against non-hatching. This must be an important consideration, given the highly perishable nature of the provisions within the crypt, which leave little or no leeway for replacement eggs to be laid should the first fail.

Complex parental care in bark beetles

The remaining subsocial beetles all breed in galleries in wood, where they form intimate mutualistic associations with fungi. These either help to break down the wood into an assimilable state or else provide the actual food itself, as in the ambrosia beetles (Scolytidae and Platypodidae). So vital to survival is this association that fungus-cultivating scolytids have evolved special pouches on the body, designed specifically for carrying the fungal spores to new trees. Although many fungus-growing scolytids live communally in complex gallery systems, there are no social interactions between the adults and their independently developing offspring.

In both *Ips* and *Gnathotrichus*, the male remains in the entrance gallery and guards his growing progeny. The most highly developed subsocial behaviour is, however, restricted to *Monarthrum*, in which the male and female work together to construct a series of short branches or 'cradles' off the main gallery. Each 'cradle' will be home to a single larva which feeds on fungus-infested wood chips. These are supplied in the form of a 'stopper' at the mouth of cradle at the time of egg-laying. Once the larvae hatch, the female pays close heed to her maternal duties, ever ready to place a fresh supply of fungal food in the mouth of each cradle as the original supply becomes depleted. She also clears up the larval faeces, which are ejected from the cradle at regular intervals through an opening which the larva makes on specific 'cleaning' days. Some of this waste is pushed out of the galleries altogether, while a portion is retained to act as fresh compost for raising new fungus. The female closes the opening with a plug of fresh food and remains in attendance until the adults emerge.

Particularly complex social relationships have developed in some Passalidae, most of which live in dead wood. The larvae of a number of species, such as *Odontotaenius disjunctus*, do not thrive on wood which they have chewed themselves, but do much better when receiving a supply of pre-digested wood in the form of the adults' faecal pellets. The larva of *Heliscus tropicus* is even more dependent, being unable to chew wood at all and relying totally on the adults. Some species of passalids have evolved a remarkably sophisticated social life, with overlapping generations living together in a colony which has a certain amount of co-operative brood care. In a behavioural trait more typical of primitively social hymenopterans, the young adult may even help its parents to perform a variety of communal duties, including the construction and repair of the complex pupal cases destined to be utilized by a new generation of its brothers and sisters (Schuster & Schuster, 1985).

DIPTERA True flies

Despite the vast number of species, any cases of parental attention to offspring are apparently absent, while there are few examples even of females staying with their eggs.

The American horse fly *Goniops chrysocoma* (Tabanidae) lays a mass of eggs on the underside of a leaf. The female possesses exceptionally hook-like claws which cleave on to the surface of the leaf and anchor her in place above her eggs. Here she remains until the larvae hatch, when she dies.

The 'social' bat fly

Although not strictly exhibiting parental care, the special behaviour of the bizarre New Zealand bat fly, *Mystacinobia zelandica*, qualifies it for a mention here (Holloway, 1976). This insect is so unusual that it merited the erection of its own family, the Mystacinobiidae. The wingless flies live in communal groups in bat roosts sited within hollow trees. The female flies lay their eggs communally and larvae and adults live together, with mutual grooming being a common routine not only between adults but between larvae as well. There has even, it seems, been the evolution of a separate caste, in that certain males live beyond their normal reproductive allotment and take on the role of sound-producing 'soldiers'. Their task is to prevent the bats from intruding destructively on the bat-fly community's affairs. The flies are wholly harmless as far as the bats are concerned, just feeding on their guano, but they do rely on the bats for transport to new roosts.

LEPIDOPTERA Butterflies, moths

Despite the vast assemblage of species within the Lepidoptera, there are only two known examples of parental care, both of which are within the single butterfly genus *Hypolimnas* (Nymphalidae).

The brooding behaviour of *H. anomala* has been closely studied on the island of Guam (Nafus & Schreiner, 1988) where females lay their eggs beneath leaves of the tree *Pipturus argenteus* (Urticaceae). The size of the egg batch is incredibly variable – anything from 10 to 1272 – but a typical clutch contains about 500 eggs. The female positions herself above the eggs, facing downwards towards the leaf tip with her wings closed. Not all females stay on guard – some abandon their clutch as soon as it is complete – but more than 50 per cent

Egg-brooding in mosquitoes. The best-known example of maternal care within the Diptera concerns the Neotropical mosquito *Trichoprosopon digitatum*. The female lays her eggs exclusively in water-filled fruit husks, usually on the ground. She oviposits her clutch under cover of darkness and then remains standing over them for the next day or so until the larvae hatch, when she departs. If disturbed by a predator she 'rows' her egg mass to the edge of the pod, where more cover is often available. However, the main benefit of these maternal behaviours is to prevent damage by the torrential downpours typical of the mosquito's breeding season. These can smash up the egg masses, or even wash them clear out of the container, a hazard which any mosquito has to overcome if it is to breed successfully in a relatively small, open pod, which is easily overfilled (Lounibos & Machado-Allison, 1987).

Similar behaviour was recently (1989) confirmed by the author in a mosquito (probably a species of *Uranotaenia*, *fide* Lounibos) from rainforest in Madagascar. Several females were found sitting astride mature (i.e. dark) eggs in the water-filled stem internodes of the giant bamboo, *Cephalostacyum viguieri*. They stayed with their eggs when disturbed and refused to fly away and desert them, despite being very 'distressed', a typical reaction in any egg-brooding insect. This observation probably amplifies earlier comments on *U. shillitonis* females as 'resting' inside cut bamboo stems in Madagascar (Lounibos, pers. comm.) The main function of guarding in this mosquito is unknown, but may be primarily to prevent slopping out of the eggs, particularly as this species is much smaller than *T. digitatum*.

There is currently a question-mark over the motive for the apparent maternal behaviour in six species of *Armigeres* mosquitoes from the Far East (Lounibos, 1983). The female attaches her eggs to her hind legs and carries them around, even seeking a meal from a warm-blooded host with her eggs still in place. *Armigeres* also nests in empty fruit husks and bamboo internodes, a habitat which seems to have stimulated the evolution of some kind of post-ovipositional investment by the females. Just why *Armigeres* sticks her eggs on her legs and carries them around (for several days?) is not known, but it has been suggested that she retains them so that she can deposit living larvae within the small body of water normally utilized.

▲ Female *Trichoprosopon digitatum* mosquitoes stand guard for a day or so over their egg rafts until the larvae begin to hatch. This species breeds in fallen fruit husks, and the main function of the maternal presence is to prevent the eggs from being washed out of the husk by torrential rain.

accept the role of sentry until the larvae have hatched. About 50 per cent of these guardians then persevere and stick with their densely packed family of baby caterpillars for the 2–3 days which they spend on their original leaf. The guardian only finally deserts her post when her progeny take it upon themselves to migrate to a fresh leaf.

The main enemies are three species of ants. The largest of these, *Solenopsis geminata*, is also fortunately the rarest. It seems unperturbed by the female's defensive tactics, which consist of spreading and fluttering her wings, and manages to spirit away all her eggs one at a time, despite the guardian's best efforts to prevent it. By contrast, the most common ant, *Monomorium floricola*, is much too small to be able to carry away the eggs. It is therefore more easily repulsed by the mother because of its need to remain alongside the egg mass while it bites open the egg shell and extracts the contents. Unguarded egg masses suffer a much higher rate of loss from this source than guarded clutches, proving the value of guarding against this common predator. With *S. geminata*, egg masses are cleaned out, whether unguarded or guarded. Survival of the young caterpillars is also increased by their mother's chaperonage and she can almost double the number of her offspring reaching their second day of life if she stays on guard until then. *Hypolimnas antilope* from the Philippines exhibits similar maternal care but less is known about its rate of success against various enemies.

HYMENOPTERA Sawflies, wasps, ants, bees

SAWFLIES

Subsocial behaviour has been observed in ten genera of sawflies in the families Argidae, Pergidae, Diprionidae and Tenthredinidae. The most detailed and recent studies concern two Brazilian argids, *Themos olfersii* and *Dielocerus diasi* (Dias, 1975 & 1976). The female of *T. olfersii* straddles a compact cluster of eggs on a leaf, while, in *D. diasi*, she sits near the base of the leaf, over which the eggs are more widely spaced. Both species are warningly colored, advertising their distasteful properties to visually-hunting vertebrate predators, and they are apparently virtually immune to attack from this source. They thus 'fly the warning flag' over their eggs where it provides visual protection against vertebrate egg-predators. The guarding female reacts to a threat with a wide repertoire of defensive measures. She will buzz her wings intimidatingly, try to lunge at the intruder and bite it, strike it with an extended forewing, spread her wings down and over the eggs in a protective umbrella, or clamp her body down on to the eggs and then perform rhythmic 'press-ups'. She will also turn around to counter an attacker head-on, presumably 'confident' that this will not leave the rearmost eggs unguarded. This could lead to losing everything and is precisely why the shield bug *Antiteuchus tripterus* stays put and refuses to budge when her eggs are attacked from the rear.

The female sawfly's countermeasures are quite effective but the eggs suffer loss rates of some 7–20 per cent. Around half of the higher figure is due to attacks by scelionid parasitic wasps. The female remains for some days with her larvae after they have hatched, during which time they feed in dense aggregations nearby, and she sometimes 'escorts' them as they move to a fresh leaf. It has been theorized (Dias, 1975) that brooding behaviour in *T. olfersii* has evolved in order to cope with the exigencies of feeding upon the thick and very tough leaves of the food plant, *Eriotheca pubescens*. To gain access to the leaf blade, the larva must first gnaw a difficult entry through the leathery leaf edge near its base. To accomplish this challenging feat, the larva needs exceptionally large jaws. These can only be accommodated in an oversized head and the newly hatched larvae are noticeably big-headed when compared with other species. Development of the large head apparently demands an extended incubation period within the egg. This can only be practicable if they have some form of long-term protection – hence the presence of the maternal guardian.

WASPS

Maternal care

Within the enormous division Parasitica, examples of parental care are very scarce, and apparently restricted to just one or two examples in each of two superfamilies. Within the Ichneumonoidea, the braconid *Cedria paradoxa* from India stands on guard by her brood as they develop upon the caterpillars of various pyralid moths. The female is believed to protect her progeny from attack by hyperparasites. Within the superfamily Chalcidoidea, maternal care has been observed in the eulophid genus *Melittobia*.

The diverse members of the Aculeata employ a wide range of parental behaviours, culminating in the highly social organization seen in many wasps and bees, as well as the completely social ants. With the exception of the Bethyloidea, most of which are parasites on other Hymenoptera, and numerous 'cuckoos', the main characteristic of the aculeate wasps and bees is the construction and provisioning of a nest in which the offspring are reared. In most cases the female seals the nest, or each cell within it, before the egg has hatched, so there is no actual contact between mother and offpring.

Within the Bethylidae, there is a range of

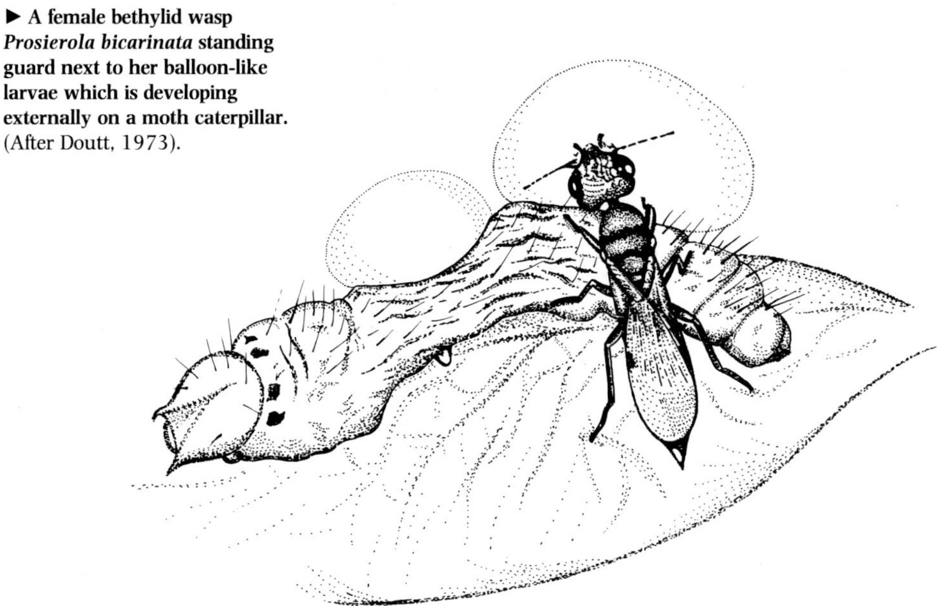

► A female bethylid wasp *Prosierola bicarinata* standing guard next to her balloon-like larvae which is developing externally on a moth caterpillar. (After Doutt, 1973).

behaviours which parallel those in unrelated groups of Hymenoptera. Thus many bethylid females lay their eggs on caterpillars, and the larvae develop as external parasites, as in some Ichneumonoidea and Pompiloidea. However, unlike in those groups, some female bethylids stay on guard during the development of their brood. The female stings the host caterpillar to immobilize it and may then feed on fluid leaking from the site of the incision.

Prosierola bicarinata from Mexico fixes 3–5 amazingly large eggs on the back of the immobile host caterpillar of *Desmia funeralis* (Doutt, 1973). The larvae hatch within a day or so and then balloon up in volume so rapidly that, within only 2 days, they have finished their feeding and resemble bloated translucent bags of liquid. The larva's long 'neck' is buried within the host's body and, if it encounters one of its siblings during feeding, it will kill it and drain it dry within a few hours. The female maintains a vigil throughout this rapid development, fiercely attacking intruders with her jaws agape and sting at the ready. She also guards the pupae, which are produced by the third day of feeding. Other parasites which utilize this species of caterpillar as a host are heavily attacked by hyperparasites, so it seems likely that, by becoming a guardian, the female bethylid exchanges a low rate of fecundity for a high rate of survival for her offspring.

Other bethylids move their paralysed host to a safer site after stinging it but then depart after laying their eggs. However, *Epyris* behave more like the solitary wasps in their habit of constructing a nest to receive the prey. *E. eriogoni* from the USA does not dig her own burrow but merely constructs a side-branch in an existing tunnel constructed by a *Bembix* wasp. The adoption of other wasps' burrows, both used and disused, is a fairly common trait in the solitary wasps (Sphecoidea). The female *E. eriogoni* employs an unusual method of dragging her prey – a tenebrionid beetle larva – to the burrow. She stands to one side of it as she grasps it in her mandibles and it may even ride up slightly on to her back, a must unusual method which has also been described in *E. extraneus* (Rubink & Evans, 1979).

Nesting in spider-hunting wasps

The spider-hunting wasps (Pompiloidea: Pompilidae) are relatively primitive, a feature which is reflected in the broad diversity of their reproductive behaviour. In some species, e.g. *Homonotus* (Old World) and *Notocyphus* (New World), the female lays an egg on the abdomen of a spider, which she has first stung in order to produce temporary incapacitation. She then departs, leaving her larva to feed externally on the spider. The host resumes its normal activities, although the unwelcome hitch-hiker meanwhile feeds on its substance, eventually killing its host before pupating.

When spiders which make their home inside a silken tube are attacked, the host's residence does duty as a ready-made nest, obviating the need for the female wasp to construct one of her own. Such behaviour is found in numerous pompilids, including some *Pepsis*. The American *Allochares azureus* works an unusual variation on this theme, being highly specialized for entombing filistatid spiders within the the intricate silken labyrinths of their webs (Deyrup *et al.*, 1988). The spider is stung out in the open, after being flushed from its lair, and the wasp then takes it back inside and wraps it up in its own silk (or sometimes in the silk of an abandoned web nearby) before laying an egg on it. The wasp prepares the egg-laying site by using her own abdomen as a scraper to brush a patch of hairs off the spider's abdomen. The wasp herself is virtually hairless and this probably prevents her from becoming trapped in the spider's sticky threads while negotiating her tricky way through the web. The wasp's larva feeds in the relative security of the tacky web and constructs a densely hairy cocoon which is unique in the family.

Most pompilids adopt a different course of action, first stinging a spider and then digging a burrow in the ground in which to store it away from possible attack by parasites. This habit of catching the prey *before* the nest is begun, and of catching just a single, large prey item before sealing up the nest and leaving, is regarded as primitive. It is typical of most pompilids and some sphecid hunting wasps. Before starting to dig her nest, the female pompilid sets the spider aside in some relatively secure spot, usually a little way up on a tuft of grass. Here it is less likely to be spotted by an ant scout, who would quickly recruit a hoard of helpers, to which the wasp would have no choice but to yield her prey without a fight. In fact, if more than a few ants turn up, the wasp will usually abandon the spider.

While busy digging its grave, she returns at regular intervals to check the spider, probably to keep tabs on the arrival of any ants and to check whether the unfortunate beast is beginning to show signs of movement. This sometimes happens when the first paralysis wears off, and the spider has to be re-stung one or more times to make sure. Pompilid burrows are very simple, consisting of a crudely finished cell at the bottom of a short tunnel. The female backs into the nest, pulling the spider behind her, lays an egg on it and then seals up the burrow. In doing this she uses the tip of her abdomen as a pile-driver, vibrating it up and down in a blur to firm down the earth of the final closure. Those species which nest in sandy ground, level off the nest by kicking sand backwards across the closure with their legs. In more gravelly substrates the female painstakingly gathers small pebbles with which to disguise the entrance.

Nesting in sphecoid hunting wasps

Ground-nesting is also the most common method in the hunting wasps in the superfamily Sphecoidea, all of which are solitary, save for a single genus, *Microstigmus*. Solitary sphecoid wasps employ three types of provisioning which are thought to form an evolutionary sequence:

(1) *Mass provisioning*, in which the nest is stocked with one or more prey items as rapidly as the female can find them. She then lays an egg and seals the cell without ever seeing her offspring. She will often stock several cells in sequence within an individual nest.

(2) *Delayed provisioning*, which is similar to the last except that the egg may hatch before the female has finished her task. However, she does not interact with the larva.

(3) *Progressive provisioning*, in which the female supplies food 'as required' to the larvae. She knows when to act by regularly monitoring the state of the larva and the nest. This mode of provisioning is thought to reduce parasitism, as the female often checks on the cell contents during provisioning trips and may eject refuse and the larvae of parasites.

Finding the nest

Most pompilids only supply a single, large prey item, so that the female does not need to take an accurate fix on the position of her nest in order to make repeated provisioning visits. The same applies to many sphecoids which also bring just a single large item. However, provisioning with multiple prey, as found in the more advanced pompilids, most sphecoids and the vespoids, has

Problems for the mass-provisioner – to close or not to close? Mass-provisioners are faced with the problem of what to do with the nest during their absence – should they close the entrance temporarily, or leave it open? Building a temporary closure involves more work and reduces the amount of time spent in hunting but has the advantage that parasites are not free to stroll right in and wreak havoc in the unoccupied nest. However, the need to open the burrow on arrival with prey might possibly give a parasite who has been shadowing the laden wasp just the chance it wants. It will have time to drop its larva upon the prey while the female is busy digging out her front door. A large item of prey is especially at risk, as the female has to dump it unattended near the burrow so that she can free all her legs for the job of digging.

Leaving the nest open means that parasites are at liberty to walk right in, but as they often locate the burrow in the first place by orienting on the female herself, leaving it open but deserted may not be too great a problem. An open burrow has the advantage that the returning wasp can run straight in through the entrance, without giving a pursuing or waiting parasite time to larviposit. However, the method of entry can actually vary with the size of prey being carried and the method of transport. In *Oxybelus strandi* from Japan, the female carries small prey (always a fly) beneath the body, gripped firmly with her legs. She rushes straight into the nest without pausing and on down to the cell below. A medium-sized fly is transported in the same way, but will be temporarily left in the entrance so that the female can reverse and pull it down backwards. Large flies are transported impaled on the wasp's sting as far as the entrance, where they are disengaged temporarily while the female enters; then she pulls them down with her front legs. These alternative methods of prey transport allow the wasp to enter the nest as quickly as possible, without having to broaden the entrance to accommodate large prey (Tanaka, 1985).

▼ Many hunting wasps make a temporary closure to the nest while they are out searching for prey. This must be dug out when the owner returns, as in this large *Sphex tomentosus* with its katydid prey in a Kenyan forest.

required the evolution of a finely tuned homing ability. This has reached such a pitch of perfection that a returning female can fly straight to her nest, even when it is set among hundreds of others in bare and apparently featureless terrain.

The key to her accuracy is her sequence of actions upon leaving the nest, especially when this is for the first time. She obtains a 'site fix' by flying a widening series of low circles or figures of eight above and near to the nest. As she flies her pattern, she fixes in her memory nearby features, such as projecting stones or tufts of grass, large pebbles or bits of animal dung, as well as more distant features, such as bushes or trees and even very far-off landmarks such as hills.

Progressive provisioning in Bembix

Progressive provisioning using flies (Diptera) is found in the closely related genera *Bembix, Stictia* and *Rubrica*. The nests are usually made in sandy soil, often in well-used paths or playing fields. Right from the start, the female manifests a remarkable capacity to tailor her actions to the changing needs of her brood by supplying the tiny newly hatched larva with a small, easily managed, soft-bodied fly for its first meal. She then maintains a regular supply of flies of varying types, gradually increasing the size of prey as her voracious larva grows strong and fat and is able to cope with even tough spiny customers such as robber flies (Asilidae). Although these flies are themselves fierce hunters, they nevertheless fall regular prey to the slick methods employed by female hunting wasps. In the Neotropics, several species of *Bembix* and *Stictia* are known locally as 'horse guards' from their habit of concentrating their lethal attentions around domestic horses, where they pounce upon blood-sucking tabanids.

The female wasp's first task each morning is to inspect the nursery chamber and assess whether there is likely to be a shortfall in food that day. In *Bembix, Stictia* and *Rubrica*, this precursory visit may also be devoted to cleaning out the cell: removing detritus, such as the sundry husks of flies which soon clutter up the nesting chamber. The female then sets off to hunt, returning regularly throughout the day with flies to replenish the stock. In some instances, she may have to slave away particularly hard to satisfy the ravenous appetites of a host of uninvited dipteran guests. The larvae of these inquilines are often found in *Bembix* nests and seem to be ignored by the female

▶ **With a horse fly (Tabanidae) slung snugly beneath her underside, a *Bembix* sp. hunting wasp returns to her burrow in a Kenyan forest. Species of *Bembix* and the related genera *Stictia* and *Rubrica* often prey on horse flies which are pestering cattle and horses, earning them the name of 'horse guards' in parts of Central and South America.**

wasp. In wasps which provision their nests with a single large prey item, the owner's larva will be quickly finished off by the invading fly larvae. However, in *Bembix* nests, the inquilines do not attack the wasp larva, with a good and perfectly self-serving reason: this would bring about the abrupt end to their own regular supply of free food. Without the presence of her own larva to provide the stimulus for prey capture on her daily inspections, the female would simply abandon the nest, leaving the inquilines to starve.

In colonially nesting species, such as *Bembix*, *Stictia* and *Rubrica* (and many other colonial nesters), the female wasp, upon returning to her nest with prey, often has to run the gauntlet of attacks by kleptoparasitic females of her own or a closely related species. These bandits do not go out and hunt for themselves, but pounce on a returning female and try to steal her hard-won fly. This is not easily done and the usual result is a hectic free-for-all on the ground, during which the opponents try desperately to bite and sting one another, the owner of the contested fly usually gripping it tightly while she battles away. These clashes can last a minute or more before the victor – usually the original owner – frees herself and scrambles into her nest, sometimes first having to retrieve the fly, which has become dislodged during the fracas.

Some females implement such acts of brigandage for several days in succession but then start to hunt on their own account. A shortage of prey during certain seasons probably increases the amount of kleptoparasitic behaviour. This also takes the form of straight nest-robbing by intruding females, who steal part of the stock of flies while the resident is out hunting. During such incursions, the owner's larva is apparently not harmed and the resident continues to provision the nest. The final closure puts an end to any such incursions and, in some species, such as *Stictia maculata* from Costa Rica, the female camouflages the site of the nest with bits of leaf and twigs (Matthews *et al.*, 1981).

Nesting in *Ammophila* sand wasps

The single genus *Ammophila* encompasses just about every type of provisioning behaviour found in solitary wasps. There are:

(1) Species which struggle mightily to hump a single large caterpillar to the nest, big enough to suffice for the entire development of a single larva. When handling such bulky and awkward items, the wasp clamps her jaws firmly around its front end. She frequently buzzes her wings to give some degree of lift when her cargo tends to drag against something on the ground. She always carries the caterpillar with its legs pointing upwards, so that the tiny hooks on its feet do not get caught in snags.

(2) Other species which mass-provision with a stock of more easily handled, smaller caterpillars, which can be rapidly air-lifted to the nest. More time spent in hunting is therefore balanced against less time and effort spent transporting the prey back to its destination.

▶ **With her jaws gaping threateningly a female *Ammophila beniniense imerinae* sand wasp from Madagascar turns on an ant which is trying to make contact with her caterpillar. Sand wasps find ants a particular nuisance, and often deal with them by picking them up in their jaws and dropping them at a distance.**

(3) Some species which commonly practise delayed provisioning with caterpillars, so that the female overlaps with her newly hatched larva.

(4) A select few species which implement progressive provisioning using small caterpillars.

Included in type 1 are many common European and North American species, as well as African representatives, such as the slim-bodied *Ammophila dolichodera*. This species is unusual in frequently making use of hairy caterpillars, rather than the smooth-bodied types preferred by most *Ammophila*. *A. dolichodera* is also unusual in catching the prey first and only then getting down to digging the nest, a trait which is more typical of the closely related genus *Podalonia* (Weaving, 1984), which also provide just a single large caterpillar to their larvae; most *Ammophila* dig the nest first, seal it temporarily and then fly off for the hunt.

Once she has stung and paralysed her victim, the *A. dolichodera* female caches it on a clump of grass or protruding stick well off the ground. On a hairy caterpillar, she prepares the actual site for attachment of her egg by depilating a small zone with her mouthparts. She then searches for a nest site, which is normally no more than 15 m distant. While digging, she regularly returns to check on the caterpillar, much as a pompilid with its spider, or indeed any wasp which temporarily stables its prey while it digs a nest. Spoil from the excavation is air-lifted some distance from the nest and then dumped, a sophisticated method which is not universal – many *Ammophila* execute this particular piece of spadework on foot. As in many large hunting wasps which are digging in hard ground, the *Ammophila dolichodera* female's labours can often be heard from a distance, due to the loud buzzing sounds which she emits. It is thought that the vibrations transmitted through the jaws help to loosen the ground, especially when small stones are being extracted from a stubborn matrix of soil. When she has finished, the female, as normal for the group, backs down into the burrow, dragging the caterpillar behind her and laying an egg on it.

When closing the nest, she uses her forelegs to rake sand across the entrance, also carrying in pebbles and bits of twig with her jaws. In common with many other solitary wasps, the material for the final closure is quarried from around the nest entrance, forming a small saucer-shaped depression which is later filled in with bits of debris. These are raked across with the legs, completely obliterating all signs of the nest. In some species, the female uses the hard flat top of her head as a pile-driver to tamp down the soil. Just before they finish the job, many species cast around near the nest in a careful search for a pebble, a piece of rabbit dung, a hard seed or pellet of compacted soil. Any of these has regularly been seen to do service as a 'tool' with which to carry out the final job of tamping. As in most solitary wasps, the *Ammophila* female performs these multifarious and demanding tasks in a manner which can best be described as 'frenziedly urgent'. This is the price to pay for having only a short life, with much to be fitted in if reproductive success is to be assured in the face of the many adversities.

Actual digging techniques can vary within a single genus. For example, *Tachysphex intermedius* first grubs up the soil with her jaws, then uses her front legs to pass it back beneath her body, where her rear legs and abdomen can act together to bulldoze it backwards out of the hole. *T. mergus*, by contrast, assembles a ball of soil between her mandibles and front legs and then backs out of the tunnel to dump her load of spoil well away from the entrance (Kurczewski & Kurczewski, 1984). This prevents the accumulation of soil in a ring around the entrance, which is typical of a *T. intermedius* nest. A similar dichotomy of methods is found in many other wasps.

Kleptoparasitism in *Ammophila sabulosa*

Kleptoparasitic attacks against the nests of other females of the same species are fairly common in solitary wasps. The *modus operandi* employed by the thieves to see through their nefarious designs has been thoroughly studied in the European *Ammophila sabulosa*. This species practises mass provisioning with small caterpillars, placing a temporary closure in the nest during each foraging trip. The female attaches her egg to the first caterpillar stored in the cell but whether or not the egg will ever be allowed to hatch is rather hit-or-miss, as nearly one-third of the eggs are destroyed by other females intent on prey theft (Field, 1989). This is a much higher rate of loss than that due to the more usual enemies, such as miltogrammine flies, which account for only 5 per cent mortality (at least in Norfolk, England).

The low level of losses from this cause may be partly because the nest-owner removes parasite eggs during her cell-cleaning exercises, a useful habit which is known from a number of different *Ammophila* species. In the American *A. dysmica*, the female wasp actually intensifies her nest-cleaning operations when she has detected a parasite in the vicinity (Rosenheim, 1987). Nest-cleaning is also found in solitary vespoids, such as *Euodynerus foraminatus* (Eumenidae), in which the female spreads out and inspects the caterpillars within the cell, spending as much as 20 minutes in a minute examination of them (Cowan, 1981). Some caterpillars may fail this testing scrutiny and be ejected, presumably because the wasp has detected a parasite's egg or larva lurking upon them.

In *Ammophila sabulosa*, conspecific brood-parasites are capable of digging out closed nests with great ease and specialize on recently stocked nests. These may be more easy to detect because of a chemical which is thought to linger temporarily around them, and also because they usually contain an egg rather than a newly hatched larva. Upon gaining illicit admittance to another wasp's nest, the female felon may proceed in one of two ways. She can drag the stock of caterpillars (usually 1–4) out of the nest, reserving her special attention for the last one, which normally holds the resident's egg. The felon drapes this looked-for caterpillar half out of the entrance while she bites her way down its length, eventually arriving at the egg, which is usually eaten. If she does not polish it off, she brushes it away or picks it up in her jaws and dumps it a short distance away. She then re-stocks the nest with the same batch of caterpillars, lays her egg on one of their number and re-seals the nest, concealing all evidence of her deed. It is vital that she destroys the original egg, as otherwise it will hatch before her own, with dire results for her larva, which, being younger, will be eaten by its larger companion.

However, there is no guarantee that her chicanery will reap the desired rewards, as there is no shortage of other females up to similar dirty tricks. So the same nest may be opened one or more times by succeeding females, all of whom take the same course of action. In fact, the very act of illicitly opening the nest seems to make this more likely, as nests which have suffered the attentions of one brood-parasite seem to be highly susceptible to further incursions, so much so that fewer than 20 per cent of the eggs laid by brood-parasites manage to survive. An alternative strategy, which may be followed by a female intent on raiding a nest, is to steal one or more of the caterpillars and use them to stock her own nest. This saves her the laborious and lengthy hassle of catching her own caterpillars. As before, the host egg is scrupulously removed from one of the caterpillars and the entire stock may be cleaned out on separate visits. On some days, several females will spend the whole of their active period checking out the status of each nest, re-sealing those which have yet to receive their provisions and stealing from, or brood-parasitizing, those nests which are stocked. Any one female does not normally restrict herself to such felonious behaviour but may also go to the trouble to provision her own nests in the conventional way.

At least two *Ammophila* species, *A. pubescens* in Europe and *A. azteca* in North America, practise progressive provisioning, not only of a single nest but of two or three simultaneously. The female remembers the location of each nest and supplies each larva with food as and when it is needed, basing her response on regular inspections (Evans, 1965).

Builders with mud

A number of genera within the Sphecoidea use mud as the sole or main material in nest construction. The conspicuous nests of the larger mud-daubers, such as *Sceliphron*, are often a familiar sight in warmer countries, as they may be plastered against the walls of houses, on furniture or even on clothes hanging out to dry. More natural locations for these nests would be around the overhanging bases of rocks and cliffs, where the nest is sheltered from direct sun and rain, or in similarly protected spots on tree trunks. One Brazilian *Pison* has the unusual habit of placing its cells in the lower regions of the nest of the reddish hermit hummingbird, *Phaethornis ruber*, but not in the nests of other hummingbird species in the same rainforest (Oniki, 1970). This would seem to be a risky undertaking, as the brooding birds occasionally catch their uninvited lodgers and feed them to their fledglings.

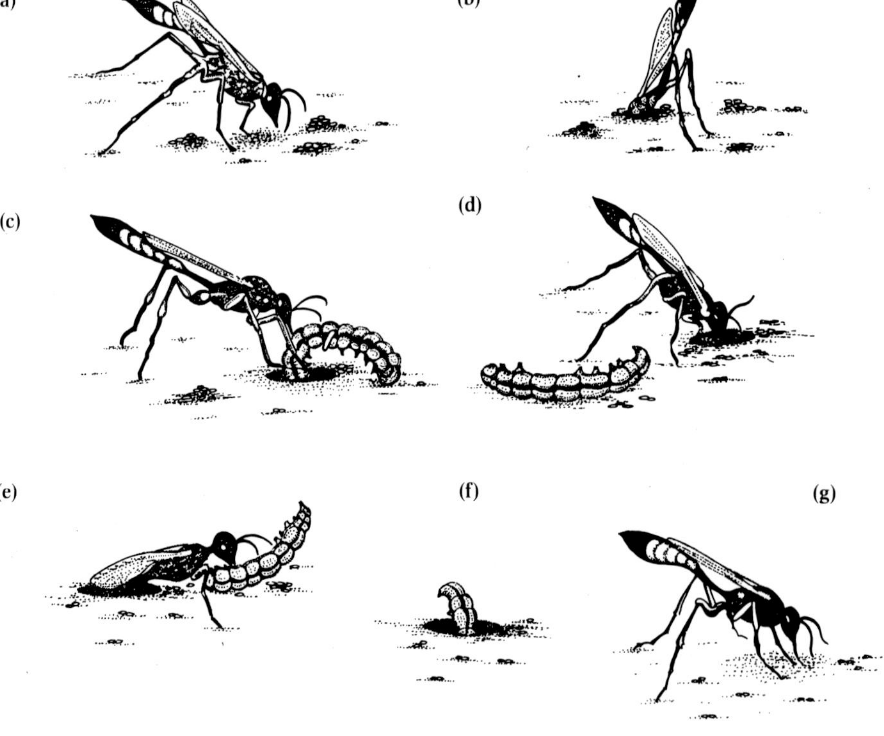

► **The events involved in a kleptoparasitic attack by a female of the sand wasp *Ammophila sabulosa*. (a) The wasp searches for a stocked and sealed burrow of her own species. (b) Having found a burrow she clears the entrance and makes her way into it to seek out the caterpillar on which the rightful owner's egg has been deposited. (c) She drapes this caterpillar across the burrow entrance and bites her way along its length until she encounters the egg, which she either eats or dumps some distance away. (d–f) She picks up the caterpillar and any others that she may have removed and re-stocks the burrow where she lays her egg on one of the caterpillars it contains. (g) The wasp now re-seals the burrow.** (After Field, 1989).

▲ With a glob of mud to the fore, a female *Sceliphron fuscum* mud-dauber wasp flies up to her nest on the underside of a tree in Madagascar. She will add this to the half-built cell on the lower part of the nest.

▼ Using her mandibles to feed mud to her front legs, a female potter wasp *Eumenes fenestralis* gathers her supplies of building material from the edge of a puddle in South Africa.

Pison normally utilizes natural cavities, such as hollow bamboo stems, beetle borings in trees, curved-over leaves and the old nests of other wasps and bees, typical habitats which are exploited by a wide variety of solitary sphecoid and vespoid wasps. Many *Trypoxylon* take advantage of similar cavities (although others build free-standing mud nests), so the bizarre habit of *T. latro* in placing its mud cell within the remarkably tiny and flimsy nests of the primitively social sphecoid *Microstigmus comes* is puzzling indeed (Matthews, 1983). In fact Robert W. Matthews, who first observed this amazing custom, describes it as 'analogous to building a house inside a laundry bag', the more so as only abandoned nests are used and these soon break away from the leaf. The reason for this extraordinary behaviour may lie in the extreme competition for natural cavities in the rainforest habitat, with the omnipresent ants snapping up most of the best properties and excluding most other cavity-nesters. Ants are, however, rarely tempted to make their homes in *M. comes* nests. The cells of *T. latro* also rarely seem to attract parasites, so by resorting to such an unusual nest site, this wasp may have succeeded in eluding two of its main adversaries – ants and parasites – at the relatively small cost of occasionally losing a nest due to accidental detachment.

Female mud-daubers gather building materials by 'mining' moist soil, often from around the edges of puddles. Even where suitably damp supplies are abundant, the females are still finicky about quality and spend some time 'testing' different spots until they hit on a medium of the right consistency and set to work. In the most common genus, *Sceliphron*, the female digs with a jerky action, her body pointed up at a steep angle as her jaws scoop out the malleable material, forming it into a ball by pushing it backwards towards the front legs. As she buckles down to her task, the wasp often emits a loud buzzing, which is not, however, accompanied by any noticeable wing movement. This 'whole-body vibration' probably functions to homogenize the soil–water mix, so that the final blob occupies the minimum possible space and is highly malleable with a glossy wet finish.

In flight, the pellet of mud is supported between the mandibles and the front legs, with the hind legs and body trailing down, presumably to act as a counterweight. *Sceliphron* nests usually consist of several cylindrical cells, built adjacent to one another, then covered with a final coat of mud to give a smooth, rounded finish. The female builds each cell 'back to front', with the rear end starting life as a crescent of mud applied to the substrate by her jaws, which exercise great precision and delicacy of manipulation. She further assures that the glutinous material runs out smoothly by 'buzzing' it on its way. She quickly lays down the whole of her load across the desired spot, then works back and forth over it several times to give it its final shape, like a baker forming dough. While working, she usually straddles the piece of work so that her legs and jaws operate from either side. This contrasts with potters (Vespidae) which normally sit at right angles to the work site and add small amounts of mud from the pellet until it has been exhausted.

The foundation crescent is expanded with the addition of several more pellets, until eventually a tube is formed. The industrious artisan has to bring in about 30–40 pellets before the job is complete, usually in one continuous sequence of collecting and building, lasting some 1–1.5 hours. Sometimes the source of the mud is changed part-way through, so that the finished cell may be rusty brown at the rear and greyish at the front. Each cell is then stocked with a number of paralysed spiders, a habit which has evolved independently from the similar spider-hunting habits of the Pompilidae. As provisioning is often a lengthy process and may be interrupted by nightfall, the female closes the cell overnight with a rather thin temporary cap of mud, fashioned from just a single pellet. Next day, the female opens this rather flimsy seal by the simple expedient of regurgitating water on to it, softening the mud so that she can cleave it open with her powerful mandibles. The final closure to the finished cell is far more robust, consisting of a convex lid formed from at least three or four loads of mud.

The habit of making mud nests has evolved independently in the mason or potter wasps (Vespidae: Eumeninae), which are non-social members of the Vespoidea. As in mud-daubers, the building material is usually collected from damp spots, although some species are able to turn to account completely dry dusty sand by mixing this with water from the crop to form their own 'plaster' on the spot. In a desert species, such as *Delta dimidiatipenne* in Israel, this *in situ* method of materials manufacture has one great advantage. The wasp is able to 'mine' the dry basics from directly beside the nest, saving her from having to fly back and forth to a mud supply, which in a desert may be quite a long way off. Instead, she only has to make the occasional trip to water to replenish her crop.

She uses her jaws to feed the dry sand into the scooped front legs, where it assumes an increasingly glossy appearance as water is fed into the mix. The nest is built on top of a stone in open desert and, when dry and bleached, the finished nests, often 3–4 placed side by side, come to resemble the stone itself. Each pellet of mud is laid down to form a circle, one upon the other, each addition being individually visible as a building-block on the crude exterior, which contrasts with the smooth finish applied to the interior. She works at right angles to the structure, 'dribbling' the clay on to the rim while her antennae bend downwards to feel the interior of the growing vault. She embellishes the entrance with a neat little lip, moulded with delicate precision as the mandibles both spread and shape the raw mud. Once the cell is complete, the female backs partly inside and lays an egg, which is suspended from the roof on a short silken thread. Only now does she set off in search of caterpillars with which to mass-provision the cell. She feeds in each caterpillar through the neatly lipped entrance, which assures a smooth passage, using a combination of her jaws and middle legs. The

▲ After completing her nest, a female *Delta dimidiatipenne* potter wasp inserts her abdomen into the entrance and lays an egg, suspended from the roof on a slim thread. This done, she will commence her search for caterpillars with which to stock the nest.

▲ The female *Delta dimidiatipenne* carefully threads a caterpillar in through the nest entrance. Note that she has built several nests side by side, all of which will eventually be camouflaged with an overall rendering of mud.

◀ With accomplished ease, a female potter wasp *Delta dimidiatipenne* uses her mandibles and front legs to convert a ball of mud into a delicate lip on her nest in the Israeli desert.

caterpillars are only lightly paralysed and are able to wriggle around somewhat within the cell, which is why the delicate egg is suspended above them on its string, out of harm's way. For the first few vulnerable days of it life, the newly hatched larva avoids the crush of bodies below by clinging to its egg shell with its tail while it gnaws on the nearest available caterpillar. Only when it is larger and tougher does it drop down to join its food. The female stocks the cell with up to a dozen or more caterpillars – depending on size – and then demolishes the neatly lipped entrance and plasters the crude exterior with a smooth coat of mud.

Most potter wasps are mass-provisioners, but a few go in for progressive provisioning. *Paraleptomenes mephitis* from Sri Lanka (Krombein, 1978) sits inside a cell which holds an egg or young larva. She closely monitors its state of development, supplying caterpillars as required. As the larva grows and needs a supply of larger caterpillars, the space inside the cell becomes crowded and the female moves outside, only popping her head inside when checking the larva's needs. Even though she builds several cells in close proximity, she only rears one larva at a time. Progressive provisioning is also seen in the African genus *Synagris*. However, *Synagris cornuta* has the remarkable custom (for a solitary wasp) of furnishing its larva on a progressive basis not with complete items of food but with a paste of chewed-up insects, a habit which is more typical of the social wasps in the subfamily Vespinae.

The road to social behaviour

There has been much discussion about the stepping-stones to true sociality within the wasps. The first step may have been the communal use of the natal burrow by members from the same 'family' This is known from the American bee-wolf *Philanthus gibbosus* (Sphecidae) in which the brothers and sisters stay together in the burrow for a while. This gives the impression of a certain touch of sociality but all the females, except perhaps one, soon leave to establish their own nests, although one or more of the males may still remain. When a female stays behind, she is usually the largest and she expands the original burrow and uses it herself. This is sometimes in temporary company with a sister, so that the two females may for a brief time provision the nest simultaneously (Evans, 1973).

A rather unconventional sphecid with social-like habits is the Neotropical *Trigonopsis cameronii*, which has reached a so-called communal level of social organization (Eberhard, 1974a). The nest is built of mud, much in the manner of a *Sceliphron*, save that the cells are constructed in the reverse sequence, i.e. with the entrance coming first and the rest of the cell then being extended out behind it. This means that, once a complete ring of mud has been laid down to form the entrance, the wasp subsequently has to insert her body through it in order to extend the walls around her, finally finishing off the rear wall with her body almost entirely inside. As in *Delta* and most other wasps, the female acts as a mobile water-tender, filling her crop at a puddle, or from droplets of dew or rain on leaves. She can then mix several loads of mud from fairly dry soil before she needs a refill.

Unlike in *Delta*, the nest exterior ends up with a rough 'as-built' finish, although the female pays great attention to smoothing off the interior, devoting as much as an hour to the job. She provisions the cells progressively with cockroaches, capping each cell with mud after every visit and then opening it again next time by softening the cap with water, before slicing it off with the mandibles. She often saves time and effort by re-furbishing old, used cells. This is much easier than building a cell from scratch, as she only needs to widen the entrance (this is left by the emerging adult with a minimum 'tight-squeeze' aperture) and then throw out all the garbage from within. This includes cockroach husks, bits of cocoon and larval faecal remains. She then re-works the inner surface to give a smooth as-new finish. The saving in time is considerable, as a renovation takes only 3–4 hours, compared with 6 hours or more for a new cell. This saving is achieved by 'borrowing' mud from other parts of the nest, rather than collecting fresh supplies from a 'mine'. Up to four females may work on the nest simultaneously, each constructing and provisioning her own cells. Such a situation is ripe for exploitation and a good deal of stealing of one another's prey could be expected, accompanied by the inevitable fights. In fact such events occur only rarely, possibly because the nest mates are closely related.

The advantage of such good communal relationships is a reduction in the rate of parasitism, there being always at least one female guarding the nest, although there is also the frequent presence of males, who guard in return for sex. Females also effect mutual cell repairs, making good any damage to caps which they happen to notice, as well as repairing holes which they themselves make while 'borrowing' mud to make a temporary closure. Two pompilids from the Philippines, *Macromeris violacea* and *Paragenia argentifrons*, exhibit behaviours which are similar in many ways.

The Stenogastrinae (Vespoidea), occurring from India to New Guinea, is a somewhat enigmatic group which encompasses virtually the whole range of social development, from solitary, through communal, to an advanced form of communal or semi-social behaviour, bordering on the truly eusocial arrangement seen in the Polistinae and Vespinae.

For example, *Parischnogaster mellyi* from Thailand is either solitary or communal (Hansell, 1982), often constructing its small nests on buildings. Maximum colony size seems to be about six, split equally between the sexes. The egg-laying procedure is complex and is probably uniform throughout the subfamily. The female threads her abdomen (gaster) between her legs, and then uses her front tarsi to 'comb' a blob of a gooey white liquid off her abdominal tip and on to her mouthparts. She places this in an adjacent empty cell, then resumes her former 'threaded' posture and lays an egg, which she picks up in her mandibles and places into the cell, on top of the blob of secretion. She then repeats the goo-producing exercise and places the resulting blob on top of the egg. This is now enshrouded in the gluey secretion, which appears to be produced by the abdominal Dufour's-glands. The same Dufour's-product later seems to comprise the main food provided to the young larvae, making this a unique method of provisioning within the wasps.

The rather primitive elements of sociality which characterize the on-nest behaviour of *Eustenogaster calyptodoma* from Malaysia are perhaps typical of the subfamily (Hansell, 1987). A colony usually contains only one or two females. If the latter, then there exists a definite division of labour, with the oldest female remaining as sentry on the nest, while her younger nest mate goes out on foraging trips. How this split is established and upheld is uncertain, as there are no overt signs of dominance by one female over the other. Young females leaving their natal nest often short-circuit the high-risk task of founding their own colony by usurping another wasp's nest, complete with its brood, which they will continue to rear alongside their own. By not committing infanticide, as would be normal practice after take-overs in other wasps, the usurper ensures that early help will be on hand to rear her own brood and protect the nest.

A female intent on having a go at making a take-over calculates the chances of success by sticking her head into the spout-like nest entrance. If the owner is at home, she will show her displeasure by loudly rattling her abdomen against the nest wall. This usually dispels any further ideas of invasion and sends the intruder packing. However, if the nest is temporarily deserted, the would-be usurper enters. When the rightful owner returns, the two wasps engage in an abdomen-rattling contest, which usually seems capable of expelling the intruder without any need for physical grappling.

When multiple females occur on a nest, there may exist a definite dominance hierarchy which approaches the truly social state. In *Parischnogaster nigricans serrei* from Java, the 'top' or 'alpha' female acts like a queen, laying all the eggs and never leaving the nest, where she performs such tasks as cell inspection and defence against conspecific intruders (Turillazzi & Pardi, 1982). The subordinate females spend much of their time out seeking food which they feed to the larvae. Subordinates are also responsible for the regular job of reinforcing the chemical anti-ant secretions above the nest, which are applied with the abdomen. The dominance hierarchy itself is decided by physical intimidation, in that the alpha female maintains her status by biting and stinging subordinates, to the accompaniment of a loud threatening buzzing. There is also a strict descending order of hierarchy among the subordinate females but how this is decided is unknown, as there are few interactions at this level. Nevertheless, the beta female 'knows' her position and duly takes the alpha role if it falls vacant. A number of males also reside on the nest, where they are fed mouth to mouth (trophallaxis) by returning female foragers. The males' main function may be to help protect the nest against attack and to mop up raindrops, which could threaten the nest's structure. This could pose a particular problem for a nest which is built upon dangling roots and twigs in damp rainforest.

The truly social wasps

With just a single known exception, all the truly social (eusocial) wasps occur in the family Vespidae, within the superfamily Vespoidea. The exception is the Neotropical sphecid *Microstigmus comes*, which is the sole social member of the Sphecoidea (Matthews, 1968). This species is of exceptional interest, representing as it does a completely independent origin for sociality, quite unrelated to that found within the Vespidae.

The colonies of this tiny wasp comprise a mother or 'queen' who is the sole reproductive, along with several generations of her offspring (Ross & Matthews, 1989). A nest may be founded by a single female or by a group of females working together. Nests are positioned beneath leaves of the palm *Chrysophila guagara*, a relationship which is more than just chance, for the palm itself provides the sole source of nesting material; a 'one-piece' construction method, unique within the Hymenoptera, is used (Matthews & Starr, 1984). The underside of the palm frond is covered with a detachable felt-like coat, which is simply 'rolled up' by a construction gang of female wasps. These beaver away in an approximate circle to detach the material with their jaws and pull it inwards into a central mound. During this gathering process, each female uses glands at the tip of the abdomen to attach silk to the material, so that eventually it hangs from the leaf as a loose downy mass.

Using a combination of chewing and silking, the wasps now lower the lump on a central stalk. This is then formed into a spiral by a single wasp, who moves up and down its length, regularly pulling it into loops with her legs. One or two females also burrow around in the upper part of the bag, eventually forming a small entrance. The process thus far takes several days, during which time the foundresses return each night to their existing nests. However, once the simple blob of down is lowered on its spiral, the builders move in and spend the next few days burrowing around, producing a series of small cells within the bag. These are quickly reinforced with silk, which is also added to the nest exterior, giving it a stronger, neater appearance.

One of the most interesting aspects of this whole complicated process is the fact that several females are able to assemble in a co-ordinated fashion on a specific area in order to follow a pre-determined 'plan' of action. This implies some manner of communication between them, capable of designating the chosen spot and 'agreeing' what is to follow. The females also co-operate in other matters, such as provisioning the cells with springtails, keeping the nest clean and defending it against enemies.

There are certain habits which characterize the social vespids. They all lay their eggs in cells and then provide food to the larva. This is never given in the form of whole prey but always consists of a paste of chewed-up animal material. Nests are normally made of carton, a paper-like material which is produced from wood fibres. These are harvested during visits to regular 'scratching sites' such as fence posts, dead branches and hollow, woody plant stems. The wasp peels away a small strip of bark with her mandibles, forming it into a small ball and combining it with salivary secretions, which act as a glue. Such freshly gathered materials are not the only ones which go to make a new nest – paper from old nests is also frequently recycled, although only rarely in some genera.

In *Polistes*, *Vespula* and most other wasps, the mastic of raw paper is added to the nest by whichever wasp has carried out its collection, whether that be a queen founding a new nest or a worker doing some building work upon a nest full of compatriots. However, in some species, the various elements of nest construction are performed by a chain of workers specializing in each task (see *Polybia occidentalis*, page 207).

In a few species, e.g. *Pseudochartergus chartergoides* and *P. fuscatus* (Jeanne 1970) from South America, and *Ropalidia opifex* from Asia, the nest is constructed of pure secretions derived from within the builders' bodies. In *R. opifex*, 2–3 leaves are glued together with an envelope of pure secretion having the finished appearance of an almost transparent, cellophane-like membrane (Maschwitz *et al.*, 1990). The secretion, which consists mainly of protein, is produced from the mandibles and takes more than a day to harden completely. The adults also provide a translucent cap for the bottom of each pupal cell, a characteristic which is also found in a number of genera of Old World polistines. Once the larva has pupated, the adults chew off the carton bottoms from the cells, dispose of the larval meconium in its sac, and then re-seal the cell with a new translucent bottom.

Construction work is undertaken by certain specialist workers, which, if accidentally lost, can only be replaced after a lapse of several days. This suggests that a considerable time is needed for the selected replacements to build up the necessary reserves of secretion. Within the humid environment of the rainforest, the envelope remains supple and remarkably durable. It is exceptionally efficient at shedding heavy rainfall from its smooth almost polished exterior. This may be an important advantage in damp tropical climates, where carton nests normally have to be carefully placed beneath overhanging leaves, rocks or trees, which give the required protection from rain. Such protected spots are often in short supply, limiting the abilities of many wasps to expand their populations, or forcing them to crowd into certain favourable areas. *R. opifex*, by contrast, can build its waterproof shelter among any convenient leaves, often high in the canopy, where the main enemies of all social wasps – the ants – are much rarer.

Nest-founding in social wasps

In many Polistinae, a nest is established by a single queen, who is often joined by several extra females (usually sisters) without the aid of an accompanying swarm of workers. This *independent-founding* behaviour is present in the Old World genera *Belonogaster* (the most primitive vespid genus) and *Parapolybia*, in the widespread New World genus *Mischocyttarus*, and in the familiar cosmopolitan genus *Polistes*, along with the majority of the Old World *Ropalidia* (Jeanne, 1980). Reproductive dominance on the nest is established by purely physical means, in that the 'queen' regularly bullies her co-foundresses into accepting their submissive status by acting aggressively towards them. Except in *Belonogaster*, there is only a single queen (monogyny).

In *swarm-founding* Polistinae one or more queens leave an existing nest, accompanied by a swarm of workers who do most of the work in swiftly setting up a new nest. This mode of nest initiation is restricted to tropical species, and is practised, among others, by the large genus *Polybia*. There are multiple queens (polygyny) who establish and maintain their status not by the crude physical 'strong-arm' methods of the independent-founders, but via the subtle persuasive influence of special pheromones. Within the Vespinae, a colony is invariably founded by a single queen, who rears the first brood of workers by herself before handing over most of the work to her offspring. Pheromones are used to mediate reproductive dominance.

One of the great characteristics of eusocial insects is the existence of castes, individual members of a nest's population who are usually specialized for a particular role. In many social wasps, castes cannot be distinguished by external appearance or by size, yet each individual certainly 'knows' its own role within the nest. In such species, the 'workers' are normally virgin females, who perform most of the daily 'housework' necessary to keep the nest going, with the major exception of laying fertile eggs. However, in the rather primitive *Belonogaster*, the dominance hierarchy is broadened at the top to include several large 'queens', who all lay viable eggs. As these 'queens' grow older, their reproductive status wanes, and they are replaced by younger nest mates, whose erstwhile jobs of foraging and nest maintenance are in turn taken over by their former bosses.

Behaviour in independent-founding Polistinae

When several females have established a nest, the emergence of one of their number as 'queen' follows a period of aggressive interactions, destined to establish rank within a dominance hierarchy. This basically boils down to the winner being the one who can be thoroughly nasty much of the time and browbeat the others most successfully. This war of nerves then continues on down the line in a descending scale so that it eventually establishes a linear 'chewing' hierarchy. This is a deadly serious business, with everything to play for in the reproductive sense, and the contestants chew and sting one another and plunge head-long into vicious grappling matches, which often result in both combatants falling off the nest on to the ground.

There is usually a close correlation between rank and the degree of ovarian development. The highest-ranking female regularly launches unprovoked assaults upon her subordinates and often eats any eggs which they lay. This rough treatment eventually promotes the atrophying of their ovaries and the final 'acceptance' of the role of 'worker'. The most dominant 'alpha' female or 'queen' now assumes the status of sole or main egg-layer, and it is her progeny which are reared by the now co-operative 'workers'. However, in at least some species, such as *Polistes annularis* in North America, the alpha female does not establish total reproductive dominance and lays an average of 80 per cent of the nest's total output of eggs. Of the other 20 per cent, the second-ranking female probably lays around half, leaving just 10 per cent of the total spread over the rest of the females

(Strassmann, 1981). If the queen is deposed in a palace coup, usually but not always by the beta female, then the queen drops to the bottom of the hierarchy, pushing all the other females up one place.

Sometimes two females in an equal state of ovarian development may found a nest. When this happens, the ferocity of their contacts may be particularly intense and if one or other fails to accept a subordinate role within the colony, then the only answer must be either expulsion or death for one of the two. Despite all the unpleasant but necessary goings-on which precede the final settling-down of life on the nest, in general there are great advantages to being a member of a multiple-foundress nest. This even applies to being a non-reproductive member, as the 'queen' will normally be a sister. The fact is that a solo foundress who lacks defenders to guard the nest is much more likely to lose her entire investment in its early stages and by then it may be too late to try again. Better by far then to share a nest and thereby contribute to its greater success, even if that means raising your sister's brood rather than your own.

As the new generation of adults begins to swell the numbers on the nest, they too begin to develop functional ovaries. This does not proceed very far, further development being prevented by the 'queen' using the same rough-house methods as before. Without warning, she lunges at her own daughters, biting them on the mandibles, eyes, neck, thorax, wings and legs. A recipient of this uncalled-for onslaught reacts by pressing herself low against the nest in a 'submissive' posture. This seldom has the desired effect of placating the queen, who often takes it as a cue to rally to the attack all the more viciously. It is only through being uncompromisingly domineering in this way that the 'queen' can preserve her exalted position, and ensure that the ovaries of the workers remain in an atrophied state.

The line of succession

The original founding queen often needs to be replaced for a variety of reasons and, in *Polistes*, the method of prosecuting the succession is different in tropical and temperate nests. In temperate species, one of the oldest dominant co-foundresses is legitimately placed next in the line to the 'throne', and normally takes over. In *Polistes exclamans* from North America, the replacement queen is usually the oldest of the foragers, and one who has performed more foraging and acted more aggressively towards nest mates than any of her competitors who did not win the succession (Strassmann & Meyer, 1983). This habit of old workers dominating young ones is not restricted to temperate *Polistes* but also occurs in *Ropalidia marginata* and *Mischocyttarus flavitarsus*, and probably in most other temperate members of these genera.

In the tropics, it is young non-foraging females who enjoy the highest rank and replace a lost or old queen. As young individuals are being constantly produced, this means that there is a constant shift in the chain of hierarchy, with workers dropping down the ladder as they age, to be replaced by newly hatched youngsters. Once a 'worker' has started foraging away from the nest, the chances of her ever becoming queen are nil, and she no longer plays any role in the dominance interactions on the nest.

At some point, all nests produce a number of females who are destined to be potential queens. In temperate species, these are the individuals who will leave the nest as it falls into disrepair in Autumn, when they will go into hibernation once they have mated. They lead a peaceful existence on the nest, not becoming part of the dominance-hierarchical rat race, and they do not perform any work during their sojourn on the nest. In the wet tropics, where nests continue throughout the year, these 'gynes' do not appear to be specially raised. Instead, they are those females who avoid being totally dominated by the 'queen', possibly because she is ageing or perhaps because there are simply so many new females being produced that they overwhelm their ruler's capacity to get around to being nasty to all of them, all of the time. If this happens, and they avoid being roughed-up on a regular basis, then there is nothing to prevent their ovaries from developing.

Usurping behaviour

A foundress will often be deposed not from within by a 'palace coup' but from without, by a usurping female from another nest. In *Polistes fuscatus*, single queens with young nests are particularly at risk from this direction, as they still lack any workers which would defend the nest (Klahn, 1988). In Iowa, nearly 20 per cent of single foundresses lost their nests to usurpers at this fledgling stage, amply demonstrating the advantage of multiple-founded nests, which only suffered losses of just over 2 per cent. The presence of several defenders on the nest was obviously highly effective in preventing take-overs in the latter case.

Usurpers are usually females who have lost their own nest to a predator and, instead of joining another nest as a subordinate, or founding a new nest, choose instead to usurp an existing nest. This saves them a considerable amount of time and work as, on an established nest, the imminent hatching of a series of helpers will be on the cards. However, a less well-developed brood, which may be destined to become gynes, will often be systematically eaten by the usurping female. This act of infanticide is necessary to ensure that it is her own gynes which will be the sole reproductives eventually produced by the nest. One drawback to taking over a nest is the lack of the correct 'nest odour' on the usurper. This deficiency is usually recognized by the newly emerged 'workers', who may, as a result, fail to co-operate with the interloper and may even try to kill her. This ability to 'sniff out' an invader is possible because nest odour is learned directly from the comb, not from nest mates. Despite this drawback, usurpation still pays better dividends than attempting to re-found a colony late in the season. Broadly similar behaviour is also seen in *Mischocyttarus flavitarsus* (Litte, 1979).

Activities on the nest

The general day-to-day activity on *Polistes* nests is particularly easy to observe, there being no protective envelope covering the cells. A 'worker' arriving back at the nest with a food bolus held in its mandibles usually proffers it first to the queen or the beta female. A large bolus may be shared with two or more individuals. The wasp normally holds the food for a while in its jaws, which malaxate the soft material, extracting much of the liquid. This is imbibed by the adult herself, she being unable to pass solid food (other than tiny particles contained in the liquid) through her mouthparts. After a period of malaxation, the wasp approaches a larval cell, sticks her head inside and beats a rapid tattoo against its walls, producing a sound which is clearly audible a metre or so distant. This vibratory announcement of incipient food is common to a number of genera. Some vibrate the antennae against the cell wall, others rattle the abdomen against the nest, while certain species

◀ Having returned to the nest after a successful foraging trip, a *Polistes townsvillensis* wasp worker shares its food bolus with two of its nest mates. Most of the food – a mashed-up caterpillar – will be fed to the larvae, possibly after the adults have extracted some of its juices for their own nourishment.

▼ Shortly after a forager returned with a bolus of food, two *Polistes myersi currasavicus* wasps feed it to larvae in their open cells. Eggs and larvae are clearly visible in some of the open cells; the white caps on some cells are silken cocoons spun by the larvae. This nest, on the island of Bonaire, is attached to a large cactus, a favoured nest site for many social wasps, possibly due to the additional protection given by the spines.

perform both actions in unison.

Some observers report that this drumming serves to announce to the larvae the arrival of the food bolus. However, in *Polistes fuscatus* it has been noted that drumming only commences *after* the adult has made a silent round of the larval cells, bearing solid food (Pratte & Jeanne, 1984). She then makes a second round, this time drumming on the nest and providing further food, but this time in liquid form. It is thought that drumming in this context advises the larva that it is about to be given a liquid meal and should *refrain* from offering a salivary secretion to the arriving adult. Such mouth-to-mouth (trophallactic) secretions are regularly supplied to the adults by the larvae. It has also been observed that adults not bearing food often walk around on the nest, performing drumming operations which seem to be aimed at *eliciting* the larval exudations. The whole subject of drumming is therefore somewhat complex and confused, possibly because we cannot detect subtle differences in drumming codes, which could convey alternative messages to the larvae. But it is certain that drumming always takes place in the context of feeding of one kind or another. Larva–adult trophallaxis is found in all independent-founding Polistinae and all Vespinae. Just why the adults should require this source of nutrition, when they are perfectly well able to forage for themselves on a variety of food sources (flowers, fruits, honeydew, soft-bodied insects) is not understood. It seems likely that the larvae are able to provide the adults with certain substances which can no longer be made in the adult stage. What these are is not presently known.

While it is normally the 'workers' or 'queen' who tour the nest feeding the larvae, males of at least six species (*Polistes dubius, fuscatus, gallicus, major, metricus,* and *snelleni*) also sometimes perform this duty. They do however malaxate the food bolus for longer than the females, presumably extracting a higher proportion of the liquid content for their own use (Cameron, 1986). Another necessary task in which the males take more regular part is controlling of the temperature of the nest. This involves standing on the nest and vigorously fanning the wings to cool the cells, a vital chore if direct sun strikes the nest for any length of time. The males may peform this task alone or in consort with females, although males generally tend to fan for shorter periods than their sisters. If the

▲ **With their nest temporarily being struck by the rays of the direct setting sun, four *Ropalidia anarchica* wasp workers fan their wings to cool it down. If the situation worsened, the wasps would take further remedial action by collecting water and applying it to the nest, cooling it via evaporation.**

temperature continues to rise, the workers of most wasps will switch *en masse* from foraging to ferrying in water. This is applied to the carton and lowers the inner temperature via evaporative cooling, assisted by multiple fanning by a substantial proportion of the available residents.

One behaviour, performed by Old World polistines (*Belonogaster, Ropalidia, Parapolybia, Polybioides)* but not by any in the New World, is the removal by the workers of the larva's peritrophic sac (meconium), containing its accumulated waste products. This is done just before the larva pupates and, in *Belonogaster griseus*, the adult is ready to receive the sac, having previously cut off the cell bottom to gain access to the larva. The waiting adult grasps the sac in her jaws and tugs it free, then air-lifts it some distance from the nest before dumping it. If the larva fails to receive any assistance, it becomes 'constipated' and dies. The opened cell is finally re-sealed by the adult with a transparent membrane secreted from the mouthparts (Marino Piccioli, 1968).

Behaviour in swarm-founding Polistinae

Sorting out the queens

In swarm-founding polistines the division of labour has been carried a step further than in the last group, so that the queens never leave the nest and are solely concerned with reproductive activities. There is also a tendency for the workers to change their roles as they age, starting out by performing domestic tasks, such as feeding the larva, and then increasingly leaving to forage as they grow older. The queen (or queens, there may be up to 15 per nest) which initially emigrated with the swarm do not all necessarily end up maintaining their exalted status. In *Polybia* and *Metapolybia*, it is not aggression between queens which decides the eventual hierarchy, but regular 'testing' by the workers of the queen's 'attitude'.

This 'testing' takes the form of 'queen dances' in which workers dart at the queen as if to head-butt her (West Eberhard, 1977). If this elicits an aggressive response, which in this situation means that the queen acts the part and 'demands' a trophallactic hand-out from the worker, the latter serves up the required liquid and then leaves. However, 'timid' queens who 'turn the other cheek' just land themselves in deeper trouble, as such timidity is an invitation to the workers to gang up on the queen and inflict severe 'punishment'. Within just a few days of the swarm first settling, the queens will increasingly be subjected to this third-degree style of inquisition, which eventually achieves its purpose of evaluating each queen for the kind of 'true grit' needed to perform her future role successfully. Once the queens have been sorted into 'resisters' and 'submitters', the latter are tormented by the workers until they either quit the nest or accept demotion to the inferior role of worker. By the end of the 'testing' procedure, only a handful of queens wiil be left in residence. These are quite capable of coping with the increased demand for egg production, due to the extreme development of their ovaries. Whether the same system operates in other genera is unknown and it is possible that, in many species, all the original founding queens remain and reproduce.

Controlling the workers

The queens keep control of their workers not via directed aggression but by the release of a pheromone from a gland which is probably located in the head. In some species, chemical measures are also accompanied by displays of aggression but these are purely ritualized affairs, not the tooth-and-nail bullying seen in *Polistes* and other independent-founding genera. However, workers on swarm-founded nests do establish and maintain a definite dominance hierarchy via direct physical intimidation. This often operates through trophallactic exchanges, in which the receiver is very much dominant over the giver – a kind of 'feed-me-or-else' system. The queens' job of ensuring that egg-laying competitors do not have a chance to develop is thus executed on her behalf by high-ranking workers, who intimidate young female nest mates to the point of ovarian atrophy. However, that the queen (or queens) still exerts ultimate control is evident from the fact that, if she is removed, the younger females soon crank up the daily level of aggressive activities, with the end result that they mate and lay eggs on their own account.

Swarming behaviour – guidance

Swarm-founding polistines have generally made an advance in nest architecture, with multiple combs enclosed within a protective paper envelope, unlike the open combs of *Polistes, Mischocyttarus, Belonogaster* etc. The construction of the nest tends to be an amazingly rapid and slickly executed operation, following immediately upon emigration from the old nest. Swarming may be induced by the destruction of a nest, forcing its occupants to undertake an emergency replacement job, or by fission of a healthy nest. In the latter case, a number of young queens leave to set up their own home, accompanied by a mass of supportive workers. Prior to this exodus, levels of aggression between the queens rises to a point where much of the brood is cannibalized. Eventually the young emigrant queens leave their normal living-quarters, deep inside the nest, and assemble on the nest envelope. The actual swarming process is co-ordinated by a series of sophisticated communication behaviours, which direct all the members of the swarm to the correct spot, this having been previously selected by a group of scouts (Naumann, 1975).

A frequent cause of swarming in Neotropical *Polybia* is the urgent requirement for total evacuation of the nest during a raid by *Eciton* army ants. The dispossessed wasps usually cluster on a leaf near the old nest, often spending a day or two in this temporary refuge. Their regular activities do not close down completely, as workers still forage for food, bringing it back in liquid form and feeding it to the more active members of the swarm. During this period, increasing numbers of trail-blazing workers leave the nest and search for an appropriate site for a new one. Once this is found, they establish a 'scent trail' between the swarm and the new site by dragging their abdomens (gasters) against objects along the route, depositing a scent secreted by the Richard's-gland near the tip of the abdomen. Initiation of the mass emigration to the new location is by so-called 'breaking runs', during which the scouts who 'know' the route run around over the cluster of resting wasps. This stimulates many of the wasps into leaving and these pioneering emigrants further reinforce the route by performing their own abdomen-dragging. Eventually the whole swarm moves in a straggling line along the intervening trail (Jeanne, 1981; Forsyth, 1981).

Nest construction in Polybia occidentalis

Polybia occidentalis is a typical swarming species, being very common in Central and South America in both dry and wet forests. It is highly unusual in having queens which are smaller than the workers, to which they also look identical save for the queen's slightly fatter abdomen, brought about by the increased development of the ovaries. When a swarm has assembled at a new site, the exact spot where the nest is to be built is marked out by several workers who apply an oral secretion. During the amazingly rapid process of construction which follows, there is generally a marked division of labour among the workers. Some act as water-gatherers, others devote themselves entirely to collecting wood pulp, while others remain on the nest and add the incoming supplies to its fabric (Jeanne, 1986).

A water-carrier brings in a load which is probably the maximum which can be accommodated in the fully distended crop. Upon returning to the nest, she dispenses her load to any builders who have run out of material. Their first requirement is to stock up with water, which can then be combined with pulp arriving with the pulp-carriers. The latter also stock up with water from a water-carrier before leaving the nest, as they will need to regurgitate water on to the source of supply (dead wood) before scraping off a wad of fibres. Upon their return, they offer their cargo to a builder who has just loaded up with water. Such a full load of pulp is a considerable oversupply for handling by a single builder, so she malaxates the material in her jaws before sharing it out with other builders, thereby reducing it to manageable proportions. She then adds the pulp to the structure, delicately forming the shape of the new cells with her mandibles.

▼ In a Trinidadian rainforest, while returning pulp-gatherers share out supplies on the top of their nest, several more workers add pulp to make new cells. In this swarm-founding polistine wasp *Polybia occidentalis cinctus* the rapid expansion of a newly founded nest is facilitated by the way in which workers tend to specialize on one task, i.e. pulp-gathering, water-gathering or building.

P. occidentalis nests are quite complex structures internally, so there is probably considerable mileage to be gained from capitalizing on the depth of experience amassed by builders, who are regularly labouring at the same work site and so 'know the ropes' of that particular job. On smaller nests, this advantage is less pertinent, leading to reduced subdivision of tasks, so that a returning pulp-collector may often add her load to the nest herself. Small nests also tend to suffer from a log-jam in the distribution of supplies to the relevant artisans, due to the lack of synchrony in arrival of the two basic materials. However, in general, this 'production-line' system is a highly efficient way of assembling and applying materials. Its great virtue is its ability to capitalize on the high load-carrying capacity of the water-gatherers during just one trip, which greatly reduces the total number of trips which are needed.

Behaviour in vespine colonies

In the Vespinae, a colony is invariably founded by a single queen toiling alone to rear her first brood of workers. Once these helpers have emerged, she rarely leaves the nest, relying increasingly on them for foraging and nest-building, while she concentrates on egg-laying. However, there are exceptions, such as *Vespa tropica pulchra*, in which the queen continues to carry out such domestic jobs as feeding the brood and constructing cells right up until her death. There is, in most species, a marked development of castes, and queens are much larger than the workers and also differ in physiology, colour pattern and the relative proportions of various body parts. As in most of the swarm-founding polistines, there is a tendency for workers to change jobs as they age: they start life as nurses tending the larvae, then leave the nest more and more to forage for pulp and liquids. A further switch of roles, to the more highly skilled duty of catching prey, then precedes a final return back to the nest as a guard.

Control of the colony, and the suppression of ovarian development in the workers, is conducted through pheromones produced by the queen. In the hornets (*Vespa* spp.), the pheromone is secreted on to the surface of the queen's body, whence it is licked off by the workers, who find this 'royal' substance highly attractive. A queen-produced pheromone also seems to be responsible for the initiation of 'queen cells' by the workers. If a queen is absent, then the workers never get around to building these special rearing units because the necessary trigger pheromone is absent. In vespines, the future queens are reared in specially large cells and fed a special 'queen' diet.

In *Vespa*, the queen's pheromones seem to exert a general pacifying influence on the whole colony, so that worker–worker aggression is noticeably rare or even absent. This contrasts with the situation in *Vespula* and *Dolichovespula*, whose workers are often very bad-tempered towards one another; this applies particularly to young females during a period when they exhibit a small but temporary development of the ovaries. The queen's control of her offspring is also mediated by pheromones but the method of transmission is uncertain, as there is none of the queen-licking behaviour seen in *Vespa*. However, in *Vespula atropilosa* and *V. pensylvanica* from the USA, if a worker chances across some of the queen's liquid faeces it will quickly feed upon them, afterwards passing some of the material on to other workers and to the larvae. As the faeces of males, gynes or workers are not treated in this way, it is suggested that it is via her pheromone-laced faeces that the queen exerts her chemical authority (Akre *et al.*, 1976).

In the early stages of founding a nest, there is often extreme competition for available sites, especially in species which utilize disused burrows in the ground. There is also a tendency for this competition to extend to the coveting of actual nests, with attempted take-overs and savage battles being the natural consequence. It is therefore not unusual to find one or more dead queens lying beneath a nest, a sign that take-over fights are usually to the death. This do-or-die approach is well illustrated by the deadly combat between queens of the Oriental hornet, *Vespa orientalis*. This is no pussy-footing skirmish – the combatants are out to kill one another. Sting after sting is rained home against the adversary's abdomen as the royal duellists frantically grapple in a tangle of legs and stabbing stingers (Ishay *et al.*, 1983). One aim of this hectic scramble is to weaken the opponent by biting off her antennae, legs and wings, but the main goal is to inflict a mortal blow by landing home a penetrating sting to the neck. Whereas multiple stings to the opponent's abdomen only exert a slow debilitating effect, a direct sting in the neck induces instant paralysis, with death following within a few hours. These desperate encounters may last for several hours and, after this kind of period, a Pyrrhic victory is often the inevitable result. By this time, neither combatant can avoid being badly injured and even the 'winner' may have been stung so repeatedly about her abdomen that she eventually fades away and dies, or else becomes infertile due to the long-term effects of the venom.

A *Vespula* queen has to work hard to establish a nest: gathering pulp, building the cells, adding the outer envelope and then catching live prey which has to be chewed up before being offered to the larvae. With workers present, the queen still continues to malaxate food brought in by the workers and feeds it to the larva. She also 'borrows' carton from the envelope and uses it to make or repair cells. In fact, it appears that the outer envelope serves not only as an all-inclusive roof but also as a storage reservoir for carton, as returning workers regularly add new pulp to the envelope, rather than using it immediately for new cells. At some future time, a worker in need of raw materials for the construction of new cells can draw her supplies from the envelope 'stock' by chewing off a spare portion. This saves her from the more time-consuming task of having to locate a source of pulp and then gather her supplies.

She masticates the dry carton, which quickly becomes soft and pliable, whereupon it can be used to build a new cell. Unlike in *Polybia occidentalis*, a returning pulp-collector does not pass her load over to a waiting worker but adds it to the nest herself. The jaws accurately apply the malleable material in a narrow strip, while the antennae measure distance and the degree of straightness relative to neighbouring cells. As in *Polistes* (and many other wasps), each cell is used for rearing a succession of larvae, being cleaned out between times before quickly receiving a fresh egg. Some species are more finicky than others about nest sanitation: *Vespula pensylvanica* queens dispose of dead larvae outside the nest, whereas *V. atropilosa* allow them to clutter up the bottom of the nest (although in many *Vespula* colonies there are plenty of inquiline fly larvae, e.g. *Volucella bombylans* (Syrphidae), ready to grow fat on the contents of the wasps' garbage dump).

Another aspect of nest sanitation is the queen's habit of regularly inspecting capped pupal cells, chewing away the cap and extracting the occupant if she perceives it to be dead or diseased. As the nest matures, the workers eventually take over this corpse-disposal role. As autumn approaches and the nest sinks into a gradual decline, they become less discriminating in this job, with the result that an increasing percentage of healthy larvae are extracted and killed, along with the sick and the dead. As long as they are not too diseased or putrid, such 'rejects' are recycled and fed to the healthy brood. On a young nest, the queen is also responsible for ventilation, standing at the entrance and fanning her wings to send a draught of cool air over the cells. Worker polybiines also adopt the same tactics in their enclosed nests and, in many species, water may be applied to the capped cells to improve the cooling effect. When temperatures are too low for pupal development, remedial measures are also available, and the workers position themselves in an empty cell next to a pupa and engage in a spell of vigorous abdomen-pumping. Remarkably enough, this is capable of producing a 10°C rise in the pupa's temperature.

As in the independent-founding Polistinae vespine adults often solicit liquid food from the larvae. In *Vespa*, the larvae themselves communicate their need for food by rubbing their mandibles loudly against the walls of their cells. This quickly prevails upon some of the workers to come running with food. Similar 'food calls' are produced by larvae of a number of other species, including *Vespula germanica, V. vulgaris* and *Dolichovespula rufa*. As in other wasps, vespine adults also announce the imminent arrival of liquid food for the larvae by drumming on the nest with the abdomen. The *Vespa tropica pulchra* queen announces her arrival, bearing a crop-load of food, by stamping her mid and hind feet.

Brood-parasites

Not all vespines rear their own brood. Some exploit the workers of another species to do it for them, acting as 'cuckoos in residence'. Such acts of social parasitism are carried out in two ways:

(1) *Facultative temporary parasites*, such as

Vespula dybowskii and *V. squamosa*, are able to found and maintain their own colonies well enough but prefer to foist themselves on another colony in early spring, killing the resident queen and taking over the job of sole egg-layer. The original brood is left intact, so it is the host's workers who emerge first and tend the parasite's brood. For a while, host and parasite workers co-operate to rear the parasite's queens and males but the host workers soon die off, leaving the nest in sole occupancy of the parasite and her offspring.

(2) *Obligate permanent parasites*, such as *Vespula austriaca*, *V. adulterina* and *V. arctica*, are totally dependent on being able to railroad the workers of the host nest into rearing the parasite's queens and males. As this requires a full complement of workers, the invader temporarily spares the life of the host queen, who is allowed to carry on with here vital egg-laying role for a while. Once her usefulness is over, she is eliminated. The parasite then takes over the egg-laying role herself but only produces male and queen eggs, which are reared by the host workers.

The methods used by a *V. austriaca* queen when usurping a *V. acadica* nest in North America are quite simple – she uses violence, intimidation and summary execution to take the throne and demoralize the work-force so that they will carry out her wishes without question. Once she has put to death the original queen, the parasite dominates the workers by submitting them to frequent maulings (Reed & Akre, 1983). She bullies them mercilessly by holding them down so that she can give them a thorough chewing over virtually the whole body. The unfortunate recipient of this brutality responds by curling up protectively and may then subserviently serve up a meal to its oppressor through trophallaxis before finally escaping. By constantly riding roughshod over her terrorized nest mates in this way, the parasite usually establishes complete dominance within 10 days, when she eases off the intensity of the maulings. Even so, forced trophallaxis with her conscript workers is still common, during which she emphasises her dominance by gripping the unwilling donor's forequarters and pulling her close until she has coughed up the required forfeit into her tormentor's jaws. The *V. austriaca* queen derives most of her nourishment from the host workers in this way, yet seldom solicits liquid food from the larvae. This is in contrast to the normal behaviour of the host queen, who relies mainly on larval trophallaxis.

Rather more subtle methods of conscripting a foreign work-force are also to be found. Once a *Vespula arctica* queen has successfully invaded a *V. arenaria* nest, she is reluctant to leave and depends on the host queen to feed her by trophallaxis (Jeanne, 1977). Once the host's workers have started to appear, they too feed the parasite and, occasionally, she will even return the service when solicited by a worker. The invader also sometimes feeds the larvae with solid food brought in by the workers and also by regurgitation, but such episodes are always brief. The mechanism by which she manages to take over the nest and then control it is far from obvious, lacking as it does the extreme and manifest application of physical menace seen in *V. austriaca*. The *V. arctica* queen probably exerts her control by pheromones, which she applies to the nest surface during frequent abdomen-dragging excursions across the comb. Another possibility is that the parasite's noticeable tendency to groom her abdomen tip during the first 24 hours on the nest spreads a chemical allomone which is responsible for producing a state of passive acceptance in the host queen. As this grooming gradually decreases over the ensuing days, it seems that this is quite likely.

It does, however, appear that this chemical governance may take a while to become established, as in the early days of a take-over the host queen may attack the parasite, biting at her wings and legs and trying to manhandle her off the comb. The invader's response is to hunker down and submit, trying to 'sit it out' by showing the minimum of reaction, with no tendency to either retaliate or sneak away. This is precisely the response of a *Vespula acadica* worker to an assault by a *V. austriaca* queen. However, it pays the *V. arctica* queen to bide her time in this way, minimizing conflict until her pheromones discharge their insidious role and present her with total control of the nest. Once this happens, the two queens will live side by side for a time in complete harmony, until eventually the host queen is slain, although whether it is the parasite or the workers who perform the deed is unknown. One important factor in all this is that the parasite should mount her take-over before the host's workers have begun to appear. They would be liable to provide the interloper with a hostile reception and by making a concerted attack they could even kill her.

The only *Polistes* known to suffer such invasions hail from Europe, where three species of parasitic *Sulcopolistes* are found, mainly around the Mediterranean basin. *Sulcopolistes* lack a worker caste and depend on the host's workers to rear their brood. The queen *Sulcopolistes sulcifer* is able to match herself against a nest defended by both queen and workers by taking on just one adversary at a time (Turillazzi *et al.*, 1990). She does this by tempting them out to intercept her as she approaches the nest. By grappling with each opponent on the ground, her superior fighting abilities enable her to win an easy victory. Although she may use her powerful jaws to amputate the odd leg or antenna, she keeps her sting strictly out of the battle. Her adversaries are more use to her alive than dead, being destined to pander to her needs as her obsequious servants. The *Polistes* defenders are hampered by no such considerations and strive to bring their stings into action as their first line of attack, although seemingly with little profit. The *Polistes* alpha queen fights longest and hardest, as she has the most to lose. The subordinate beta and gamma gynes give in more easily. The usurper finally prevails and is able not only to land on the nest but to maintain a foot-hold there, although she may still be obliged to cross swords once again with some of the defeated residents as they straggle home.

Sulcopolistes atrimandibularis queens are unique in taking over a number of *Polistes biglumis bimaculatus* colonies, several of which serve solely as food stores. These are plundered regularly by the *Sulcopolistes* queen during foraging trips from her main 'nursery' nest (Cervo *et al.*, 1990). Her own brood is reared by the host workers in the nursery nest, while the *Polistes* brood in the other nests functions as a reliable and easily obtainable source of food with which to provision the developing *Sulcopolistes* males and queens. Initially it is only the *Polistes* pupae which are taken, thus allowing the larvae to develop fully into a food supply for later exploitation. The parasite has to perform this body-snatching herself, as she has few workers at her disposal on the nursery nest, due to the low production of workers by her host.

ANTS

Social behaviour evolved in the ants a very long time ago and today all of them are eusocial, although a few species may now be secondarily eusocial, having replaced the queen with fertile workers. As a result,

► In founding a new colony, the first thing that a newly mated queen ant must do is prepare a nest. The first rains which break the long dry season are the trigger for the massed nuptial flights of the leaf-cutting ant *Atta bisphaerica* in the Brazilian *campo cerrado*. After mating and shedding their wings, the queens rapidly dig a hole in the ground to establish the new nest; few such nests will succeed in the long term.

▼ One of the principal hazards facing an *Atta bisphaerica* queen leaf-cutting ant in her efforts to establish a new nest in the Brazilian *cerrado* is the hostility from members of her own species occupying nearby nests. Here a queen is gradually overcome and killed by a swarm of workers, helped by one or two soldiers with their grossly enlarged heads and formidable mandibles. Such problems also have to be faced by the young queens of other ant species.

there are no examples of any of the apparent stages between the solitary and eusocial way of life which are to be found in the wasps and bees. From the point of view of their classification, ants are usually discussed in terms of their family, subfamily or tribe. This approach is, however, not totally practical in terms of their behaviour, since particular life-styles may be spread across the various taxonomic subdivisions, so within this section the ants will be dealt with in terms of their particular way of life, i.e., whether they are army ants, harvesting ants, fungus ants etc.

Army ants and nomads

In order for a species to qualify as an army ant it has to conform to a number of specific behaviour patterns. Typically, army-ant colonies are very large with many thousands or even millions of workers accompanied by a single queen. Army ants do not have permanent nests but use temporary bivouacs whose sites are changed at frequent intervals. Some species do not build any form of nest at all. Army ants are predacious, seeking out their prey in mass raids during which a continuous column of individuals is maintained between the raid area and the bivouac, communication being based on mass recruitment. Army ants are able to handle food items with greater efficiency than most ants. With such large numbers of workers available, large prey items can be carried by groups of them instead of having to be cut up on the spot and carried in small pieces by individual workers. A fair amount of food is stored temporarily along the foraging trails in protected caches until it can be transported back to the nest. The two subfamilies which best fit these requirements are the Old World army ants, the Dorylinae, also called driver or legionary ants, and the New World Ecitoninae.

Whereas the dorylines are mainly hypogaeic, a number of the ecitonines are epigaeic and perhaps understandably a lot more has been written about their behaviour, especially that of the Neotropical species *Eciton burchelli*. Ecitonine colonies pass through alternate statary and nomadic phases during their life cycle. During the statary phase, the ants' activities are centred on a fixed bivouac site for a period of about 3 weeks. During the first week of this phase, the queen grows at an enormous rate as her ovaries develop and become swollen with tens of thousands of eggs (physogastry). Then, over a period of several days, she lays upwards of 300 000 eggs. These eggs hatch into larvae towards the end of the statary phase and this is followed by the emergence of callow workers from the cocoons.

Callows of *Neivamyrmex nigrescens* do not take part in any raiding activities for at least 3–7 days after they eclose but they do emigrate to the new nest site (Topoff & Mirenda, 1978). Newly eclosed callows are found packed together in the centre of the bivouac surrounded by a mass of adults

Sociality in ants. Like the termites, sociality in ants is based on the caste system but whereas termites have only one queen, some ant colonies may contain more than one. In some ponerine ants, however, morphologically distinct queens have been secondarily lost and have been replaced by mated, egg-laying workers. Obligate parthenogenesis is found in the Japanese queenless ant, *Pristomyrmex pungens*. Most but not all ant colonies have males present at some time during their existence and all but some slave ants also produce at least one type of worker. Division of labour in the workers may be based on morphology, i.e., each different type of ant within the colony has a specific role or number of roles and this may also relate closely to their size and structure. The role of defending the colony, for example, may rest with the largest of the workers, the majors or soldiers, who may rarely also perform other duties. In *Camponotus fraxinicola*, the major workers, once the nest has been breached, are no better at defence than the minor workers (Wilson, 1974). They do, however, perform another role within the colony, namely that of food storage (Eisner *et al.*, 1974). Not only do they have very large fat bodies but they are also able to store large amounts of sugary liquid in their crops, which they can regurgitate and feed to other colony members when it is needed. Such repletes are also to be found in other ants and reach their maximum development in the honey-pot ants. The abdomens of these ants increase to many times their normal volume as their crops distend with stored sugary fluids. They are unable to move around in such a state and remain suspended from the roof of the nest galleries, regurgitating the food when required. Division of labour, as in hive bees, may also relate to age. In *P. pungens*, for example, there is strong evidence (Tsuji, 1988) that egg-laying and brood care, and other in-nest activities, are carried out by the younger workers which, as they get older, lose their fertility and become foragers outside the nest. All but the most primitive ants engage in a process called trophallaxis (mutual feeding of regurgitated food), which takes place between workers and is also the way in which the larvae receive their food.

▼ The nest of the weaver ant *Oecophylla smaragdina* is a pouch formed of one or more leaves folded over and held together with silk. As the adults themselves cannot produce silk, they use a larva for the purpose. Most of the work in sealing up the nest is performed from the safety of the interior. Exterior jobs are usually performed at night, and usually consist of sealing up any extra large holes with a dense curtain of silk, which may be 3–5 cm across. While performing this task, the adult worker holds the larva in her jaws and moves it back and forth across the cavity like a shuttle; if necessary, more workers temporarily hold the leaves in place until the work is finished.

with, between the two of them, the food collected from the previous raid. At the beginning of the next raid, many workers leave the nest but there are still enough left in the nest to keep the callows confined. As the emigration commences, however, more and more of the adults are recruited until eventually the callows are exposed, mill around the nest and then eventually join the mass of workers leaving. As each day passes, the callows develop their darker adult coloration and show less tendency to cluster in the centre of the bivouac. By the fifth day of emigration, the callows are beginning to take part in raiding activities.

The emergence of many thousands of new *Eciton* workers stimulates more intense raiding activities and the beginning of the nomadic phase, when the statary bivouac is deserted and a new bivouac is formed at the end of each day's raiding. During a day's raids, the ants may travel anything between 100 and 200 m. The workers carry and feed the developing larvae and surround them each night in their temporary bivouac. The queen, by this time much slimmed down so that she is only about twice as long as the largest worker, marches in the column accompanied by a retinue of workers. The relationship between the queens and their retinues has been studied for a number of species, including *E. burchelli* (Rettenmeyer *et al.*, 1978). At the beginning of each day, the queen is in the bivouac, usually at the centre surrounded by a tight cluster of mainly minim workers. At the beginning of each day, in all of the New World species surveyed, the queen does not usually leave the bivouac until about 75 per cent of the workers have left. From laboratory studies, it seems that, at least for *Neivamyrmex nigrescens*, it is worker activity that determines precisely when the queen moves off. The adults immediately around her buffer her, as they do for the callows, from the increasing activity of the rest of the ants until they too become aroused and begin to disperse. At first, the queen's movements may lack direction but eventually she is drawn into the column of workers leaving the nest. Queens typically emigrate when it is dark or gloomy and only those of *Eciton rapax* are known to do so in bright sunlight. It would appear, however, that the queens are not necessarily shunning the light but that emigrations are timed to take place in the dark.

An *Eciton* queen may leave the old bivouac site in the company of a retinue of between 100 and 500 workers stretched out between 0.5–1.0 m in front of and 1–2 m behind her. If she leaves without a retinue she soon acquires one, though this may not contain many workers if she happens to be particularly athletic. Retinues generally contain a higher proportion of large workers and soldiers than those parts of the column ahead of or behind the retinue which, when well-developed, is two to three times wider than the rest of the column, though one *E. hamatum* retinue was recorded at about 25 ants wide. The longer the ant colony has been migrating, the more guard workers there are to be found along its margins and it is these individuals who often join the retinue as it passes. The retinue is not constant but changes all the time, with some ants leaving and others entering as the queen moves along. If she stops for any reason, the retinue accompanying her may move on without her and a new one forms around her as she moves off again. If the retinue is disturbed for any reason, even by a puff of wind, then the guard ants may stop and adopt a defensive attitude, rearing up with their mandibles open and their antennae vibrating rapidly. If the disturbance is a major one, then the queen stops and the workers cluster around her in a tight mass and the rest of the column may also stop, or continue past, depending on the nature of the interruption. The queen may stop for anything

▲ Bivouac of the army ant *Eciton burchelli* in a hollow under tree roots in Trinidad. The whole mass of ants is suspended from those workers which are clinging onto the surface of the tree by their jaws.

between 10 minutes and, rarely, 1 hour before she moves on. Only part of the cluster moves on with her, the rest of them staying behind for up to 10 minutes before they too move on. The ants which form the retinue are older workers while young workers and callows form bridges, allowing the queen and her retinue to pass across gaps in the trail.

There seems to be a general relationship between retinue size and life-style, with bigger retinues forming in epigaeic species. Of the five *Neivamyrmex* species observed during the present investigations, all were hypogaeic and had retinues similar in size to those of hypogaiec *Eciton* spp. The main difference was the lack of soldiers in the former and the lack of continuous rows of guard workers along the column margins. *N. cristatus* workers seem to sense the imminent passing of the queen for they show a great deal of excitement for up to 2.5m in advance of her migration. In the laboratory, the retinues of majors around the *N. nigrescens* queen increased in size the further she advanced towards the new bivouac site. In this species, the retinue tends to move ahead of the queen, occasionally extending for just a few centimetres behind her. These migrations continue for as long as the larvae are growing but, as they reach full size and begin to pupate, so the intensity of the daily raids begins to decrease and the colony goes into the next statary phase of the life cycle.

The way in which colonies of *Eciton* multiply was first worked out for *E. hamatum*. For most of the year during the colony's alternating statary and nomadic phases, the queen remains the centre of attention. This situation changes, however, when early in the dry season the sexual brood is produced. In *E. hamatum*, the sexual brood contains about 1500 males and only six new queens but no workers. At an early stage of development of this brood, a large worker force becomes associated with it when the colony splits. The remaining workers maintain their association with the colony queen. As the larvae approach maturity so the bivouac is found to consist of two clear zones, one of which contains the sexual brood and its workers, the other the queen and her workers. From the sexual brood, it is the young queens who hatch first, followed a few days later by the winged males. The emergence of the latter stimulates great activity in the colony, which sets off on an intense raid, followed by the usual emigration but this time also by splitting of the colony. These raids are carried out along two main trails leading away from the old bivouac site and, as the day progresses and they become more intense, the young queens and their attendant workers move out along one trail while the old queen and her retinue depart along the other. Only one of the virgin queens is allowed to reach the new bivouac site, the others being restrained by the workers and then left alone to die. Occasionally, perhaps because her health and attractiveness are waning, it is the old queen who is prevented from entering her new bivouac and, instead, it is a virgin queen who continues this half of the original colony. Some days after they emerge, a number of the males may fly from the colony, presumably in search of virgin queens from others, though some may possibly remain to mate with their virgin sisters. Whatever the situation, the young queens have usually mated within a few days of hatching.

Although, as already mentioned above, army ants traditionally come from the subfamilies Dorylinae and Ecitoninae, there are ants from other subfamilies which could also be called army ants. *Leptogenys diminuta*, for example, is a ponerine ant which hunts and brings back prey co-operatively by swarm raiding and also lives in temporary nests. It does not, however, fit the army-ant description completely since food is first sought by scout ants who then recruit workers to it, whereas army ants forage *en masse* without any direction from scouts.

In *L. diminuta*, there is also no continuous column of workers between the nest site and the foraging area. A recently discovered (Maschwitz *et al.*, 1989) *Leptogenys* sp. from Malaya does, however, seem to be an adequate army ant. It forms single-queen colonies consisting of tens of thousands of workers who carry out co-ordinated raids but without the initial activity of scout ants. Like army ants, they remain in the same nest site for just a few days and produce no real nest structure as such. Unlike the ecitonines, however, *Leptogenys* does not exhibit alternate statary and nomadic phases and the eggs, larvae and pupae are carried during the periodical migrations. Colony migrations in this ant follow one or sometimes two raids and the new nest is always in a site which has not previously been occupied. Emigration commences during the hours of darkness, usually not long after the last workers have returned from the preceding raid. The pupae are the first to be taken from the bivouac, during the first third of the emigration, and the workers carry them in single file held beneath their body. Movement of the larvae follows somewhat later and reaches a maximum in the second third of the emigration, while the callows are among the last to leave the old bivouac site. Queens have been seen within the columns in both their ergatoid state, when apart from their lighter colour they resemble workers, and their physogastric state. The queen is not accompanied by a specific group of workers when moving to her new nest site but, at intervals, groups of workers stand at the side of the trail, facing inwards with their mandibles open, checking each passing ant with their antennae. Males are carried by workers and at dusk they may be seen flying off and leaving their colonies. Like the columns of the 'typical' army ants, those of *Leptogenys* are accompanied by a variety of ant guests, including staphylinid beetles, Collembola, the highly adapted myrmecophilous isopod *Exalloniscus maschwitzi* (which often hangs on to a pupa), a ptiliid beetle (also on pupae) and several mites.

Army-ant-type behaviour has also been found in the myrmecine ants. Two species of *Pheidologeton* from South-East Asia show behaviours which converge with those of the true army ants (Moffett, 1984 & 1988). *P. silenus* workers, for example, forage *en masse* along narrow columns which fan out into swarm raids up to 3 m across. This species takes mainly animal food, typical of army ants, though the closely related, swarm-raiding *P. diversus* also harvests seeds directly from plants, not at all a habit of army ants. Although the raid front of these species is fairly rapid it is not as fast as that of most true army ants. *P. silenus* is also somewhat limited in its ability to handle prey for it is not very successful at catching large or agile species, whereas true army ants can handle almost anything. Like army ants, both *P. silenus* and *P. diversus* indulge in group transport of large prey items but only *P. diversus* has been found to store food in temporary caches.

Although the term nomad often occurs in a description of their emigrating habits, army ants are not nomads in the strict sense, in that they do not move their animals to new pastures like human nomads. There is, however, an ant which does just that. The dolichoderine ant *Dolichoderus cuspidatus* from Malaya forms an obligate symbiotic relationship with the pseudococcid *Malaicoccus formicarii* and its close relatives in the same genus (Maschwitz & Hänel, 1985). Like the army ants, *D. cuspidatus* forms temporary bivouacs which house the single, ergatoid queen, in excess of 10 000 workers and upwards of 4000 larvae and pupae. In a typical bivouac, there is no nesting material and the workers cluster around the central mass containing the queen, eggs, larvae, pupae and large numbers of mealy bugs. The bugs reproduce viviparously within the nest. Nests may be found underground, on the ground or in trees but more than three-quarters occur in the herb layer of the forest, at heights below 2 m, and in bamboo stems. The workers carry the mealy bugs along an established trail system between the nest and their feeding sites on a wide variety of plant species. Since they feed only on fresh growth, the ants continually have to shift them to new feeding areas on the plants.

Moving of the nest tends to correlate with an increasing distance between it and the foraging site and it may take place in nature anything from once or twice a week to once in 15 days. The process of nest-changing is similar to that in army ants, with the workers transporting the brood and the mealy bugs and also forming living bridges across gaps in the vegetation. During nest-moving, no mealy bugs are carried to feeding sites but, once the new bivouac site is established, then feeding recommences. When comparatively short distances are involved, the move takes 2 to 3 hours but over longer distances it naturally takes more time and intermediate depots may be formed.

On the basis of the distribution of existing colonies, which occur in clumps, often with gaps of hundreds of metres or even some kilometres between clumps, new colonies are clearly formed by budding from old ones. During investigations of this ant in the field, one colony was found to have divided into two halves, though the actual process of budding was not observed. Budding is an obvious answer to the question of how the ant ensures that each new colony ends up with a retinue of mealy bugs from which to feed. Male ants are on the wing twice a year, once in the dry season and again, some months later, in the wet season.

Ant–ant relationships

In the way that other insects, for example staphylinid beetles, have evolved close relationships with ants, so many ants have developed various associations with other ants. These vary from situations where two ant species have adjacent, interconnecting nests and forage in the same habitat, the phenomenon of parabiosis, to the situation where one species is totally parasitic on another, the state referred to as inquilism. As might be expected, not all of these behavioural subdivisions are clear cut and there are species which show intermediate behaviours. These are probably a reflection of the lines of evolution which have led from one particular type of behaviour to another. It is possible, for example, that the ants who are permanent parasites on other ants, the inquilines, could have arisen independently along three different pathways, from the xenobionts, from the slave-makers or from temporary parasites.

Parabiosis

Parabiosis is characterized by having two ant species living in very close association without either of them apparently taking more from the relationship than the other. The Colombian rainforest ants *Crematogaster limata parabiotica* and *Monacis debilis*, for example, build their nests adjacent to one another, with connections between them, rearing their broods independently within them. The workers pass to and fro along common foraging trails and, in some parts of their range, the two have been seen collecting honeydew together from membracid bugs. Similar relationships have been noted between *Crematogaster* ants and other ant species, even to the extent of there being interspecific greetings between workers when they meet. As long as the relationship remains a neutral one, then parabiosis is maintained but, once one ant starts taking more from the other than it gives in return, then the relationship becomes xenobiotic.

Xenobiosis

More so than for parabiosis, there are different degrees of xenobiosis. In the northern USA and southern Canada, for example, *Leptothorax provancheri* forms an association with *Myrmica brevinodis*. *L. provancheri* builds its own nests, close to those of its host, in which it rears its own brood. It then makes connections between the two in the form of short galleries through which its workers, but not those of *M. brevinodis*, can pass. The relationship is xenobiotic rather than parabiotic because the *L. provancheri* workers obtain food from the host workers by trophallaxis.

The relationship between the North American ant *Leptothorax diversipilosus* and the western thatching ant, *Formica obscuripes*, is an example of extreme xenobiosis, since the former species is unable to survive in the absence of the latter (Alpert & Akre, 1973). The *L. diversipilosus* workers move around freely within the host nest, though the brood is reared separately from that of the host. The *F. obscuripes* workers seem to accept their tenants, only occasionally showing any hostility towards them. The *L. diversipilosus* workers obtain their food by trophallaxis from the *F. obscuripes* workers. A xenobiotic relationship between two species which does not involve food stealing by trophallaxis is that between the myrmecine *Megalomyrmex symmetochus* and the fungus ant *Sericomyrmex amabilis*. *M. symmetochus* rears its brood in the host's fungus garden and feeds on its fungus.

The pathway into xenobiosis, which involves food-stealing or trophic parasitism, may well well be via those ants which follow the workers of other species to their feeding grounds and use them for their own foraging. In Europe, for example, *Camponotus lateralis* workers follow *Crematogaster scutellaris* foraging trails and then make use of the same food sources. The two species nest separately and the relationship is not obligatory. Similarly, in Trinidad, workers of *Camponotus beebei* follow the trails of *Azteca chartifex*, though they use them at different times of the day, presumably because, on the rare occasions that they do meet, the *Azteca* show extreme hostility to the trail-followers. It has also been found that, whereas *C. beebei* is able to recognize and follow the trail pheromone of *A. chartifex*, the latter is unable to recognize the former's scent.

Slave-making ants

Slave-making species are those which raid the nests of other ants, capture their pupae and return with them to their own nest. When these pupae hatch, they accept the nest as their own and act as workers for the slave-making ant, foraging for it and caring for its brood. Ants which exhibit slave-making, or dulosis, do so to different degrees and there appears to be a fairly well-marked progression from the facultative to the obligate slave-maker.

At the bottom of the ladder are facultative slave-makers, such as *Formica sanguinea* from Europe, whose behaviour has been intensively studied over the years. This species is able to survive quite happily in the absence of slaves and as few as 2 per cent of colonies have been reported as containing them in some areas. Slave-making raids take place in colonies from which the sexual individuals have just departed and only the old queens and workers remain. It is likely that, as for other slave-makers, the nests to be raided are found by scout ants who then lay down a pheromone trail back to their nest. The *F. sanguinea* then move out on a broad front, following the scout trails, for distances as much as 100 m, until they reach the nest to be robbed. When sufficient workers have accumulated, they then enter the nest, remove as many larvae and pupae as they can find and transport them back to their own nest. Any defenders who attempt to stop the pillaging are killed by the raiders. The American species *F. integra* raids its hosts more often than *F. sanguinea* and workers will remain in the raided nest overnight, returning home the next day, if a raid has extended into the hours of darkness. The behaviour of another American ant, *F. wheeleri*, is somewhat unusual in that it uses slaves of two different species and there is division of labour between them. Workers of the slave *F. neorufibarbis* go with the slave-makers on their raids and assist them in nest-plundering while, back at the *F. wheeleri* nest, they assume a defensive role. The second slave ant, *F. fusca*, on the other hand, remains in the nest where it apparently specializes in food storage and undertakes the care of the *F. wheeleri* brood.

The most highly developed of the formicine slave-makers are the Amazon ants of the genus *Polyergus*, from North America, Europe and Japan and eastern Siberia. The Amazon ants are obligate slave-makers whose workers are adapted solely for attacking and plundering the nests of slave species. They are unable to construct nests and therefore they use the nests of other ants and they have to be fed by the slaves. Nests of the slave ants of the genus *Formica* are found by the scout Amazons who lay a recruitment trail back to their home nest. The recruited workers then make their way at speed to the *Formica* nest and attack without any hesitation, rapidly dispatching any of the nest's residents who get in their way with their sharp, piercing mandibles. The *Formica* cocoons are then transported back to the home nest where a proportion are allowed to hatch to produce more slaves while the remainder serve to feed the colony. From field and laboratory investigations, it seems that new colonies are usually formed by budding, when newly fertilized queens return to the home nest and then take a proportion of the workers on a raid, taking over the raided nest as the basis of a new colony. In the laboratory, a queen of *P. rufescens*, the European species, was seen to make her way into a *F. fusca* nest, eventually killing its queen and using the resident workers to rear her own workers.

Slave-making has evolved independently in the myrmecine ants of the genera *Harpagoxenus*, *Leptothorax* and *Strongylognathus*. Experiments using slave-deprived colonies of three species of *Harpagoxenus* in the laboratory (Stuart & Alloway, 1985) revealed that each of them retained varying degrees of self-sufficiency from their presumed free-living ancestors. The three species investigated were *H. canadensis* and *H. americanus* from North America and *H. sublaevis* from Europe. In the presence of slaves, the *Harpagoxenus* did little in the form of domestic duties but, when deprived of them, *H. canadensis* especially exhibited a wide range of normal ant abilities both in and out of the nest.

The slave-making activities of some North American *Harpagoxenus* and *Leptothorax* have been studied in some detail (Alloway, 1979; Stuart & Alloway, 1985; Bourke, 1988). A comparison of the life-styles of two species, *Harpagoxenus americanus* and *Leptothorax duloticus*, revealed some similarities as well as some important differences between them. Both species use scouts to locate slave-ant (*Leptothorax* spp.) colonies. Scouts consist of either single individuals or groups of two or more ants who always remain in fairly close proximity to one another. Single scouts sometimes join these groups when they meet up and individuals may leave groups and begin scouting alone. Unlike the formicine ants, the initiation of raids in these two species seems almost infinitely variable, though at least four main raiding patterns are discernible:

(1) A lone scout finds a target colony, seeks out and finds its entrance though does not try to enter. The scout then returns to its own nest where its arrival engenders some degree of excitement. The scout then returns to the target colony, dragging its sting along the substrate as it does so, followed by a single column of slave-makers with sometimes some slaves as well. Trail-laying *L. duloticus* scouts tend to move more slowly and haltingly than those of *H. americanus*. The target colony is attacked by the whole group on their arrival.

(2) A group of scouts finds a target colony, some attack immediately while one or two return to their home nest to recruit further workers, and perhaps some slaves, to the attack using the trail-laying method described in (1).

(3) A group of scouts, perhaps with one or more slaves, finds a target colony, vanquishes it and then recruits reinforcements.

(4) A scout returns but, instead of leading workers back to the target colony, it stimulates formation of a scouting group which then goes out and, by random seeking, discovers the target colony, attacking it without further recruitment.

Once at the nest, the two species carry out their raids in a similar fashion. While a number of slave-making workers guard the nest entrance to prevent the occupants from escaping with their brood, the remainder, as well as accompanying slaves, attack or intimidate the resident workers. There is a noticeable difference between the two species with regard to the way that the battle is fought. Target-colony workers are very hostile to the invasion and consequently *L. duloticus* may kill as many as 75 per cent of them in the battle, though those which offer the least resistance and are not carrying brood are permitted to escape from the nest. The response of target-colony workers to *H. americanus*, however, is exactly the opposite, for they offer no resistance, though they may be attacked by accompanying slave ants. If one of them attempts to escape with a portion of the brood it drops it immediately on encountering one of the slave-makers. As a result of this benign attitude of *H. americanus*, all but a few, i.e. all except those attacked by the accompanying slaves, of the target-colony's residents escape from the attack with their lives. Once the raid is over and the target-colony workers have been killed or dispersed, the raiders carry the brood back to their nests.

This has always been observed to be the case for *H. americanus* but, on a number of occasions, *L. duloticus* colonies have emigrated to the raided nest. Brood transport is facilitated by the recruitment of further slave-makers, always apparently with some slaves in *H.americanus*, sometimes without slaves in the other species. Prior to an emigration, all of the *L. duloticus* workers, but no slaves, from their nest are recruited to the raided colony where the brood is examined thoroughly, with both mandibles and antennae, for a period in excess of 1 hour. When such a nest is taken over, the *L. duloticus* workers go back to their old nest and return with their brood to the new one. This activity seems to excite the slaves in the old nest and, although they show no hostility, they do seem to impede this brood transport. In order to facilitate matters, the *L. duloticus* workers lay aside brood transport for a while and instead pick up and carry their slaves to the new nest until there are no longer enough of them at the old nest site to be a nuisance. On the occasions when this nest-moving was observed, the whole process took about 90 minutes. Captured brood is treated differently by the two species. *L. duloticus* eats large numbers of eggs and young larvae from the booty and all sexual pupae. *H. americanus* also eats the eggs and small larvae as well as some of the sexual pupae. A number of the latter, however, are allowed to develop into adults, thus alates of the slave species leave the nest without being molested.

Colonies of *Harpagothorax sublaevis* may be based on a single queen or on a group of egg-laying workers. Within a subset of the workers in a single-queen colony or in queenless colonies of this species a particular dominance order has been found to exist (Bourke, 1988). This dominance order is independent of worker body size but is correlated with the development of the ovaries, frequency of trophallaxis and the amount of time spent in the nest. Removal of high-ranking workers from queenless nests indicates that they inhibit the ovarian activity of the workers below them, removal of a queen having the same effect upon the high-rankers. The latter tend to spend significantly less time scouting and raiding, reducing the risk involved in being outside the nest. Since they spend more time in the nest, this also gives them a greater chance of feeding, resulting in greater egg development. Thus, should the queen be lost for any reason, they are in a position to take over the role of egg-layers.

Within the Palaearctic genus *Strongylognathus* occurs the final transition from slave-making to complete inquilism. *S. alpinus* wages total war against the closely related slave species, usually *Tetramorium caespitum*. During a raid, it is accompanied by slave workers and they, along with the *S. alpinus* workers, kill off the queen and all reproductives and carry all of the brood and surviving workers back to the slave-makers' nest. In the parasitic species *S. testaceus*, however, the *Tetramorium* queen lives alongside the slave-making queen. Only the latter produces eggs resulting in both workers and reproductives, for somehow the reproductive ability of the *Tetramorium* queen is limited to that of giving rise to workers. In the mixed colony, the slaves considerably outnumber the slave-maker workers, who not only do nothing within the nest but also no longer go out on slave-making raids. Thus one final step, that of losing the workers altogether, will take this species into the state of extreme inquilism.

Temporary parasitism

Temporary parasitism involves the queen of one species entering the nest of another species, killing the host queen and then taking over egg-laying. For a while the host workers care for the parasite brood until its own workers can take over the role. This has resulted in a variety of ploys which are used by the queens to gain entry into the host nest. *Formica rufa* queens, for example, just dive straight into the host nest and are frequently killed by the workers. However, the success of this ant as a species clearly indicates that, often as not, she is successful in her attempt to usurp the host queen. Queens of *F. exsecta* enter the host nest by stealth, or else allow themselves to be carried in by workers, while the queens of *F. pressilabris* use the latter ploy, pretending to be dead in order to encourage the host workers to pick them up. Newly fertile queens of *Lasius umbratus* kill a worker of a host ant and run around with it held in the mandibles, presumably to assume the host's nest smell before they attempt to enter the host nest. The invading *Formica* and *Lasius* queens somehow are able to eliminate the host queens. The queens of two *Bothriomyrmex* from North Africa, *B. decapitans* and *B. regicidus*, allow the workers of their *Tapinoma* hosts to take them into their nest where they then hide in the host brood. Eventually, the queen settles on the back of the host queen, cuts off her head and takes over the colony. A similar behaviour is found in another North African species, the inquiline *Monomorium santschii*, though in this case it is the host workers who are somehow induced to kill their own queen. Ants of the genus *Epimyra* are well on their way to complete inquilism, for though they retain a worker caste in small numbers, they never assist the host workers, who carry out all the colony's tasks. Once the *Epimyra* queens enter the host nest, they quickly kill the host queen or queens with their sharp, pointed mandibles. Deprived of their queens, some of the host workers take over the role of egg-laying to ensure further workers for the colony. One species, *E. ravouxi*, has taken the final step to inquilism and allows the host queen to live and continue producing workers.

Inquilines

We have already seen some examples of inquilines at the ends of the apparent evolutionary lines from xenobiosis, temporary parasitism and dulosis, when queens of individual species have ended up living in the nest of another ant species, relying partly or completely on the host workers to maintain their brood. True inquiline species show a number of behavioural and caste changes, as well as morphological changes, in association with their life-style. In inquilism at its extreme, ants:

(1) Lose the worker caste.

(2) Replace the queen with an egg-laying individual resembling a worker and called an ergatogyne. In some cases there may be a continuous series of intergrades between the queens and the ergatogynes.

(3) Have more than one egg-laying queen in the same nest.

(4) Have nuptial flights which are reduced or non-existent so that mating occurs between closely related males and females. Dispersal of the queens is very limited.

(5) As a consequence of reduction in their mouthparts, are able to feed only on liquid food regurgitated by workers.

(6) Have small reproductives; in some inquilines, the queen is smaller than the host workers.

(7) Have queens, especially in the physogastric state, who are very attractive to host workers. These queens appear to produce some substance which the workers lick avidly from their surface.

An excellent example of the change from the free-living to the fully inquiline way of life is shown in three species of *Plagiolepis*. *P. pygmaea* is a free-living ant with winged queens and males and abundant workers who appear before the reproductives. *P. grassei* is an intermediate parasite on *P. pygmaea*. The *P. grassei* queen is winged but smaller than the host queen and the male is somewhat female-like and occasionally has only rudimentary wings. Workers of *P. grassei* are rare, with only one to each ten mated queens; furthermore, they appear after the reproductives. *P. xene* is fully parasitic on *P. pygmaea* and its queen is only one-third the size of the host queen. Wing development of the queen is variable and wings are often rudimentary, as they always are in the male, who is very like a female. There are no workers.

The fact that some parasitic queens destroy the host queen while others let her live seems to be related to two possible reproductive strategies. In the absence of a queen, the workers supply all of the food to the developing reproductives of the parasite, so that as many as possible are produced in a short time before the now unreplaceable host workers die. Alternatively, the host queen can be left intact to produce workers, and thus parasite reproductives, over a longer period of time but at a slower rate. Which strategy is adopted by a particular species probably relates to the ecology of their habitat.

BEES

The overwhelming majority of the 25 000 or so known species of bees are solitary. This means that the female works alone to construct and provision a nest for rearing a brood which she will never see. Advanced eusocial behaviour has evolved only in some members of the Apidae, especially the honey bees (Apinae) and stingless bees (Meliponinae), all of which are highly social.

Lack of sociality does not, however, imply lack of either commitment or success, and the measure of care and 'ingenuity' which the mother bee devotes to the preparation of her brood chambers is often quite extraordinary and goes far beyond the mere placing of an egg upon a suitable food source, as seen in most other insects. Thus, although relatively few bees are counted even as subsocial, with the mother not overlapping with her offspring, they are still counted as having advanced maternal care. Like all insects which invest a considerable amount of time and effort into each offspring, their fecundity is quite low, with only a relatively small number of well-cared-for larvae being produced during the lifetime of each hard-working female.

Unlike most insects, solitary bees are less common in the tropics (especially in the rainforests) than in temperate zones and, indeed, it is the deserts of North America and the Middle East which are the richest habitats in terms of species. This is probably because of the dangers in damp tropical climates posed by moulds, which are prone to infecting the larval food stores of pollen and honey. Unlike some other subsocial insects which nest in the ground, such as earwigs, certain crickets and beetles, the female solitary bee will not remain within the nest and be on hand to give direct protection from developing moulds. Solitary wasps do not have this problem, as their larvae are supplied with prey in a living but paralysed 'stay-fresh' state. The bees have attempted to overcome this fungal problem by applying waterproofing to the inside of their burrows but this has apparently been effective on a large scale only in relatively damp cool-temperate climates. Bees which nest in the damp tropics are still relatively few in number, although some may have solved the mould problem by turning to nesting in fairly dry 'artificial' environments, such as the insides of termite nests.

Solitary bees

General nesting habits

Solitary bees can be conveniently divided up according to their nesting habits:

(1) *Miners*, for those species digging nests in the ground.

(2) *Masons*, for any species which use earth or other malleable materials to build durable cells within existing cavities, or in the open.

(3) *Carpenters*, for those species whose powerful mandibles enable them to excavate their own tunnels, even within quite sound timbers.

Mining bees generally line their cells with secretions manufactured within their own bodies. This is usually applied as a waterproof lining covering the soil enclosing the nest cell. In six families of bees, this secretion is produced by the Dufour's-glands in the abdomen. The chemical composition of the secretion varies from family to family: terpenes in the Andrenidae; alkanyl butanoates in the Melittidae; macrocyclic lactones in the Halictidae, Oxaeidae and Colletidae; oily triglycerides in *Anthophora*. Once applied, the material soon dries to a glossy finish; in Halictids, it oozes meanwhile into the soil to bind with the cell lining. Mason bees utilize for this task naturally-occurring materials such as leaves, mud or the resin bleeding from wounded trees.

Mining bees include some large and familiar temperate genera such as *Andrena* (Andrenidae). The other families of specialist miners are the Melittidae, Oxaeidae and Fidelidae, while members of several other families nest in the ground as well as in other substrates. The nest architecture of mining bees varies enormously, depending on the type of ground. Single-celled nests are typical in unstable ground where multi-celled structures would be threatened by collapse. Above ground, some nests are marked merely by a small tumulus of loose earth, while the entrance to others is furnished with a turret whose design is normally species-specific. The function of these turrets has been the subject of much debate but is still unknown as they do not appear to provide any extra degree of protection against parasites.

Nesting procedures – coping with moulds

Dufourea novaeangliae (Halictidae), which nests in sandy banks in the eastern USA, has a primitive nest architecture, in which single egg-shaped chambers are situated at the ends of extended side-tunnels, branching off the main gallery (Eickwort *et al.*, 1986). The cells are lined with Dufour's-gland secretion which permeates the adjacent soil and binds with it so that the lining cannot be peeled away. The lining is applied directly to the unmodified cell interior and is completely waterproof. The female bee takes out some extra insurance against moulds by enclosing the spherical food mass in a waterproof coating from the Dufour's-gland. Similar waterproofing of the vulnerable provisions is quite common among solitary bees and is presumably effective, even though the larva's feeding will broach the coating to an increasing degree as time goes on. However, the female halictid cannot waterproof the entire cell as she is unable to apply the secretion to the final closure, so there is always some degree of humidity entering the cell.

Measures to prevent the food stores becoming mouldy in incompletely waterproofed cells are also important in many other bees. In order to minimize the chances of infection, the 'bee loaf' itself may be shaped in a species-specific manner, which is designed to reduce the amount of contact between food and cell to the minimum. Halictids and andrenids produce more or less spherical bee loaves, while the Eurasian melittid *Dasypoda altercator* goes some way to curing the contact problem (which may be particularly acute in its completely unlined cells) by furnishing the lower part of its spherical brood ball with a trio of short legs.

Colletes has solved this problem in two ways. Firstly, by a complex and ingenious piece of engineering, the female is able to close the cell completely with a waterproof lining. Secondly, the cell is fumigated internally with a powerful fungicide and bactericide called linalool. This is released as a spray from minute glands at the bases of the mandibles. As she squirts away, the female champs her mandibles rapidly so that they function as a pump (O'Toole & Raw, 1991). However, keeping the cell too dry can result in a new difficulty – the pre-pupa may become *too* dry and dehydrate in the totally waterproofed cell. The solution to this problem, and indeed the whole process of nest-building and provisioning in these bees, is so amazingly complex and sophisticated that it merits description at length.

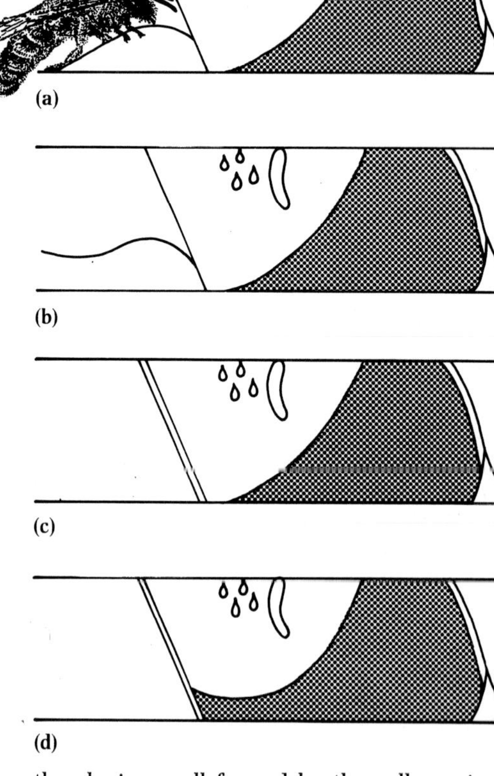

▶ The final steps involved in the provisioning of a cell by the female of the colletid bee *Colletes kincaidii*. (a) The female bee sits on the cell platform, which she is about to use to close off the cell. Before the next stage she secretes Dufour's-gland fluid around the cell rim. (b) The cell platform is detached from the ventral cell lining and is then pushed up by the bee until (c) it closes up the original opening into the cell, which was through the centre of the cell rim. The cell platform becomes glued to the cell rim by the Dufour's-gland secretion. She then worries away at the cell cap, near its base, with her jaws. The eventual result of this activity is that the cell provisions slump into the position shown in (d), where they form a depression in which the larva will eventually lie and feed. (After Torchio *et al.*, 1988).

Nesting in Colletes kincaidii

Colletes is a large genus, rich in species, found throughout the world except Australia. *C. kincaidii* builds its burrows in vertical sandstone cliffs in the eastern USA and its nesting procedures are probably broadly typical for the genus (Torchio *et al.*, 1988). The female's first job is to apply the waterproof lining. She does this by laying down a coat of saliva with her bilobed tongue, which then does service as a brush to combine droplets of Dufour's-gland secretion with the saliva. She also mixes Dufour's-gland secretion with saliva by performing frequent half-somersaults, during which her tongue and anal gland come into contact. By itself, the Dufour's-gland secretion remains in liquid form for a long period but, combined with saliva, it polymerizes and rapidly forms a quick-drying waterproof polyester (Albans *et al.*, 1980).

The bee now begins to build the 'cell platform' by nodding her head while releasing copious saliva. This soon forms a series of lumpy projections from the floor at around a bee's length from the cell's back wall. She then draws a series of salivary strands upwards from this base, forming a lattice of hardening strands. She adds more saliva to finish with a thin platform jutting up at a 60–70° angle from the floor and leaning towards the entrance. Now she extends the platform from its free leading edge down on to the floor so that it forms a hump. This she ties into the lining on the cell wall, using yet more saliva. She coats the whole lot with further saliva, then prescribes a circle above the platform, laying down a line of secretion around the cell's interior. This will form the foundation (the 'cell rim') upon which the cell cap will eventually be glued.

This complex enterprise takes her about 2 hours. Then she rests before heading out to forage for provisions. These consist of a mixture of pollen, scraped off the scopae of her hind legs, and nectar regurgitated from her crop. She uses the furry tip of her abdomen to sweep up any loose pollen, which she has previously preened off her body, and to firm down the mass of provisions, which constitute a sloping mound twixt floor and roof at the rear of the cell. Egg-laying takes place with the bee lying atop of her platform, placed very precisely so that she can swing the tip of her abdomen against one wall and attach an egg to its upper surface, just below the roof. The egg is accompanied by a considerable quantity of Dufour's-gland fluid. Then she quickly dabs some 15 droplets of this fluid on to the wall beside the egg.

She closes the cell by first rotating her body around the cell rim while applying a fresh coat of Dufour's-gland fluid. This rim is now 'pre-glued' and ready to receive the cell platform, which she bites free from its lower attachment. Pushing mightily with her head, she bends the platform upwards until its edges contact the cell rim. The previously applied secretion acts as a contact adhesive, forming an instant bond which, with the application of some additional 'glue', seals the cell completely.

With the waterproof cap safely in place, it could perhaps be assumed that the hard-working artisan's job was complete. But she has yet more work to do in order fully to ensure the welfare of her larva within the closed cell. She begins to 'worry' away at the cell cap just above its base, biting and pushing at it with her head, now and again turning around to smear the area of attack with Dufour's-gland fluid. Under this constant physical and chemical pressure, the cap gradually beings to soften and warp inwards. Finally, its inner surface contacts the sloping wall formed by the cell provisions. This causes the semi-liquid provisions to flow downwards from their highest part (at the rear of the cell) and upwards along the lower part of the cell cap. This slumping movement finally stops when surface tension pressure on the semi-liquid mass has become stabilized.

This prolonged effort has now changed the shape of the provisions *after* the cell has been closed. There now exists a dished hollow on the lowermost surface of the food mass, while its upper part has been correspondingly reduced, pulling it down short of the egg, thereby giving the newly hatched larva more freedom of movement. The reduced degree of slope also enables the small larva to slide down safely into the dish-shaped hollow, where its C-shaped body can securely float on a pool of food. However, the female's complicated machinations have also benefited her larva in another vital aspect. Inside its waterproof cell, the larva may indeed be safe from flooding, or from having its food supply go mouldy. However, it now may be *too* dry for its own good, which could spell disaster for the overwintering pre-pupa (larva in an immobile 'resting' stage).

The female secured the correct degree of future humidity when she released the gush of *pure* Dufour's-gland fluid along with the egg, plus the several spots of the same liquid, which she so meticulously dabbed on to the cell wall nearby. In its pure state (i.e. unmixed with saliva), this product remains semi-liquid for some time, only gradually solidifying and, while so doing, releasing water as a by-product. Some bee pre-pupae are thought to absorb atmospheric water and thereby gain weight during their pre-pupal stage. In view of this, it seems likely that the *C. kincaidii* female's elaborate preparations for maintaining humidity in a waterproof cell must be vital for her larva's welfare. That such a complex suite of behaviours should have evolved in *C. kincaidii* (and other *Colletes*) is quite staggering and illustrates the lengths to which even non-social insects may go in order to ensure the successful development of their brood.

As already mentioned, bees in several families prevent premature mould-induced spoliation of the loaves by covering them with a thin protective membrane secreted by the female. One bee has solved the problem of nesting in a damp habitat in a truly remarkable way, by actually harnessing the humidity prevalent in its tropical environment to help convert a rather poor-quality food source into something far better (Roberts, 1971). This paragon among bees, *Ptiloglossa guinnae*, builds its branched nests in the ground in the cool, rain-drenched environment of the Costa Rican cloudforest. Each membrane-lined cell is provided with a copious supply of extremely thin, runny liquid, in which pollen is only sparingly present. However, yeasts seem to thrive in this otherwise marginally nutritious pabulum and it is probably these yeasts, bubbling away in their miniature subterranean vats, which yield the main food supply for the bee larvae. In a unique departure from normal practice, the cell is left open, allowing the gaseous by-products of fermentation to escape safely and preventing the watery supplies from drying out by admitting the humid forest air.

The yeasts probably wage their own chemical warfare, which would exclude the more unwelcome destructive kinds of moulds, thereby solving that particular problem. Also, by entering into a mutually beneficial alliance with yeasts in this way, the bee is able to capitalize on the numerous species of Solanaceae (plants of the potato family) which thrive in these forests. Few bees extensively utilize pollen from plants of this family, as it is low in nutrition – unless of course it is 'enriched' by fermentation.

Perhaps even more remarkable is the edible nest-lining produced from the Dufour's-gland of *Anthophora abrupta* (Norden *et al.*, 1980). The female applies a dense coat of this lining to cells, which are constructed in clay banks. The substance has a cheesy odour and is produced from the hugely developed Dufour's-gland. Once the larva has polished off its normal provisions of nectar and pollen, it satisfies its remaining needs by tucking into the cell-lining over the next 2 days (Batra *et al.*, 1980).

Building and provisioning in mason bees

Most mason bees avail themselves of a ready-made home by merely adapting some existing cavity for their specific purposes. Unless the nests happen to be made in a series of 'tenements', such as a beetle-riddled tree trunk, nests are usually far more scattered than in the miners, many of which nest in dense 'villages' when suitable ground is in restricted supply. Some masons are familiar to everyone, due to their presence around houses, where they nest inside cavities in walls and fences.

In Europe, the red mason bee, *Osmia rufa* (Megachilidae), often nests in old nail-holes in garden sheds or the decaying mortar of old walls. Like many mason bees, the female uses mud to form the partitions between cells, collecting her supplies from a miniature 'mine' where the earth is damper than elsewhere. Up to ten or more *Osmia* females often use the same 'mine' and may excavate tunnels down to a depth of several centimetres in order to gain access to soil of the required malleability. The little miner works quickly, hacking out the mud with her mandibles and then fashioning it into a glob against the only handy solid surface – the underside of her abdomen. She bends this forward for the purpose, pushing the finished glob against her mandibles, where it is carried while in flight, often being held fast with her front legs, which are pressed against the rear end of the cargo.

Back at the nest, she applies the mud to the inside of the tube by using her front legs to feed it forwards through her champing jaws. As the semi-liquid material flows from her mouthparts she nods her head up and down and from side to side, applying a thin strip of mud around the inside of the nest as she prescribes a circle with her body. She then builds upon this foundation, using mud from several more collecting trips, expanding the ring inwards until it meets in the centre to form a solid partition. After adding more mud to reach the required thickness, she smooths down the work surface, using the special horns on her mandibles as 'floats' as she moves her head up and down across the entire surface. This smoothing process is important for the later survival of the emerging adult. As the adult, once it has emerged from the pupa, has insufficient room to turn around inside the cell, it is vital that the larva (which can turn around) should make certain that it pupates with its head facing the front (exit) partition. It manages to figure out which way is which by nosing against a partition to detect whether it has a rough or a smooth finish. A rough surface (because the female could not reach it, as it was inside the cell) indicates that this is the rear surface of the partition at the front end of the cell and is therefore the way to face. A smooth finish indicates the front smoothed-down side of the partition at the rear end of the cell.

Not all *Osmia* species build mud partitions. *O. bruneri* from the USA is one of many masons which utilize a pulp of chewed-up leaves (in *O. bruneri* the leaves are of evening primroses, *Oenothera* spp.) for construction of the cell partitions (Frohlich, 1983). Upon returning to her nest in a hollow stick, the female *O. bruneri* deposits her blob of pulp on the floor at the back of the nest. Then, bit by bit, she chews pieces off and adds them to a growing circle of pulp, which she forms into a rim inside the tunnel. Further layers are added and this eventually becomes a substantial disc, which is finally reinforced with bits of pith bitten off the insides of the tunnel and pressed into the still-damp work surface.

The finishing touches are more complex, for the bee now converts the flattened disc into a concave dish, spreading pulp outwards to the rim and using her front legs to impress a dimple in the centre. After finishing off by coating the whole structure with a secretion brushed on by the mouthparts, the bee prepares the ground for the final sealing of the cell, and the beginning of the next one, by turning around and constructing a 'threshold'. This is actually destined to be the foundation for the next partition. She measures out where this should be placed by lying full-stretch with her head against the newly completed partition. Then, keeping the tip of her abdomen firmly planted against the floor, she curls inwards and places a blob of pulp on the floor next to her abdominal tip. It is upon this tiny 'marker' wad that she now builds a narrow rim around the tunnel. This will be her constant reminder of the size of the cell and consequently how much food will be needed to provision it completely. This 'full-length' measuring technique is only used for the inner cells, which are destined to contain the larger females. The smaller males in the outer cells will need less food, so the bee lies on her side and bends into a tight C-shape before placing the wad of pulp next to her abdomen. A similar threshold is also constructed by *O. rufa*, although of course in mud, and, indeed, thresholds are typical of many members of the genus, including *O. lignaria, O. pumila, O. marginata, O. californica* and *O. georgica*.

O. bruneri provisions her cells with a 'loaf' of pollen and nectar, brushing her pollen loads off the scopal hairs beneath her abdomen with her back legs as she reverses into the cell. She then regurgitates her cropful of nectar and uses her mandibles to knead the combined ingredients into a moist dough. In between foraging trips, she grooms herself and, in so doing, seems to ingest a droplet of abdominal liquid which is passed to her mouth via her legs. This behaviour seems to be common within the family, and may be intended to add a glandular agent to the provisions, probably to inhibit the growth of moulds. The female knows that she must begin to build the final closure once the provisions have reached the threshold, although she first deposits an egg on the store of food.

O. californica also utilizes leaf pulp, usually from mallows, *Malva* spp., or evening primroses (Torchio, 1989). She nibbles a rounded bolus of pulp from the edge of a leaf, grips it in her jaws and then flies off, not directly back to the nest but to a patch of bare earth. Here she uses her mandibles and front legs to coax the ball across the soil, where it picks up loose particles until it is liberally coated with a gritty crust. Back at the nest, she uses the same techniques as *O. bruneri* to construct a soil-pulp partition, whose final strength is augmented with a brushed-on coating of regurgitated nectar, collected on a special foraging trip. Funnily enough, individual females dispense with the soil-collecting phase and use pure pulp instead.

After putting in place her threshold, the *O. californica* female provisions the cell with the normal pollen-nectar 'bee loaf' in a broad mound occupying almost the entire cell behind the threshold. It is interesting that, during this phase, the female does not slavishly follow an invariable routine but can, in fact, act in a way which indicates that she 'knows' where she is at any one point in the general scheme. Thus, when setting up the threshold, she may temporarily 'borrow' material from the completed cell partition – and then go out and gather material specifically to effect the necessary repair, not 'forgetting' to brush the finished patching-up job with its coat of nectar. She may also extend the threshold somewhat during provisioning, presumably because she made a mess of measuring-up in the first place.

When she has finally filled the cell with provisions, the female bee is faced with a steeply sloping, relatively smooth wall of 'bee loaf', which has no crevice in which an egg could be securely placed, nor any safe lodging place for the larva. Her next task is therefore to modify the loaf's shape to accommodate these requirements, which she does in a quite remarkable piece of behaviour. She digs and bites a large cavity in the loaf, then saturates the newly exposed surface with nectar, which she kneads thoroughly into the pollen. She now

Snail-shell nesters. Many *Osmia* species construct a linear row of cells inside twigs or old beetle borings, using methods similar to *O. montana*. As pre-existing cavities are always in short supply, it is not surprising that some species have evolved the habit of utilizing some unconventional nesting sites. Many Eurasian species nest inside empty snail shells, an otherwise underexploited waterproof home, which can be very abundant in habitats such as coastal sand dunes and limestone grasslands. Numerous species in at least five genera nest solely in snail shells, although there are only two examples in the USA. In Eurasian species, such as *Osmia bicolor* and *O. aurulenta*, the bee partitions the cells with chewed leaf pulp; *Hoplitis spinulosa* utilizes sheep or rabbit dung for this purpose, while *Rhodanthodium septemdentatum* uses resin. In many of these species, the female adds fragments of snail shell to the building materials to form the final closure. Usually the shell is left *in situ* when the female has completed her work and *Osmia bicolor* conceals it with pine needles or bits of grass. The sight of a little bee flying off with a huge 'moustache' of long grass many times her own length trailing along behind her is somewhat surprising. *O. rufohirta* and *O. rufigaster* often carry out the job of concealment before they even start nesting, dragging the shell under cover of a stone or grass tuft. As the smooth shell alone would provide inadequate purchase for her feet, the labourer eases her task by cementing a few 'foot-holds' of mastic to the shell. *O. tunensis* saves itself the bother of attaching these foot-holds by selecting only muddy shells, the mud itself providing a natural grip-tight surface.

▼ Numerous species of mainly Old World bees nest in disused snail shells. This *Hoplitis lhotellieri* female is closing the mouth of her nest with tiny fragments of chewed-up shell combined with a mastic of masticated leaves.

backs into the cavity and deposits her egg on the wet provisions. Yet her final objective is still not met, for she must now seal the egg inside the provisions by converting the open cavity into a closed chamber. She executes this seemingly intractable task by moulding the edges of the cavity upwards and inwards with the tip of her abdomen, forming a circular lip which eventually meets in the centre. She uses the remainder of her final pollen load to apply a dry coating to the outer wall of the loaf, then seals the cell with pulp.

The female of *O. montana* uses broadly similar methods, which nevertheless differ in a number of details. When collecting leaf pulp, the female sits astride the leaf and, bending forwards, makes a line of perforations in the shape of an arc, her tool being the sharp tips of her mandibles, which impinge on the leaf from either side like clippers. She is now able to twist the leaf and tear off the perforated section, as if detaching a postage stamp from a sheet. She then reduces the material to tatters by worrying at it with her mandibles and front legs, and carries it back to the nest. When adding it to the partition, she feeds the ragged leaf through her rapidly opening and closing mandibles, after which her tongue flicks the pulped material forwards on to the work surface. Full maceration is thus delayed until the material is actually being used. No threshold is constructed before provisioning begins, and the shape of the pollen loaf is radically altered to accommodate the egg in a chamber, as in *O. californica*.

However, the mechanics of this undertaking differ. While regurgitating nectar to soften the loaf, the *O. montana* female uses her mandibles to slice away material from the front of the pile and pass it backwards beneath her body, using her front legs as rakes. She is now standing on a layer of displaced food, facing a substantial cavity in which the egg is laid. This done, and with her back to the provisions, she rakes food off the floor and passes it back to her pulsating abdomen, which rams it into a pile more than half the height of the cell. Legs and abdomen together then push this up further to connect with the original loaf on the roof, forming an ovoid chamber within the loaf.

Leaf-cutters

Leaves are used as nest materials by the leaf-cutting bees in the large (1500 spp.) cosmopolitan genus *Megachile*. Most species compete with *Osmia* for nest sites in similar cavities but a few nest in the ground. Many Eurasian and North American species employ similar methods of building and provisioning, which can be described in general terms. After choosing a nest site, the female seeks out a plant with suitable leaves. Some species have definite preferences, e.g. in Europe, *M. willughbiella* chooses a fairly narrow range of smooth leaves such as rose, willowherb, lilac and beech; while the smaller *M. versicolor* (also common in the USA) restricts its attentions almost entirely to roses.

The female may land briefly on several leaves in turn, often rejecting tough old leaves in favour of younger ones, at least until the supply starts to run out. She settles astride the leaf margin and quickly uses her powerful jaws to snip away (audibly) a section of leaf. Just before the final bite, the cut section becomes unstable and wobbles from side to side, causing the bee to buzz her wings and take the weight as it finally comes free. However, she only flies a short distance before landing to adjust the leaf's position within her mandibles, folding a long section lengthwise so that it forms a half-tube, with its outer convex surface facing upwards towards her. The female can alter her technique to cut sections of at least three basic sizes and shapes. She also 'knows' where she is at any point in the building programme, so that if a piece of leaf does not fit or is dropped, she returns to the source and cuts the correct shape again. Semi-rectangular sections of about the bee's own length do service as overlapping panels to form the end of the cell; longer oval sections make up the sides, while the cap of the fully provisioned cell is made out of several almost perfectly circular leaf discs. The sections of leaf are not glued firmly together, but adhere lightly to one another by their own sap, which is deliberately 'bled' from the leaf by the bee, who chews the overlapping edges, possibly adding saliva as she does so.

There are only a few variations on this general very widespread theme. The Eurasian *M. analis* substitutes strips of birch bark for the more usual leaves, while a number of species use petals rather than leaves for the cell cap. Some cut entire sections of leaves, take them to the nest, then chew them into a mastic and use this in at least part of the building process. The North American *Megachile (Eumegachile) pugnata* uses mastic to build a ring inside the tube. This serves as a foundation which receives complete oval sections of leaf, constituting the back end of the cell (Frohlich & Parker, 1983). Four such sections are overlapped to form the complete cap, the bee chewing each one to fix it to the mastic foundation ring, then tamping it home with a rapid pile-driver action of the head, using the outer surfaces of the mandibles as the hammer. Once all four sections are in place, they are secured more firmly by the addition of more mastic around their perimeter and then across the whole surface. Quantities of specially collected soil are then pressed into the surface of the soft pulp with the mandibles and tamped firmly home with the head. This final layer benefits from the incorporation of an abdominal secretion applied by the legs and mouthparts, probably strengthening and waterproofing the structure.

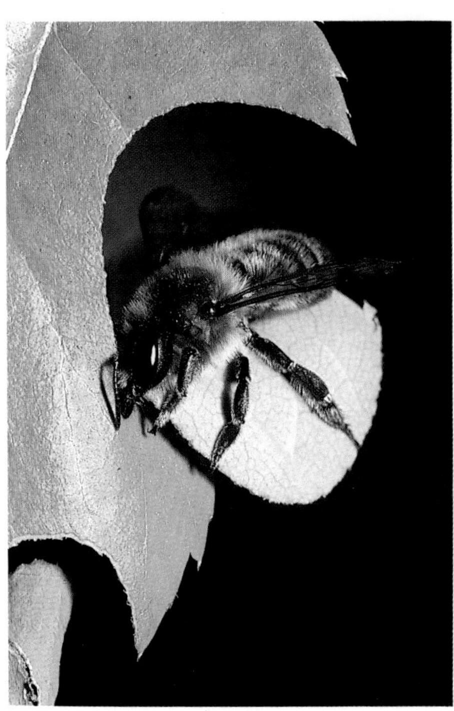

▲ The favourite nest-lining material for the European leaf-cutting bee, *Megachile willughbiella*, is a rose leaf. Using her powerful jaws, the female quickly snips away a section of leaf. She will cut different shapes according to the various stages of nest construction, 'knowing' which point she has reached at any one time.

Carder bees

Some of the most unlikely nesting materials are seen in the *Anthidium* carder bees, which line their cells with fluff derived from various sources. The common European *Anthidium manicatum* nests within readymade cavities and lines the interior of its cells with flock collected or 'carded' from plants having densely pubescent leaves. When harvesting her supplies, the female bee edges backwards along the leaf, shaving off the down with her mandibles, her progress being marked by a narrow line of depilation. Using her front and middle legs, she collects the fibres together to form a fluffy ball, nearly as big as the bee herself, and flies back to the nest. Using her head as a press, she applies the stock of down to the interior of the cell, spreading the fibres out across the surface with her jaws. She needs 6–8 loads to finish the job, after which she provisions the cell and closes it with a plug of down. The final closure of the entire nest is often completed with a covering of twigs, soil or pebbles. Remarkably enough, the female seems to be aware if any part of the nest is left exposed and can respond accordingly to remedy the situation in a way which is beyond the 'normal' sequential line of events. A nest in the author's garden had been built between the overlapping slats of a fence and, when complete, the line of cells was clearly visible from above. This defect was noticed by the female, who responded by collecting pebbles, which she carefully dropped (along with quantities of loose earth) through the offending gap until the nest was completely covered.

The European *Anthidium florentinum* lines her cells with the fluffy parachutes attached to dandelion (*Taraxacum*) seeds, while the African *Serapista denticulata* incorporates animal hairs into the felty densely-woven nest envelope.

Using resin

Numerous bees around the world utilize resin, rather than mud or leaves, in cell construction. The resin may serve just as a waterproof lining, as in *Trachusa* spp. (Cane, 1981), or may be the main construction material, as in some *Chalicodoma*, certain euglossines and the stingless bees. Resin has certain distinct advantages over other materials, mainly its durability and natural waterproofing properties. The bee can also make a small amount go a long way, due to the resin's natural ability to be spread very thinly compared with other materials. Once applied, its sticky surface is ideal for the reception of additional reinforcement in the form of sand or small pebbles. However, resin may also have another highly desirable property. It releases highly volatile, anti-bacterial compounds which effectively fumigate the nest, keeping it free of bacteria and moulds which could infect the provisions (Messer, 1985). This would be particularly important in the tropics, the main home of the resin-using stingless bees. The big puzzle about resin is how the bees which use it manage to work such a sticky material without themselves becoming entangled or having their jaws put out of action.

Resin-using bees are also widespread in drier areas, such as the Mediterranean and the deserts of North America, a habitat in which aromatic resin-producing plants are particularly abundant. The American species of *Chalicodoma* all employ resin for their nests within dead wood, while species of *Dianthidium* utilize resins within a variety of natural cavities, or even for making nests placed on exposed surfaces. *Dianthidium ulkei* constructs her cells within hollow stems (Frohlich & Parker, 1985). She collects resin oozing from pine trees, using her mandibles to slice away a ribbon of it as she edges backwards, rolling up the material into a ball. She spreads this smoothly over the centre of the back wall of the cell, using the mandibles and front legs. To this sticky foundation, she then adds a fine powder comprising chewed fragment of pith from the inner walls. An additional layer of resin forms the base for a layer of pebbles, brought in during collecting trips which alternate with resin-gathering forays. She devotes much time and skill to moulding the final interior into the correct shape and finish, finally adding two different abdominal secretions, applied by the legs and mandibles. It appears that these, mixed with the resin by chewing, may soften it and make it more easily worked. Her final task before provisioning is to build a thick, pebble-reinforced cell cap, in which she leaves an aperture just large enough to admit her body. This aperture is surrounded by pure resin, which she will use for the final closure. The provisions are highly liquid, such that the egg floats on the surface and the finished cell is capped with resin and pebbles.

The ability to scrape the sticky resin off the tree is obviously well developed in all bees which use this material, but in Wallace's giant mason bee, *Chalicodoma pluto*, the female's mandibles and labrum are grossly enlarged for the purpose. This huge bee gleans resin from rainforest dipterocarp trees, using the double-toothed tips of the mandibles to prise free a few scraps, then scraping these up with the tongue until a ball of resin 10 mm across is formed (Messer, 1984). The material is used in combination with plant fibres to line the cells, which are placed in galleries within the arboreal carton nests of *Microcerotermes amboinensis* termites. In this potentially hazardous environment, the tough fibre-reinforced resin cells of the bee are well able to resist invasion by the involuntary and unwelcoming landlords. Galleries are regularly re-used after a new lining has been fitted. Several bees usually occupy a single nest, possibly with a communal form of social organization. This, along with the habit of using cells over again, is probably dictated by the extreme scarcity of suitable nesting sites within the forest.

A number of other tropical bees also foist their nests upon termites. These include several species of stingless bees, the euglossine *Eufriesea laniventris* and at least four species of *Centris*: *C. derasa* in nests of *Microtermes arboreus*; *C. sponsa*, *C. bouvieri* and *C. thoracica*. As a genus, *Centris* exploits particularly diverse nesting sites. Some go for the abandoned nests of other bees, while *C. labrosa* and *C. analis* re-use the tough abandoned mud nests of *Sceliphron* wasps. In the USA, these same nests are also utilized by the female of *Hoplitis biscutellae*, who inserts her urn-shaped cells into the abandoned cavity.

The durability of the earthen nests made by *Sceliphron*, *Eumenes*, *Delta* and other mud-using wasps is matched by mason bees, such as *Chalicodoma* and *Hoplitis*, which construct free-standing earthen cells on rocks, cliff-faces or houses. For sheer rock-like quality of final product there is little to rival the nests of *Chalicodoma siculum* from the Mediterranean region. The nests of this bee are so hard that to scratch one with a fingernail merely leads to frayed nails, yet not a single grain of material will have been removed. This remarkably concrete-like finish is probably due to the inclusion of a Dufour's-gland extract in the building-clay. This is a valuable toughening and water-resisting ingredient which cannot be added to the mix by mud-daubing and mason wasps, which lack any such gland.

The nests of *Chalicodoma sicula*, which are usually fully exposed on flat rocks (although occasionally in bushes), are therefore resistant over many months to the worst that wind and rain can throw at them. They remain in good enough condition to be re-used the following spring by other smaller species of bees. Unlike many masons, the female *C. siculum* scorns the ready-made mud around puddles and mixes her own mortar on the spot by combining saliva with fine silty material. She can often gather this close beside the nest, so she saves considerable travelling time. Back at the nest, she works the glob of mortar on to the rock with her jaws and front legs,

▲ **Backing into her mud nest, a female mason bee *Chalicodoma sicula* in Israel uses her back legs to scrape pollen off her abdominal scopa.**

▲ **A female mason bee *Chalicodoma sicula* caps a finished nest on a large rock in Israel. These nests are so durable that they are often used again in the following year.**

building a cone-shaped structure upon a circular foundation by adding successive layers of mortar. Once she has filled the cell with nectar and pollen, she caps it with mortar and starts to build a new cell alongside, thereby saving on mortar by using part of one wall for two adjacent cells. Once she has constructed and provisioned several cells, she covers the whole group in a smooth coat of mortar, disguising their outline so that the whole nest cluster resembles a smooth bump on the rock.

Chalicodoma parietinum builds a similar nest, but incorporates several small pebbles within the mortar, giving it a rough exterior. The interior, by contrast, is painstakingly smoothed off by the female so that no pebbles protrude inwardly. As in the previous species, the finished group of nests is finally covered in a neat top coat of mortar. Their pebble-and-mortar composition is so resistant to wear that the same nests are often used again the following year, after a thorough spring-cleaning by the new occupant. Another highly skilled mason which also incorporates pebbles into its structures is the Eurasian *Hoplitis anthocopoides*, one of several European bees which have recently arrived by accident in the USA. In nesting habits, there are many parallels with *Chalicodoma parietina*: the nests are often placed in exposed positions on rocks; mortar is made from dry sandy soil mixed with saliva, and is combined with pebbles to give a durable finished product which is difficult to penetrate, even with a knife. These nests too can stubbornly resist the worst winter weather, and can be re-used the following year (Eickwort, 1975).

The female *H. anthocopoides* returns to the nest site from her quarries (one for silt, a different one for pebbles) carrying in her mandibles a single pebble, along with sufficient mortar with which to fix it in position. Once the foundation layer is in place, she adds more pebbles from the inside, leaning in over the newly laid wall and pulling the pebble up into position on the rim with her front legs. This done, she cements it in place with a dribble of mortar. As the nest grows, numerous chinks are often left between the stones. The female often closes these with mortar collected specially for the purpose, although she will always eventually plaster over the entire interior surface to give a final smooth finish. This mortar lining is extended upwards over the rim into a delicate lip (as in *Delta* wasps, but unlike in *Chalicodoma* spp., which do not form a rim). After provisioning with pollen and nectar, and laying a single egg, the female closes the nest with two sets of mortar. The first is derived from the material of the lip, which is bitten off in the mandibles, softened with saliva and then applied to inner rim, while the rest is fresh material from her quarry. As in *Chalicodoma*, an entire group of cells is finally topped off with mortar.

The social bees

Unlike the ants, which are all eusocial, in the bees it is still possible to find species with habits ranging from the solitary to the highly social. As a result, we get an insight into the pathway along which full eusociality may have evolved, at least in the bees, for the phenomenon may well have evolved along different lines in the ants and wasps. Unfortunately, the overall picture is somewhat complicated by the fact that some bees pass through different stages of social development during the formation of a colony. As a result, a species showing a fairly high level of social development early on may revert to a simpler level once the colony is established.

Communal nesting, quasisocial and semi-social bees

The minimum requirement for sociality is for two or more females to tolerate each others presence by, for example, sharing a common nest entrance but having no other form of co-operation. Communal nesting of this kind is found in a number of solitary bees but is not necessarily obligatory. Some species may be solitary or communal while others are always communal. It is possible that factors such as the availability of suitable nest sites dictates the behaviour of facultatively communal species. In the South American halictid *Augochloropsis diversipennis*, communal behaviour results from females staying within the burrow system within which they hatched and then extending it for their own use.

The next level of sociality within the bees involves the beginning of co-operation between females. Quasisocial females may help each other to build and provision their cells, so that the number of females in a given colony usually exceeds the number of open cells in that colony. All females, however, retain the ability to lay eggs. The Indian nomiine bee *Nomia capitata* appears to be quasisocial in that sisters both forage and lay eggs in co-operatively provisioned cells. In this species, however, it appears that one favoured individual, perhaps the original foundress or an older sister, remains within the colony and oviposits (Batra, 1966). Quasisociality is only rarely reported and it may be that it represents a stage in the development of a colony with a higher level of sociality (O'Toole & Raw, 1991).

Within the semi-social bees may be found the beginnings of the caste system which characterizes the truly social insects. In a semi-social nest, some of the females have large, active ovaries and are the egg-layers while others have poorly developed ovaries, seldom lay eggs and may not even have mated. The latter are the colony's foragers while the egg-layers do little foraging. The Neotropical halictid bee *Augochloropsis sparsilis* is an example of a semi-social bee. Nests of this species consist of burrows dug into banks, the burrow ending in one or more groups of cells (Michener & Lange, 1958). Each nest contains several females, who often, though not necessarily, are sisters. Nests containing unrelated bees appear to result when females who have emerged from different nests join another female who has excavated a new burrow. The males leave the nest soon after emerging and probably fly off to mate with females from another nest. As expected, there is some division of labour within the nest, with bees with poorly developed ovaries foraging and the others remaining in the nest and producing eggs. In nests containing three or four bees, only one seems to be an active forager but as numbers increase, so the numbers of foragers increase. Usually, only one cell is provisioned at a time.

Subsocial bees

As so far described, parental care in the bees has been restricted to providing a fixed supply of food on which a single egg is laid before the cell is sealed. The developing larva is then left to get on with it with no further assistance. In a subsocial bee colony, the female looks after the larva once it has hatched and may provision it progressively. The female of the Japanese anthophorid bee *Ceratina japonica*, for example, performs a number of maternal duties (Sakagami & Maeta, 1977). She breaks down the cell partitions at intervals, removes the faeces and then re-builds the partitions; she guards the nest entrance, preventing the entry of parasites; she associates with young adults in the autumn and the latter take pollen from her. Some females have an extended life-span and may rear young the following year having overwintered. Cells are arranged in rows down the hollow centre of plant stems so that, as the first young adults emerge, they have to make their way past their younger siblings in order to get to the outside. They do this by progressively breaking down the partitions and re-building them as they make their way up through the row of cells. At the same time, they transfer each pupa and/or larva back one place so that, as hatching proceeds progressively, so the row of occupied cells gets shorter. *C. japonica* is unusual in that, within a single species, a range of behaviours from the solitary to the subsocial may be found.

This range of behaviours is also exhibited by another anthophorid tribe, the Allodapini. The subsocial members of this tribe show a seeming advance on *Ceratina japonica* in that the females rear their brood progressively in an unpartitioned chamber, a situation not found in other non-apid bees. *Allodape mucronata* from Africa is fairly typical of these bees (Skaife, 1979). The female founds her nest in the form of a burrow in the hollow stem of a suitable plant. She lays a number of eggs loose in the bottom of the burrow and then guards them until they hatch. She then feeds them on demand with a mixture of pollen and nectar and arranges them so that the oldest larva is nearest the entrance. Eventually, she may be engaged in feeding six or seven larvae at any one time and, when she is not doing so, she is either on guard at the nest entrance or keeping the burrow clean. Once the larvae have pupated, she remains in the nest until her offspring hatch. A number of young remain behind in the nest with their mother and help her to raise a further generation; the rest leave to found their own nests. A nest may thus contain anything between two and six adult females, one of which is the principle egg-layer while the others forage and help to look after the nest, the beginnings of eusociality.

Eusocial bees

A typical eusocial bee colony contains an egg-laying queen with at least two, and often more, generations of mature females, the workers. As with other social insects, their roles include nest construction, brood care, nest-guarding and foraging. At certain times, males also may be present within the colony.

Bees in existence today which show a primitive level of eusociality are the sweat bees, the Halictidae. Sweat-bee colonies are founded by a single fertile female, who may be joined by other females of her generation; thus the colony passes initially through solitary and subsocial phases. In the North American sweat bee *Lasioglossum lineatulum*, for example, which has two generations per annum, the spring nests are founded by from one to six mated females (Eickwort, 1986). Where the nest is founded by more than one female, which happens for between two-thirds and three quarters of nests, all of them are potential egg-layers but the largest of them does not forage and is fed by her nest mates. Whether or not this female is the only, or at least the dominant, egg-layer is conjectural. Mated adult females are the only stage that overwinters and they emerge in the spring and spend a few days imbibing nectar from the available flowers. Nests are dug anew in the soil though, on occasions, old nests may be utilized. Side-burrows are dug and lined with waterproofing material and these are then provisioned with pollen and nectar. Foraging females seek out their own pollen and nectar sources and do not communicate this information to foraging nest mates as do hive bees.

A single egg is laid on the food mass and the burrow is then sealed with soil. Such cells may occasionally be examined by adult bees but usually the developing young receive no further care. Adult bees emerge from these cells in June of the same year and roughly one-quarter of them are males, who then leave the nest to seek out and mate with between one-quarter and one-half of the newly emerged females. The females which mate have been found to have well-developed ovaries; in the remainder, of which only a small proportion mate, the ovaries are undeveloped. Thus, at this stage of the life cycle, this bee has two castes: queens and workers. All of the females return to the nest and begin to increase the number of cells and forage for the next generation. The colony remains eusocial for a short time but then the original foundresses die and it becomes semi-social in character, with only one generation of adults present. One or more of the fertile females takes over the role of queen from the original foundress, though there is usually some form of dominance hierarchy, with the first-emerged of the females taking over as the major queen.

Foragers at this time are those females with less well-developed ovaries and a low likelihood of mating, though they do not differ significantly in size from their sisters. The nest entrance is guarded by one of the workers, who blocks the narrow entrance to the nest burrow with her head and only allows bees whose smell she recognizes to enter. She usually attempts to bite intruders and, if this fails, she then turns round and blocks the entrance with her abdomen. Which bee fulfils the role of guard in *L. lineatulum* is not certain.

The next generation emerges in August and contains proportionally more males then the spring emergence. All of the females mate and then return to the nest where they subsequently hibernate to become the foundresses for the spring nests of the following year. As is usually the case, the males soon all die. A closely related sweat bee, *L. zephyrum*, has also been studied in great detail and shows some subtle differences in its behaviour. The queen of this species is known to direct some of the activity within the nest for, when a forager returns, she meets it and leads it to whichever particular cell she feels requires provisioning (Breed & Gamboa, 1977). Should the worker lose contact with the queen for any reason, then it becomes lost and confused and cannot carry out provisioning of the correct cell until it once again meets the queen. Dominance hierarchy is determined behaviourally, the bee which nudges the others most with her head becomes the queen though, among bees of the same age, this role is normally filled by the largest individual. The degree to which the queen influences the other workers is correlated with colony size (Buckle, 1985). In large colonies, the dominant workers show more queen-like characteristics than those in smaller colonies and, furthermore, a higher proportion of the workers in the larger colonies are willing to mate. In this species, it is the second female in the dominance hierarchy who normally guards the nest entrance.

In the widespread Eurasian sweat bee *Lasioglossum malachurum*, the foundresses are somewhat larger than their worker daughters and there may be two, or even three, generations during the summer before the queens for the following year are produced. The foundress remains in the nest for the whole summer, dying a flightless, worn-out shadow of her former self in the autumn. In this and another Eurasian species, *L. marginatum*, the cells, unlike those of the American bees above, are kept open during larval development. Those of *L. malachurum* are then closed up for the pupal stage but in *L. marginatum* they are kept open even at this time. The foundress of the latter species is long-lived; 5 or 6 years is the norm. By the end of her first season in the nest she can no longer fly and is a full-time egg-layer, eating food that the workers have placed in the cells. All of her offspring, who are still alive at the onset of winter, then hibernate and are thus available to continue their labours at the commencement of the next season.

Towards the end of the colony's fifth or sixth year, the males are produced and at this time a number of the young females leave the nest to mate and then hibernate as the foundresses for the following spring. This bee is one of the most advanced socially of the sweat bees since the nest is perennial and the queen maintains her individuality throughout the life of the colony. Despite this, there appears to be none of the communication between individuals that is found in the highly social bees, workers just carrying out tasks that they find need doing.

The second major group of primitively eusocial bees are the familiar bumblebees, which are typically northern temperate in their distribution, though a number occur in the Amazon tropical forests. In the temperate species, colonies are founded in the spring by queens, who mated in the previous autumn and then hibernated, relying upon food reserves laid down while feeding on late-summer and autumn flowers. As with the sweat bees, the early stages of the bumble-bee colony exhibit first solitary and then subsocial characteristics before becoming eusocial in the later part of the season. Having revitalized herself on pollen and nectar from the spring flowers, the foundress seeks out a suitable nest site, sometimes underground in a ready-dug burrow, such as that of a mouse, or, in other species, above ground but using dry plant material to cover the cells. The queen lays eggs and forages to feed the developing larvae until the first generation of workers is produced. These workers are often quite small, sometimes too small to leave the nest and forage, in which case the queen has to continue with this task until she can rear some larger workers. The production of small workers usually correlates with cold springs and a paucity of spring flowers. As the seasons progress and the number of flowers and availability of food increases, so the number and size of the workers increases. As it does so, some of the workers begin to lay eggs and, as long as they remain undetected by the queen, they eventually develop into the colony's males. When the queen does find a worker egg in a cell, she destroys it and replaces it with an egg of her own. In *Bombus lapidarius*, it has been reported that the workers may then retaliate by eating the queen's egg, this aggressive behaviour on their part correlating positively with the degree of their ovarian development. The males which eclose are then available to mate with the young queens as they emerge in the late summer and autumn.

Honey bees

The seven known species of the genus *Apis* are collectively called honey bees and include one of the best-studied of all animals, the hive bee or western honey bee, *A. mellifera*. A queen of this species lives for between 3 and 5 years and, during that time, she produces many generations of workers who are all her sisters. The queen does almost nothing but lay eggs, as many as 2000 per day during the foraging season. All of the rest of the hive's tasks are performed by the workers, who may number anything between 40 000 and 80 000 in a well-established colony.

The worker-bee's behaviour within the colony is age-related so that, as she ages, so her roles change. The first half of her roughly 6-week life she spends in the nest, the rest she spends foraging. For the first day or two of her life, she does very little but feed on pollen and honey and, during this time, her head glands and wax glands are developing. Once this development is complete, she is able to produce royal jelly, which she feeds to the young larvae, and wax, used in the production of new comb. As well as these tasks, she also tends the queen, keeps the nest interior clean and receives nectar from returning foragers. She processes the nectar that she receives by sitting and repeatedly opening and closing her mouthparts. This exposes the nectar to the air and, as water evaporates from it, it becomes thicker and thicker. Once it has been partially ripened in this way, it is on the way to forming honey and is placed into a storage cell. The honey is not fully ripe for several more days and the process of evaporation is hastened by workers standing and creating a stream of dry air over the honeycomb by fanning with their wings. Fanning is not age-restricted and may be carried out by any worker. If by chance no cells are available, she retains the sugary liquid in her alimentary canal and eventually absorbs it. This increases her wax production so that she is able to build more cells to receive incoming nectar. There is thus a built-in mechanism within the colony to ensure that, when storage space is in short supply, it is rapidly made available. In the same way, nurse workers collect pollen to feed the larvae from pollen-storage cells but, if these are empty, they take pollen directly from the returning foragers.

Other tasks performed in the nest include capping of the cells once the larvae are pupating, cleaning of old cells for re-use, and removal of nest detritus, including old bees that have died in the nest. This particular behaviour seems to be restricted to about 2 per cent of the workers in the colony (Visscher, 1983). This fact was discovered when bees which had been killed by freezing were allowed to thaw out and were then placed in an observation hive. Of the worker bees, 10 per cent ignored the corpses completely while another 75 per cent of them showed some interest, which varied from a perfunctory antennation to antennation accompanied by licking with the proboscis. Of the remainder, some workers grasped the corpses and pulled them around for a short distance or time but only the seemingly special 2 per cent inspected the corpse briefly, picked it up in the mandibles and carried it away for disposal.

Having fulfilled her role as a house bee for the first half of her short life, the worker then becomes a forager. What stimulates a worker to leave the nest to forage is not known but it may relate in part to what happens when she returns with a load of pollen after a foraging trip. If she can get rid of her pollen load quickly, either by having it taken from her by a hungry house bee or by placing it in a pollen-storage cell, then she may well be encouraged to depart immediately on another pollen-collecting trip. If she fails to off-load her pollen, because there is already plenty available in the nest, then she may be delayed in departing to collect more. In this way, demand stimulates collection.

Honey bees have, of course, evolved a special form of communication system, which is used to inform other workers of the precise position and abundance of flowers at a foraging site, the well-known 'dance of the bees'. An examination of the system in three species of honey bee gives us an insight into how these dances may have evolved from the simple to the more complex in terms of information transmitted to fellow workers.

The simplest dance is that of the South-East-Asian bee *Apis florea*. A scout worker, having found a foraging source, returns to the horizontal platform above the single vertical comb which forms the nest. Then, surrounded by other workers, she walks in a straight line across the platform, waggling her abdomen; she repeats this several times. The direction in which she dances is the direction in which the food source lies.

Scouts of a second South-East-Asian bee, *A. dorsata*, land on the vertical, sun-facing surface of their single vertical comb to perform their dance. If they walk vertically upward, then the foraging source is towards the sun. If the food is at any angle to the left of the sun then, instead, they walk upwards at that angle to the left of vertical and, if it is to the right of the sun, then they do the same to the right of vertical.

The common honey bee or hive bee, *A. mellifera*, which lives in hollow trees or other suitable cavities in the wild, has the most sophisticated form of communication for not only can the scout inform other workers of the direction and approximate distance in which the food source lies but she can also inform them of its quality. What is more, this information is all imparted in the darkness of the nest or hive, where it is presumably perceived by touch and hearing, since it consists of both movements of the body and vibrations of the wings. If a scout finds food within 20 m of the nest, then she performs a round dance in which she walks in a series of circles, changing direction frequently. The more sugar there is in the food, the more often she changes the direction of the dance. A food source between 20 m and 100 m is indicated by means of a dance intermediate between the round dance and the waggle dance, the latter consisting of a compressed figure of eight, which is used on its own to tell of food at distances of more than 100 m. Again, the dances are performed on the vertical face of the comb, with the vertical representing the direction of the sun at that moment. The direction in which she dances the cross-bar of the figure of eight represents the direction of the food in relation to the sun as seen from the nest entrance. The distance to the food is indicated by the distance for which she moves and the number of body-waggles she makes as she walks the cross-bar of the figure of eight. Not only can the bee perform the dance accurately on entering the nest but, even after 30 minutes, she can still impart the correct information, for she has the ability to compensate for the few degrees of arc moved by the sun over that time. The quality of the food source is probably imparted by the frequency of the waggles and the buzzing of the wings which accompany the dance.

In order for the nest to operate successfully with so many occupants in an enclosed space, *A. mellifera* has to be able to control its internal temperature and humidity. The larvae, for example, are unable to survive at temperatures much below 32°C or above 36°C. If the temperature in the nest falls, therefore, the workers shiver their wing muscles to generate heat. If the nest becomes too warm then they either fan cool air into the nest with their wings or they collect water and spread it on the surface of the comb to induce evaporative cooling.

The integrity of the honey-bee's social system is basically maintained by the presence of the queen and the size of the colony. Although all of the workers are females, their egg-laying ability and their drive to rear new queens is inhibited by the secretion of queen substance by the queen, thus she remains the sole egg-producer in the colony until circumstances dictate otherwise. The queen is constantly surrounded by a small but ever-changing entourage of workers, who pick up queen substance when they make contact with her. By standing still so that workers can make a good contact with her, and by periodically changing her position within the nest (Seeley, 1979), the queen is able to ensure that sufficient queen substance is distributed among all of the workers. The importance of worker–worker transmission of the inhibitor may be appreciated from the fact that the queen only contacts about 35 per cent of her workers over each 10-hour period. This 10-hour period is crucial, for if there is no contact with a queen for this length of time, inhibition is lost and the workers begin to rear new queens. The most intensive contact with the queen is made by workers who are 3–9-days-old and, having contacted her, they then move rapidly around the nest antennating and being inspected by their nest mates. In the 30 minutes following their contact with the queen, these workers perform fewer nest tasks than other workers. Each worker needs a minimum quantity of queen substance in order to maintain the status quo of the colony as a whole. As the season progresses, however, and the size of the colony increases, there comes a point at which more and more workers begin to receive less than their daily requirement of queen substance and this then stimulates them to begin rearing new queens. This requires a change in the behaviour of the nurse workers, who now have to feed the larvae which are going to be queens exclusively with royal jelly rather than with pollen. At this time, the attitude of the workers towards the old queen changes and they begin to act aggressively towards her; she also lays fewer eggs at this time.

The first new queen to emerge normally kills off the other young queens before they emerge and she then leaves the nest on a mating flight. On her return, the old queen leaves the nest with a large swarm of workers to found a new colony, leaving the new queen in control of the old colony. Sometimes more than one young queen emerges and, if they do not fight to the death with other young queens, they may also leave the nest with some workers to found a separate colony. As with food sources, new nest sites are found by scout bees who then return to the swarm, where they perform a dance to indicate its position. The whole swarm then flies off to the new site. The death from old age or ill health of a queen stimulates the rearing of young queens but it also allows a certain number of workers to begin egg-laying as well. These workers lay eggs in drone cells and the males which are eventually reared are larger than the drones produced from queen-laid eggs. Thousands of worker-derived drones may be produced in queenless colonies. A small number of workers in queen-right colonies also lay eggs but the numbers are kept at a very low level, due to agonistic behaviour towards them by other workers; worker-derived drones are thus much rarer than queen-derived ones (Page Jr & Erickson Jr, 1988).

Stingless bees

The second group of bees to have evolved a high level of social behaviour are the stingless bees of the apid subfamily Meliponinae (Wille, 1983; O'Toole & Raw, 1991). They show enough differences in their social structure to be discussed separately from the honey bees. One important difference between the two groups is that, whereas the latter use progressive provisioning for their larvae, the stingless bees use mass-provisioning in the manner of solitary bees. It takes several workers to fill a cell with the necessary food and their gathering attracts the attention of the queen to it. Once the cell is full, one of the workers lays an egg in it, whereupon the queen eats both the egg and a little of the food before laying her own egg in place of the original one. This is the queen's main method of feeding, though she does get some food directly from the workers as well. Not all stingless bees lay eggs which are then eaten by the queen, i.e. trophic eggs, while, in some species that do, the eggs are actually bigger than those laid by the queen herself. Having laid her egg in the cell, the attendant workers then cap the cell. Division of labour within stingless bees seems to be much the same as that in honey bees, relating to the development of the royal-jelly-producing head glands, though the timing of each age-related function differs somewhat. The age order of tasks for stingless bees in general goes from incubation and repairs of the brood chamber through cell construction and provisioning, nest-cleaning, feeding of queens and young adults, reception of nectar and guarding the nest entrance and, finally, to foraging. Stingless bees as a whole are less capable of controlling nest temperature than the honey bees but, in those that can, the same methods, of wing-shivering to elevate temperature and wing-fanning to lower it, are employed.

Communication by foraging stingless bees of the sites of food sources differs markedly from that of honey bees, based as it is on scent. The worker of a typical stingless bee, on finding a suitable food source, makes several forays between it and the nest, seemingly checking that what it has found is indeed worthwhile returning to. Then, using secretions from its mandibular glands, it marks a trail between the food and the nest, stopping every few metres to place a spot of scent on a leaf, stone, twig or other suitable object. On arriving back at the nest, other workers then follow the scent trail to the foraging site, though it is believed by some researchers that the scout worker actually leads the foragers back and forth for a number of trips. Distances between scent marks vary between species, with a noted range of from <2 m to >30 m, with some indications of a greater amount of scent being deposited nearer to the food source. The advantage of scent-mediated communication is that it allows the foragers to assess the height of a food source, information not available through the honey-bee system. In some species, workers are alerted to food sources when the returning scout makes strong sounds and runs around in zigzag fashion, frequently butting other bees. Stingless bees of the genus *Melipona*, however, do not use scent-mediated communication but, instead, use a method considered by some to be intermediate between the other stingless bees and the honey bees. On returning to the nest, the *Melipona* worker makes a buzzing sound whose length indicates the distance of the food source from the nest; the longer the buzz the further away the food. She then flies part of the way towards the food, to give the other foragers the direction, before returning to the nest. Reduce this short flight to a walk upon the comb, pointing in the direction of the food source, and one has the dance of the honey bee.

▲ A battle between workers from two different colonies of *Trigona* stingless bees in Trinidad; here the workers are actually killing one another.

The reproductive cycle of stingless bees shows some differences within the group as a whole, as well as from honey bees. For all but the *Melipona* bees, the rearing of queens and workers is carried out by the workers, with the production of queens determined by the workers. In *Melipona*, however, up to one-quarter of the females reared are queens. In both *Melipona* and some species of *Trigona*, virgin female queens are kept trapped, either within a small chamber or roaming free within the nest. This is apparently an insurance against the death of the old queen. As new queens are raised, so the workers kill the imprisoned ones. The founding of a new stingless-bee colony is initiated by scout workers who first find a suitable site not too far from the parent colony. Workers then carry building materials and honey from the old colony to this site, where they build the protective covering for the nest and construct a few food pots and some brood cells. A virgin queen from the old nest now leaves for the new nest, accompanied by a small swarm of workers. Males from surrounding colonies then fly around the new nest and the virgin queen leaves it on a mating flight, before returning to begin laying. Thus, in stingless bees, the new nest is founded by a young queen, a consequence of the fact that the old stingless-bee queen, unlike her honey-bee counterpart, has lost the ability to fly.

Feeding Behaviour

The term 'feeding' encompasses a number of different activities. The animal has first to find its food, either by living on it, by seeking it actively, by waiting for it to approach or, in some instances, by letting an animal of a different species provide it. At some time during this stage, the palatability of the food may need to be assessed by smelling, tasting or even handling it. Once the food is identified as edible it then has to be dealt with in some way in order to render it into a state suitable for ingestion into the alimentary canal. It is the first of these stages which involves some extremely interesting behavioural adaptations.

◀ Crab spiders do not produce a web but are 'sit-and-wait' predators and, unlike web-building spiders, they do not store food for later use. This means that whereas a web-builder will desert its meal to rush off and kill a newly arrived insect, adding it to a 'larder' within the web, any insect which shows up while a crab spider is feeding will be immune to attack. This male *Eristalis arbustorum* hover fly is therefore in no danger of being killed as he walks around, and even over, a female *Misumena vatia* spider (Thomisidae) in England. He is attracted by the death whines emitted by a conspecific female pierced by the spider's fangs.

PLATYHELMINTHES Flatworms

Terrestrial planarian worms are predacious, feeding on other invertebrates, such as slugs and earthworms. In 1990, a new species of planarian from Kenya was described (Jones *et al.*, 1990) which takes an unusual prey. *Microplana termitophaga*, as its name implies, feeds upon termites. The worm lives at the mouth of the ventilation shafts on the termite nests and stretches itself down into these, waving its head end around until it contacts a termite. The latter sticks to the mucus produced by the worm, which quickly retracts its head end, forms a loop around the prey, and then begins to feed; the husk of the prey is discarded when feeding is complete.

▼ Terrestrial planarian worms are predacious, feeding on other invertebrates, such as slugs and earthworms. Illustrated is an unnamed species of flatworm from Trinidad, which is feeding on the eggs of a water snail, themselves unusually laid on land.

CRUSTACEA Woodlice, crabs

Platyarthrus hoffmannseggi are small, blind isopods which live almost exclusively in ants' nests. A study of their behaviour indicates that they rely upon the ants as a source of food, at least in part, for when ants are supplied with a diet containing the dye neutral red, the latter appears in the guts of the woodlice within 3 days (Williams & Franks, 1988).

▼ Like most of their kin, terrestrial crabs are scavengers, eating anything edible that comes their way. *Coenobita compressus* is a terrestrial hermit crab from Costa Rica, here seen feeding on driftwood.

ONYCHOPHORA Velvet-worms

Macroperipatus torquatus forages nocturnally on the rainforest floor on the island of Trinidad where its main prey are crickets and a few other invertebrates (St J. Read & Hughes, 1987). As it forages across the forest floor, it continuously sweeps the head from side to side, normally reacting to the presence of prey at about 1 cm up to a maximum of 4 cm. By means of slow, steady movements, the peripatus is able to approach most prey items undetected. Once contact is achieved, the prey is examined by delicate applications of the antennae, which are withdrawn immediately each time they touch it, thus leaving the prey undisturbed. Examination of the prey varies from the perfunctory to the prolonged before either it is attacked or the peripatus moves on. In attacking prey, *M. torquatus*, like other Onychophora, first enmeshes it in a sticky glue squirted from the oral papillae. The squirting action is very abrupt, the streams of glue being invisible to the human eye, and the prey is suddenly seen to be covered in an entangling mesh of threads. *M. torquatus* normally releases the glue at a distance of about 0.5 cm from immobile or slow-moving prey though, when artificially starved, it may be prepared to attack moving prey at distances of up to 4 cm; at longer distances, attacks tend to be less successful. One squirt is usually sufficient to immobilize the prey though extra squirts may be directed at the legs of more active individuals. When the most dangerous of prey, i.e. spiders, is attacked, up to 30 extra squirts of glue may be employed, with a number of them directed directly at the spider's jaws. Harmless prey, such as woodlice, are sometimes just grasped in the peripatus's jaws without the deployment of the glue. Once the prey is trapped, it is immobilized further when the peripatus bites through one of the arthrodial membranes and injects it with saliva. While the peripatus is waiting for the saliva to digest the prey, it feeds on the strands of glue, from both the surface of the prey and the substrate around the prey. The glue is a protein and its total loss would mean that an unacceptably high price would be paid by employing it in prey capture.

DIPLOPODA Millipedes

In Paraguay, a chelodesmid millipede has been recorded as associating with the leaf-cutting ant *Atta sexdens rubripilosa*. The millipede apparently feeds on the discarded debris from the ant's fungus gardens and uses the ant's trail pheromones to locate new nests (Fowler, 1981).

ARACHNIDA Scorpions, spiders, mites, etc.

SCORPIONES Scorpions

Observations on the hunting behaviour of scorpions pose difficulties since the majority of species are nocturnal while those that are known to be diurnal tend to live in the gloomy depths of tropical forests. From a number of purely qualitative studies made over the years, it would appear that scorpions either use a 'sit-and-wait' method of hunting or wander around actively and catch anything that comes within reach. In heavily built species with large pedipalps, the prey is usually subdued physically, with the sting only employed occasionally if the prey struggles a great deal. Slimmer species, with lightly built pedipalps, seem always to sting their prey, as do the earlier developmental stages of some of the heavily built group.

The hunting behaviour of the vaejovid scorpion *Paruroctonus mesaensis* has been studied both in the wild in California (Polis, 1979) and in the laboratory (Brownell & Farley, 1979). Scorpions have the advantage that they fluoresce under ultraviolet light so that they can be observed during their nocturnal foraging with little disturbance of them or their prey. *P. mesaensis* is a 'sit-and-wait' predator which emerges at dusk and then generally travels a short distance before adopting a motionless stance at its chosen foraging station. The greatest numbers of individuals are on the surface early in the night but, as time passes, numbers fall and, when dawn comes, the stragglers that are still present retreat into their burrows. *P. mesaensis* can detect and orient to small arthropods when they are at a distance of up to 0.5 m. The scorpion responds to the presence of prey by turning towards the direction of the stimulus and then advancing with its pedipalps held out in front. Once the prey is encountered, it is grasped in the pedipalps; amazingly, aerial prey is also taken in mid-flight. The researchers were able to prove that the scorpion oriented and moved towards the prey as a result of vibrations transmitted through the sand substrate of the desert floor. Of the prey animals taken, 80 per cent were cursorial, 10 per cent fossorial and 10 per cent aerial.

One of the fossorial animals taken by this scorpion is the burrowing cockroach, *Arenivaga investigata*, which moves around a few centimetres below the desert surface. Having detected movement beneath the sand, *Paruroctonus* manoeuvres until it is above the cockroach and then pushes its pedipalps down until it contacts and is able to grasp the cockroach. Once prey has been caught, the way it is handled varies somewhat. Small, helpless items are consumed immediately whereas active, struggling prey is usually stung. When an animal such as a thick-cuticled beetle is caught, the scorpion probes around with its sting until it finds a soft area, such as a joint, through which to inject the venom. All prey is normally eaten head-first. Interestingly, whereas adults usually consume the prey where they catch it, immatures often carry it up on to adjacent vegetation before feeding. The percentage of the diet formed by a particular group of prey organisms tends to change in year groups. (*P. mesaensis* matures at between 19 and 24 months.) For example, scorpions (including other *P. mesaensis*) are a major part of the prey of adults, making up 24 per cent of the diet, whereas in 1–2-year-olds they make up 18 per cent and in 0–1-year-olds only 8 per cent of the diet.

The feeding behaviour of two other species of scorpion have been studied in the laboratory. The giant among scorpions, *Pandinus imperator*, a west African rainforest species, orients to and catches its prey (Casper, 1985) in much the same way as outlined for *Paruroctonus mesaensis* above. The adult scorpion is so large (up to 17 cm from tip of chelicerae to end of sting) that it is apparently able to subdue all prey with its stout pedipalpal chelae, the sting never being used. Observations on members of a growing family of this species indicate that, when they are small, they tend to use the sting to subdue prey but, with increasing size and age, the sting is used less and the pedipalps more and more.

In a study (Bub & Bowerman, 1979) of the prey-catching technique of the desert scorpion *Hadrurus arizonensis*, the animals were maintained in a terrarium on a 15 cm deep substrate of sandy soil. The prey offered was either the American cockroach, *Periplaneta americana*, or the house cricket, *Acheta domesticus*. When hunting, the scorpions take up an alert stance with the body slightly off the substrate but with the tips of the pectines maintaining contact with it. The tail is curled up over the body and the pedipalps held forwards with the tip of the movable finger of the chela in contact with the sand. In this stance, *Hadrurus* is able to detect moving prey and orient towards it; if, however, the prey stops moving, then the scorpion will freeze also. Further movement of the prey close to the scorpion will elicit a lunge towards it and, often as not, the former is successfully grasped in the chelae. Each time *Hadrurus* captures a prey item, it always stings it at least once, a soft part of the body normally being searched out into which to inject the venom. This relates well to the fact that this species has relatively lightly built pedipalpal chelae.

SOLIFUGAE Sun spiders or solifuges

Our knowledge of the feeding behaviour of this arachnid group is somewhat limited, probably because the majority of them are nocturnal, spending the heat of the day in their burrows in their mainly desert habitats. Some species stalk their prey, especially where this is at a high density, e.g. around termite nests, while others are known to climb trees, with the aid of special suckers on the pedipalps, and hunt their prey. In Texas, two species of *Eremobates* often enter people's homes to hunt, possibly drawn there by light, and a number of African desert species have been reported as feeding on insects drawn to the light of camp fires. They are voracious hunters and some of the larger species have been seen to capture scorpions, large spiders and even vertebrate prey in the shape of small birds, rodents and lizards.

One diurnal species, *Metasolpuga picta*, has been studied in detail (Wharton, 1986) in the field in the Namib Desert of South-West Africa/Namibia. Solifuges are very active animals and most of the time when they are out of their burrows is spent in foraging, though males may spend most of their time searching for mates. Even in daylight, they are not that easy to observe because they move very rapidly and are very cryptic. Wharton was able to monitor the foraging activities of 65 *M. picta* in the field, 15 of which he was able to follow continuously for at least 15 minutes.

It turns out that females and immature individuals move rapidly around their habitats in a zigzag fashion, investigating all places where prey might lurk, such as burrows, pebbles, clumps of dead grass and larger stones. An immature individual, observed for as long as 12 hours, used its second pair of legs for digging and the palps and fore legs for sensing as it foraged. During its foraging, it dug up four small invertebrates adjacent to four different pebbles and ate them. The immature and female *M. picta* either eat or attempt to eat every invertebrate that they encounter whereas males often run into potential prey but ignore it. Prey is located by direct contact, either when it is run into or during the investigations of pebbles etc. outlined above. It was not possible to determine whether any prey is taken when they enter the odd burrow, though it seems possible, and indeed only a single tettigonid was positively identified as prey in the field. The tettigonid in question was a *Comicus* sp., which is nocturnal, spending the day concealed in a burrow. One *M. picta* discovered the general area in which one of these crickets was hidden and commenced to dig a series of holes around it. Before the solifuge could catch it, the cricket emerged from its burrow and jumped away to safety. Holes were similarly dug around another cricket but, in this instance, it was somehow detected and the solifuge hauled it out of the burrow using its chelicerae and palps. It then chewed the cricket for about 15 seconds before taking it into a hole to consume it.

Since there were no visible signs to the human observer that the crickets were present beneath the surface it would seem that *M. picta* is able to detect its prey by vibrations and possibly by chemoreception. The malleoli are known to be chemoreceptors and they are kept in contact with the ground as the solifuge forages.

ARANEAE Spiders

The Araneae are a group of exclusively predatory animals and, as a result, they have evolved some very sophisticated methods for prey capture, notable among these, of course, being the familiar silken snare, which, as we shall see, can manifest itself in many different ways. In dealing with the finding and capturing of prey by spiders, we find that there are a large number of generalists, who share common techniques to achieve this end, and a number of very interesting specialists with unique behavioural adaptations.

MYGALOMORPH SPIDERS

The more primitive mygalomorphs can basically be subdivided into two groups: those that sit in wait for and ambush their prey and a lesser number which move around and actively seek prey. Of the ambushers, many sit at or inside the entrance to their burrow, which may have a lid in the form of a trapdoor, hence their common name of trapdoor spiders. Sitting in this position, they are able to detect the soil-borne vibrations of approaching prey or, in those species whose front legs protrude from the entrance, air-borne vibrations also. When the prey is within reach, they dart from the burrow and pounce upon it though such spiders seldom completely vacate the burrow to catch their prey, which gives them a limited feeding area directly round the entrance. Studies made on the American trapdoor spider *Ummidia carabivora*, which sits beneath its fractionally open trapdoor with its legs in contact with the undersurface, indicate that the prey may, in some instances, actually need to walk on the trapdoor before the spider can respond, 95 per cent of possible prey which walks past being completely ignored. Once detected, however, prey is captured very efficiently with only a small percentage of the spider's strikes missing their target once the prey has touched the trapdoor (Coyle, 1981). At all times, the spider's abdomen remains within the burrow, with the third and fourth pairs of legs anchoring it to the wall. The prey is grasped by a combination of the pedipalps and the front legs which are then used to hold it as the spider backs into the burrow. In such spiders as *Ummidia*, it would seem that the choice of burrow site is of great importance, since it needs to be in a position where there is a good chance that it will be stumbled upon by passing prey.

One family of mygalomorphs, the Theraphosidae, which includes the so-called 'bird-eating spiders', have overcome this seeming aversion to leaving their burrows and will do so to hunt in the area immediately around them. Many of them do still construct a burrow; others live in tunnels under stones while some of the tropical forest species have taken to living above the ground in the trees, where they occasionally take young or sitting adult birds from their nests.

The limited catchment area of some of the trapdoor spiders has been considerably increased in those species which set up a radiating system of trip-lines around the entrance to the burrow. In some species, these consist of lines of silk but the twig-line spider, *Arganippe raphiduca*, from Australia, uses small twigs, leaf stalks and other debris to form the trip-lines. As with *Ummidia*, these spiders sit at the burrow entrance but, in this instance, with their legs in contact with the inner ends of the trip-lines. When a prey item stumbles upon one of the lines, the spider rushes out and grabs it and then returns with it to its burrow, presumably finding the entrance by making its way back along the trip-line.

In theory, single trip-lines could be bypassed by any prey which happened to walk in the gaps between them, thus, the more trip-lines, the more efficient the system. It is only a short step from this to the full funnel or sheet web built by other mygalomorphs, such as the notorious Sydney funnel-web spider, *Atrax robustus*. Arthropods which walk on to the silken sheet are immediately detected and captured by the spider.

A group of specialists within the mygalomorphs are the so-called 'purse-web' spiders of the family Atypidae. Like the trapdoor spiders, the atypids construct a burrow which they then line with silk but here the similarity ends, for the latter extend the tube beyond the burrow entrance across adjacent surfaces. In the European *Atypus affinis*, the extension is relatively short, a few centimetres or so across the surface of the ground and the end is sealed to form the 'purse'. In the American *Atypus* (*Sphodros*) species, the tube is continued vertically up the trunk of an adjacent tree and, in both these and in *Atypus*, the aboveground portion of the tube is camouflaged with suitable material obtained from the surroundings. These spiders lie in wait within their 'purse' until a wandering arthropod walks across its surface whereupon it is pierced in the belly through the wall of the tube by the spider's formidable fangs. The basal segment of the jaw is furnished with a set of sharp teeth and these are then used to rip a hole in the wall of the tube and the prey is dragged inside to be subdued and eaten. The spider usually repairs the damage before feeding commences.

▲Harvestmen (Opilionidae) are voracious carnivores, feeding on any manageable prey which comes within their reach. This individual is feeding on a fly in forest in the Smoky Mountains of Tennessee.

ARANEOMORPH SPIDERS

The araneomorph spiders, like their main prey, the insects, have colonized most of the earth's available ecological niches, though the exploitation of freshwater habitats by spiders has never approached that of insects. Accordingly, it is within the araneomorph, or true, spiders that the greatest number of adaptations for prey capture are to be found and it is fashionable to group the various spider families into a number of divisions on the basis of their prey-capture techniques. Thus, there are:

(1) *Short-sighted hunters*, small-eyed spiders many of whom are nocturnal or live under stones, bark etc., which wander around their habitat stumbling upon prey by accident as much as by design.

(2) *Long-sighted hunters* which have traditionally included the wolf spiders, the jumping spiders and the lynx spiders and have always been thought of as actively going in search of prey using their large, forward-looking eyes. There does, however, appear to be a progression in the degree to which hunting actively takes place within these families, with the wolf spiders being more inclined to sit around and wait for prey to come to them while, at the other extreme, the jumping spiders actively wander around searching for prey.

(3) *Ambushing spiders*, typified by the crab spiders, the Thomisidae, which sit motionless, often camouflaged, in an exposed situation such as a flower and wait for prey to come to them.

(4) *Web-building spiders*, all of which build some form of web whose primary function is to capture prey.

Within the active hunters, there is obviously plenty of scope for the evolution of various prey-catching behaviours. Within the web-builders, however, the web catches the prey and the spider then has simply to run and subdue it. In a number of species, however, the web has been secondarily reduced or modified so that there is a useful saving on silk, but this means that the spider has had to change its behaviour accordingly. Thus, in the discussion on web-building spiders (page 235), the emphasis will be upon these particular species.

The hunters

Spiders of the family Lycosidae, usually referred to as wolf spiders, can be diurnal or nocturnal and may be tied to a burrow or may be free-living. Of the species which dig a burrow, some hunt from its entrance, pouncing on passing prey in a manner akin to that of mygalomorphs, while others leave their burrows to hunt, using them as a retreat during periods of inactivity. Lycosids have keen eyesight which, especially in diurnal species, enables them quickly to detect moving prey and they also possess a tapetum which presumably improves the visual capabilities of those which are active nocturnally. Wolf spiders of the genus *Pirata* live in the vicinity of water and are able to run on its surface and take prey in much the same way as some Pisauridae.

Although diurnal wolf spiders are traditionally thought of as running around actively in search of prey, from personal experience this does not seem to be the case for most British species. Rather, they seem to be wait-and-see hunters, resting quietly in one place until prey comes within their visual range and then pouncing on it. Quantitative evidence for this type of activity is available for the very common British wolf spider, *Pardosa amentata* (Ford, 1978). In a laboratory set-up, Ford found that this species does indeed adopt a sit-and-wait strategy with a periodic change in its foraging site. The frequency at which a new site is adopted increases with increase in ambient temperature and, what is more, the time spent at the site increases when food has been consumed there. In fact, the longer the time taken to digest the food, the longer the time spent at a particular site. The total time actually spent in motion is, however, very small for, even at the highest temperature used in the experiment, *P. amentata* was active for a mere 0.0032 per cent of the day.

Since they are diurnal hunters, spiders of the family Pisauridae are sometimes also referred to as wolf spiders. *Pisaura mirabilis* is very much a wait-and-see predator, leaping upon any suitable prey that comes within reach of its perching place, usually the leaf of some suitable herb or shrub. Occasionally, it may actually leap into the air and take an insect in mid-flight (pers. obs.). Some pisaurids are, however, specialists and live on floating vegetation or along the sides of ponds, lakes or streams where they feed upon water-borne prey. The European spider *Dolomedes fimbriatus* sits on a floating leaf with its front legs touching the surface of the water. In this position, it is able to detect the vibrations of insects which have fallen in and are struggling on the water surface. The spider is then able to run across the surface of the water to grasp the prey, returning to a leaf to feed.

The American species *Dolomedes triton*, the fishing spider, is so-called because of its ability to take insect larve, tadpoles and fish from below the water. That this species uses surface waves on the water to detect and home in on submerged and floating prey has been adequately demonstrated (Bleckmann & Lotz, 1987). They found that *D. triton* is able to pinpoint precisely the centre of a series of concentric waves, produced by a fish at the surface, up to a distance of 18 cm or so. Despite this, the spider is not that efficient at catching its prey, with only a 9 per cent success rate when responding to a surface-wave stimulus. Accidental contacts, can be even more successful, with a 16 per cent success rate. Some fishing spiders have been seen to dabble their front legs below the surface of the water to act as a lure to small fish. Whereas *Dolomedes* is diurnal, the closely related *Trechalea magnifica* from Costa Rica is nocturnal and has been reported as feeding upon freshwater shrimps, which it takes from small streams (Van Berkum, 1982).

The most active of the spiders are the jumping spiders (Salticidae) and most, but not all, of the lynx spiders (Oxyopidae), the majority of both families having a tropical or subtropical distribution. Just how active the jumping spiders are may be ascertained by simply watching one of the common zebra spiders, *Salticus scenicus*. These little spiders, with their characteristic black and white striped abdomen are common around human habitations in both Europe and the USA. They are active on warm summer days when they are in almost constant motion, interspersing periods of walking with quicker spurts, stopping every now and then to turn their heads from side to side in search of a likely victim. The only time they are likely to be found motionless is when they are consuming a prey item. This sort of behaviour is true for most salticids wherever they are found. Despite their small size (most are less than 15 mm in length) they are able to form clear images

▼ The jumping spiders (Salticidae), like this *Hyllus* sp. feeding on homopteran prey on Mount Kinabalu in Borneo, are active hunters. They use their large, forward pointing eyes to focus on their victim and can then jump several body lengths onto it.

◀ **Although spiders normally feed on other arthropods, some, including of course the bird-eating mygalomorph spiders, will also take vertebrate prey. The ctenid *Cupiennius coccineus* from Costa Rica, seen here with the tree frog *Hyla ebraccata*, habitually takes small frogs as prey, as well as the usual range of insects such as katydids.**

of prey, using their large main eyes, from several centimetres away. Once the prey is focused upon, the spider can then make its characteristic leap upon it, though it always runs a dragline of silk out behind it as a precaution should it miss its target and fall. Their ability to jump long distances, as much as 40 times their body length, appears to be due to their small size and correspondingly light weight rather than to any special development of the legs, such as is found in the Orthoptera. Salticids are capable of capturing prey many times heavier than themselves, even managing to subdue large jumping insects, such as katydids, and this can only be the result of a very potent, quick-acting venom. Some South-East Asian species have perfected a special prey-capture technique: they leap on to orb-web spiders as they sit at the centre of their webs and consume them on the spot.

There are, however, a number of salticids which build large webs for catching prey. Two such salticids, *Simaetha paetula* and *S. thoracica*, come from Queensland and their prey-catching behaviour has been studied both in the field and in the laboratory (Jackson, 1985). *Simaetha* constructs a space web similar to that produced by spiders from other families and Jackson himself admits that these webs were often mistaken for those of dictynids when they were noticed for the first time in the wild. Behaviourally, *Simaetha* is similar to such spiders in that, when prey lands in its web, it moves rapidly across the silk to grasp or jump on to it. Two species of *Spartaeus*, *S. spinimanus* from Singapore and Thailand and *S. thailandicus* from Thailand, construct large sheet webs on the surface of trees (Jackson & Pollard, 1990). Their main prey is moths which they capture either on or close to the web by walking up to them and then lunging at them. Their success in prey capture was found to be enhanced by the presence of the web.

On the basis of observations upon the behaviour of the Oxyopidae, which are generally active hunters on vegetation and do not use a web in prey capture, the hypothesis was put forward that they have evolved from a web-building ancestor. Such a web-building oxyopid has now been discovered in Costa Rica (Griswold, 1983). This spider, *Tapinillus longipes*, builds a web not unlike that of some linyphiids, with a space web below which there is a sheet web. The spider hangs below the sheet web and runs to the underside of any insect which lands upon it. It then bites the prey through the sheet, the duration of each bite relating to the size of the prey, i.e. larger prey was bitten for a longer period. The prey is then pulled through to the underside of the sheet, grasped in the chelicerae and dragged off to the spider's retreat where it is fed upon without wrapping and then discarded.

Web-building spiders

Although it is now believed that, primitively, silk was used by spiders to line their burrows and to protect their eggs, its use in the making of prey-capturing webs has become of paramount importance in many spider families. Ground-dwelling species live under stones or within burrows and thus any silken trap tends to be of a two-dimensional design; those spiders which live above the ground on vegetation, however, often produce highly three-dimensional webs.

Despite the plethora of individual aerial web types in existence today, they are all based upon three general designs; thus there are three-dimensional space and scaffold webs, two-dimensional sheet webs, and orb webs, all with their many modifications and derivatives. The space web is the most primitive of these and, in its simplest form, consists of a three-dimensional tangle of threads whose purpose is to entangle flying or jumping insects for a long enough time for the spider to get to and immobilize them, by biting, throwing silk over them or swathing them in silk. This type of web has been improved in the Theridiidae by the addition of viscous lines. In other spiders, such as the Linyphiidae, the addition of a sheet of silk below the space web serves as a platform upon which insects fall and where they can be pounced upon by the web's occupant. It is almost certain that the orb web is a further sophistication of this horizontal platform, since the more primitive orb-weavers usually have horizontal webs whereas they are vertical in the more advanced species. The incorporation of spirals of sticky silk in the ecribellate spiders, and of hackled-band silk in the cribellate spiders, has improved the catching properties of these types of web.

There has been a good deal of argument over the years as to why it is that some of the orb-web builders, both cribellate and ecribellate, incorporate a stabilimentum into the centre of the web. One idea that has been put forward is that it enables flying birds, which would otherwise wreck the web, to avoid it, thus preventing the spider from losing an important resource. More recent investigations (Craig & Bernard, 1990), however, point to a different role for the stabilimentum. Primitive spiders produce a silk that has a high ultraviolet reflectivity whereas that of the modern orb-web builders has a low ultraviolet reflectivity. The silk that is used to manufacture the stabilimentum, however, has a high ultraviolet reflectivity. Experiments revealed that webs decorated with a stabilimentum and containing a spider intercept more prey per unit time than those which lack a stabilimentum and from which the spider has been removed. Furthermore, more insects are trapped on the side of a web containing stabilimentum silk than on the side lacking it. Many insects are attracted to flowers on the basis of their high ultraviolet reflectivity; thus, it is possible that the incorporation of the stabilimentum is effectively to mimic a flower and to attract insects to it.

The way in which these normal types of web function has been described many times in the literature so that the intention from now on is to describe how these basic web patterns have been further modified in some species to make them either prey-specific or perhaps to increase their efficiency, if that is possible.

Many of the modifications to the web relate to a reduced material investment in terms of the protein utilized. Clearly it is of advantage to a spider to be able to reduce the size of its web and thus the amount of silk it needs to invest in it, as long as its overall catching efficiency is not sacrificed. In some species, this apparent reduction in the use of silk may in fact represent the primitive state where full silk production to form an advanced web structure has not yet evolved. This may be the situation in the scytodid spider *Drymusa dinora* from Costa Rica. This spider dwells exclusively under logs where it makes use of natural tunnels and crevices in which to manufacture a simple space web consisting of just a few tangled threads. The spider sits in wait, roughly centrally in the web but towards the upper side. Small prey crawling beneath the log is attacked directly and bitten until it is immobilized. It is then carried to the spider's resting site and is consumed without prior wrapping. The spider is apparently able to appreciate the entry of larger prey into the web area for it responds by rapidly spinning a barrier across the tunnel or crevice at the end opposite to where the prey entered. Assuming the prey is unable to cross this barrier and turns back on itself, the spider then spins a barrier at the point of entry, thus trapping the potential victim. *Drymusa* attacks by making several bites with its chelicerae and then waits for the prey to become immobile, whereupon it wraps it and carries it to the resting place for consumption. Very large or strong prey is allowed to proceed unchallenged through the web area.

Whereas *Drymusa* perhaps represents a primitive level of silk usage, within the orb-web-building cribellate family Uloboridae are two genera which have also adopted a secondary reduction in protein use. The triangle spiders of the genus *Hyptiotes* employ just a 50–60° sector of the original orb web, while *Miagrammopes* has reduced its web to a single line. *Hyptiotes* sits to form a bridge between to separate lengths of silk. One length passes from her spinnerets to the twig behind her while the other is attached to the apex of her triangular web with the whole held taut by the front legs and usually a loop of slack held by the third legs. When an insect hits the triangular web, the spider allows it to relax and lets go of the loop held by the third pair of legs, while at the same time releasing more silk from the spinnerets. This act projects her towards the web and, as she moves, she now takes up any slack and then repeats the process, each pull and release effectively snapping the web shut and further entrapping the insect. Eventually, her movement into the web proper brings her into contact with the trapped insect, which is then wrapped in silk for later consumption. *Hyptiotes* has to replace her web completely once it has been used to take prey.

Miagrammopes sits in the same position as *Hyptiotes* but with just a single horizontal thread, the centre section consisting of combed-out cribellar silk. Again, the spider holds the thread taut with a loop of slack held by the legs and waits, in this instance, for an insect to select it as a perching place. The spider immediately releases the slack, trapping the insect's feet in the fluffy silk,

Variations in space webs. Whereas the space web originally evolved for the capture of flying insects, it has also been adapted for capturing cursorial prey on the ground. Webs of this type are constructed by theridiids, such as *Theridion saxatile* and *Steatoda bipunctata*. The space web is attached to low-lying vegetation and, from it, tightly stretched sticky lines run vertically down to the ground. When a walking insect crawls into one or more of these, it sticks fast and its struggles then cause the silk to break from its moorings, with the result that the prey is lifted bodily off the substrate. How effective this is may be ascertained from the report that the web of one of the house-dwelling theridiids, *Achaearanea tepidariorum*, was actually seen to lift a small mouse off the ground where, unable to escape its bonds, it died. The spider's reaction is to approach the prey, haul it up into the space web and then throw silk over it and wrap it before biting it.

The European *Episinus* species have reduced the type of web just described to almost the minimum possible, namely two vertical strands of sticky silk attached to the substrate. The spider sits in a position where it can grasp and subdue any prey which walks into the viscid lines. In this instance, the spider has reduced its outlay of protein in producing the silk but has reduced its overall prey-trapping area. The Australian and South American theridiids of the genus *Phoroncidia* have gone one step further in keeping just a single sticky trap-line which is suspended horizontally between two vertical structures and is used in much the same way as that of the cribellate *Miagrammopes*. It has been suggested (Eberhard, 1981a) that a sciarid fly may be attracted to the web of the Colombian species *Phoroncidia studo*. Theridiids of the genus *Euryopis* build no permanent web and, instead, hunt nocturnally, trapping prey by throwing silk over them and then wrapping them.

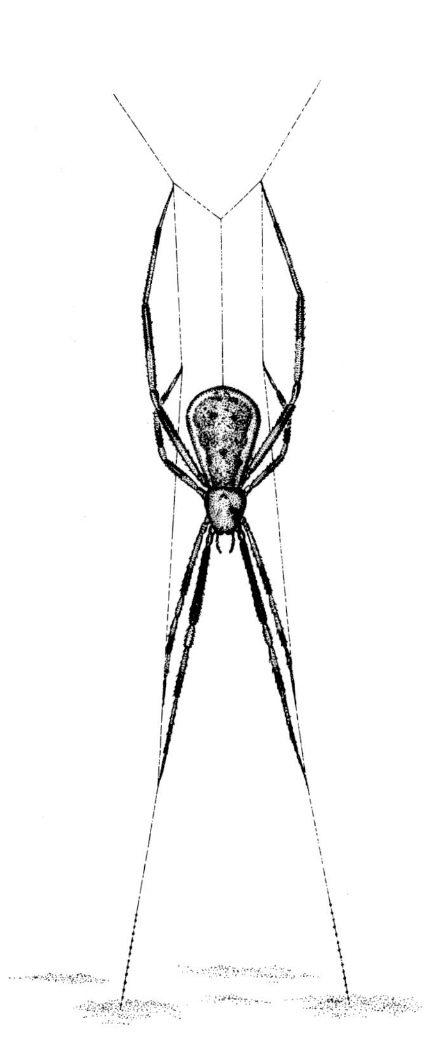

▲ ***Episinus angulatus* is a theridiid spider in which the web has become reduced to the bare minumum. The spider hangs head downward and grabs any walking insects, such as ants, which walk into and become stuck on either of the two lengths of viscous silk.** (After Bristowe, 1971).

whereupon she rushes over to it and wraps it. In a study (Opell, 1990) of the web of *M. animotus*, a close association was found to exist between the size of the spider and the web's effective trapping surface area. Older spiders tend to produce looped threads with a greater surface area and thus a greater area for prey retention. It was also revealed that cribellar-thread stickiness increases with spider size and that larger spiders tend to capture larger prey.

Araneid orb webs and their variations

The typical araneid orb web appears in a number of guises and, in some species, shows modifications similar to those already described for the cribellates and the theridiids. An orb web effectively gives a very large catching area for flying insects but at the expense of a large outlay of protein in the form of silk; therefore, it has to be used as efficiently as possible. There is evidence pointing to the fact that at least some flying insects can actually see spider webs quite clearly and can accordingly take steps to avoid them. Evidence for this has come from videotaping fruit flies and mosquitoes flying towards webs suspended across frames in a wind tunnel. This research has revealed that the insects are able to detect the webs from about 100 mm distance and individual threads at distances of 6–12 mm. If there is an unrepaired hole in the web then the insect will fly through it. Presumably as a response to the behaviour of these flying insects, orb-web spiders tend to make some positive choice as to the positioning of their webs. Diurnal species, which are active in good light, usually build their webs close to bushes and other structures, which act as a background, reducing the silk's visibility. Nocturnal species, or those which live in gloomy forests where light levels are low, do not have to build against such a background and accordingly can exploit a much greater volume of prey-trapping air-space.

Many, but not all, orb-web spiders sit in the centre of the web as they wait for some flying insect to collide with it. Sometimes the web is asymmetric with the hub above the web's centre and it has now been indicated that this is for the perhaps obvious reason that it takes the spider longer to run uphill to the area above the hub than it does for it to run downhill to the area below the hub. Once prey is trapped, orb-web spiders often vibrate the web to further entangle their victims. What happens to the prey after it has been captured and wrapped varies within the araneid orb-web spiders. Sometimes the prey is consumed immediately or is left stored at the capture site. Alternatively, as in the case of members of the genus *Nephila*, the prey is always removed to, and stored at, the hub of the web. It seems that this aspect of their behaviour may relate to the common presence of kleptoparasitic spiders within the web and is an attempt to remove the prey from their reach.

There are two araneids from Australasia with reduced orb webs which are believed to occupy a half-way point between the typical orb-web spiders and the bolas spiders. *Poecilopachys australasia* constructs a web at an angle to the vertical which at first sight appears to be a normal, if somewhat untidily formed orb web, but Clyne (1973) has observed that the spirals are not laid down as such, but are formed in single sections, the direction in which they are actually inserted often being reversed. When the sticky spiral is put into a normal orb web, it is always spun in the same direction from the centre out, i.e. clockwise or anti-clockwise, without any change in direction. *Poecilopachys* leaves a short length of dry silk where the viscid lines

▼ **The araneid spider *Pasilobus* sp. from New Guinea builds a web which is a reduced-area modification of the normal orb web. When an insect flies into one of the hanging loops it breaks off from its outer support thread leaving the insect dangling. The spider then rushes up and hauls it in to be dealt with.** (After Robinson & Robinson, 1975).

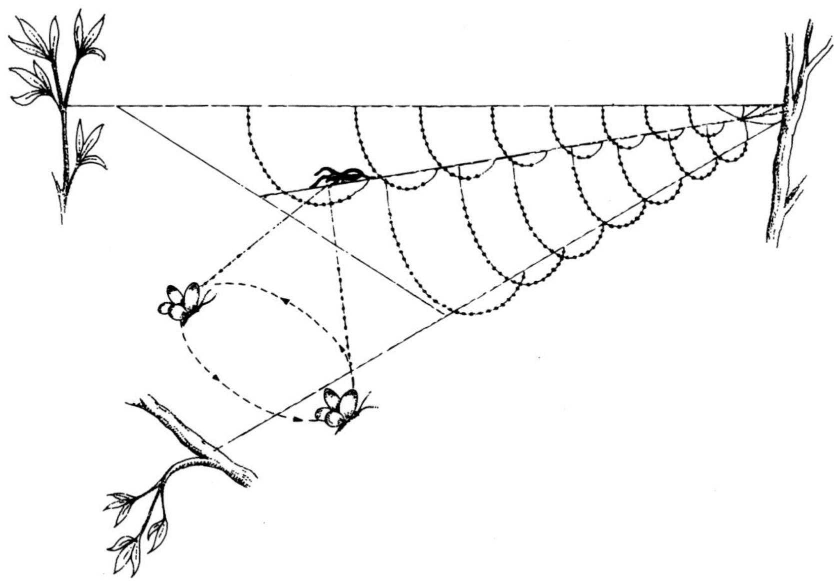

attach to the radii. When one of these viscid lines is hit by an insect, it may break at one of the radii, leaving the insect dangling but still capable of becoming further entangled on adjacent spanning lines if it attempts to fly. The spider takes its prey by hauling it up on the line on which it is trapped.

In the web of *Pasilobus* from New Guinea (Robinson & Robinson, 1975), there has been a reduction in the amount of silk used so that only three radii are present and thus it resembles a sector of the *Poecilopachys* web. Again the viscid spanning threads are laid down separately and they loop downwards below the plane of the roughly horizontal web. The spanning threads are held tightly only at their junction with the mid-line thread; at the other end, there is a less secure joint which breaks when an insect contacts the line, leaving it dangling in mid-air below the mid-line thread. The spider sits at the hub of the web, facing along the mid-line, and when an insect becomes trapped she runs along this line until she contacts the viscid line to which the prey is attached. She then pulls the prey up to within contact distance and bites it. It was found that the spanning threads were usually close enough to each other to ensure multiple entanglement of a strong-flying insect.

Wixia ectypa from the USA has been described (Stowe, 1978) as having a reduced web but of a different nature from that of *Pasilobus*. In this spider, it is the viscid lines that are lost and the radii which are retained and act as trip-lines. *Wixia* constructs its web at all heights in the outer branches of trees. The web is set up with eight radii as the norm and it is usually constructed near the horizontal in the fork of a branch. Stowe reports never having seen a web during daylight hours and assumes that the spider, which incidentally is highly cryptic, dismantles it at the end of its night's hunting. One interesting aspect of this web is the multiple attachment points of the radii but their function becomes obvious when considering the way in which the spider uses it. The spider sits at the hub of the web, which functions as a trip-line snare for arthropods walking along the branch on which it is built. The likelihood of prey triggering a trip-line is presumably increased by the multiple attachments of the radii. The normal response of the spider is to rush from the hub down the stimulated trip-line to the branch. She then runs backwards and forwards along the branch until she contacts the prey. Since the prey is not suspended in the web but is on a solid substrate, the spider now has first to immobilize it by rapidly circling the branch and tying it down and then by biting it. As soon as the prey has succumbed to the venom, it is cut free of its bonds, wrapped in more silk and then taken to the hub to be consumed.

Members of the family Theridiosomatidae have, in the past, been included in the Araneidae but they have certain features unique to them as a group. The European species *Theridosoma gemmosum* constructs an orb web but then bites out some of the meshes and strengthening spirals before the web is completed. The final product resembles a half-open umbrella with a thread leading away from the ferrule to an anchoring point. The spider sits on this tightly stretched thread, facing away from the web, and, when an insect becomes entangled, the spider suddenly lets go of the thread, further entrapping the victim. This method of prey capture is somewhat reminiscent of *Hyptiotes* but, whereas the latter has to rebuild her web after each take, this is not true of *Theridiosoma*. Within the same family are spiders of the genus *Wendilgarda* and their unique method of prey capture has been described (Coddington & Valerio, 1980) for three species found in Costa Rica, Guatemala and Honduras. The webs are built above moving streams and the viscid lines are connected to the water surface with each under some tension. Insects which have fallen into the water and are moving downstream on the water surface are trapped by these floating threads.

The specialists

A number of spiders from various families have developed prey-capture techniques which lie outside the general classification outlined above, so that they might be referred to as the specialists. Unique among spiders, since except during dispersal it spends its entire life submerged below the surface of static freshwater ponds and lakes, is the water spider, *Argyroneta aquatica*. *Argyroneta* constructs a thimble-shaped, silken lair attached to water plants and filled with air, which is replenished at intervals by fresh bubbles brought down from the water surface. The spider hunts by darting out on passing insect larvae or other arthropods, which are grasped in the front legs, bitten and then taken into the lair to be consumed. They will also take insects struggling on the water surface, indicating that they are able to detect the formers' vibrations, and, at night, they may leave the lair to hunt small arthropods, such as shrimps, on the bottom of the pond.

While some tropical members of the family Scytodidae build small sheet webs to trap prey, the family encompasses some species with a unique method of prey capture which has earned them the name of spitting spiders. The spitting spiders are nocturnal hunters which wander around their habitat taking any suitable prey which they encounter. *Scytodes thoracica* is one of those species associated with human dwellings and, as a result, it has spread over much of the world. *Scytodes* has a characteristic domed cephalothorax containing large glands which produce both gum and venom and are connected as usual to the fangs. Once it is within range, the spider squirts over the victim a double stream of gum which effectively anchors it to the substrate. As the spider spits the gum, it oscillates its head rapidly from side to side in order to spread the lines of gum across the prey. *Scytodes* usually then advances upon the victim with some caution, delivering a bite in one of its legs if it struggles and not commencing to feed until it is motionless.

Rather than waiting for prey to come to their webs, the net-throwing spiders of the family Dinopidae take their webs to the prey. These spiders manufacture a rectangular web on a scaffold of dry silk in a stretched state. When completed, it is cut free, whereupon it collapses down to about one-tenth of its original size. It is then held at each corner by one of the front two pairs of legs. Members of the genus *Dinopis* then make use of their net in one of two ways: they either hang head-down on a dry-silk web and drop it over insects walking beneath, or they can sit in the web and throw the web over insects as they fly past. Once an insect is caught, the web may be shaken to further entrap it before it is wrapped and consumed. *D. guatamalensis* in Costa Rica, according to observations made by one of the authors, may use some form of chemical signal to attract flying prey. The author set out early one evening with the intention of photographing the spiders at work. Eventually he found a spider of this species setting itself up in readiness for a night's snaring on its dry-silk web about 30 cm above the ground. No sooner had the spider positioned itself with its web than a moth appeared, where none had been seen before, and the spider captured it. Because the author had taken insufficient photographs, he took this moth away from the

▲ The ogre-faced or stick spiders of the family Dinopidae have evolved a novel method for catching prey. Instead of waiting for their victims to come to their web they take their web to the victim. This *Dinopis guatemalensis* in Costa Rica is hanging from its loosely-built scaffold web and is holding its 'net' of elastic silk in its front legs. From this position it can either scoop up any insect walking across the ground beneath it or enmesh it in mid-air as it flys past.

▼ The web of the araneid bolas spiders, such as *Mastophora*, has been reduced to the absolute minimum of a simple 'trapeze', from which to hang, and a bolas. The latter is a single line, swung by a front leg, on the end of which is a sticky globule. The latter elongates as the bolas is swung and sticks to any moth which passes by, effectively trapping it. The spider then hauls it in and feeds on it.

spider, which immediately re-positioned itself in its web with its net at the ready. Once again, a moth appeared seemingly out of nowhere and was caught by the spider. Since no moths appeared to be flying around before the spider was ready, but they turned up as soon as it was in the trapping position, it could be that the author's hypothesis of a chemical attractant has some basis and is clearly a case for further research.

Observations on *D. subrufus* and another dinopid, *Menneus subrufus*, in Australia (Austin & Blest, 1979) reveal a somewhat different catching technique for the latter species. Instead of taking insects from the ground (apparently the exclusive habit of the *Dinopis* which was never seen to take prey in mid-air but always from the ground), *Menneus* aligns itself so that it faces vertical stems of grass and twigs and captures prey walking up and down these. In the habitat studied, these two species were sympatric and the researchers were able to compare the prey taken by the two different techniques. The main prey of *Menneus* turned out to be blatellid cockroaches (92 per cent) whereas, although *Dinopis* took blatellids (23 per cent), it also took ants, other species of spiders and a range of other arthropod groups. One interesting aspect of feeding seen by the authors was when a male *Dinopis* opportunistically shared with a *Menneus* female a large cricket, which had been caught and was being consumed by her. In comparison with the Australian dinopids just described, *D. longipes* in Panama is more of a specialist, stationing itself exclusively above the trails of leaf-cutting attid ants (Robinson & Robinson, 1971).

The somewhat bizarre prey-catching behaviour of the bolas spiders of the genera *Mastophora*, *Dichrostichus* and *Cladomelea* was first described early this century but since then a great deal more has been learned about them. The behaviour of the Colombian species *Mastophora dizzydeana* (Eberhard, 1980) is representative of these spiders as a group though there are minor differences between this and other species. Hunting takes place from a horizontal trapeze line, spanning an open space, from which the spider hangs by the hind two pairs of legs. The bolas consists of a sticky ball on the end of a short length of silk which is held by one of the outstretched front legs. On the approach of a flying moth, the bolas is swung by the spider so that the sticky ball on the end comes into contact with the moth. Roughly 50 per cent of strikes are successful, in spite of the spider's very rapid reactions and the relatively slow approach of the moth. If the strike is successful, however, the moth inevitably sticks to the ball, whereupon the spider descends the bolas line, bites the prey and wraps it. The spider will then ascend the bolas line and, on reaching the trapeze, will begin to feed. On occasions, however, the spider hangs the prey on the trapeze without feeding and recommences hunting with a newly spun bolas. When hunting, the spider always positions itself with the underside of the abdomen downwind and, if the wind direction changes, then the spider will re-orient itself accordingly. That an attractant, volatile chemical is produced by the spider can be concluded from the behaviour of the moths which are caught (Eberhard, 1977). In more than 100 observations, it was found that the moths always approach the spider from downwind with their antennae extended; often several passes are made at the spider by an individual moth. In the wild, two species of moths, both noctuids, constitute the main prey of the spiders investigated and these are always males. One moth species in particular, *Spodoptera frugiperda*, is known to have a female which produces pheromones that attract males of the same species. It would appear, therefore, that here there is sound evidence to support the theory that *Mastophora* uses chemical attractants.

Production of the sticky ball on the end of the bolas takes between 1 and 2 minutes. This involves a line of viscid material being extracted from the spinnerets and compacted into a ball by the hind legs, with 80–100 pulls sufficient to complete it. Individual pulls are rapid at first but they slow down as the ball accumulates, completion of the ball apparently coinciding with exhaustion of the supply of viscid silk. Used balls, or those which have accidentally adhered to some object other than prey, are always eaten, as are those which have been held in readiness for prey which has failed to appear; the time during which a bolas appears to remain viable varies between 14 and 41 minutes. There is usually a delay of up to 1 hour before the spider manufactures a new bolas and resumes its hunting. The need periodically to renew the sticky ball relates to its structure. Internally it contains a mass of folded threads and these are then surrounded by layers of jelly-like material of differing viscosities, the outermost layer being very liquid. It is the drying-out of these outer layers which determines when the bolas needs replacing. Experiments showed that, when a newly made sticky ball is taken away from a spider and allowed to hang free for 90 minutes, it loses approximately 40 per cent of its volume and also its stickiness.

▲ A number of salticids mimic ants and use this as a ploy to get them close to their prey, the ants themselves. This is the case with this African *Cosmophasis* sp., seen here with its prey, which it resembles fairly closely.

At least two other araneids closely related to the bolas spiders are believed to produce chemical attractants but have dispensed altogether with the use of silk to capture their prey. The bird-dropping spider, *Celaenia excavata*, from Australia, sits on a leaf or on its egg sacs with its first two pairs of legs held out in a manner akin to that of ambushing thomisid spiders. Several observers have noted moths, always males, circling round and then homing in on these spiders, only to be grasped by the raptorial front legs and bitten. Similar behaviour has been noted (Eberhard, 1981) for a species of *Taczanowskia* in Colombia, except that, in this case, the spider hunts by hanging down from one or more horizontal lines, reaching out and grasping any moths which come within range.

Another aberrant araneid which has almost dispensed with a web is *Arcys nitidiceps* from Australia. In daylight, this spider uses a different hunting technique from the one it uses at night (Main, 1982). During the day, it sits crab-spider-like exposed on a leaf, where it has been seen feeding on flies. At dusk, *Arcys* constructs a trapeze line and then hangs from it by its hind pairs of legs with its front legs again held like those of a crab-spider. Spiders on their web have been seen with prey but there is no evidence to indicate that they catch these other than opportunistically, i.e. no pheromone attractant seems to be involved.

Ant-hunting spiders

A spider, when specializing in hunting ants as prey, has to take into account their pugnacity. A number of different techniques are employed by spiders to overcome this problem. Various species of jumping spiders mimic ants well enough to be able just to walk up to them and capture them without any resistance from the ant. *Cosmophasis* from Africa is a good example (pers. obs.), its abdomen bearing the same banded pattern as the ants on which it feeds. *Tutelina similis*, from the USA, is a

salticid which mimics ants of the genus *Camponotus* both morphologically and behaviourally (Wing, 1983) . Thus, apart from resembling the ant in its appearance, the spider moves its front legs in an antenna-like manner, this being especially marked when it is stationary. *T. similis* stalks its *Camponotus* prey while it is engaged in tending scale insects, the spider attempting to remain behind and out of the ant's visual field at all times. At an opportune moment, the spider rushes in, bites the ant and then immediately retires to a safe distance. The spider then waits until the ant is motionless from the effects of the venom before picking it up and dragging it away to a safe place to feed.

The spider *Aphantochilus rogersi* (Aphantochilidae) from Brazil is a very accurate mimic of cephalotine ants and is often to be found in the neighbourhood of their nests. This spider has been studied both in the field and in the laboratory (Oliveira & Sazima, 1984). In the area studied, the models were ants of the genus *Zacryptocerus* and, on two occasions, the spider was seen feeding upon these ants in the field. In the laboratory, *A. rogersi* would only take *Zacryptocerus* ants as prey, rejecting all other types of insect offered and, thus, it seems that this spider is possibly an obligate ant-predator. Two-thirds of attacks by the spider on ants in the laboratory were from the rear, the ant's petiole being grasped in the spider's jaws. The remainder were frontal attacks with the ant being grasped by the head or neck, the spider keeping the front legs well out of reach of the ant's jaws until it became immobilized by the spider's venom. The Paraguayan clubionid *Corinna vertebrata* is a good structural and behavioral mimic of the grass-cutting ant, *Acromyrmex landolti fracticornis* (Fowler, 1981). The spider lives within the ant's nest and also leaves with the ant when it goes foraging, picking off stray foragers as prey when it does so.

Salticids which are not ant-mimics also specialize in ant prey but it has been found (Jackson & Van Olphen, 1991) that their catching techniques show some variation. *Corythalia canosa*, from Florida, when hunting ants, displays a specific behaviour pattern which it does not use when taking other insect prey. Once the spider has caught sight of its intended victim, it becomes activated, running in short spurts to enable it to come head-on to the ant. It then either makes a short jump or lunges at the ant and grasps it by the thorax or head. During this time, the salticid arches its legs, keeping its body two to three times higher than usual above the surface on which it is hunting. If the ant is large, the spider will back off immediately after its first contact, attacking again a few seconds later. *Pystira orbiculata* employs two different methods of capturing ants, both of which are less specialized than that used by *Corythalia*. *P. orbiculata* either pursues its prey and leaps directly on to it from 4–10 body lengths distant or it stands head-down on the trunk of a tree and ambushes ants passing beneath it, keeping a dragline attached to the tree as it does so. *Pystira* also takes insects in the wild by the pursuit method but, interestingly, only ants are taken by ambushing, as far as has been observed.

Robinson and Valerio (1977) describe an unidentified salticid taking *Pseudomyrmex* ants from an acacia tree. This ant, when in danger, produces an alarm pheromone which attracts other members of the colony to its assistance. To overcome this problem, the salticid merely dropped in on the ant from above, grabbed it and then jumped off again, to remain suspended on its dragline while it dispatched its prey. Despite being attracted to it, the other ants were unable to help their captured fellow because they were unable to descend the dragline.

Whereas members of the Salticidae often seem to have associations with ants, it is rare to find theridiids which prey upon them. Two examples of ant-feeding theridiids have, however, been described from the USA, each employing a different trapping behaviour. The theridiid *Euryopis coki* is found on the mounds of the harvester ant *Pogonomyrmex owyheei*. *E. coki*, unusually for a theridiid, does not use a web to catch its prey but just attacks an ant directly (Porter & Eastmond, 1982), first anchoring it to the ground with silk and then biting it in the leg to subdue it. During the time that it takes for the poison to take effect, the spider may be attacked by other worker ants but it is able to entangle them with silk and drive them off. Once the ant is comatose, it is dragged away from the mound, attached to the spider's spinnerets, to be consumed in a safe place.

A second species from the USA, *Euryopis funebris*, has been recorded (Carico, 1978) as also feeding upon ants. This species, however, specializes in *Camponotus castaneus*, which forages on trees. The spider catches the ants by throwing silk over them and fastening them down to the tree bark.

Another theridiid, *Steatoda fulva*, has been recorded (Hölldobler, 1970) as feeding on another species of harvester ant, *Pogonomyrmex badius*. In order to avoid being attacked by its aggressive prey, the spider has adopted a behaviour pattern which relates to the ants' daily activity. During hot weather, when they are active, the ants emerge from the nest at about 08.00 hours and reach a peak of activity between 11.00 and 12.00 hours. Activity then begins to decrease so that, by 14.00 hours, it has virtually ceased, to recommence at about 15.00 hours with a further activity peak at 17.00 hours. It is during the ants' period of inactivity in the early afternoon that *S. fulva* approaches the depression at the centre of the nest area to construct its web, which it attaches to grass stems around the edge of the depression. The first ants to emerge become entrapped in the web but, as they struggle, they seemingly release their alarm pheromone, which attracts the immediate attention of other ants, who attempt to release their fellows from the silk. On finding this to be an impossible task, the ants then retire into the nest and seal the entrance, whereupon the spider, which up to now has stayed well out of harm's way, emerges, wraps its prey and begins to feed. It would appear that *S. fulva* can make use of the original entrance only once, for the ants then keep it closed and open up a new one some distance away. Initial research indicates that the spiders are able to detect the nests of the ant by smell.

Sociability in spiders

Sociability has evolved in a number of spiders from different families and, among the advantages, are joint efforts at prey capture. The degree of sociability varies from species that are facultatively social to those that are obligately so. The former exhibit communal behaviour in that two or more adults live together in a semi-permanent group but do not co-operate with each other in any way. Co-operative behaviour, on the other hand, is exhibited by those species which join in with each other in prey capture, food-sharing, parental care and construction of a communal web. In the uloborid genus *Philoponella* all stages from solitary web through to obligately communal are to be found within the individual species. *P. oweni* is a facultatively communal species from the southwestern USA and, within one habitat, females may be solitary or may live communally. There is no indication in this species as to what makes them choose the solitary or group existence, though there are differences between the two life-styles (Smith, 1982) which may help to provide an answer. In the same habitat, both solitary and communal females produce the same number of egg sacs though the latter do produce more eggs per egg sac. To counter this, the egg sacs of the communal females suffer significantly more from the attentions of the parasitic pteromalid wasp *Arachnopteromalus dasys*. The sum total of these observations is that the reproductive success resulting from the two life-styles is approximately the same. One further inference is that, since the communal females produce more eggs, they may be better at prey-catching than the solitary ones.

The Mexican araneid spider *Metapeira spinipes* is also facultatively communal and some of the factors which dictate whether a spider adopts a solitary or shared existence have been investigated (Uetz *et al.*, 1982). In this species, the abundance of prey and the severity of the environment seem to dictate whether or not a female remains solitary or joins up with other females. The general picture is that, where environmental conditions are harsh and prey availability is as a consequence low, the spiders tend to be solitary or to form only small groups. Where environmental conditions are more favourable and prey is more abundant, the spiders tend to be more sociable. In moist, warm, tropical sites, for example, where prey is abundant all year round, the colonies tend to be very large.

Experimental moving of colonies from areas of relative prey abundance to areas poor in prey indicate that the spiders make a deliberate choice as to whether they remain solitary or become communal. Thus, when a number of small colonies were moved in this way, the average group size fell significantly from 10 in the relatively prey-rich area to 7.75 in the prey-poor area and the nearest-neighbour distance nearly doubled. In a separate experiment, the move to a prey-poor habitat was compensated for in half of the colonies by placing mounds of cow dung near to them in order to attract flies. In the groups without cow dung there was again a significant decrease in colony size and nearest-neighbour distance but, with cow dung, there were no significant changes. *M. spinipes*, therefore, probably represents an intermediate in the evolution from the solitary species to the fully integrated co-operative species, with individuals able to choose whether they remain single or join up with other spiders as a response to environmental factors.

Members of obligately social species, in which many spiders occupy the same web in a co-operative fashion, have to be able to recognize each other's vibrations as they move around on the silken lines. It has been demonstrated (Burgess, 1979) that the web of the Mexican social cribellate spider *Mallos gregalis* (Dictynidae) attenuates vibration frequencies below 30 Hz and above 700 Hz. Frequencies within this range, which are ably transmitted by the web, relate to vibrations produced by the spiders' major prey items, flies. Vibrations transmitted through the web by a trapped fly elicit a response from all available adults who then immobilize it. Once the fly is comatose, it can then be fed on both by adults and by immatures incapable of subduing large prey on their own. The spiders themselves, and larger prey, tend to produce vibrations within the attenuated frequencies and thus do not draw attention to themselves.

In a study of another social spider, the Neotropical theridiid *Anelosimus eximius* (Christenson, 1984; Nentwig, 1985), it was found that this species, with its communal web, when compared with non-social web-builders in the same habitat, caught an unusually high number of large flying insects, which included cockroaches and grasshoppers. As with *M. gregalis*, several spiders are attracted to and help in the subduing of prey and the larger the insect and the stronger the vibrations it produces, the more spiders that attack it. Attacks are usually carried out by any adult females and a few older immature females within about 40 cm of the prey, as it falls from the upper strands of the web on to the sheet below. Silk is thrown over the victim as the spiders attempt to bite it and, in the case of very large prey, or prey such as ants, which are capable of biting back, the silk that the females use is stickier than that used on less agile or less dangerous prey. In both *Mallos* and *Anelosimus*, therefore, the social option allows spiders to exploit prey of a greater size than they could manage on their own, thus giving them access to a much greater spectrum of prey in the habitat in which they live. Although social spiders co-operate in prey capture, there may be some competition between individuals when it comes to feeding.

Individuals of the highly social southern African eresid *Stegodyphus mimosarum* neither defend any particular nest nor retreat within the communal web system and assist each other in prey capture. Once the prey has been subdued, however, struggles may occur between spiders to pull it into their own particular retreat (Ward & Enders, 1985). One consequence of this is that smaller prey may be torn apart in the process. Spiders which had not participated in the capture of prey tend to induce brief fights with the catchers as they try to muscle in on it.

Observations on the Indian species *Stegodyphus sarasinorum* (Bradoo, 1980) point to there being 'pilot spiders' within the community, individuals who have not fed for a while and who wait on the web for the arrival of prey items. Once prey has landed in the web, they rush to subdue it. As the prey struggles, they repeatedly tap and pluck the web with their hind legs to recruit other helpers to the prey, this behaviour being especially noticeable when pilot-spider numbers are low and they are a long way from the main nest area of the web. With large numbers of spiders attacking it at once and injecting venom, the prey is subdued very rapidly and there is no need for any wrapping, an overall saving in dietary protein for the spiders. Subdued prey is moved towards the nest area during the hours of daylight but, late in the day, it is consumed at the capture point in the web

▲ As well as males of her own species, the female of the large, Neotropical araneid spider *Nephila clavipes* may also entertain other guests in her large orb web. In this illustration, an individual of the kleptoparasitic theridiid *Argyrodes elevatus* has crept stealthily up and is sharing the host spider's prey.

area. Female spiders catch small insects and transport them to the nest area where they are fed upon by the colony's spiderlings.

Kleptoparasitic spiders

The members of one particular genus of theridiid spiders, *Argyrodes*, are recognized as being mainly kleptoparasites in the webs of other spiders, though some of them have retained the ability to construct and use their own webs if necessary. In order to avoid becoming a meal for their host spiders, these kleptoparasites have evolved a series of special behaviour patterns which allow them access to the host's web and food items. The foraging behaviour of the Central American *Argyrodes elevatus* has been studied in detail (Vollrath, 1979). *A. elevatus* is most commonly to be found in the webs of the araneids *Nephila clavipes* and *Argiope argentata*. These two spiders employ somewhat different tactics in subduing and dealing with prey items which may be related to the relative success that the kleptoparasite has in the respective webs. *A. argentata* wraps prey and either suspends it from the web at the capture site on the orb or, in some instances, transports it to the hub; *N. clavipes* always transports its prey to the hub.

A. elevatus does not construct a web of its own but instead hangs outside the prey-catching orb of the host's web, connected to the radii by fine signal threads which it has previously laid down. Vibrations from the host's web are transmitted to the kleptoparasite via these threads, because prey-wrapping movements by the host have been seen to initiate raids by *A. elevatus*. The latter seems to be aware of whose web it is in for, in a *Nephila* web, it searches at the hub of the web whereas, in an *Argiope* web, it searches both the hub and around the orb. In searching for a prey item, *A. elevatus* moves along very slowly and waves its long front legs both forwards and sideways whilst continuously tapping the host threads. These slow movements are clearly to reduce the likelihood of it being detected by the host but, occasionally, it is attacked, in which case it drops out of the host's web on a dragline. Although movement is slow when the host is motionless, if the host is actively attacking and wrapping prey, *A. elevatus* will move around much more rapidly. Once a wrapped prey packet is discovered, it then has to be removed from the host's web and even this is done with great care so as not to attract the host's attention. Incredibly, as the kleptoparasite cuts a prey packet out of the web, it holds the cut line in its front legs and only very slowly lets off the tension in the line. Presumably if it did not, the line would twang and alert the host. *A. elevatus* is seemingly attracted to still-living wrapped prey or newly trapped prey by its vibrations or by those of the attacking host spider. Once in the general area, it detects the prey by touch rather than scent. It can take *A. elevatus* as little as 1–2 seconds to find stored prey at the hub and only a further 4–10 seconds to cut it out and carry it away; prey up to ten times its own weight may be removed. On occasions, the kleptoparasite does not remove the prey packet but instead sits at the hub of the web and shares it with the feeding host spider.

A. elevatus is much smaller than its hosts and its daily food requirements could almost certainly be fulfilled by a few small insects trapped in their webs but these are probably ignored as being below their prey-size threshold. Despite this, the kleptoparasite removes far more food than it possibly needs and if, for example, a packet containing undigested prey is stolen, then it is abandoned and an alternative is searched for. There is some evidence that the host spiders are aware of the disappearance of prey for they start moving around the hub in what appears to be a search for it. *N. clavipes* has several times been seen to climb into the barrier web outside the hub and shake it and, as a result, has been able to recover a prey item which has been set swinging by this action.

The behaviour of the New Zealand species *Argyrodes antipodiana* shows both similarities to and differences from that of *A. elevatus* (Whitehouse, 1986). For a start, the New Zealand species can build a sticky space web of its own and use it independently to catch prey. Most of the time, however, it is dependent upon its host, the araneid *Araneus pustulosa*, as a source of food. When this type of space web is attached to a host's web, it is referred to as a support web and it plays an important role in the kleptoparasite's foraging behaviour. Like *A. elevatus*, *A. antipodiana* takes insects to which the host has not responded, as well as stealing wrapped prey and feeding with the host spider. One important difference, however, is that *A. antipodiana* indulges in a degree of araneophagy, by using aggressive mimicry to attract the host's spiderlings within catching range and by actually feeding on the host itself when it is at its weakest during moulting. (The Japanese species *A. fissifrons* also preys upon its main host, *Agelena limbata*, while it is moulting (Tanaka, 1984).) *A. antipodiana* also habitually eats silk, both from its own web and the host's web, often prior to the construction of a new web or the extension of an old support web. In approaching the host in order to feed, the kleptoparasite moves in a typical slow fashion, returning at intervals to the support web, laying a line of silk behind it as it does so. These lines serve as a reinforcement of the support web. It locates the position of the feeding host by probing gently with a front leg, the host rarely responding to being touched, and then, while itself feeding, the kleptoparasite always arranges to be on the opposite side of the web to that of the host. Prior to starting feeding, however, *A. antipodiana* always runs a strand of silk back to the support web and then attaches a dragline to this about 2 cm from the feeding site. If disturbed by the host during feeding, it can swing away out of danger on this line. Like *A. elevatus*, *A. antipodiana* also moves food items to other parts of the web. In catching host spiderlings, *A. antipodiana* uses aggressive mimicry in the form of tapping or plucking at the web in the manner of a trapped insect. The spiderlings respond by moving towards this disturbance, behaving as if they were searching for the supposed insect but, once they are within range, *A. antipodiana* lunges forwards and pulls the spiderling victim into its waiting jaws.

The minute symphytognathid spider *Curimagua bayano*, which does not exceed 1.3 mm in length, has been reported as living in association with a mygalomorph spider of the genus *Diplura* in Panama (Vollrath, 1978). The latter, which grows up to 4 cm in body length, builds large sheet or funnel webs for trapping its prey. Within a short time of capture, any prey is reduced to a pulp by the action of the spider's powerful chelicerae and the digestive enzymes poured on to it. *Curimagua* often rests on the cephalothorax of its host but, when the latter is feeding, it makes its way over the front of its head and down on to its jaws. Here it sucks up the fluids enveloping the digesting prey, its abdomen swelling visibly as it does so. Other individual *Curimagua* moving about on the host's web are also attracted to the feast, presumably by vibrations produced by the masticating *Diplura*. Also associated kleptoparasitically with diplurid mygalomorphs are two African mysmenids: *Isela okuncana* from southern Africa and *Kilifia inquilina* from Kenya.

Certain species of sparassid spiders of the genus *Olios* occupy the webs of social spiders in Australia and Sri Lanka (Jackson, 1987). In Queensland, for example, *O. diana* is found in the highly adhesive communal web of the cribellate spider *Badumna candida* (Amaurobiidae). It is not as well adapted to moving around on its host's web as many of the previously described kleptoparasitic spiders and tends to stay around the periphery, keeping its legs as much as possible on the leaves and twigs to which the web is attached. These spiders will, however, move on to the silk of the web to take prey if necessary, using their large size and brute force to break free of the entangling threads. The sparassids construct their own lairs within the host-web complex. All stages of the life-cycle of the sparassid have been observed within the host's web.

Araneophagic spiders

Quite a number of spiders from various families indulge in araneophagy, i.e. they feed on other species of spider. One family in particular, the Mimetidae, have become specialist araneophages, which has earned them the common name of pirate spiders, although, since pirates are not usually cannibals, the term is something of a misnomer. Mimetids normally enter the webs of other spiders, especially those of theridiids and araneids, rather than feeding upon free-living types, which presumably would be much more difficult to sneak up on than the sedentary web-dwellers. A study of two *Mimetus* species, one from Australia and one from New Zealand (Jackson & Whitehouse, 1986), has revealed that a combination of stealth and aggressive mimicry accounts for the success of these spiders. The two mimetids were studied both in the laboratory and in the wild, where it was found that they did indeed specialize upon theridiids and araneids, feeding mainly on smaller individuals which posed a lesser physical threat. Both species, on entering an alien web, indulge in aggressive mimicry by vibrating the web in much the same way as that of a struggling insect. The prey spider usually responds accordingly by running towards the mimetid which then attacks its victim, immobilizing it with a powerful venom while at the same time holding it in its spiny front legs. As a general rule, the *Mimetus* act as aggressive mimics only when their potential prey is quiescent; as soon as it becomes active, the pirate spiders remain still, with the first two pairs of legs held ready to grasp the prey in typical, ambushing, spider fashion. On occasion, if the prey approaches too rapidly or is large, the mimetids will drop out of the web.

Whereas araneophagy is the norm for the mimetids, it occurs only in the odd species in other spider families. That cosmopolitan species, the daddy-long-legs spider, *Pholcus phalangioides*, often found around human habitations, both catches prey in its own untidy, non-sticky web and invades the webs of other spiders and preys upon them. It has been found (Jackson & Brassington, 1987) that its capture efficiency varies with the type of web that it is in. As might be expected, it is most efficient when capturing insects in its own web with a fall-off in success as it enters the webs of other spider species, least success being achieved in those of cribellate spiders. When in an alien web, *Pholcus* attracts the occupant by use of aggressive mimicry in the form of various types of so-called 'vibratory behaviour'. When a spider approaches, *Pholcus* often elevates its body out of harm's way on its very long legs, remaining motionless until the prey makes contact with a leg.

This contact stimulates an attack from *Pholcus*, during which silk is thrown over the prey to immobilize it, an action which it shares with the Theridiidae. In fact, when *Pholcus* attacks a theridiid, the latter normally retaliates by throwing silk at the same time, usually without doing any harm. When in an alien web, *Pholcus* sometimes bites out some of the lines and replaces them with some of its own, especially in cribellate webs. When walking in sticky alien webs, its legs sometimes become caught. The reaction of *Pholcus* to this is to bite the silk away and then groom the leg clean before cutting away the sticky lines and replacing them with some of its own on which to perch.

Araneophagy has been closely studied (Jackson & Blest, 1982) in a primitive, web-building jumping spider *Portia fimbriata* from Australia. This spider has the ability to invade the webs of other spiders, whether sticky or non-sticky, cribellate or ecribellate, or two- or three-dimensional, without any apparent difficulty. In an alien web, *P. fimbriata* employs aggressive mimicry in the form of vibratory behaviour to attract the prey spider to it. Sometimes the potential victim responds with only a localized movement but this is enough to help *Portia* get a fix on it so that it can move towards it. *Portia* will also pursue an individual of another species of free-living salticid. In this case, it moves very slowly forwards and then freezes immediately if the prey turns to face it. It would seem that these prey salticids are unable to appreciate the danger that *Portia* poses towards them. As might be expected, vision plays an important part in the hunting technique of this species. *P. fimbriata* also feeds on the eggs of other spiders when the opportunity arises, opening up the egg case with its chelicerae and raking the eggs towards its mouth with its front legs. A close study has been made on the way in which *Portia fimbriata* captures another salticid, *Euryattus* sp. (Jackson & Wilcox, 1990). *Euryattus* females, unusually for salticids, make their nests inside suspended rolled-up leaves and, in order to prey upon them, *Portia* has to employ a unique form of prey-specific hunting behaviour. Upon finding a nesting *Euryattus* female in its suspended leaf, *Portia* moves down either the existing guy-line or its own dragline and positions itself at one of the leaf openings. It then uses vibratory behaviour to attract the *Euryattus* out of its nest. The unique feature of this behaviour is that it appears to mimic the courtship display of the male *Euryattus*. This prey-specific behaviour is only demonstrated by populations of *P. fimbriata* in the rainforest area of northern Queensland, where it lives sympatrically with *Euryattus*. *P. fimbriata* individuals from populations where *Euryattus* is not present fail to display such behaviour, as do other *Portia* spp.

Like *Portia fimbriata*, the New Zealand gnaphosid *Taieria erebus* is a highly accomplished araneophage and its behaviour has also been studied in detail (Jarman & Jackson, 1986). As with other spider-eating spiders, *Taieria* employs aggressive mimicry to attract its prey when it is in an alien web. Once again, this involves vibratory behaviour including, in this instance, fluttering of the palps against nests or webs and stamping on a strand of silk using all eight legs, giving the appearance of the spider 'running on the spot'. *Taieria* also plucks at the alien web, either with its palps or legs, most commonly the front two pairs of legs, and shivers by placing the tarsus of a leg on to a line of silk and moving it rapidly up and down for periods of 1–3 seconds. In the wild, *Taieria* is found in areas in which the webs of spiders, such as the amaurobiid *Badumna longinguus*, theridiids of the genera *Achaearanea* and *Steatoda* and the araneid *Araneus pustulosa* are plentiful, as are good numbers of various free-living hunters such as salticids. As well as observing *Taieria* in the wild, the researchers also set up experiments in the laboratory in which its approach to the webs of alien spiders with their occupants was studied. Its success in various web types is opposite to that of *Pholcus* (page 244) in that *Taieria* achieves its greatest success in densely woven cribellate and non-sticky webs and least success in sticky, ecribellate webs. Unlike *Pholcus*, it is unable to handle the problems associated with walking on webs containing sticky silk. Away from webs, cursorial spiders are usually attacked when they accidentally stumble into a stationary *Taieria*, though, occasionally, the latter will make a sudden rapid attack upon another spider from as much as 4 cm distant. It is unlikely that *Taieria* would be able to see the prey in this instance, air-borne vibrations being almost certainly the stimulus for the attack. *Taieria* was also seen to feed on spiders residing in silken nests. It enters the nests by biting a hole through the silk with its chelicerae, placing its front legs into the hole and then forcing its way in. Alternatively, when there is a nest opening, it will enter through this. Often as not, the occupant tries to escape and is attacked by the intruder as it makes the attempt.

Members of the theridiid genus *Argyrodes* are commonly kleptoparasites in the webs of other spiders and some are known to be araneophagic. *A. attenuatus* from Colombia constructs a small web of dry silk which it uses not as a snare but as a resting place both for itself and its prey (Eberhard, 1979). This spider merely sits in its web and waits for other spiders to wander on to it, whereupon they are wrapped in wet silk and devoured. *A. attenuatus* also catches nematocerous flies which have the habit of hanging on the spider's web and using it as a perch. Spiders of the closely related genus *Rhomphaea* from New Zealand capture other species of spider which enter their web, as well as entering the webs of other spiders to prey upon them. In an alien web, *Rhomphaea* employs aggressive mimicry to attract its prey but, unlike the previously described araneophagic species, it uses a special technique for trapping it (Whitehouse, 1987). Prior to launching an attack, it spins a length of viscid silk which is held in the form of a triangle. As the prey comes within reach, the attacking *Rhomphaea* rotates its legs forwards, collapsing the triangle of silk around the victim and effectively scooping it up. Related spiders of the genus *Ariamnes* use a similar technique but in their own webs. *Argyrodes antipodiana*, another closely related theridiid, does not employ a net but instead uses the technique of an ambushing spider (Whitehouse, 1986) by grasping the prey with the first two pairs of legs and then drawing the spider towards it to deliver the fatal bite.

Tasting by spiders

Having trapped its prey, the next thing that any spider does is to test its palatability simply by tasting it. There is some evidence that those warningly coloured insects which are unpalatable to vertebrates may actually be fed upon by spiders. Members of the families Thomisidae, Theridiidae, Araneidae, Salticidae and Agelenidae have all been found with their chelicerae sunk into insects such as coccinellid and cantharid beetles and zygaenid moths (pers. obs.). Whether the spiders actually consume these species, or digest them externally and then reject them once they begin to ingest the liquefied body contents, has never been proven but they are certainly not rejected immediately. Quantitative and qualitative data on the discrimination of unpalatable butterflies by the tropical araneid *Nephila clavipes* has, however, been provided (Vasconcellos-Neto & Lewinsohn, 1984). Various unpalatable (at least to vertebrates) butterflies were tossed into the orb webs of this spider and the response noted. Trials in the field in Brazil, employing 27 butterfly species, revealed that *N. clavipes* responds consistently to a particular butterfly species. Members of the subfamily Ithomiinae and some Danainae are almost always set free while Heliconiinae, Nymphalinae, Acrainae, Pieridae and Papillionidae are normally consumed. As might be expected for a short-sighted spider, the butterflies' warning coloration is not involved in the spider's response since the latter eat the palatable models of the unpalatable butterflies. It is most likely that distastefulness is indicated by the presence of certain noxious chemicals present in the butterflies which are rejected or released.

The insects in the exopterygote division of the insects, both adults and nymphs make use of the same food source and their feeding activities can be considered together. In the endopterygotes, the adults and larvae utilize different food sources and, in many of the examples, it is the feeding behaviour of either one or the other that is described, e.g. most adult butterflies have simple feeding strategies, whereas this is not the case for a number of butterfly caterpillars. Consequently more attention is paid in the text to variations in feeding behaviour of the latter.

ODONATA Damselflies, dragonflies

Dragonflies and damselflies habitually take their insect prey while on the wing and hold it in their prehensile legs whilst feeding. A number of Anisoptera are known to or thought to indulge in 'accompanying' behaviour, i.e. they fly in the company of herds of vertebrates moving slowly through grassland, or occasionally shallow water, attacking any suitable flying insects that are disturbed. This type of behaviour has been shown to be true (Corbet & Miller, 1991) for the pan-African sympetrine *Brachythemis leucosticta*; accompanying behaviour is shown by both solitary males and females, immatures and groups. An unusual form of feeding behaviour, at least for Odonata as a whole, has been noted in the giant tropical damselfly *Megaloprepus coerulatus* in Costa Rica (Young, 1980). Predominantly males approach and hover before the large orb webs of *Nephila* spiders. They then pluck small spiders, presumably kleptoparasitic theridiids and not the web's owner, from the web and then alight on nearby vegetation to devour the prey. Of 52 visits to *Nephila* webs by *M. coerulatus*, successful captures of a spider were made in 25 per cent of cases observed.

BLATTODEA Cockroaches

The cockroaches owe their success in part to their somewhat catholic diets in that they seem to make use of whatever food is available at the time. Some species are more specialized; the three British species, for example, seemingly are pollen-feeders while, in Mexico and South Africa, cockroaches of two different species have been observed feeding upon ripe berries (pers. obs.).

Feeding behaviour in females of the forest-dwelling Costa Rican cockroach *Xestoblatta hamata* relates to different phases of the ovarian cycle. Early in the cycle, the insects feed on the shed bark of the legume *Inga coruscans*; chemical analysis of sequestered materials has indicated that this food source is low in nitrogen but high in lipids. Following this, the cockroaches tend to take in protein followed by carbohydrate before oviposition. Interestingly, one important source of nitrogen for the females comes from the males. Following mating, the males void the content of their uricose accessory sex-glands and the females feed upon it. If labelled uric acid is injected into a male, labelled compounds are found in the female's eggs after she has mated with him, an example of a male contributing directly to the development of his offspring. More of this male-derived uric acid is taken in by females on low-nitrogen diets than by those on a high-nitrogen diet.

ISOPTERA Termites

All termites feed on dead plant material either directly or indirectly by utilizing it as a culture medium for the fungi on which they feed. Termites of the genera *Macrotermes, Longipeditermes* and *Hospitalitermes* are blind but have returned to above-ground foraging in the rainforests of South-East Asia. A study of a number of species from within these genera (Jander & Daumer, 1974) has provided an insight into how these sightless creatures are able to make their way over distances as great as 300 m from their nests to their foraging sites. Two species, *Macrotermes carbonarius* and *Longipeditermes longipes*, have similar foraging techniques. Both of them forage on the ground, travelling seldom more than 10 m along trailways, which they build by removing small particles of debris hindering their passage and then make smooth by filling any cracks and depressions with small pieces of soil. They both forage for leaf litter, particularly rotting leaves from which small sections are cut to be carried back to the nest, though *M. carbonarius* workers also cut up and collect fallen green leaves. The numbers of workers leaving and entering the nest at any one time are approximately equal, indicating that individual workers make a number of trips during each foraging session. Whereas *M. carbonarius* and *L. longipes* forage along cracks and depressions in the ground, *Hospitalitermes umbrinus sharpi* follows crest-lines along the tops of sticks, roots, leaves and other elongate objects above the

▲ **Workers of *Hospitalitermes* termites collect lichens from the surface of forest trees and transport balls of them, clearly visible here, back to their nest. Amongst the workers may be seen a few nasute soldiers, with their long 'snouts' standing guard over the workers.**

ground, moving over the ground itself only when absolutely necessary.

Like those of the other two species, foraging expeditions of *H. u. sharpi* are intermittent affairs, taking place at intervals of up to 1 week. Foraging columns are ordered affairs, with the lead being taken by soldiers, who also mount a guard on either side of the column as it lengthens. At the foraging site on a particular tree, each worker collects its load of food and then makes its way back to the nest along the same trail as it followed outwards. Whereas out-going workers move in the two outer lanes of the column, those returning do so along the centre lane. Should a termite start going the wrong way, it soon turns round, seemingly as a result of the large numbers of head-butts it receives in attempting to go the opposite way to that in which the lane is moving. In this species, each worker makes only one foray on a particular expedition. These forays usually start during the second half of the night, reach maximum intensity from mid-morning to mid-afternoon and then rapidly decline. A group of soldiers is always last back into the nest.

The form of *H. umbrinus* from Sarawak, however, emerges in the evening, forages at night and then returns to the nest the following morning, the columns forming and behaving in much the same way as those of the Malaysian subspecies (Collins, 1979). Workers of the Sarawak subspecies were found returning with two different forms of food ball: dark ones consisting mainly of crustose lichens, presumed to be from the leaf canopy of the foraging site and light ones containing bark cells, sclerenchyma cells, xylem vessels and filamentous and colonial blue-green algae, presumably collected from the tree's bark. Foraging is highly organized, with rows of workers cutting material away from the tree and passing it back to waiting ball-makers who then transport it back to the nest. A cutter might then replace a departed ball-maker and itself be replaced by a newly arrived worker, their being no observable morphological differences between individuals indulging in the respective 'trades'.

Termites from the subfamily Macrotermitinae, including *Macrotermes*, *Odontotermes*, *Microtermes*, *Ancistrotermes* and other genera, do not feed directly upon the food that they have collected but instead use it to culture fungus gardens in a manner akin to that of fungus-cultivating ants. The way in which these termites make use of the fungus has been elucidated for *Macrotermes* (Leuthold *et al.*, 1989). Initially, the collected food material is consumed by termites who are <30 days from ecdysing from the final instar into a worker. The ingested food passes fairly rapidly through them and the primary faeces produced are then used as the basic material for the fungus garden. The fungus which develops is a *Termitomyces* sp. and it is known that the termites are able to control or prevent the growth in the garden of any unwanted species of fungi.

Field and laboratory observations on the Indian species *Odontotermes gurdaspurensis* (Batra & Batra, 1966) have shown that soil which has been processed by workers, and also the defensive oral secretions of the soldiers, seem to stop growth of undesirable fungi. The fungus is cultured on a fungus comb from which eventually grow spherules, which serve as a source of food for the <30-day-old workers and which they then pass on to the developing nymphs. In *Macrotermes*, at least, the spherules contain digestion-resistant fungal conidia which then mix with the ingested foraged food and serve as an inoculant for the new areas of fungal comb. Workers >30 days old act as foragers outside the nest and feed on the fungus-comb material once it has produced its spherules. These workers produce the final, true faeces, which then forms carton, which is used as the basic building material for the manufacture of the fungus combs within the nest.

MANTODEA Mantids

Mantids are the archetypal sit-and-wait predator and they are not particularly fussy about what they eat, apparently relishing anything they can get hold of. Apart from taking cursorial insects, the nymphs of the African flower mantid *Pseudocreobotra* have been seen to snatch bee flies out of the air and the larger species have, on occasion, been known to take frogs, lizards and other small vertebrates as prey. The African species *Tarachodes afzelii* lives on tree bark and, as a consequence, tends to be surrounded by ants on their way to and from the forest canopy. It is not surprising, therefore, that ants form a major prey from the third instar to the adult of this mantid. Rather than sitting and waiting for an ant to pass, the mantid, on seeing one, chases it, catches and immediately consumes it. When preying on other insects, however, they adopt the usual sit-and-wait procedure.

ORTHOPTERA Crickets, locusts, grasshoppers

Members of the family Tettigoniidae, the bush crickets or katydids, are either herbivorous, omnivorous or carnivorous. Within the herbivorous members of the family are examples such as the African Copiphorinae, which use their large jaws to crack open tough seeds. Other copiphorines and the rather ferocious-looking members of the Saginae, however, use their large jaws for a different purpose and, along with the European oak bush cricket, *Meconema thalassinum*, are obligate carnivores.

The Saginae will attack almost any arthropod up to their own size, though animals less than one-tenth of their size are not often taken as prey. They prefer other orthopterans as prey but also feed on cicadas and caterpillars, and *Cloniella praedatoria* has been found catching the butterfly *Danaus chrysippus* from the tops of tall flowering grasses. Although exclusively carnivorous, they do drink the fluid from fresh fruits without consuming any of the pulp. The Saginae are highly active hunters who walk about their habitat detecting prey by the vibrations which they produce. They leap on to their victims, grasp them in their spiny front legs, which are not always too effective at maintaining a hold since prey often escapes, and then kill them with a bite in the throat. Another European species, *Tettigonia cantans*, is also carnivorous and has been seen to catch a smaller

▲ (Above) Mantids are the archetypal 'sit-and-wait' predators of the arthropod world. This female *Acontista* sp. from Trinidad is here feeding on a male coolie butterfly, *Anartia amathea*.

◀ Many katydids are opportunistic feeders and will seize any suitable prey which comes their way. This male *Tettigonia cantans* from Europe is starting to feed on a smaller species of katydid that was unfortunate enough to have jumped straight into the larger individual's jaws.

▲ (Above right) Tropical grasshoppers are often very selective about which part of a leaf they eat. Fresh young growth is often neglected, in favour of unappetizing-looking withered areas on the leaf edge, as in this *Poeciloclоeus* sp. (closely related to *P. bullatus*), busy chewing away in a Peruvian rainforest. Such discrimination in favour of the unhealthy parts of leaves is probably due to the lower concentrations of toxic defensive compounds in damaged areas.

species of tettigonid in its powerful front legs and dispatch it very rapidly with its powerful jaws (pers. obs.). The cone-head *Pyrgocorypha hamata* is able to attack and demolish tough-shelled, tropical chafer beetles.

The crickets of the family Gryllidae tend to be very omnnivorous in their diets and many of them are scavengers, taking what they can get, including animal faeces. This is also true of the family Raphidophoridae, which includes the cave crickets and the sand-treader crickets. A large species of sand-treader, a *Macrobaenetes* sp. from California, normally feeds nocturnally upon organic debris, including seeds and dead insects, in its sand-dune habitat. At times, these dunes are subject to high winds which scour the surface clean of such edible debris and, at this time, the sand-treaders resort to feeding upon lizard and kangaroo-rat dung (Rentz, 1970). In order to prevent the prize from being blown away by the strong wind, the cricket actually encloses it in a 'cage' formed by its legs. Once the cricket has finished, it merely raises its head and some of its legs and the wind whips away the remains of the food. As well as taking plants and fungi in their diet, New Zealand wetas commonly feed on a particular species of bug and fly as well as other raphidophorids (Richards, 1962). Rat faeces also form part of their diet.

Grasshoppers of the family Acrididae are exclusively vegetarian in their diet, though some of the most brilliantly coloured eumastacids obtain their vegetable matter second-hand, i.e. they feed upon the plant content of faeces. Most grasshoppers live on or close to their food plant, although at least one exception, the American species *Cibolacris parviceps*, passes the hours of daylight on the desert floor where its coloration renders it cryptic. Towards dusk, the grasshoppers commute to nearby creosote bushes and commence feeding. Some species have adapted their feeding behaviour in order to feed upon noxious leaves, such as those of the cassava. The oldest cassava leaves contain the most toxins; therefore the West African elegant grasshopper, *Zonocerus variegatus*, feeds selectively on the young leaves, diseased areas or yellowing leaves, all of which pack less of a toxic punch.

The choice of a food plant for the North American desert-dwelling grasshopper *Ligurotettix coquilletti* is closely tied up with territorial defence (Greenfield *et al.*, 1989). This grasshopper is an oligophagous species found mainly on creosote bush, *Larrea tridentata*, a plant whose leaves contain the defensive phenolic compound nordihydroguaiaretic acid (NDGA). The levels of NDGA differ from plant to plant and males 'kept' on high-level plants suffer high mortality rates while females and nymphs in the same situation feed normally but show low growth rates, seemingly as a result of a lowered food-conversion ratio. Consequently, males tend to defend and feed on plants with low levels of NDGA in the leaves, thus attracting females to them.

Myrmecophila is a genus of wingless crickets which live in the nests of different ant species. *M. manni* comes from the semi-arid and arid regions of the USA and Mexico and is mainly associated with the ant *Formica obscuripes*. The ants are very aggressive towards the crickets, who have to use their superior agility to avoid being attacked; despite this, the crickets maintain a close company with their host ants. These crickets appear to have broken the code which ants use to induce trophallactic feeding from another ant (Henderson & Akre, 1986). They do this by duplicating the antennal drumming used by one ant to obtain food from another. Under laboratory conditions, the crickets are able to elicit trophallaxis not only from *F. obscuripes* workers but also from workers of other species and from ant larvae. The crickets also strigilate their host ants, both living and dead.

DERMAPTERA Earwigs

Earwigs are basically omnivorous, with a marked tendency to prefer food of animal origin. The common European earwig *Forficula auricularia* is well known by gardeners for its habits of feeding on the petals and young foliage of garden plants while, in captivity, *Labidura riparia* prefers animal food in the form of flies and other small insects. Two arixeniine earwigs from the Malayan region are know to associate exclusively with bats of the family Molossidae. These two species, *Arixenia esau* and *Xeniaria jacobseni*, live in bat roosts and the former feeds mainly upon the body exudates and skin debris of the naked bat.

THYSANOPTERA Thrips

Thrips feed mainly by penetrating plant tissues with their piercing mouthparts and then sucking up the sap, though a number are also known to feed in a similar fashion upon other small arthropods. Kirk (1984) recorded eight species of thrips from three different families feeding upon individual pollen grains. These insects suck out the contents of the pollen grains but leave the cell walls intact. In some instances, pollen is consumed at a prodigious rate; one thrips was seen to imbibe the contents of more than 100 pollen grains in less than 15 minutes.

HEMIPTERA True bugs

This group of insects is characterized by its sucking mouthparts. Of the two groups into which the Hemiptera is divided, the Homoptera are exclusively plant-feeders while the Heteroptera contains plant-feeders, predators on other invertebrates and blood-suckers of vertebrates.

▶ A large proportion of the Hemiptera, including many of the Heteroptera, are plant-feeders. The homopteran leaf-hopper *Rhaphirrhinus phosphoreus* (Cicadellidae) is, like all members of its sub-order, a plant-feeder. The individuals in this group are sucking the sap from a rainforest leaf in Peru.

▼ Adult booklice or barklice (Psocoptera) scraping the surface of a leaf as they feed in rainforest in Peru.

HOMOPTERA

Most aphids just sit and suck the phloem contents of their host plants and undergo their life cycles without any complications. Aphids of the genus *Eriosoma* from the Holarctic region, for example, make galls by rolling the leaves of elm trees (Ulmaceae), their primary hosts. Some species may then alternate elm trees with a secondary host or hosts. Fundatrices of *E. yangi*, however, do things differently (Akimoto, 1981). They do not form galls of their own but wander around on the young elm leaves until they moult to the third instar. Then they enter the galls of other species of *Eriosoma* aphid and take it over, killing the rightful owner of the gall in some way during the process. On occasions, the latter is not killed and both species of aphid then feed and reproduce within the single gall though they remain spatially separated.

HETEROPTERA

The plant-feeding heteropterans normally live upon their host plants and exhibit little specialized feeding behaviour. Those species which feed upon other arthropods are often as not opportunistic or sit-and-wait hunters, though there are some interesting examples of special prey-catching behaviour to be found within this group. Not least of these is the bait-and capture technique of the Costa Rican reduviid assassin bug *Salyavata variegata*. The life-style of this interesting species has been described in detail (McMahan, 1983). Both adults and all nymphal stages of *S. variegata* live on the carton nests of *Nasutitermes* species in Costa Rica. Adults and nymphs opportunistically snatch their termite prey from openings in the nest but only the third, fourth and fifth instars have been seen to use bait to attract prey. Baiting involves the use of the carcase of a previous victim as a tool to entice another termite out of an opening in the nest to within the bug's reach. In feeding experiments, both nymphs and adults showed some choice in their prey, preferring large worker termites to small workers and small workers to soldiers. Whereas the adult bugs are cryptically coloured, the

► Many bugs which feed on seeds often have the habit of clustering around a single food item. This is probably because each bug injects its own dose of digestive enzymes, making its own contribution to the combined attack necessary to the successful pre-digestion of the seed's kernel. This is the cottonstainer, *Dysdercus flavidus* (Pyrrhocoridae), feeding on kapok-tree seeds in Madagascar.

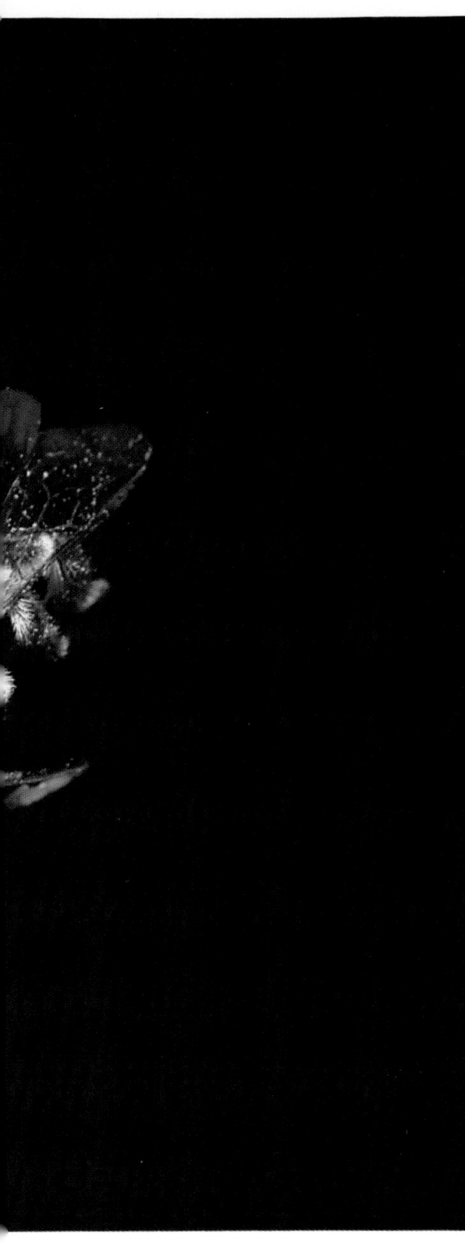

▲ **The assassin bugs are aptly named for they are a major predator of other insects, especially in the tropics, where many species are brightly coloured. *Apiomerus flaviventris* is here seen feeding on a bee in Tamaulipas State, Mexico.**

nymphs are not and they camouflage themselves by gluing to their body surface crumbs scraped off the carton nests.

Some reduviids use another form of tool to help them catch prey, in this instance tree resin. For example, the South-East Asian resin bugs *Ectinoderus longimanus* and *Amulius malayus* dip their hairy front legs into tree resin and use it to help them trap *Trigona* bees which are attracted to it. The same habits are reported to be true for the Neotropical species *Beharus lunatus*. A somewhat different, though related, behaviour has been reported for another Neotropical species, *Manicoris rufipes* from Brazil (Adis, 1984). Nymphs of *M. rufipes* lie in wait for returning *Trigona* bees on which

▲ ***Salyavata variegata* is a Neotropical assassin bug (Reduviidae) which 'fishes' for its supper. It lives on the nests of *Nasutitermes* termites, feeding on its host. In order to attract new prey, it dangles the empty husk of a previous termite meal on the end of its proboscis. The covering of detritus on its body presumably hides it from its enemies.**

they feed. The bees often land on trees in the vicinity of their nest, where the bugs are waiting. As a bee lands, the nymph moves towards it and tries to touch it with its front legs, which are covered with a yellowish coating to which the bees adhere. Newly moulted bugs lack the adhesive and have to catch the bees in their first two pairs of legs but, as they catch more bees, so the material, which the bees are carrying back to their nest, accumulates on the bugs' hairy fore-tibia until there is sufficient to trap all further bees it contacts. Examination has shown the adhesive substance to be a mixture of pollen, nectar and resin. Also in Brazil, nymphs of an *Apiomerus* sp. of resin bug have been seen to feed upon termites (*Termes* sp.). They search for damaged termite galleries on the trunks of trees and then place their resin-covered front legs on to the termites as they pass over the bark.

Whereas the previous species specialize in *Trigona* prey, the emesine assassin bug *Stenolemus lanipes* from the southern USA is a specialized predator on spiders (Snoddy *et al.*, 1976). The adult bugs live in crevices and other protected areas in the immediate vicinity of the webs of their prey, the common house spider, *Achaearanea tepidariorum*. The bug has both structural and behavioural adaptations to allow it to feed upon this web-building theridiid spider. The micro-spines on the legs and tarsi, for example, are covered with fragments of spider web or some silk-like material which appears to act both as camouflage and as an anti-adhesive agent allowing the bug to move freely on the spider's web. During hunting, *S. lanipes* makes a slow approach to the web, first striking it with its long, arched antennae. On entering the web, it walks with a distinct vibratory movement as if copying the effects of a piece of leaf caught in the web and blowing in the wind. The oscillations produced in the web by this type of movement do not disturb the web's occupants and allow the bug to approach within striking distance. The young of *A. tepidariorum* remain in the web with their mother for some time before they disperse and it is the former which are the bug's main prey; adults are seldom taken. In fact, when both the adult female and the young are present in the same area of web, the bug seems to move unerringly towards the latter.

An assassin bug which is not an assassin is the highly unusual coprophagous reduviid *Lophocephala guerini* from the Indian subcontinent and Sri Lanka. In the western Ghats of India, both adults and nymphs live in crevices beneath stones, always in an obligate association with the ant *Anoplolepis longipes*. The ant may be found without the bug but the bug is never found without the ant. All stages of the bug emerge from their daytime hiding places at dusk and make their way to their feeding sites, guided and accompanied by the ants. The bugs feed on the liquids exuding from masses of fermenting cow dung, which has to be at the

▲ Shield bugs (Pentatomidae) feed on both plants and animals. *Troilus luridus* is a European species which is commonly found feeding on caterpillars and other insects. In this instance, an adult is busily engaged on feeding on a 'distasteful' seven-spot ladybird, *Coccinella 7-punctata*. Many invertebrate predators will feed on prey which would be rejected by vertebrates such as birds. Pentatomids are particularly noted for their lack of fussiness on this point, and will even dine on the body-contents of such highly poisonous prey as *Zygaena* spp. burnet moths in the larval, pupal and adult stages.

correct stage of fermentation before it is acceptable. As a result, the distances which may have to be travelled by the bugs can be quite considerable. Since acceptable dung masses are few and far between, all of the bugs in one particular micro-habitat may attend the same feeding site at the same time. At dawn, they retire to their original hiding places. There are indications that the ants benefit from the relationship by feeding upon the bugs' faeces.

Both adults and nymphal stages of the European pentatomid *Picromerus bidens* prey upon the larval stages of Lepidoptera and chrysomelid leaf beetles. In areas where they are abundant, the nymphs will feed avidly upon all stages of the distasteful burnet moths of the genus *Zygaena*. Up to three nymphs have been seen feeding upon a single pupa of one of these moths (pers. obs.). The yellow and black assassin bug *Ectrichodia crux* lives under stones where it specializes in another chemically protected group, millipedes, as prey. It is possible to find a whole family of these bugs, adults and their bright-red nymphs, assembled around the body of a millipede and feeding upon it.

Bugs of the family Termitaphididae live in an obligate symbiotic association with fungus-cultivating termites. The bugs feed on fluid which they suck from the fungus and, in return, they produce from abdominal pores a secretion which is taken up by the worker termites. Another obligate association has developed between mirid bugs of the genus *Ranzovius* and certain spiders, though whether this is an example of commensalism, kleptoparasitism or egg-sac predation is not yet clear. *R. moerens* (= *R. californicus*) has been studied in California (Davis & Russell, 1969) where it is associated with the webs of the agelenid spider *Hololena curta*. The mirid moves around diurnally on the spider's sheet web, with apparent ease and on both upper and lower surfaces, in search of food. It feeds on both animal and plant material which is caught in the web and it seems likely that it is able to kill small insects itself. For no obvious reason, the spider completely ignores the bug's presence, even when it is within touching distance. Further knowledge of the relationship between these bugs and their spider hosts has come from a study (Wheeler Jr & McCaffrey, 1984) of two other species from the eastern USA. *R. contubernalis* lives in the webs of the subsocial theridiid spider *Anelosimus studiosus* and also in those of the agelenid *Agelenopsis pennsylvanica*, while the latter also hosts *R.*

◀ **(Top) The larva of the green lacewing, *Chrysopa vulgaris* (Neuroptera), is a voracious predator and is likely to consume this colony of aphids on a nettle leaf in double-quick time.**

◀ **(Bottom) Pond skaters (Heteroptera: Gerridae) normally feed by sucking the body fluid from luckless insects which have fallen into the pool on which they live. In some instances, however, they are known to take small fish. Here, two individuals of the European *Gerris lacustris* are feeding on a small minnow.**

agelenopsis. From this latter study it appears that *R. contubernalis* feeds mainly on insects that have been caught on the web but are too small to stimulate the host spider's own feeding response. For example, large numbers of aphids often become trapped in the web of *Anelosimus studiosus* and are ignored by the spider, although they form a substantial supply of food for the bugs; the story is the same for the boxwood psyllid, *Psylla buxi*, when the webs are constructed on boxwood.

NEUROPTERA Lacewings, mantispids

The Neuroptera have biting mouthparts and feed upon other insects, both as adults and as larvae, the latter including the well-known ant-lions. Supreme amongst the Neuroptera in their hunting behaviour are the members of the family Ascalaphidae, their abilities comparing favourably with those supreme aerial hunters, the dragonflies. The North American species *Ululodes mexicana* flies with great speed and precision for long periods (Henry, 1977), hovering and darting erratically as it pursues small, flying insects. Adults are active for just a short time each day, from 15 to 20 minutes after sunset for a period of around 45 minutes. Initially, they hunt at heights of around 10 m but, as it gets darker, so they fly at progressively lower levels until, as darkness finally descends, they are down to as low as 15 cm before hunting activity finally ceases.

Adult mantispids are Neuroptera with raptorial front legs which resemble those of the mantids and are used in a similar fashion to capture insect prey. The North American polistes-wasp-mimic *Climaciella brunnea occidentalis* was found (Batra, 1972) to forage on a variety of flower species, including thistles, from one of which they were seen to lick and chew gummy exudates from various wounds on the plant. In captivity, they drink droplets of honey and water and also capture a variety

► The scorpionflies (Mecoptera) are basically scavengers, feeding upon dead and dying invertebrates and even the occasional small vertebrate. Here three individuals of the European *Panorpa communis* are feeding on a hedgehog killed in a road accident.

of the insects offered to them, including aphids, bugs and ladybirds.

Larval mantispids, however, prey upon the eggs of spiders within the spider egg sac and they employ either one of two basic behaviour patterns to gain access to their prey. In the first instance, *Mantispa virida* is an example, they may penetrate an already formed egg sac; alternatively, as in *Climaciella brunnea*, they may board a female spider and wait until she produces her egg sac before they enter it. Larvae of the latter are known (Redborg & Macleod, 1983) to rear up on their tails and often sway back and forth, legs out-stretched when in search of a host. The larvae of some species are facultative penetrators or boarders. One such is *Mantispa uhleri*, whose larvae actually maintain themselves by feeding on the blood of their spider host, *Lycosa rabida*, until it lays its eggs (Redborg, 1982).

MECOPTERA Scorpionflies, hangingflies

Of the three families within the Mecoptera, the Panorpidae or scorpionflies are scavengers, often kleptoparasitically, the Bittacidae or hangingflies are predatory while the Boreidae are plant-feeders.

Although the panorpid scorpionflies are known to feed on almost any dead animal that comes their way, and this includes mice (pers. obs.), it has been known for some time that they are able to take either freshly caught or wrapped prey from spiders' webs. During observations on the panorpids of southeastern Michigan in the USA, for example, individuals of all of the nine resident species of *Panorpa* were seen to raid spiders' webs of one kind or another, though some species did so more often than others (Thornhill, 1975). On locating prey in a web, a panorpid may either fly directly to it or land on adjacent vegetation and walk on to the web from it. Normally they feed on the prey *in situ* but, on occasions, scorpionflies may fly off with small flies held in their jaws. The scorpionflies move around on the spider silk with little apparent difficulty and, if they do occasionally get enmeshed in the web, they have a special means of escape. They exude a brown liquid from the mouth on to that part of the web in which they are trapped and this dissolves the silk, so releasing them. If they are artificially milked of this fluid then they are unable to release themselves from the silk. Scorpionflies enter spiders' webs quite readily and, as long as the occupants are small they are relatively safe. In the study area, however, over 50 per cent of panorpid deaths occurred at the hands of web-building spiders.

The hangingflies are highly adapted aerial predators with, somewhat unusually, a pair of powerfully raptorial hind legs. *Harpobittacus nigriceps* and *H. australis* from Australia exhibit similar hunting techniques (Bornemissza, 1966). Observations have been made on males hunting for nuptial gifts for their females, who apparently rely completely upon these as a food source. *Harpobittacus* males catch their prey by flying slowly up the vertical stems of grasses and other herbage, or by raking their legs across the surface of flower clusters. The elongated, prehensile tarsi vibrate rapidly as they contact the surfaces of the plants over which they are hunting. The hind tarsi immediately coil round any soft-bodied arthropod which they contact and hold it firmly. Flying insects, such as flies or bees, however, are often held by the wings, which are stretched out between the hangingflies' tarsi; the saltatory legs of grasshoppers and crickets are also effectively immobilized by the coiled tarsi. *Harpobittacus* hangs by its front legs and then uses its middle legs to assist its hind legs to position captured prey before killing it with its biting mouthparts; the rapidity of the prey's death indicates a possibly toxic saliva. From the fact that these Australian hangingflies take a wide range of prey, both mobile and immobile, obvious and camouflaged, it is concluded that hunting depends entirely upon tactile stimuli received by the tarsi of the hind legs. Prey includes adult and larval insects and spiders.

Three North American hangingflies of the genus *Bittacus*, namely *B. apicalis*, *B. pilicornis* and *B. strigosus* from southeastern Michigan, have the same habitat preference and show a seasonal succession with some overlap of the species (Thornhill, 1977). They do, however, show marked differences in their foraging behaviour as adults. Of the three, *B. apicalis* is diurnal whereas the other two are crepuscular and nocturnal. *B. strigosus* forages to some extent in low vegetation at dusk and at dawn but forages in the trees during darkness. *B. pilicornis*, however, forages nocturnally in the herb layer, as does a fourth species, *B. stigmaterus* (Thornhill, 1978). *Bittacus* employ three different methods for catching prey:

(1) They fly from a hanging position and grasp prey from an adjacent plant or from mid-air.

(2) They use their prehensile hind tarsi to grasp prey that comes within range of where they are hanging.

(3) They sweep vegetation, as described for the two Australian species of *Harpobittacus*.

The diurnal *B. apicalis* and the crepuscular *B. stigmaterus* use mainly the first two methods and only occasionally the third, whereas the latter method is employed primarily by the mainly nocturnal *B. pilicornis* and *B. strigosus*. There is a significant difference in the type of prey taken by males and females of *B. stigmaterus*. Males tend to concentrate on Diptera and Lepidoptera within a particular size range whereas females, who in any case receive most of their food in the form of nuptial gifts from the males, primarily catch Homoptera, Diptera and Hymenoptera. Females, in fact, tend to catch what happens to be the most abundant prey available at any one time. Palatability and size are also of importance in determining the prey taken by males of *B. apicalis* and *B. pilicornis*, both of whom present nuptial gifts to the female, but males of *B. strigosus*, who do not present such gifts, show less selectivity in their prey choice.

▼ This female *Panorpa communis* scorpionfly is feeding kleptoparasitically on a crane fly (Tipulidae) hanging in a spider's web in England. Such habits are common in the group, especially in the males, which frequently trespass in spiders' webs to steal prey which will serve as nuptial gifts. This entails an increased risk of mortality amongst the males, compared with the females, arising from attacks by the resident spiders.

▲ **Butterflies feed on a range of liquid foods, including nectar, sap and urine, and males especially will take up water rich in dissolved salts from streamsides and puddles. Feeding from damp ground in Peru is a *Heliconius erato favorinus*. Female *Heliconius* have developed the ability to exploit protein-rich pollen as a food source.**

LEPIDOPTERA Butterflies, moths

Adult butterflies and moths use their long proboscis to take up liquid food in the form of nectar, fruit juices, sap, urine and the liquid on the surface of fresh dung. There have also been a number of observations of heliconiids taking the liquid from the corner of the eye of vertebrates such as caymans in South America. In South-West Africa/ Namibia, Büttiker & Bezuidenhout (1974) collected a total of 66 specimens of eye-frequenting moths from seven different localities. The moths were collected from the eyes of domestic cattle and goats and were the three noctuids *Arcyophora longivalvis, A. zanderi* and *A. patricula*. The large geometrid moth genus *Scopula* contains a number of species whose males are either lachryphagous and/or zoophilous; these include *S. lacriphaga, S. haematophaga* and *S. malayana*, all from South-East Asia (Bänziger & Fletcher, 1985). Apart from tears, their food sources include sweat, droplets of blood exuding from insect bites and wound fluids, all of which they take from mammals, including man.

A number of butterflies indulge in special forms of behaviour in order to assure themselves of a supply of food. Antbutterflies, for example, follow columns of army ants in order to find their principle food source, the droppings of antbirds which themselves follow the army ants, feeding upon insects that the ants displace from the forest floor. Ray & Andrews (1980), working in Costa Rica, found that females of three ithomiines, namely *Mechanitis polymnia isthmia, M. lysimnia doryssus* and *Melinaea lilis imitata*, associated with the raids of the army ant *Eciton burchelli*. Where the ant swarms are dense, between 8 and 12 of these female butterflies may be seen dipping down for a moment into the midst of the ants, occasionally stopping to feed on a bird-dropping. In some instances, the mimics of the antbutterflies accompany the models and also feed at the bird-droppings (pers. obs.).

An unusual form of foraging behaviour has been noted (Gilbert, 1976) in the small lycaenid *Megalopalpus zymna*, which lives in rainforest from Ghana to Gabon. This butterfly was noted feeding from the posterior nectar gland of a lycaenid caterpillar which, at the time, was being attended by ants. The butterfly was apparently able to wade unharmed on its rather long legs through the ants to get at the caterpillar. As it feeds, it holds its abdomen at such an angle as to minimize its accessibility to the surrounding ants.

The diet of the average butterfly is very low in protein. As a result, the protein that is required for egg production has to be laid down during the larval feeding stage and passed on to the adult. *Heliconius* butterflies live for months and lay a few eggs each day over their long life. They do this as a result of a modification in the normal butterfly feeding habit of taking low-protein liquid food. When they visit a flower, as well as taking up nectar, they scrape their proboscis across the anthers and, consequently, pollen adheres to it (Gilbert, 1972). Nectar is then regurgitated on to the pollen, which accumulates at the base of the proboscis. Over a period of several hours, this mass of pollen is then 'kneaded' with the end of the proboscis by coiling and uncoiling it. The result of this activity is that the pollen grains release amino acids and other important soluble food substances which are then re-imbibed along with the nectar; the empty husks of the pollen grains are discarded. In a study of ten species of *Heliconius* butterflies in Trinidad and Costa Rica (Boggs *et al.*, 1981) under greenhouse conditions, various aspects of pollen usage were elucidated. Females tend to collect significantly more pollen than males and older females collect significantly more pollen than younger ones. Because the different species have significantly differing abilities to utilize small or large pollen grains, this results in significantly different mean pollen collections for the same species in different sites. As expected, pollen collection is also affected by the availability of flowering plants of favoured species at different times of the year. Pollen loads have also been reported on the proboscides of three papilionines from Costa Rica: *Parides childrenae*, *P. arcas mylotes* and *Battus crassus lepidus* (DeVries, 1979).

At first glance, the larval lepidopteran with its efficient cutting and chewing jaws would seem to have few in the way of problems in obtaining food, since female moths and butterflies surely select their egg-laying sites with great care, ensuring that they contain an adequate supply of food for their offspring. One problem that has been overcome by some caterpillars is that of the presence of defensive trichomes, which have extremely sharp points, capable of piercing their soft skin, on the leaf surface. The South African shrub *Solanum coccineum* bears leaves which are covered in stellate trichomes but, despite this, they are still fed upon by caterpillars of a noctuid moth of the genus *Pardasena* (Hulley, 1988). These caterpillars are apparently able to walk upon the trichomes and they normally began to feed near the distal end of the leaf. Initially, they use their mandibles to 'mow' the trichomes from a narrow strip, just a few millimetres long, close to the edge of the leaf. Trichomes are cut off close to the surface of the leaf and these accumulate beneath the labrum in front of the mandibles. At intervals, the clumps of trichomes are tossed away with a flick of the caterpillar's head. Having cleared an adequate area of trichomes, the caterpillar now begins to feed on the leaf tissue, inclining its head so as to avoid the thick mat of trichomes on the leaf's lower surface. These are left dangling in space as the tissue is eaten away from above them and they eventually fall away under their own weight.

In Central America, the 'maler mujer' or 'evil woman', *Cnidoscolus urens urens* (Euphorbiaceae), presents two lines of defence against grazing animals; it is covered in urticating trichomes and contains a sticky latex which flows out of any break in the surface of the plant. Caterpillars of the sphingid moth *Erinnyis ello* have evolved ways of overcoming both problems (Dillon *et al.*, 1983). The effectiveness of the defences of this plant can be appreciated by the numbers of dead butterflies, moths, katydids and other insects which may be seen caught on the leaves. Despite this, late-instar *E. ello* caterpillars are able to defoliate a whole plant up to more than a metre across in a single day. When feeding, the caterpillar initially clears the leaf petiole of urticating trichomes by the simple expedient of grazing them down. It then moves into the cleared area and continues grazing, gradually moving towards the leaf base. A small amount of latex is released and this is removed from the body with the mandibles. The caterpillar now begins to crush and constrict the petiole 1–2 cm from the leaf base without completely severing the leaf. This has the effect of cutting the lactifers which supply the leaf and run in the region of the phloem. Next, the caterpillar moves on to the leaf surface, mowing the trichomes as it goes, and then begins feeding on the leaf edge close to the base of the leaf. It is usual for the caterpillar to consume the whole leaf before it moves on to the next one. On occasions, the caterpillar constricts a small stem rather than a single leaf and then eats all of the leaves on that stem.

It has been reported (Rathcke & Poole, 1975) that the caterpillar of the ithomiid butterfly *Mechanitis polymnia isthmia* avoids the trichomes of its *Solanum* host plant by spinning a silken scaffolding on the tips of the trichomes and then feeding from the safety of this platform. Caterpillars of the same butterfly from Panama, however, when placed directly upon the leaves of their solanaceous food plants are not adversely affected by the sharp, recurved trichomes. The same appears to be true for the caterpillars of *Heliconius* spp., which feed on *Passiflora adenopoda*, which has similar trichomes (Benson *et al.*, 1976).

Most larval butterflies feed on vascular plants, though a few are known to feed on lichens, lycopsids and mosses. In Central America, their exists a riodinine, *Sarota gyas*, whose caterpillars feed on the epiphylls which typically grow on the leaves of a wide range of tropical forest plants (De Vries, 1988). The caterpillars feed on the foliose liverworts and blue-green algae on leaves of at least four different plant species, though the leaf tissue itself is never eaten.

Whereas the caterpillars of most butterflies are phytophagous, a number are carnivorous for at least some part of their lives. Perhaps the best-documented of these is the European large blue butterfly, *Maculinea arion*, whose early instars feed on the flowers of *Thymus* sp. The fifth instar, however, induces ants of the genus *Myrmica* to carry it into their nest where it completes its growth by feeding on their larvae and pupae. In South Africa, early-instar caterpillars of two lycaenids, *Lepidochrysops methymna methymna* and *L. trimeni*, feed in their early instars on the flower heads of *Selago spuria* (Claassens, 1976). These caterpillars are milked for their honey by *Camponotus maculatus* ants who stroke and drum with their antennae in the vicinity of the honey gland. Until they are ready to enter the ants' nest, the lycaenid caterpillars hang on to the host plant if an ant tries to remove them. Caterpillars which are ready to enter the nest are carried there by *C. maculatus* workers. Initially, these caterpillars feed on the silk which surrounds the ants' cocoons but, once they reach the approximate size of a cocoon, they begin to feed on the ant pupae and any available ant larvae. *Lepidochrysops* caterpillars reared in a formicarium indulge in apparent 'begging' behaviour when ant broods are in short supply. Whether the ants feed them has not been observed, though feeding of lycaenid caterpillars by ants has been recorded in a number of species, including the Eurasian alcon blue, *Maculinia alcon*.

Caterpillars of the South-East Asian lycaenid butterfly *Anthene emolus* are obligate symbionts with the weaver ant *Oecophylla smaragdina* (Fiedler & Maschwitz, 1989). Females oviposit on the larval food plant, *Saraca thaipingensis* (Caesalpiniaceae), bearing *O. smaragdina* pavilions. When the egg hatches, the caterpillar wanders around until it is discovered by an ant worker, which palpates it with its antennae, then picks it up in its jaws and carries it into the pavilion. These first instar caterpillars feed upon the host-plant leaves within the pavilion and they are, in fact, unable to survive outside it. Second instars are carried to and from the pavilions, back and forth from their feeding sites on the young shoots, by the ants. Palpation by the ant's antennae causes the second instar to raise its head and thorax, making it easy for the ant to pick it up. Both first- and second-instar caterpillars feed by scraping holes in the leaf surface. Third and fourth instars are seldom carried by the ants, partly because they are too big but mainly because they rarely raise the head end when palpated by the ants; rather than biting holes in the leaf surface, they eat the whole leaf. The relationship is not perfect as, occasionally, an ant may transport a caterpillar from an acceptable food plant on to one on which it cannot feed. There is, however, a tendency amongst the caterpillars to refuse the invitation to be picked up when they are already feeding at a suitable site.

Caterpillars of a number of lycaenid species from different parts of the world are known to feed on Homoptera, such as aphids and coccids. Females of *Allotinus major* from Sulawesi, for example, lay their eggs on or near to females of a membracid bug and the caterpillars which emerge then feed on the membracid nymphs (Kitching, 1987). Both the caterpillars and their prey are protected by the presence of the ant *Anoplolepis longipes*. The Australian lycaenid *Jalmenus evagoras* is often found in association with myrmecophilous membracids. The caterpillars of this species usually make good use of the bugs' presence by feeding upon the honeydew which they secrete (Pierce & Elgar, 1985). First- and second-instar lycaenid caterpillars may actually ride on the backs of adult membracids in order to feed but, unlike those of *Allotinus major*, they have never been seen to feed upon the membracids themselves.

The only pre-social insect which is known to make use of chemical recruitment trails is the North American eastern tent caterpillar, *Malacosoma americanum* (Fitzgerald & Peterson, 1983). Groups of between 50 and 300 sibling caterpillars live in a communal tent which they build in spring on the branches of rosaceous trees. When they are hungry, the caterpillars leave the tent and search for a group of leaves on which to feed. Initially, they may follow existing trails but, as they move on to new areas of the tree, they lay down strands of silk and an exploratory trail pheromone. Once they have fed, successful foragers return to the tent along the same trail, this time reinforcing it with a chemical recruitment marker secreted from a site located between their anal prolegs. Hungry tent mates then preferentially follow these marked trails directly to the food source and they also enable the caterpillars that laid them down to re-locate it next time that they are hungry.

A number of moth caterpillars have been noted as being kleptoparasitic. Caterpillars of the nolid moth *Nola innocua* from Japan, for example, bore into the galls of the aphid *Nipponaphis distyliicola* on the shoots of *Distylium racemosum* (Itô & Hattori, 1982). Here they feed on the internal tissue of the gall, sometimes penetrating two or more galls to complete their growth. *N. innocua* caterpillars feed on these galls during July and August but, in September, when the supply of *Nipponaphis* galls is exhausted, they switch to the galls of another aphid, *Monzenia globuli*.

Caterpillars of the noctuid moth *Neopalthis madates* associate with the colonial theridiid spider *Anelosimus eximius* in Panama (Robinson, 1977). At all times of the day, the moth caterpillars feed upon insect remains within the web and, when they are not feeding, they either rest beneath plant debris trapped in the web or assume cryptic stick-like postures. Caterpillars of a number of other moth species have also been reported as living in spider webs.

▲ Beetles may be carnivorous, herbivorous or omnivorous as well as, in for example the fungus beetles (Erotylidae), fungivorous. Whilst mating, this female *Scaphidomorphus quinquepunctata* is feeding on a small fungus on the surface of a log in rainforest in Peru.

COLEOPTERA Beetles

Beetles, both adults and larvae, have well-developed jaws which are used to effect upon a very wide range of food types,

▼ Chafer beetles (Scarabaeidae) may often be found in large numbers feeding on pollen. This *Amphicoma aleppensis* is the primary pollinator of *Anemone coronaria* in Israel.

▲ Rapidly rising morning temperatures on the South African savannahs can soon make life uncomfortable for insects feeding openly on fast-withering flowers. The answer employed by this blister beetle, *Decapotoma lunata rhodesiana*, is to pull down the *Ipomoea* flower over itself as a sunshade. Hooking its tarsi into the edges of the flower and pulling inwards, the beetle gradually works its way round until the whole of the flower's perimeter has been tucked inwards. Then, reaching up, the beetle pulls the whole flower down on to itself to form a closed tent in which it can feed in relative comfort and security.

depending upon species. Exceptionally, a number of meloid beetles have mouthparts which have evolved to suck nectar and pollen from flowers. In the South American honey beetles of the genus *Nemognatha*, a sucking tube is formed by extension of the galeae while, in the genus *Leptopalpus*, the maxillae are elongated instead. Of beetles with normal biting jaws, some are purely carnivorous, others are herbivorous, many are omnivorous and others specialize on such things as dung or man's stored foods. Most of their feeding strategies are straightforward although, as might be expected in the largest order in the animal kingdom, there are some interesting specialists.

The tiger beetles of the family Cicindelidae are highly active hunters and are some of the most voracious of all insects. The majority are daylight hunters on the ground, especially on sandy soils where their larvae, which are also predatory, can burrow easily, although some hunt on trees and a few apterous species are nocturnal. In the Neotropics, the tiger beetle *Cheiloxya binotata* hunts its prey along rivers by running on flotsam and skating on the water surface. Individuals of another Neotropical genus, *Oxycheila*, forage on underwater prey in fast-moving mountain streams. A number of tiger beetles live along pond and lake margins and on mud-flats where they employ their mandibles to probe into the soft mud in a search of prey.

The family Carabidae contains beetles most of whom are essentially predacious upon a wide range of other small invertebrates. Two species from southeastern Arizona, however, *Helluomorphoides latitarsis* and *H. ferrugineus*, have adopted what seems to be a very dangerous life-style, for they plunder the raiding columns of the army ant *Neivamyrmex nigrescens* apparently with impunity (Topoff, 1969). The ants forage nocturnally along scent trails laid down initially by scouts, and the beetles, in their turn, forage along these trails. When a beetle comes into contact with an ant worker in possession of a food item, such as the larva of another ant species, the ant worker immediately drops it. The beetle then picks it up and consumes it on the spot or may carry it a short distance away from the ant trail before it does so. At times, the beetles may forcibly remove food items from the ants' jaws. Not only do the beetles eat the food collected by *N. nigrescens* but they also make considerable in-roads upon the latter's larvae whenever they came across a cache of them. The beetle, in this situation, actually feeds to the point where its abdomen becomes visibly swollen and protrudes beyond the end of the elytra.

The tropical forest-dwelling staphylinid beetle *Leistotrophus versicolor* resembles a bird-dropping and is a specialized, obligate predator on adult Diptera (Forsyth & Alcock, 1990). Its normal method of prey capture is to sit on vertebrates' dung or carrion and then to ambush any suitable flies which are drawn to these. Where such materials are absent, the beetles resort to a different form of attraction. Under these circumstances, they sit around on leaves and rocks and use chemical secretions to lure small flies towards them. The chemicals involved are laid down from the tip of the abdomen on to the substrate, with the beetle then positioning its head above the point of application. Alternatively it may wave the tip of its abdomen in the air towards small flies that approach it.

Lycid beetles are normally protected from predation, at least by vertebrates, on account of their unpalatability. In Arizona, however, there are four sympatric species of *Lycus* which are preyed upon by sympatric, but edible, cerambycid mimics of the genus *Elytroleptus* (Eisner *et al.*, 1962). In the habitat investigated, the model exceeds the mimic by about 100:1, though the cerambycids have only a small effect upon the lycids. In a sample of 1181 of the latter, only 18 bore cerambycid-inflicted injuries. Attacks by *Elytroleptus* vary in their degree of severity. Some observations revealed that they cause superficial damage to the base of the lycid's elytra which results in bleeding and they then suck up the blood. In other cases, the cerambycid bites its way into the thorax and feeds upon its internal contents; in some instances the now-dead lycid is, at this point, discarded but, in others, the cerambycid may consume the whole of the lycid's internal contents. The feeding relationships within this group of beetles is in itself of interest. In this particular habitat, there are two distinct mimetic associations, one consisting of two yellow-orange lycids mimicked by *E. ignitus* and the other comprising two similarly coloured lycids, with black tips to the elytra, mimicked by *E. apicalis*. Both field and laboratory investigations showed that neither of the cerambycids was particularly fussy about which of the four species of *Lycus* presented itself as a meal.

The North American colydiid beetle *Lasconotus subcostulatus*, in its later larval stages, is a predator upon the larvae and unpigmented pupae of the scolytid beetle *Ips paraconfusus*. *L. subcostulatus* females lay their eggs in concealed sites in the galleries of their host beetle. First- and second-instar larvae feed upon fungi but the third stage is the first predatory one. It burrows through the frass-filled galleries, feeding upon both fungi and *Ips* larvae and pupae as it does so.

In the laboratory, it was observed that, once a larva began to feed on an *Ips* larva, other *Lasconotus* larvae were attracted to it so that as many as nine would be feeding together at any one time (Hackwell, 1973).

Loricera pilicornis is a European carabid beetle which lives on damp soil where aggregations of its main prey, springtails (Collembola), are found. Springtails are able to jump actively to avoid capture and this beetle has evolved a special method of trapping them. On the underside of its head and on the basal antennal segments are a series of long, stiff hairs which form a cage in which the springtails are trapped. During the capture sequence, the basal segments of the antennae strike downwards, trapping the prey between the antennal hairs and those on the base of the head, preventing escape in any direction.

A unique form of predatory behaviour is to be found in fireflies of the genus *Photuris* from the New World. It has been found (Lloyd, 1975; Lloyd & Wing, 1983) that females of several *Photuris* species are light-seeking, aerial predators, i.e. they are led to their prey, males of other species of firefly, by the signals that they emit to attract females of their own kind. Female fireflies will also use aggressive mimicry to attract the males of other species of *Photuris*, as well as those of the genera *Photinus*, *Pyractomena* and *Robopus*. In this instance, they perch on a leaf and flash the female signal of their prey species. This attracts the male of that species who, on contacting the mimic, is immediately devoured. Females of *Photuris versicolor*, for example, are able to attract the males of four other species of firefly with distinctively different flashing sequences. In order for *P. versicolor* females to attract males of other species, they have to be able to copy not only the flashing sequence of the latters' females but also the delay which occurs after the male has signalled.

In field observations, two different *Photuris versicolor* females answered males of both *Photinus tanytoxus* and *Photuris congener* and another female answered simulated flashes of *Photinus macdermottii* while she was in the process of eating an unidentified *Photuris*. The ability of the females to change their flashing pattern in response to a male stimulus is quite remarkable. In the 1975 study, 11 females responded correctly to a *Photinus macdermottii* simulation and were then able to change to the correct reply to a *Photinus tanytoxus* simulation. Although some females could sometimes make the change immediately, in others, the male flash had to be repeated a number of times before the new response commenced. Although the mimicry is not perfect, it is effective enough to provide the *Photuris versicolor* females with sufficient prey, for each seldom answered more than ten males without one flying to her and ending up as a meal.

The paussid beetles as a whole show structural modifications which fit in with their normal life-style as kleptoparasitic inquilines in the nests of ants. Their enormously developed antennae, along with areas of the body wall, produce secretions which are readily licked up by the ants in return for the latter feeding the beetle. *Edaphopaussus favieri* lives in nests of the ant *Pheidole pallidula* in North Africa and parts of southern Europe. Like other paussids, it is readily licked by the worker ants but, in this case, it is not fed by them. Instead it is a predator on *P. pallidula* eggs, larvae and workers. In order to pass from one ant nest to another, the adult beetles follow the ant recruitment trails (Cammaerts *et al.*, 1990).

Like the paussids, many members of the family Pselaphidae are kleptoparasites in ant nests. The feeding behaviour of the European species *Claviger testaceus* is well documented but the similar *Adranes taylori* from the Pacific Northwest of North America will suffice as an example. Like other members of the subfamily Clavigerinae, oily golden hairs, or trichomes, which are highly attractive to ants, are present on various parts of the body (Akre & Bruce Hill, 1973). Observations in the laboratory indicate that the beetles spend a lot of the time standing amongst larvae and groups of adult ants. The beetles find the ant larvae highly attractive and the latter feed them by trophallaxis at frequent intervals. In general, the younger ant larvae more often supplied food than the older larvae. *Adranes* also feeds on dead larvae and dead workers when they are available. Although the beetle's trichomes are only moderately attractive to the adult ants, the younger larvae find them highly desirable. A single beetle may walk around with as many as four larvae hanging on to the trichomes with their jaws.

Adults of the North American scarabid genus *Cremastocheilus* are known to associate with ant colonies; the western species, *C. armatus*, is found in the nests of various *Formica* species, especially *F. obscuripes*. During spring, the adult beetles feed upon both ant larvae and pupae; they pierce the latters' cuticle with their sharp maxillae and slowly suck them dry (Alpert & Ritcher, 1975). *Aphodius rufipes* and *Geotrupes spiniger* are European scarabid beetles which feed on horse dung. Typical of many dung beetles, females of *G. spiniger* excavate chambers in the soil beneath the dung, fill them with the dung and lay their eggs on these brood masses. Larvae of *A. rufipes* live in short, vertical shafts in the soil below the dung and emerge from these to feed on the main dung mass. Third-instar larvae of the latter burrow deeper where they may, accidentally it seems, penetrate the brood chambers of *G. spiniger*. Under these conditions, the *A. rufipes* larvae become facultative kleptoparasites by feeding upon the *G. spiniger* brood masses (Klemperer, 1980). *A. porcus* goes one step further, for the females purposefully enter the brood chambers of *G. stercorarius* and lay their eggs directly on the dung placed there by females of the latter species; thus they are kleptoparasitic by choice rather than by accident.

Staphylinid beetles of the genera *Idioptochus* and *Leucoptochus* are found in the arboreal nests of *Graallatotermes* termites in Mozambique (Kistner, 1973). In order to obtain food, the beetles approach termite workers with their antennae spread out and touch the worker's antennae with them. This causes the worker termite to move forward and touch mouthparts with the beetle and, following some movements of the mouthparts of both participants, the termite regurgitates a drop of fluid which is taken up by the beetle. Three further species of staphylinid, *Trichopsenius depressus*, *Xenistusa hexagonalis* and *Philotermes howardi*, are associated with the termite *Reticulotermes virginicus* in North America (Howard, 1978). All three species have feeding behaviours which effectively mimic those of their host. In order to obtain stomodeal fluids, the beetle approaches from the front and uses its mouthparts and antennae to tap vigorously the mouthparts, labium and frons of the termite host. The termite then usually regurgitates some fluid which is taken up by the beetle. Alternatively, and in laboratory observations more often, the beetle approaches from behind and vigorously taps the lateral edges of the tip of the termite's abdomen with its antennae and then with its mouthparts. This usually results in the termite exuding a drop of proctodeal fluid which is then taken up by the beetle. The beetles also allogroom their hosts, possibly obtaining food in the form of surface lipids as they do so.

▲ **A group of *Orthoschema* sp. longhorn beetles (Cerambycidae) feeding at a sap-flow on a tree in Brazil.**

Many tenebrionid beetles are scavengers and none more so than the denizens of the Namib Desert dune systems with their ephemeral supplies of wind-blown food materials. Even more of a problem for them, however, is obtaining sufficient water from the fogs which are so characteristic of the area. Two genera at least have solved the problem by adopting specialized behaviour patterns. *Lepidochora* sp. dig small trenches across the path of the prevailing wind in which the moisture that the wind carries is caught and condensed. The beetles later retrace their original route and drink the water that has collected. As the soaking fogs blow in, *Onymacris unguicularis* climbs to the top of a dune and stands head-down with its abdomen elevated and directed into the wind. The fog droplets condense on the beetle's body and run downwards to the mouth to be imbibed.

HYMENOPTERA Sawflies, wasps, ants, bees

WASPS

Wasps of the large genus *Tetrastichus* are mainly primary and secondary parasites and egg-predators of a variety of arthropods. One species, however, has been found (Hawkins & Goeden, 1982) to have developed an association with gall midges (Cecidomyiidae) of the *Asphondylia atriplicis* complex on various species of *Atriplex* (Chenopodiaceae) in arid regions of California. The female wasp lays her eggs in the inner wall of the *A. atriplicis* gall and, within this tissue, develop endogalls, stimulated by the wasp larvae. Growth of the endogalls often, but not always, results in the crushing and killing of the midge larva. The *Tetrastichus* larva is therefore, in a way, a phytophagous kleptoparasite of the gall midge; the relationship also seems to be an obligate one for no gall produced by the wasp has yet been found away from the midge gall.

While most adult hunting wasps utilize nectar, sap, honeydew or other exudations as a food source, some extract part of the body fluid from the insects which they catch to provision their larvae. Alternatively, they may themselves feed upon a proportion of the food they catch, rather than supplying it all to their young, or they may even catch different prey for their own consumption, e.g. the Amazonian nyssonid wasp *Stictia signata* feeds on the mosquito *Aedes aegypti* while provisioning its young with tabanids (Diptera). In the more primitive hunting wasps, the prey is captured first and then the burrow is dug afterwards to receive it. In more advanced species, the burrow is dug before the prey is captured. Again, in more primitive hunting wasps, mass-provisioning is practised, i.e., all the prey necessary for the larva's development is packed into the burrow at one go and the latter is sealed and never re-opened by the female. More advanced species practise progressive provisioning, re-opening the burrow at intervals to supply the larva with freshly caught prey. Progressive provisioners also take the opportunity to clean the nest of any uneaten prey each time the burrow is opened.

Wasps of the family Pompilidae prey upon spiders, which they paralyse with their sting before transporting them to their nest where they lay a single egg upon them. Wasps of the genus *Pepsis* specialize in large mygalomorph 'tarantulas' as prey for their developing larvae. These wasps are able to handle their victim without any problems, despite the fact that the spider is also a ferocious predator and is often considerably larger than its attacker. *Pepsis* are very specific in the spiders that they choose and, to ensure that she has found the correct species, the female wasp must first of all check it with her antennae. During this inspection, the spider allows the wasp to walk all over it without making any attempt to defend itself or to escape. Having completed her check and determined that it is the right species, the wasp then proceeds to dig a burrow a short distance away, while the spider just sits awaiting its fate. The burrow completed, the wasp now goes back to the spider and, with a considerable amount of manhandling but with no response from the spider, finds a soft spot between the joints of the spider's exoskeleton through which to insert her sting. At this juncture, the wasp grasps one of the spider's legs, to which the latter responds by at last trying to defend itself and the two

Opportunistic predators. Many large sawflies (Symphyta) are opportunistic predators. Although somewhat clumsy in their technique – they simply jump on top of any likely insect – they may succeed in overcoming even large predatory prey. Soft-bodied prey is taken mainly, the sawfly's jaws being unable to cope with hard-bodied shiny prey such as ladybird beetles.

▶ This *Tenthredo mesomelas* in England has killed a white-legged damselfly, *Platycnemis pennipes*.

may struggle for a short time before the inevitable happens and the wasp pushes her sting home. The effects of the venom are virtually instantaneous and the spider rolls over on to its back in a state of paralysis from which it is destined never to recover. The wasp usually now grooms and malaxates the site of entry of the sting before dragging the spider into the burrow where a single egg is attached to its abdomen.

A study in southern Colorado in the USA (Gwynne, 1979) has provided an insight into the different ways in which pompilids are able to deal with the same type of prey, namely burrowing lycosids of the genus *Geolycosa*. The area used in the study had a substrate of sand in which spiders dug burrows which either had open tops or were closed off with a silken collar. All three wasp species studied hunted in the same way, by moving rapidly, and seemingly erratically, over the sand, continually flicking their wings and palpating the ground with their antennae. The behaviour of the two species of *Anoplius* studied, namely *A. marginalis* and *A. cylindricus*, is very similar except that, being somewhat smaller, the latter takes smaller spiders as prey. If the wasp finds an open burrow containing a spider, she just walks straight into it; if, however, the burrow is closed with a collar, then the wasp has to dig her way into it, not always successfully. Laboratory investigations reveal that, within the burrow, there may be a brief struggle before the wasp stings the spider. From the spider's vertical burrow, the wasp then digs a horizontal side passage, at the end of which she constructs a cell into which the prey is placed. The side passage is then filled with sand and the main burrow is at least partially filled as the wasp emerges. The behaviour of the third species, *Pompilus scelestus*, differs from that of the other two in that the prey is not left in its own burrow. The wasp either lures the spider to the burrow entrance or enters the burrow to sting it. If the spider manages to close the collar, then this species will also attempt to dig it out. Once the prey has been subdued, it is dragged by the wasp to a suitable piece of vegetation and hung up out of the way while she searches for a suitable spot in which to dig a burrow of her own. Of these three species, *A. cylindricus* restricts itself to *Geolycosa* spiders as prey and it also has a restricted habitat preference. *A. marginalis* and *P. scelestus*, on the other hand, take other lycosids as well as *Geolycosa* and have wider habitat preferences. A fourth species from the same area but not included in this study, is known to take agelenids as prey, as well as the lycosids taken by the other species.

The North American pompilid *Anoplius depressipes* specializes in the pisaurid spider *Dolomedes triton*, which itself hunts on the surface of ponds and lakes. The wasp normally hunts the spider amongst the emergent vegetation on and around such habitats. Remarkably, however, when a *D. triton* dives below the water surface to escape the wasp's attentions, the latter is able to dive also and sting and capture the spider. The wasp is then able to bring the spider to the surface and drag it across the water to dry land, where it employs the normal pompilid behaviour of dragging it to its burrow (Roble, 1985).

Tiphiid wasps of the genus *Methoca* show pronounced sexual dimorphism since the females lack wings and are smaller than the winged males. They are parasitoids of larval tiger beetles (Cicindelidae) and have special morphological and behavioural characteristics to help them to handle their highly carnivorous prey. On finding a beetle larva in its burrow, the female wasp arranges it so that the larva can grasp her in its jaws around her thorax or waist but cannot crush her. As the larva rears up to grasp the wasp, it exposes its lightly sclerotized neck region to the wasp's sting. The wasp stings the larva several times and then drags it to the bottom of its own burrow where she lays a single egg on it. Finally she carefully fills the larval burrow by carrying pieces of soil into it.

Wasps of the sphecid subfamily Larrinae

specialize in supplying various kinds of Orthoptera as food for their developing larvae. The foraging behaviour of two species of *Tachytes*, *T. mergus* and *T. intermedius*, both from the USA, has been studied in some detail (Kurczewski, 1966; Kurczewski & Kurczewski, 1984)). Both species prey on the nymphal and adult stages of the pygmy mole cricket, *Neotridactylis apicalis*, and the way in which the two wasps hunt for and capture them is very similar. In a suitable habitat, the female wasp runs around on the sand in zigzags, with frequent changes in direction, flying up every now and again and moving anything from a few centimetres to several metres to a new search area. As she runs around, the wasp keeps her wings folded flat over her back and her antennae stretched forwards, the tips continually tapping the soil surface. Every now and again, she stops and digs a few millimetres into the sand with her mandibles before moving on again. Once a pygmy mole cricket has been located, she begins to dig in earnest, employing her front legs as well as her mandibles. On returning to the hole, the wasp extends her antennae into it and pauses for a moment, presumably checking that the prey is still there, before she continues with her digging. As digging proceeds, she becomes more excited, often backing out with a load of sand and then circling round before re-entering the burrow. As the wasp approaches the prey, digging seems to continue with the jaws alone. When finally the mole cricket is reached, the wasp immediately flies up with it clutched in her mandibles around the sides of its head or neck. She then lands with it a few centimetres away and, holding it dorsal side up, bends her abdomen round and stings it in the underside of the thorax. If it attempts to escape by digging its way out, the wasp follows the pygmy mole cricket and, as it takes off, grasps it in mid-air before landing with it and stinging it.

The females of the two species then fly with the immobilized prey to their burrows before returning to the same area to continue their foraging. Whilst in flight, both species carry the prey, usually ventral side up or on its side, held in their legs, pressed tightly against their underside and with one or both of its antennae held in the mandibles. At the burrow, *T. mergus* holds the prey with her hind legs, supports herself on her middle legs and opens the entrance with her front legs. *T. intermedius*, however, who does not temporarily close her burrow, flies straight into the entrance with her prey. If the prey is large, she may leave it at the entrance, enter the burrow, turn round inside and then return to the entrance to pull in the prey by holding on to its antennae with her mandibles. Prey-stealing is not uncommon in *T. intermedius*, especially where large numbers of the wasps nest in close proximity to one another. Stolen prey is usually re-stung before it is removed to the thief's nest.

Whereas *T. mergus* and *T. intermedius* specialize in pygmy mole crickets, two other North American species, *T. parvus* and *T. obductus*, sympatric throughout much of their range, prey upon Tetrigidae (groundhoppers, grouse locusts or pygmy grasshoppers). Prey is sought while the female is in flight, *T. obductus* employing low, rapid flights 1–4 m from the nest while *T. parvus* flies more slowly and sinuously, sometimes hovering, at distances between 1 and 5 m from the nest. On finding prey, both species dive down on to it, land on its dorsal surface and then bend the tip of the abdomen beneath its thorax to sting it. The return to the nest with the prey differs: *T. obductus* adopts a rapid, sinuous flight and *T. parvus* a slow, waggling flight. Both enter the unplugged burrow entrance directly. The style, direction and duration of this return flight may be modified by the attentions of kleptoparasitic flies.

Females of the genus *Philanthus* in the sphecid wasp subfamily Philanthinae provision their subterranean burrows with bees. Most species, such as the well-known European bee-wolf *Philanthus triangulum*, which preys on hive bees, forage on flowers. Having caught and stung the prey, the wasp then squeezes the nectar out of the bee's crop and imbibes it. Like the bee-wolf, the American species *P. crabroniformis* will often chase almost any small insect that

▼ The beautiful metallic hunting wasp *Chlorion lobatum* specializes in capturing crickets with which to provision her nest. The prey is flushed out from beneath stones or other cavities, and stung once it is in the open. This female is administering a final sting to the cricket's neck region, after which she towed it the short distance into her burrow. However, the battle is not always in the wasp's favour. The crickets use their powerful back legs to good effect. If they can kick their attacker away to a distance of some 20 cm, then she will probably fail to locate her prey again, and will finally give up after scurrying around for a minute or two in ever-increasing circles. Surprisingly enough, a cricket can be stung many times in several parts of the body, including the bases of the back legs, and still kick out with sufficient power to send the wasp flying in a final life-saving gambit. Once the wasp has disappeared, the cricket can recover completely within minutes, seemingly unaffected by the repeated stings and the frantic nature of the battle, which may have lasted many minutes. Nepal.

enters its visual field, though it does not necessarily attack it. A study of this species in the field (Alcock, 1974) reveals that females often visit flowers and occasionally dart at halictid bees, their main prey, but apparently do not catch them there. Instead, the *P. crabroniformis* females forage by searching in open areas with dried mosses where halictid bees nest. The wasps sit on a suitable perch and make short foraging flights lasting just 5–20 seconds, followed by a short pause after which another flight is made. The wasps attack the halictid bees as they make their final descent to their burrow entrance after a foraging flight. On seeing the flying bee, the wasp follows it from just above and behind and then suddenly accelerates and dives on it. The two collide and fall to the ground where the victim is manipulated so that the wasp's jaws grasp the bee behind the head; the wasp then stings the bee in the thorax. Occasionally, the wasp may take bees at rest on vegetation or on the ground and, on one occasion, a wasp was seen to actually take one out of the air without falling to the ground. Sympatric with *P. crabroniformis* is a second species, *P. gibbosus*, which also feeds on halictid bees but seems to be more adaptable in its foraging behaviour (Lin, 1978). Not only does it take prey in a manner similar to that of *P. crabroniformis* but it more often takes bees in flight, both on arrival at and departure from their nest, and also from the ground when the bee is at the nest entrance. Similar behaviour is exhibited by the African larrine wasp *Palarus latifrons*. Females of this species pursue flying worker hive bees and swoop on them in mid-air, forcing them to the ground where the wasp then stings them and milks them of their nectar before carrying them off to their nest.

ANTS

Apart from the different patterns of foraging behaviour described below, ants have some interesting feeding behaviours. In common with other hymenopterans, they perform trophallaxis, i.e. they pass food from mouth to mouth between all members of the colony. A very unusual form of feeding behaviour, referred to as 'larval haemolymph feeding' (LHF), has been described for the ants *Amblyopone silvestrii* (Masuko, 1986) and *Leptanilla japonica* (Masuko, 1989), both from Japan. In both species, LHF is the only way in which the queens can feed. Prior to LHF in *A. silvestrii*, the queen handles a fifth-instar larva and strokes it with her antennae. She then grasps it between her sharply pointed mandibles at one of the intersegmental grooves between segments two and four and squeezes. The mandibles pierce the cuticle, releasing droplets of haemolymph which she then laps up. She continues feeding on the larva's haemolymph by pumping it directly from its body cavity. The larvae of *L. japonica* are more highly developed in that they possess a pair of duct organs, the 'larval haemolymph taps' used specifically for LHF. On the fourth abdominal segment of the larva, there are a pair of slits leading into a short duct which opens internally into the body cavity. Before feeding, the queen again strokes the fifth-instar larva with her antennae before grasping it in all three pairs of legs while lying on her side. She then searches with her mandibles until they close on the fourth abdominal segment and she begins to feed, apparently using her pharyngeal pump to suck haemolymph up directly, since no lapping action akin to the initial feeding activity of the *A. silvestrii* queen was observed in *L. japonica*.

Many ants are totally omnivorous and will forage for any form of animal or plant life, living or dead. Tropical army ants, for example, are the scourge of the forest floor and will capture any small animal which is foolish enough not to move out of harm's way when they are on the rampage. At the other extreme, there are ants which depend entirely upon plant material as a source of food. Harvester ants of the genus *Messor* collect seeds, remove any chaff and then bite off the radicle to prevent germination before storing them in their 'granaries' for later consumption by the colony. The myrmecine leaf-cutting ants of the tribe Attini cut pieces off leaves and petals and transport them back to the nest where they form the basis of their fungus gardens.

When watching ants scurrying to and fro from their nest, it is not always easy to

▲ This female *Palarus* sp. hunting wasp from Kenya has caught and paralysed a hive bee, *Apis mellifera*. Before carrying the bee back to her nest, she is taking a meal by drinking nectar from her victim's mouth.

▼ Army ants carry out swarm raids with large numbers of individuals fanning out over a wide area and catching anything edible which they encounter. In rainforest in Trinidad, these workers of *Eciton burchelli* are transporting a large centipede back to their bivouac.

appreciate that there is a definite order built into their behaviour. Details of the foraging activities of a large number of ant species have now been elucidated and the concept of the recruitment system figures high in these. In its most primitive form, the recruitment system involves a single ant calling on the services of a single nest mate, so-called 'tandem-running'. As the system becomes more efficient, so trail pheromones are laid to and from the foraging site and greater numbers of workers are recruited. In the most advanced systems, it is the amount of trail pheromone exuded that establishes the number of workers recruited.

The attinine ants all feed upon fungi, which they culture in their nests in cavities beneath stones, under bark, in rotting wood or in the soil. The more primitive species use insect faeces, the remains of insects and various forms of vegetable matter as a substrate on which to cultivate the fungus. The higher forms have come to rely on vegetable matter as a compost.

The ants of the genus *Atta* have gained the common name of leaf-cutting ants from their habit of cutting up leaves and flowers and transporting them to their vast underground nests. The ants chew the plant tissues into small pieces and use them as a substrate upon which they grow a fungus, their main food source. Initiation of a new colony of attine ants (Weber, 1972) begins when a newly fertilized queen leaves her parental nest with fragments of its fungal mycelium stored in her infrabuccal pouch, a blind pocket branching off the proximal end of the pharynx. Having found a suitable site for a nest, she digs a hole in the ground with her mandibles and then ejects the contents of her infrabuccal pocket in the form of a pellet, which she then manures with droplets of faecal material. Once the fungus has established its growth, she lays eggs on it.

The ability of leaf-cutting ants to maintain a pure culture in their fungus gardens is quite amazing and owes a great deal to their behaviour. For a start, the ants groom themselves and each other constantly, so that they are always clean, and 'licking' with the mouthparts applies salivary exudates which may be fungicidal or bactericidal in their action. Also, as cut pieces of plant material are brought back to the nest, the smallest of the workers usually attach themselves to them and begin to swab the surface, the process continuing even after the leaf has arrived at the fungus garden.

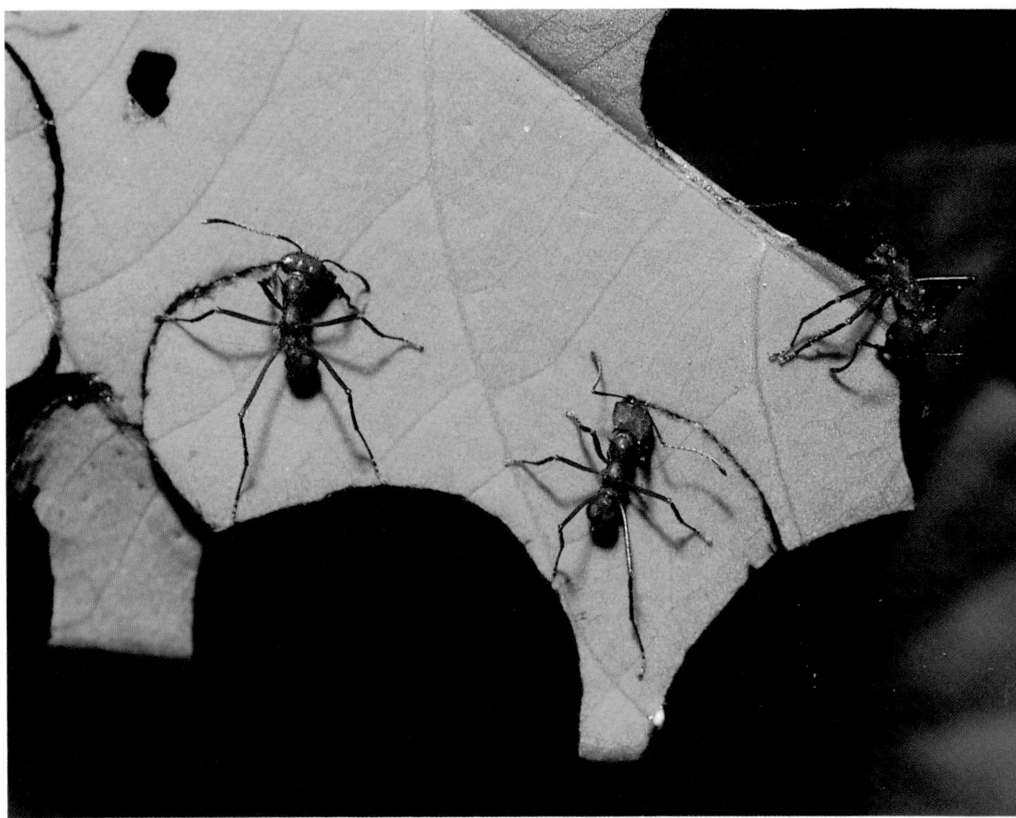

▲ Workers of the leaf-cutting ant *Atta cephalotes* cutting portions of leaf in rainforest in Costa Rica. These pieces of leaf are carried back to the nest and are incorporated in the ants' fungus garden, the fungus providing food for the ant larvae.

The ants then add their excretions to each new piece of plant material and then inoculate it with tufts of hyphae. The fungus, once it has been inoculated on to the chewed-up leaves, grows rapidly in the warmth and humidity of the nest and eventually produces swellings which are cut off by the ants and fed to the larvae. The fungal protoplasm normally supplies the ant colony with both food and water but, in the dry season, or when added plant material consists of dead leaves and flowers and is therefore low in water, the workers will drink from external water sources.

The recruitment system for foraging in the leaf-cutting ant *Atta cephalotes* has been investigated in the laboratory (Jaffe & Howse, 1979). A food source is initially located by scouts who, on discovering it, touch it with their antennae and mouthparts. They then return quickly to the nest without any food but laying a recruitment trail behind them by touching the tip of the abdomen against the substrate. On reaching the nest, the scout then returns to the food, still laying a trail. Often the scout meets other ants before arriving at the nest, in which case it turns back again immediately. Under normal circumstances, the scout makes several trips back and forth from food to nest or to where it first met other ants. During its journeying, the scout frequently stops by other ants and they then palpate it with their antennae before rushing off to the food source; others make for the food source as soon as they encounter the scout's trail, without first making contact with the scout. The numbers of ants recruited to a new food source depend upon the quality of the source and this is indicated to other workers not by the numbers returning from the initial foray but by the concentration of the pheromones in the recruitment trail.

The honey ant *Myrmecocystus mimicus* is a scavenger which not only captures termites as prey but also collects nectar from flowers and tends homopteran bugs. As with leaf-cutting ants, the food sources are initially discovered by scouts who then return to the nest, laying down a recruitment trail by alternately streaking the ground with the tip of the abdomen and then raising it for 2 to 3 seconds as they run (Hölldobler, 1981). At fairly regular intervals, the scout stops and looks back the way it came, sometimes climbing an adjacent object before doing so, and then continues its journey back to the nest. If, when the scout arrives in the vicinity of the nest, some of its fellow workers are present, it may run around randomly from one to another, performing a rapid jerking display to each of them. This display seems to affect the other workers who then start their own jerking displays, the speed at which they run around noticeably increasing.

Within a few minutes of her arrival back at the nest, the scout ant heads off again in the direction of the food source, once again applying trail pheromone and stopping every now and again to look back, towards the nest this time, before continuing. This time she is accompanied by the workers that she has recruited, which may number anything from between 20 and 300 individuals. The site of trail-pheromone production seems to lie in the hind gut for, if artificial trails of hind-gut extract are laid down, the ants are recruited to these and will follow them, but only as long as they have first been excited by a scout's jerking display. If an artificial trail is laid across a natural one, a number of the workers following the genuine trail will deviate on to the artificial one. No such response is obtained if extracts of the poison gland or Dufour's-gland are used in place of hind-gut extract. Artificial trails do not, however, cause the scout ants to deviate from their own trails. Although extracts from the poison gland do not produce recruitment, they do induce excitement in the workers. If, a stick is dipped in poison-gland extract and presented to workers then, in their excited state, they will follow the stick if it is subsequently moved along just above an artificial hind-gut-extract trail. It would seem, therefore, that the poison gland produces an excitatory substance which stimulates following and foraging behaviour and the hind gut produces the trail pheromone, at least in this species of ant.

Ants may exhibit what appear to be carefully organized raids against animal prey, such as termites. *Pheidole titanis* is a termitophagous desert-dwelling ant which comes from the southwestern USA and western Mexico; its foraging behaviour has been carefully documented (Feener Jr, 1988). The ant's main prey is a *Nasutitermes* sp. which builds carton nests in trees and forms protected runways down the tree trunks to its main food source, leaf litter beneath the tree. Termite foraging parties consist of between 200 and 2000 workers, surrounded by a smaller number of soldiers, and it is these parties which form the main focus of attack for the ant. Initially, the termite foraging parties are encountered by a scout ant, which immediately returns to its nest, laying down a recruitment trail as it does so. Within a minute of entering the nest, workers begin to appear round the nest entrance and, once 100 to 200 are mobilized, they begin to follow the scout's chemical trail. The raiding party may eventually number anywhere between 200 and 2000 workers and soldiers, who follow the scout's trail in a well-organized group until they are within about 10 cm of the termite foragers. The perimeter of the termites' foraging area is determined by a stealthy approach of the ants' vanguard so as not to disturb their prospective victims. At this point, the ants form a tightly packed group and then attack in one of two ways, depending upon the nature of the termite group. If this group is roughly 25 cm or less in diameter, the ants perform an encircling movement and overwhelm the termite soldiers; each then rushes in and attempts to capture a single termite by grabbing it around the neck from behind. More diffuse termite groups are attacked by a fanning-out of the ant raiding party and the termites are captured on the run. Once an ant has a termite firmly in its grasp, the original outward trail is then followed back to the nest.

Foraging behaviour of the large, termitophagous ponerine ant *Pachycondyla commutata*, in Brazil, follows a similar pattern (Mill, 1984). Scout ants wander around the forest floor at random, searching for foraging parties of *Syntermes* termites. On finding one, they return to their nest along a more direct route, with the gaster bent downwards and the partially extended sting touching the ground, presumably laying down a recruitment-trail pheromone from the hind gut. Workers recruited by the scout leave the nest within about a minute and follow the scout to the termites' foraging site. The ants spread out in search of prey and then return with it, in a column, to the nest.

The behaviour of the West African ponerine, obligate termite-predator *Megaponera foetens* differs in some ways from that of the Brazilian species described above (Longhurst *et al.*, 1979). This ant feeds on species of *Macrotermes* and *Odontotermes*, who forage under the cover of soil sheeting constructed by the workers. As with the ants already discussed, the termites are first

▲ A number of ponerine ants specialize in taking termites as prey. Here a *Termitopone* sp. ant is returning to its nest with the body of a *Syntermes* sp. soldier termite in Peru.

found by scout ants which then return to the nest, laying down a recruitment trail which, in this species and also in a number of other ponerines, comes from the poison gland. The scout ant then recruits a group of sister workers who follow the pheromone trail to the termites' foraging area. Here the workers begin to dig into the soil sheeting in order to extract the termites. When an ant comes across a group of termites beneath the sheeting, it releases a secretion from the mandibular glands which attracts more workers to the site, so that the maximum number of termites can be captured.

Ants of the ponerine tribe Cerapachyini specialize in taking other ants as prey. An Australian species of *Cerapachys* was found in laboratory investigations (Hölldobler, 1982) to prey mainly on *Pheidole* spp. Prey ants are initially found by scouts who then return to the nest and recruit other workers. These then follow the scouts along a previously laid pheromone trail back to the prey's nest. *Cerapachys* workers are able to attack, sting and immobilize *Pheidole* workers with impunity, even when heavily

outnumbered, since they are protected by their heavy sclerotization. *Cerapachys* raiders returning to their nest with booty are able to recruit other workers to come back with them to the *Pheidole* nest; following behaviour is initiated by a raider repeatedly raising its gaster. Larvae and pupae are stung briefly before being transported and, where prey is abundant, brood alone is carried back to the *Cerapachys* nest. In the laboratory nest, stung *Pheidole* larvae remained alive but comatose for up to 2 months, providing a ready supply of food. That the *Cerapachys* are able to sting and kill the adult *Pheidole* with ease and yet the tender larvae and pupae are simply narcotized points to a very carefully controlled stinging ability. Adult *Pheidole* are only used as prey items when brood is not available.

Members of the genus *Amblyopone* from as widely separated regions as Australia, Africa and North America appear to specialize in geophilomorph chilopods as prey. Colonies of the cryptobiotic, hypogaeic African species *A. pluto* are small, not exceeding 100 individuals, and the workers forage alone (Gotwald Jr & Levieux, 1972). Prey is first paralysed with the sting before being dragged back to the nest and placed close to the brood. The workers then feed by cutting holes in the prey's intersegmental membranes and inserting their mouthparts into them. Larvae are placed as close as possible to the prey and they then find the holes in the membranes on their own, put their heads in and begin to feed. The behaviour of these primitive ants in capturing and feeding on paralysed rather than dismembered prey parallels the behaviour of many hunting wasps and may point to a common ancestor for the two groups.

Army ants of the Neotropical genus *Eciton* are well known for their highly organized foraging behaviour. Foraging columns up to 100 m in length extend in several directions from the bivouac, the workers taking any small invertebrate prey that comes within their clutches. Field observations on *E. hamatum* have revealed that, on finding prey, a scout worker returns to the nearby column, laying down a trail by dragging her abdomen on the substrate. On reaching the column, she runs both up and down for 5–10 cm in each direction, touching other workers momentarily with her antennae and body. She then runs back and forth from the column to the prey site, continually making contact with other workers and seemingly reinforcing the pheromone trail. Within half a minute of the scout reaching the column, recruits may be diverting towards the prey, with 50 to 100 individuals leaving the column within the first minute. Workers coming into contact with prey will return to the column and recruit further workers. *E. hamatum* uses hind-gut exudates to mark the normal foraging trails but the recruitment trails leading to prey are marked with some other substance.

Pheidologeton diversus is an Asian myrmecine ant which, like some doryline army ants, employs swarm raiding (Moffet, 1984). Colonies of this species normally have one or two main foraging trails, often of long duration, which extend for anything between 5 and 100 m from the nest. These main pathways act as a starting point for foraging raids, which can originate at any time and from any point on them. Raids begin abruptly when a group of workers move out from some point along the foraging trail, or from the nest, and advance initially at speed. As the raid advances, it begins to fan out but, as it does so, the speed of advance begins to slow down. At the leading edge of the raid is a narrow band of swarming ants and trails lead back from this to a single main pathway which gets longer as the raid advances. Any invertebrate, whether large or small, is captured, the larger ones being borne down by sheer weight of numbers. They are cut up on-site and the pieces carried back to the nest. Unlike the dorylines, which are strictly carnivorous, *P. diversus* also collects seeds, fruits and nuts on its raids.

A number of ant species have developed special relationships with other organisms to ensure a source of food for themselves and their offspring. Obligate associations between ants and 'ant-plants' (myrmecophytes) have now been found in most parts of the tropics. These associations are usually symbiotic in nature in that the plant supplies the ant with a source of food while the role of the ant is to protect the plant from other herbivorous animals. An alternative term to describe the relationship is *mutualism*. Ants of the genus *Pseudomyrmex* live on Central American species of 'swollen-thorn' acacias (Janzen, 1966). The ants are tenants in the swollen bases of the acacias' stipular thorns and the plant provides them with all of their food; in return, they keep the foliage free of herbivorous animals and the area around the plant free of other competing plant species. Sugar for the ant colony is provided in the form of nectar secreted by foliar nectaries, which on the swollen-thorns are larger than on

► **Ants of the genus *Pseudomyrmex* are tenants in the swollen bases of the stipular thorns on acacias, the plant providing them with all of their food. In return, the ants keep the foliage free of herbivorous animals. Protein is furnished for the ants in the form of special Beltian bodies which are the modified tips of leaflets.**

other acacias. Protein is furnished in the form of special Beltian bodies which are the modified tips of leaflets. The Beltian bodies are made up of large, thin-walled cells full of nutrients containing protein, fatty substances and vitamins which are harvested by the ants not long after they are expanded on the new leaves. The worker ants cut the Beltian bodies up and feed them to the larvae and it is thought that the workers probably take up some of the fluid released from damaged cells during this process. Just occasionally, the ants may catch a small insect and use it to feed the larvae. At least one species of ant, *P. nigripilosa*, has, as it were, reneged on its deal with the acacia for it lives in the swollen-thorns and feeds on the nectar and Beltian bodies but does not protect the plant (Janzen, 1975). *P. nigripilosa* does not, however, get things all its own way for it is regularly evicted from its nests by obligate acacia-ants.

The ant *Crematogaster nigriceps* indulges in a similar sort of life-style on the 'whistling-thorn' acacia, *Acacia drepanolobium*, in Africa. The ants in this instance feed from foliar nectaries but obtain proteins and oils not from Beltian bodies but from small outgrowths on the anthers. In South-East Asia, *C. borneensis* forms an obligate association with various myrmecophytic trees of the genus *Macaranga* (Filia & Maschwitz, 1990). As with the acacia–ant relationship, the tree provides the ants with proteinaceous food, but in the form of small bodies on the stipules, while the ants ward off phytophagous insects. Soluble carbohydrate comes, however, not from nectaries but from scale insects which are kept by the ants in the hollow stems in which they also have their nests. A similar relationship exists between the ant *Myrmelachista*, the Central American lauraceus tree *Ocotea pedalifolia* and two *Dysmicoccus* mealy bugs (Stout, 1979). In this case, the tree provides nesting sites for the ants and feeding sites for the mealy bugs, from whom the ants obtain carbohydrate. In Africa, two species of ant, *Pachysima aethiops* and *P. latifrons*, associate with the myrmecophyte *Barteria fistulosa* (Janzen, 1972). Ant colonies live with a single queen in hollow, living branches where scale insects provide carbohydrate and fungus gardens cultivated by the ants provide other nutrients.

Many ants have developed associations with Hemiptera and the caterpillars of lycaenid butterflies but without the necessity of the accompanying myrmecophytic relationship in the above examples. The herding of aphids (Hemiptera: Homoptera) by ants as a source of honeydew is well documented but ants also form feeding relationships with other homopterans. In these relationships, the ants obtain mainly sugars from the honeydew which is used to supplement their normal diet. Whereas ants often tend these homopterans at their normal feeding sites, protecting them where possible from attack by predators, in some instances special provision is made by the ants to protect their food sources by building special structures. Two species of ants from South-East Asia, *Polyrhachis arachne* and *P. hodgsoni*, build their nests in the hollow stems of broad-leaved bamboos on which they forage (Dorow & Maschwitz, 1990). Although they have not actually been seen to feed on other arthropods, it is assumed that this is the ants' protein source, since remains of these have been found in the ants' nests. Honeydew is provided by pseudococcids for which the ants build special structures to serve as a protection. These consist of pavilions made of silk secreted by the larvae, attached to the underside of bamboo leaves and then covered with a layer of detritus.

The Malayan ant *Dolichoderus cuspidatus* has an obligatory feeding relationship with the pseudococcid *Malaiococcus formicarii* as well as with other members of the same genus. The mealy bugs are carried in well-organized processions to new feeding sites when older ones become exhausted and they are protected by the ants even to the extent of the latter forming a living umbrella over them when it rains. When the distance between the ants' nest and the foraging stations becomes too great, the whole nest and its occupants move to a new site nearer to the farmed mealy bugs. Neither the ants nor the mealy bugs flourish in isolation, so it would appear that the mealy bugs provide a large part of the ants' food supply though, when artificially presented with dead insects, the ants do make use of them (Maschwitsz & Hänel, 1985).

Honeydew produced by a number of species of scale insects and aphids forms a major part of the food of the large, diurnal ant *Camponotus detritus* from the Namib Desert of southern Africa (Curtis, 1985). Workers of this species seen to visit flowers of the succulent *Trianthema hereroensis* were presumed to be collecting pollen and nectar. This plant, which flowers throughout the year, apparently constitutes a major source of food for the ants where homopteran bugs are absent. Workers also scavenge for other edible items, such as seeds, dead arthropods and even bird and lizard faeces. Those nests of *C. detritus* which were examined did not contain any food stores nor does this species have a replete caste.

Although relationships between ants and homopteran bugs are common, there are few records of associations between ants and Heteroptera, though they do occur. In Malaya, two phloem-feeding plataspid bugs, *Tropidotylus servus* and *T. minister*, are farmed by the ant *Meranoplus mucronatus*. The bugs feed on the trunk of a broad-leaved tree where they are found in either naturally protected sites or pavilions constructed by the ants. The larval bugs defaecate continuously and the attendant ants palpate their anal region and feed on the glucose-rich droppings. The same species of ant also feeds from the nymphs of the coreid bug *Hygia cliens*, which lives on the same trees. In the same area, a similar feeding relationship was found to exist between a *Crematogaster* sp. ant and the coreid *Chloresmus* sp., but this time on bamboo. In this instance, the bug larvae defaecate when their anal regions are stroked by the ants' antennae (Maschwitz *et al.*, 1987).

Close relationships have been developed between the caterpillars of many lycaenid butterflies and ants, sometimes but not always to the latters' benefit. Some, but not all, ant-associated lycaenid caterpillars have a honey gland on the tenth abdominal segment from which ants can solicit honey by stroking the area with their antennae. Overzealous solicitation by the ants results in the caterpillars extruding a pair of dorsolateral tubercles on the eighth abdominal segment, which causes the ants to back off. In exchange for this secretion, the ants protect the caterpillars while they are feeding and allow them the protection of their nests at night or, in some cases, during the caterpillar's overwintering phase.

The North American ant *Formica integra* actually constructs 'byres' out of fine detritus in which third and fourth instars of the lycaenid *Satyrium edwardsii* spend the hours of daylight guarded by the ants (Webster & Nielsen, 1984). At night they leave the byres, again guarded by the ants, to feed on the foliage of their host tree. The fourth-instar caterpillars produce secretions of honey when solicited to do so by the ants. This is also true of the third and fourth instars of the South-East Asian lycaenid *Anthene emolus*, which live in an obligate association with the weaver ant *Oecophylla smaragdina* (Fiedler & Maschwitz, 1989). These caterpillars make a considerable donation in terms of energy in return for the care which they receive from the ants, the figure of 200 J per individual probably being an underestimate. From this, it has been calculated that the amount of energy donated by a single caterpillar over a 3-day period is sufficient to satisfy the resting metabolic rate of 55 ants. Furthermore, the total secretions of 50 caterpillars should be enough to support the production of 165 worker ants.

Although feeding relationships between ants and Homoptera and ants and lycaenid larvae are fairly common, the discovery of ants feeding from larvae of a tortricid moth was something unusual (Maschwitz *et al.*, 1986). The larvae of the tortricid moth *Semutophila saccharopa* live in silken shelters on their food plant, the giant bamboo, *Gigantochloa scortechini*, in Malaysia. Ants from at least seven different genera attend the larvae in their shelters. When an ant palpates a shelter with its antennae, the larva extrudes up to one-third of its rear end and produces a droplet of clear fluid from the anus. If the liquid is not taken up by the ant, it is withdrawn back into the body. The larva draws it rear end back into the shelter a few seconds after being stimulated by the ant. Larvae which have not been stimulated by ants for a period of time spontaneously extrude their rear ends and forcibly eject droplets of liquid for several centimetres. At least one ant, an *Iridomyrmex* sp., lays recruitment trails to the larval shelters. An analysis of the fluid exuded by the larvae indicates that it contains both carbohydrates and amino acids and/or protein.

Ant nests serve as homes for a great diversity of inquilines, not least of which are other ant species. *Leptothorax diversipilosus* is an ant which shares the nest of the western thatching ant, *Formica obscuripes*, in the USA. Although their broods are reared separately, adults of *L. diversipilosus* are able to move freely around the nest of the host ant. In laboratory nests, *L. diversipilosus* was observed to obtain food from host workers, either directly or, more often, when two workers were passing food between each other (Alpert & Akre, 1973).

What is termed as 'tool use' has been described for at least three genera of ants. The Florida harvester ant, *Pogonomyrmex badius*, for example, uses grains of sand to absorb liquid foods and transport them back to the nest and similar behaviour has been noted in a number of *Aphaenogaster* spp. In the laboratory, the fire ant, *Solenopsis invicta*, also uses tools to collect liquid food (Barber *et al.*, 1989). If starved *S. invicta* are presented with a pool of honey, they feed on it directly. If, however, they are well fed then they return to their artificial nest with various articles, such as sand grains and pieces of dry grass, which they place around the edge of the pool. These articles, plus the honey on their surface, are then transported back to the nest.

BEES

Most adult female bees feed upon nectar and forage for nectar, pollen and oils with which they provision their nests. They may then practise either mass provisioning, where all of the food the larva needs for its development is provided at one go before it is sealed in its cell, or progressive provisioning, where the larva are fed on demand. Polylectic species show little in the way of morphological or behavioural specialization and feed upon a wide range of flower types, seasonal availability often dictating the food source at any particular instant. Oligolectic species are limited to a few or even a single flower species and may accordingly have evolved morphological and behavioural specializations.

The North American bee *Diadasia bituberculata* is an oligolectic species which throughout its range utilizes different species of *Calystegia* or *Convolvulus* as a source of nectar and pollen (Schlising, 1972). This bee possesses elongated mouthparts which allow it access to the nectar deep down in the tubular corollas of these flowers and it also has both a diurnal and seasonal synchronization with the flowering of these plants.

Whereas many flowers produce abundant pollen from anthers which undergo complete dehiscence, there is a large number of angiosperm genera from more then 70 families which dehisce through short slits or true rounded pores in the anther apex. Such flowers require a special pollination technique and, therefore, a special form of foraging behaviour from the bees that visit them. Almost without exception, such poricidal flowers limit their rewards to pollen, the exception being some Ericaceae. In order to gain this reward, a large number of bee species from at least seven families have evolved the technique of 'buzz', vibratory or vibratile pollination. In order to obtain pollen, the female bees have to vibrate the stamens, which they do by rapid buzzing using their indirect flight muscles. This buzzing results in the expulsion, from the anthers' apical pores, of a stream of pollen which strikes the underside of the bees and sticks there by electrostatic attraction. The bees then gather the pollen together by grooming before packing it into their own particular transport apparatus.

▲ **Having evolved alongside one another for millions of years, flowers and insects have developed feeding relationships. Thus some flowers have short tubes and are fed from and pollinated by short-tongued insects while others have longer tubes and their nectar can only be obtained by longer-tongued insects. Some short-tongued bees, such as this early bumblebee *Bombus pratorum*, have, however, learnt the art of robbing long-tubed flowers, here a comfrey, by forcing apart the bases of the petals to get at the nectaries.**

An example of such foraging behaviour is exhibited by the Central American euglossine *Euglossa imperialis* (Buchmann, 1980). Females of this species land on flowers of a Panamanian bloodwort, (*Xiphidium caeruleum*: Haemodoraceae) and, by shivering their indirect flight muscles, produce an audible buzz which consequently vibrates not only the stamens but in fact the whole flower; the result is a rapid release of pollen from the apical slits in the anthers.

A number of bees have found ways of getting round the need to 'buzz' a flower and they are able to obtain pollen from one without facilitating pollination. The North American *Cassia fasciculata* (Leguminosae: Caesalpinoidaea) is visited by a number of bees of different species (Thorp & Estes, 1975). The most common visitors are females of *Bombus americanorum*, which collect pollen both by vibration and also by milking the anthers; this behaviour is also true of *B. fraternus* and the green sweat bee, *Augochloropsis metallica*. Two anthophorids, *Xylocopa virginica* and *Svastra atripes*, employ buzzing alone, while the smallest of the bees, female *Dialictus* spp. just milk the anthers. In field experiments which excluded the larger bees but permitted entry of the smaller halictines, fruit was set in less than 1 per cent of the flowers visited, though masses of pollen were removed.

Pollen- and nectar-robbing is by no means uncommon amongst bees and is so-called because it involves taking the plant's products without a repayment by the insect in the form of pollination. The proboscis of the hive bee, *Apis mellifera*, is, for example, too short to reach the nectary of the flower of the bluebell, *Endymion non-scriptus* (Liliaceae). The bee, therefore, forces apart the bases of the petals and obtains the nectar thus (pers. obs.). Worker hive bees have also been seen to rob pollen from the scopae of other bees in California (Thorp & Briggs, 1980). Observations on sunflower heads revealed that the hive bees quite often steal pollen from *Diadasia enavata* and, occasionally, from *Halictus ligatus*; the latter also robs the former species on the odd occasion. The act of robbing involves the hive bee either standing or hovering in a latero-posterior position and scraping off the pollen from the victim's pollen-filled scopae with the mandibles and front legs. It appears that, once a worker hive bee has learned to rob other bees, she is reluctant to resume foraging in the normal manner and hovers around waiting for a new victim.

Stingless meliponine bees of the large tropical genus *Trigona*, as well as robbing flowers of their nectar, have also evolved pollen-stealing behaviour in some species. In Brazil, they rob flowers of plants of the family Melastomataceae by initially probing for pollen through the apical pores of the tubular anthers for as far as their tongues can reach. At this point, they cut the top off the anther to expose more pollen; this act may be repeated several times (Renner, 1983). Not only do the stingless bees rob flowers but they also rob one another, not only of food but also of the materials from which their nests are constructed (Wille, 1983). This pillaging activity tends to take place between nests which are in close proximity to one another and it can be anything from slight, when the attacked colony is able to resume its normal life, to extreme, when the attacked colony is either obliterated or sometimes occupied by the attackers. Attacks can last anything from hours to weeks; one colony has been seen to attack two others simultaneously and a single colony has been seen to be attacked by three others simultaneously. Colonies have also been seen robbing one another at the same time. Honey is stolen as well as pollen and the latter, instead of being carried on the hind legs, as when collected from flowers, is softened and carried in the crop instead.

Neotropical stingless bees of the genus *Lestrimelitta* are known to be obligate robbers of other meliponine bee colonies. Workers of the Central American *L. limao*, the lemon bee, have no pollen-carrying apparatus but possess enlarged mandibular glands from which they secrete a substance with a very strong smell of lemons. Once a *L. limao* has breached the defences of the colony to be plundered, it releases this strong scent. This overwhelms the normal colony odour so that further *L. limao* workers are able to enter unhindered and remove food from its stores. They may even take over the nest entrance, preventing the rightful owners from entering or leaving while the food stores are plundered.

The African robber bee *Cleptotrigona cubiceps* uses a different ploy to obtain its food. On finding the nest of one of the small *Trigona* species, between 40 and 60 of the robber workers congregate around its entrance. They then break down the tubular entrance to the nest, whereupon the occupants pour honey through the damaged opening as a temporary repair. This honey is then collected by the attackers who, at the end of the day, return to their own nest. During the night, the attacked workers repair the nest tube but, the next day, the robbers resume their offensive and the honey begins to pour from the nest once again. This goes on for hours or even days, but eventually the attackers lose interest in it and, instead, attempt to get into the nest. Finally, the honey stops pouring out and *Cleptotrigona* workers and workers from other nests of the attacked species collect all of the honey from around the attacked nest entrance. After this, the robbers are able to breach the nest and they then plunder it of all of its contents, both food and building materials.

Trigona bees have also been found in association with membracids where they feed upon the honeydew produced by these bugs. In Guatemala, for example, *Trigona amalthea* feeds from the membracid *Antianthe expansa*, which may simultaneously be attended by *Crematogaster* ants (Schuster, 1981). In order to feed, the bee first palpates the nymph's abdominal tip with its antennae. The nymph then raises its abdomen and releases droplets of fluid from its anus, which the bee then imbibes. In Brazil, the same bee is known to associate with membracids of the genus *Aethalion*.

A number of genera from several plant families have evolved structures referred to as 'elaiophors' which secrete a mixture of lipids, amino acids and carbohydrates, seemingly as a reward for pollinators. A number of bee genera from different families have evolved structurally and behaviourally to exploit this rich food source. Unlike nectar, which is imbibed and carried in the crop, these secretions are normally collected by hairs on the front tarsi and they are then passed via the middle legs to the scopal hairs of the hind legs, where they are amassed for subsequent transportation to the nest. Perhaps the most extreme example to date of the relationship between an elaiophoric plant and a bee is found in South Africa between the plant *Diascia longicornis* (Scrophulariaceae) and the melittid bee *Rediviva emdeorum* (Vogel & Michener, 1985). The flowers of *D. longicornis* have two very long (25 mm), oil-producing spurs. The front legs of *R. emdeorum* females are similar in length to the spurs and, when they are inserted into these spurs to collect the oil, they simultaneously pollinate the *D. longicornis* flowers. Other species of *Diascia* occur with different spur lengths, each associated with its own particular *Rediviva* bee with front legs of corresponding length.

Bees of the genus *Ctenoplectra* employ a somewhat different method for collecting secretions from elaiophors found in the flowers of certain Old World Cucurbitaceae (Vogel, 1981). Females of *Ctenoplectra* bear, on the fourth and following metasomal sterna, dense fringes of hairs which form a brush, superficially resembling the pollen brush of megachilid bees. Each hair bears two rows of overlapping leaflets which, with the brush as a whole, have a strong tendency to hold fluid by capillary action. The bee walks on the flower and swings her abdomen laterally across the elaiophors to soak up the secretion. Following each swing, the abdomen is cleaned of accumulated fluid by alternate cleaning movements of the hind legs. The liquid is collected in the scopae on the hind legs where it may be mixed with pollen, which is collected separately, to form a soft paste for carrying back to the nest. Both female and male *Ctenoplectra* also feed upon the nectar secreted by these cucurbits so that there is a complete interdependency between bee and plant.

At least one other species of bee, the exomalopsine *Paratetrapedia calcarata* from Panama, collects both pollen and elaiophor secretions from its host plant. Females of this species visit *Mouriri myrtilloides* (Melastomataceae) and vibrate its anthers to release pollen, as well as collecting oily secretions using the front tarsi (Buchmann & Buchmann, 1981). The intrinsic food value of such secretions may be gleaned from an analysis of that produced by *M. myrtilloides*. It was found to contain at least 13 fatty acids, glucose, amino acids, carotenoids and a number of other organic components.

Any animal which lives in a desert or semi-desert habitat has to overcome the problems of water balance. The foraging pattern of the xerophilic megachiline bee *Chalicidoma sicula* from the Mediterranean region, for example, has to take into account the need to forage for cell-provisioning and to take in water to balance water loss. A study of this species in Israel (Willmer, 1986) indicates that the need to collect adequate amounts of water from the flowers it visits actually over-rides the need to obtain energy and that, as might perhaps be expected, water requirement is dictated by ambient temperature. Smaller bees, with a greater surface area/volume ratio might be expected to lose more water and this is reflected in the fact that they do indeed make more nectar-collecting trips than larger bees with their smaller surface area/volume ratio. Female bees construct sand nests which they fill with nectar and pollen collected mainly from *Lotus creticus* (Leguminosae). The nectar of this plant is fairly dilute and is preferred by the bees but, when it is not available, another leguminous flower with a more concentrated nectar is utilized. In order to obtain sufficient water, this is supplemented with the more dilute nectar from a sympatric labiate flower. Prior to sealing the cell, the female *C. sicula* makes a number of nectar-only forages and uses what she collects to adjust the water and sugar content of the stored larval food.

In honey bees, whose colonies operate as a highly integrated unit, foraging must relate to the needs of that unit at any one time. It has been found (Seeley, 1989) that the types of flowers visited by a colony are dependent on the current status of the colony. If the colony is well fed, then the tendency of the workers is to exploit only high-quality food sources but, if conditions are bad, then both high- and lower-quality sources are used. Roughly 20 per cent of the colony, i.e. workers between 12 and 18 days of age, act as food-storers and receive the food from incoming foragers. The foragers are able to appreciate the colony's nutritional status by finding out how difficult it is to find a food-storer when they return to the nest. The longer the forager has to queue to find a food-storer then the better the food stores within the colony and, therefore, the less time she needs to spend on lower-quality nectar and pollen sources. This automatic feedback system differs markedly from the signalling system used to convey the position and quality of food sources, the 'dance of the bees'.

It is perhaps to be expected that, within a group so large and diverse as the bees, one of them at least should use something other than pollen as a provisioning source for the larvae. One such is the stingless bee *Trigona hypogea*, which is an obligate necrophage (Roubik, 1982), protein for its developing brood coming from the corpses of dead animals. An animal carcase is detected within hours of its death and the scout bees then lay down an odour trail between it and their nest. Recruitment then results in hundreds of workers making their way to the dead animal. Other competing insects, such as wasps and flies, at the carcase are attacked and driven off by the bees.

Of the roughly 20 000 species of bees so far described, about 3700 species do not forage for themselves but let other bees do it for them; they are the so-called 'cuckoo bees'. The cuckoo habit is found in five of the 11 bee families. In order to feed their young, they have to track down the nest of another bee and lay their egg or eggs in it so that their offspring develop rather than the host's. In what is effectively foraging behaviour, the females use olfactory cues to detect host nests. Having found a nest which is being actively provisioned by the host, the cuckoo then has to choose the best time to enter and lay her own egg, preferably when the host is absent. First-instar larvae of many cuckoo bees are furnished with large, sickle-shaped jaws, which they use to destroy host eggs or larvae in order to eliminate feeding competition.

DIPTERA True flies

As with other endopterygote insects, adult and larval flies of the same species forage for different foods. Adult flies tend to be liquid-feeders, the source of the liquid coming in the form of nectar, other plant exudates, animal exudates and animal body fluids, while some flies do not feed at all as adults. Fly larvae of one species or another feed on almost anything that is edible. Between the adults and larvae, therefore, some interesting foraging behaviours have evolved.

Larvae of the fungus gnats *Leptomorphus bifasciatus* and *L. subcaeruleus* (Mycetophilidae) in North America construct silken sheets beneath rotting logs bearing fungal sporophores or beneath the sporophores of bracket fungi (Eberhard, 1970). On the silken sheet, there is always a network of slime trails which have been laid down by the larva and along which it always moves. Sheets with living occupants are always found beneath active sporophores from which spores fall on to the silk and are caught. The larva moves its head from side to side in an arc and consumes an area of the silk sheet and its accumulated fungal spores. Moving its head in a similar way, it then replaces the silk in the area which it has just consumed. Larvae of the Japanese mycetophilid *Ceroplatus nipponensis* have a similar life-style but for some reason they are faintly luminous.

The larvae of the Central American mycetophilid *Orfelia aeropiscator*, as the name implies, also fish from the air but for flying insects not fungal spores (Jackson, 1974). Originally described from a cave in Belize, in which the air was still but moist, they have also been found in Costa Rica under large leaves in the rainforest understorey where there is little wind and the air is very humid. The web contains three structural components. There is a horizontal strand, which averaged 38 cm in length in those measured, which is suspended by supporting lines, normally two to five in number, from the leaf, leaf petiole and/or twig to which the leaf is attached, at various points along its length. From the horizontal line, the larva suspends between seven and thirty 'fishing' lines, 5–7 cm in length. The web, and in particular each fishing line, is covered in a sticky mucus and there may be a blob of mucus at the bottom end of some of the latter. The *Orfelia* larva rests on the horizontal line, fully exposed, when not feeding. In Costa Rica, the only prey seen attached to the fishing lines were nematoceran flies. Large flies immediately attract the larva's attention and they are hauled in on the fishing line and consumed straight away. Tiny flies, however, sometimes remain uneaten for hours after they have become trapped. Not only does the larva haul in and replace tangled lines but it also randomly replaces lines to which prey is seemingly not attached.

Larvae of *O. fultoni*, from the Appalachian Mountain area of North America, build a sprawling web in small cavities in soil, amongst mosses or in dead wood, or in gaps between stones. The web is attached to the substrate by means of adhesive spindles to which prey adhere. Unlike the previous species, larvae of *O. fultoni* are luminous, with the light organs located in the head and tail of all instars producing a fairly constant output of blue light during the hours of darkness. The light attracts mainly small Diptera as prey and seemingly has no effect upon small, wingless, soil arthropods (Sivinski, 1982).

The larvae of fungus gnats of the genus *Arachnocampa*, from Australia and New Zealand, construct similar webs in caves and under rock overhangs. They are highly luminous, possessing, at the swollen distal ends of the Malpighian tubules, special areas for light production which attracts prey to their webs. The fishing lines, instead of having a single sticky droplet at the lower end, are liberally supplied with them along the whole length.

In a field experiment conducted in Florida (Sivinski & Stowe, 1980), discs, to which were attached either a partially digested insect from a spider's web or a freshly killed insect, were suspended along with a blank control disc in the vicinity of webs of the spider *Nephila clavipes*. The cecidomyiid *Didactylomyia longimana* was the most abundant kleptoparasite recorded: in some locations nearly every *N. clavipes* had the company of as many as six of the flies sharing its prey. This particular fly was also seen in association with *Argiope aurantia*, *Eriophora ravilla* and *Mastophora bisaccata*, all araneid spiders. It was noticed that the flies were more commonly found feeding on fluid-covered prey being consumed by the spider than on wrapped, stored prey. The spiders are obviously aware of the flies' presence for they bat them away with their forelegs as they fly in. In the disc experiment, only the partly digested prey attracted the fly, seemingly indicating that they are attracted by some product of the process of digestion. A number of milichiids also attended the feeding spiders, notable among which was *Paramyia nitens*. This fly has mouthparts nearly as long as the body which, it is surmised, allow them all the more easily to feed through the silk on wrapped prey or to imbibe fluid from the dangerous areas around the spider's mouthparts.

The ceratopogonids of the genus *Culicoides* feed as adults on the blood of homoiothermic animals, including man to

▲ Ants are able to 'milk' small homopteran bugs by stroking them; this gives them direct access to honeydew. A mosquito is unable to apply the correct stimulus to a homopteran, so if it wants a drink of honeydew, the only way to obtain it is from the ants. This tiny mosquito (possibly a *Wyeomyia* sp.) regularly dips its slender proboscis into honeydew glistening on the jaws of formidable *Ectatomma* ants in Trinidad. Why the rewards of such behaviour are worthwhile, compared with what must surely be the considerable risks taken, are unknown.

whom they can be a great nuisance. *C. anophelis*, however, has a somewhat more endearing habit since it chases after mosquitoes that have fed upon the blood of human beings or other animals and feeds from them by piercing their abdomens.

The family Tabanidae contains the so-called horse flies, well known by many people for the unpleasant blood-feeding habits of the females. The males, however, feed only from flowers but are also recorded as drinking from open water by swooping down on to it from some height, striking the surface and then flying off again. The assumption is that this unusual behaviour helps them escape the attentions of robber flies and hunting wasps. Females of one subfamily of tabanids get the best of both worlds for they have a proboscis which is twice the length of the body and they hover in front of long-tubed flowers and sup the nectar, much like a hummingbird hawk-moth. They can take a blood-meal, however, by biting with their very much shorter stylets, at the same time pushing the long labial proboscis to one side. Tabanid larvae are both carnivores, herbivores and feeders on vegetable detritus, though the most bizarre feeding behaviour of all is shown by larvae of *Scaptia muscula*. These larvae lead a very dangerous life by living in the pits of ant-lions, where they share the victims of the host larvae.

The snipe flies of the family Rhagionidae contain a subfamily, the Vermileoninae, whose adults have a long, slender proboscis used for feeding on nectar. The larvae of, e.g. the genera *Vermileo, Vermitigris, Chrysopilus* and *Lampromyia* are, however, carnivorous. They live in steep-sided conical pits in sandy substrates and feed on any small animals which fall in and are unable to escape. In order to dig its pit, the larva lies on its back just below the sand and moves around in a circle, constantly jerking its head up and throwing sand outwards to initially form a circular groove. It continues this rotation, the head flicking out all the time and, as the groove gets deeper, so the larva reduces the size of the circle until finally it comes to lie at the bottom of the pit, either completely submerged in sand or with the head and part of the thorax exposed. The waiting larva accelerates the descent of any luckless victim which falls in by flicking sand in its direction, thus precipitating an avalanche. The larva whips blindly around trying to wrap its body round the prey and, once it succeeds, it drives home two pointed hooks on the mouth which are used to inject poison. The immobilized prey is then dragged below the sand and eaten. In many ways, therefore, their behaviour is convergent with that of the better-known neuropteran ant-lions and has earned them their family name Vermileoninae and their common name of worm-lions.

There is a remarkable degree of conformity in the prey-catching behaviour of robber flies (Asilidae) from many different genera though each species tends to show more in the way of specialization in the prey taken. When foraging, most species sit in a sunny position on the ground, on rocks, on dead trees or on vegetation and they orient to face potential prey as soon as it is sighted. They then fly up towards the prey, catching it and killing it in mid-air, usually by plunging the rigid proboscis into a soft area of membrane between head and thorax of the insect victim. When taking larger prey, both parties may end up on the ground while the asilid struggles to push its proboscis into the victim (pers. obs.). Feeding takes place back at, or close to, the original foraging site and the prey may then be manipulated several times during the process. In some species, this involves the fly taking to the air again and manipulating the prey in mid-air with its legs. Once feeding is completed, the remains of the prey are usually discarded at the foraging site. Unusually for robber flies, members of the genus *Leptogaster* are not sit-and-wait predators but instead they indulge in scouting flights and pounce upon stationary prey, such as spiders and insects. Their ability to pick their prey out against a confusing background of stems and leaves is quite amazing.

Asilids take a wide range of flying insect prey though, as might be expected, there is generally a positive correlation between the size of the asilid and the size of its prey. Data from a range of North American species indicate that other Diptera are likely to be the most commonly taken prey, and this is usually the case for British species (pers. obs.), with Hymenoptera likely to be the second 'choice'. Robber flies of the genus *Stichopogon* from North America have been seen to take prey from surfaces as well as from the air. Two species, *S. catulus* and *S. trifasciatus*, were studied along a stream in the Chiricahua Mountains of Arizona (Weeks & Hespenheide, 1985). *S. catulus* was seen on a number of occasions to fly,

▲ **The robber flies (Asilidae) are adept aerial predators, taking off from their perches and snatching their victims from the air as they fly past. This European robber fly, *Dioctria rufipes*, had caught a warningly coloured sawfly but it soon dropped it with obvious distaste and 'wiped' its proboscis on the leaf for several minutes in obvious distress. Invertebrate predators do not respond to warning colours, such as those of this sawfly, which are aimed at vertebrate predators. An invertebrate attacker, such as a robber fly, must therefore first take the plunge and taste the victim before discovering its mistake, possibly at some cost to the predator if the victim is particularly well protected chemically. To withstand such 'taste it and see' treatment by both invertebrate and vertebrate enemies, warningly coloured insects are often exceptionally tough. This sawfly rapidly recovered from the robber fly's probings and flew off, apparently none the worse for its experience.**

from its foraging position on a rock, along the stream and capture first- and second-instar nymphs of the water-surface-dwelling hemipteran *Gerris*.

Traditionally it has been accepted that bee-fly (Bombyliidae) adults, with their long, slim proboscis, feed exclusively on nectar and that the larva is the main protein-assimilating phase of the life-cycle. As early as 1910, however, it was noted that certain bombyliids visited nectarless flowers and, in the intervening period, evidence has been accumulating which points to the possibility that the majority of bee flies, especially the females, feed on both nectar and pollen.

Investigations of the foraging behaviour of *Poecilognathus punctipennis* (Deyrup, 1988), recorded in the USA from the states of Florida and Georgia, have confirmed pollen as a food source for a female bombyliid. Females of this species feed upon flowers of *Tradescantia roseolens, Cutherbertia rosea* and *Commelina erecta*, all members of the Commelinaceae which typically produce no nectar. Because when feeding each fly needs the exclusive use of a single anther, the number of flies on a flower at any one time is limited by the number of available anthers. On alighting on the flower, the fly performs rapid raking movements with its front legs across the petals, glandular hairs and other flower parts until it reaches the stamen. The inference here is that the fly is able to detect the presence of the pollen with its front tarsi. In order to forage, the fly rakes these tarsi across the anther's pollen-bearing surfaces. Its collection is possibly assisted by the presence of modified hairs, which are weakly adhesive, on the ventral and lateral surfaces of the first four tarsal segments. In order to transfer pollen to the proboscis, the leg is raised and the tarsus is wiped across its tip. The wiping action takes place at irregular intervals and both tarsi are used alternately if both have retrieved pollen, which is not always the case. Microscopic observations of feeding flies indicate that the pollen is transported up the proboscis in the form of a slurry mixed with liquid, which has presumably first been regurgitated down the proboscis. During this investigation, both sexes of another bombyliid of the genus *Geron* were captured and their stomach contents examined. Pollen was found in the stomachs of all of the females but not in those of the males.

Larvae of the small family Cyrtidae are parasites of spiders while the adult flies have long, slim proboscides with which they feed at flowers. The females may lay hundreds of eggs at one time and from these hatch a larva which then lies in wait for a passing spider. The larvae can jump distances up to 6 mm by bending and then abruptly straightening the body and they can also move along spider silk in a looping motion. The larva penetrates the body of its host and feeds on it internally.

Adult flies of the family Empididae have a life-style somewhat similar to that of the asilid robber flies in that they pounce on their arthropod victims and suck out their body fluids with their sharp proboscis. A number of them, however, fly little if at all and, instead, run after prey on tree trunks and vegetation, grasping their victims with their legs. In order to facilitate this the femur and tibia of at least one pair of legs are armed with spines which oppose each other as they grasp their victim in a vice-like grip. Adult flies of the genus *Clinocera* exploit a source of food that is rare in flies, namely insects which have fallen into water and are trapped on its surface, although the habit has also evolved in at least two Dolichopodidae and also in the Ephydridae.

Hover flies (Syrphidae), after bees, are the secondmost important pollinators of flowers in many parts of the world. As with the bees, they carry out the pollination as they forage for nectar and pollen, in this case to feed themselves and not their larvae, which instead are predators on other invertebrates. With their soft, tubular proboscis, nectar-feeding is straightforward but feeding on pollen requires some more-specialized behaviour.

It was revealed during a study on the feeding of adult hover flies in New Zealand (Holloway, 1976) that there tends to be a relationship between the morphology of the fly and the type of pollen consumed. Small species with a short proboscis, sparse unbranched hairs and short, simple bristles, on internal examination, proved to contain almost exclusively anemophilous (wind-distributed) pollen in their guts. Larger, hairy species with a long proboscis, pollen-collecting hairs and long, spirally grooved bristles, on the other hand, contained pollen taken almost exclusively from nectar-producing, insect-pollinated flowers. Pollen-feeding was closely observed in one of each group during the study.

For the entomophilous pollen-feeder, the widespread drone fly *Eristalis tenax* was chosen. Pollen which adheres to the body while the drone fly is on a flower is frequently combed off. The front legs are used to comb the head and its appendages, the middle legs and one another, while the hind legs also comb the middle legs, as well as all the remaining parts not reached by the front legs. At intervals, general combing is interrupted by pollen concentration during which the hind legs are stretched backwards off the substrate and the apex of the tibia and the tarsus of one leg is scraped against those of the other. This action transfers pollen from the tibial combs on to the pollen-retaining bristles on the tarsi. The tarsi are then tapped on the substrate to remove any oversize particles or pollen which have not been trapped by the bristles. A similar sequence of events is followed during combing by the front legs, which are also kept clear of the substrate; then they are bent at the femoro-tibial joint, the mouthparts are extruded and the pollen on the fore-tarsi is consumed. In hovering flight, the hind legs make movements similar to the pollen-concentrating movements when at rest but, in a second activity, all three pairs of legs are dangled below the fly's body and pollen on the bristles of the hind tarsi is then transferred to those on the fore-tarsi.

Aerial leg movements similar to these were also recorded in two other syrphids, *Melangyna (Austrosyrphus) novaezealandiae* and the bulb fly, *Merodon equestris*. *E. tenax* was never observed to feed directly from the anthers on *Ranunculus sardous* pollen, either in the laboratory or in the field, though pollen grains of this species were found in the gut, indicating that they had used the feeding method outlined above. *E. tenax* was, however, seen taking the pollen directly from the anthers of two species of *Raphanus* in the field situation.

Although the anemophilous pollen-feeding *Melanostoma fasciatum* carried out the same body-grooming patterns as *E. tenax*, both at rest and in the air, the amounts of pollen which collected on the tarsi were minimal and at no time were the flies seen to feed from it. Instead, this species feeds directly from the anthers of plants, such as *Plantago* spp. and grasses, holding them in its front tarsi as it does so. When feeding on *Plantago*, *M. fasciatum* pushes both lobes of the labellum into a ripe anther and consumes all of the pollen it contains before going on to the next one, and so on.

The larvae of syrphid flies of the genus *Microdon* are inquilines in the nests of ants and other social insects. A typical example is *M. piperi*, which lives in the nests of its primary host, the ant *Camponotus modoc*. The fly larvae are predacious and feed upon the ant brood. They appear to be near-perfect chemical mimics of the host ants, which ignore them completely, despite their depredations on the ant larvae and pupae. The life cycles of the two insects are closely synchronized (Akre *et al.*, 1988).

Milichiid flies are known to be associated with the prey of other arthropods, for example, they attend feeding asilid robber flies (pers. obs.). In Panama, a number of species of small flies from different families have been seen feeding kleptoparasitically on the prey of spiders and an assassin bug (Robinson & Robinson, 1977). Females of a milichiid fly of the genus *Phyllomyza* were seen in the company of the araneid spider *Nephila clavipes*. The flies are tiny and sit on the spider in an inactive state for long periods and they are equally unmoved when the latter goes through the process of catching and wrapping a prey item. The spider then secretes enzymes into the prey and digestion commences. After a while, the surface of the prey becomes covered in a film of liquid and, at this point, the flies move on to it and begin feeding until their abdomens are completely distended. The same researchers also saw two unidentified flies feeding on the prey of the spider *Argiope savignyi* as well as some chloropids of the genus *Conioscinella* feeding at the prey of an *A. argentata*. Another milichiid, *Neophyllomyza* sp., was seen apparently feeding on the *Trigona* prey of the reduviid assassin bug *Zelus trimaculatus*. In Texas, gravid adult females of the milichiid fly *Pholeomyia texensis* hop on to pieces of leaf being carried by the ant *Atta texana* and ride on it into the nest to lay their eggs; their larvae feed on detritus in the nest (Waller, 1980).

▼ At first sight, this appears simply to be an *Apiomerus* sp. assassin bug (Reduviidae) feeding on a worker hive bee, *Apis mellifera* in Mexico. Closer examination, however, reveals the presence of a number of milichiid flies who feed kleptoparasitically on the fluid oozing from the wounds made by the bug's feeding activity.

Defensive Behaviour

In defending themselves against the attentions of predators or competitors, invertebrates use a number of different ploys. They may, for example, use some form of physical defence, such as that employed by spiders, who fall to the ground on the end of their life-line when disturbed. Alternatively, they may employ some form of active chemical defence by squirting some noxious substance from a gland or by regurgitating unpleasant-tasting or unpleasant-smelling gut contents. In all cases, the examples of defence described in this section involve the animals in some action; just being a stick mimic or walking around warningly coloured does not count.

▲ Along with some amphipod crustaceans, a number of millipedes are able to roll themselves up into a ball when molested. This *Sphaerotherium* sp. millipede, from the Madagascan rainforest, is also distasteful, advertising this fact with its warning coloration.

◀ When disturbed, this female *Micrathena schriebersi* araneid spider from Brazil arches her body away from her web and bobs it up and down. She thereby exposes an underside which appears to be warningly coloured. The display probably therefore serves a 'keep-off' function' similar to the 'flash' displays of insects.

▲ **When mygalomorph spiders find themselves under threat away from their burrows, they adopt the defensive pose illustrated here. This male Australian trapdoor spider has spread his legs out, presumably to make himself appear as large as possible, and is also baring his large fangs.**

PHYSICAL DEFENCE

When caught out in the open or at the mouth of a burrow which cannot be closed, a number of mygalomorph spiders react by adopting an aggressive threatening posture. They drop back on their haunches, raise their front legs and open their fangs wide, making themselves appear as large as possible. Many of these mygalomorphs are also covered in a pelt of fine hairs, a cloud of which they can scrape free by vibrating their hind legs against the abdomen. These hairs are extremely sharp and barbed and also appear to be coated with toxins so that, if they are breathed in, by a hunting mammal for example, the effects can be extremely unpleasant. It has been shown that these hairs are capable of piercing human skin to a depth of 2 mm. One final form of active defence which is employed by these large mygalomorphs is to turn the rear of their abdomen to the source of any danger and squirt a clear liquid from the anus.

The North American fishing spider, *Dolomedes triton*, although sometimes exceeding 3 cm in body length as an adult, is able to run on the surface of still or slow-moving waters in search of prey. During observations on this species on a day of light winds (Deshefy, 1981), it was noticed that, if they found themselves on areas of open water, with no form of protective cover, they would elevate their second pair of legs at roughly 70–90° to the water surface. The wind then caught them and blew them across the water, even against slight currents, until they reached another solid substrate. If the spiders were placed on open water in still air, they showed no leg-raising response until air was lightly blown over them from a cardboard tube. If, however, they were placed on a floating leaf or other debris, leg-lifting did not take place; instead, the spider crouched low against the substrate when blown. It was concluded that the spiders use this behaviour as an anti-predator device and also to minimize the amount of energy required for them to move over expanses of open water.

A very unusual form or defence has been observed in females and immature males of the North American western black widow spider, *Latrodectus hesperus* (Vetter, 1980). If threatened, the spider produces a very sticky silk from the aggregate glands, which it extracts from the spinnerets with its fourth pair of legs. It then spreads these legs so that the silk is held across its vulnerable abdomen and it can also place the silk on to the attacker if necessary. In laboratory experiments, the silk was found to be capable of deterring the attentions of *Peromyscus* mice, simply, it seems, because of its unpleasant stickiness, since the mice find it quite palatable.

Apart from the fact that they may be cryptic or can simply jump or fly away when threatened, some katydids have no form of defence. *Ancistrocercus inficitus* is a relatively defenceless katydid from Costa Rica, which may be found roosting behind loose bark. More often than not, however, it is found roosting in association with nests of the wasps *Polistes, Polybia, Stelopolybia, Synoeca* and *Myschocyttarus*. The wasps apparently provide the katydid with a certain degree of protection against predators and parasites and, unless disturbed, they remain faithful to a particular nest.

Whereas, in the classic description by Petrunkevitch (1952), the tarantula stands and lets itself be stung by the hunting wasp *Pepsis*, grasshoppers are less tolerant and attempt either to flee or to fight back. During observations of interactions between the sphecid wasps *Prionyx parkeri* and *Tachysphex* spp. (Steiner, 1981), various forms of physical defence were shown by their grasshopper prey. One obvious line of defence that was observed was that of jumping or flying away, although this did not always deter the wasps, who occasionally clung to the grasshopper in mid-air. The wasp normally grasps a grasshopper in its jaws in such a way as to be able to get its sting into the latter's throat to paralyse the jaw muscles and prevent the regurgitation of noxious fluids, another line of defence employed by the victim. Immobilizing the grasshopper's powerful jaws is quite important to all of the wasps that use them as larval food, for in at least one instance a female wasp let herself get bitten and died the next day, presumably from the effects of the bite. As the wasp attempts to get its sting into the victim's throat, the latter does its best to dislodge her, pushing with its powerful hind legs against the attacker's head and grasping front legs. This may be successful against the lightly built *Tachysphex* and its cousins but it seldom succeeds against the powerful *Prionyx*. In the wild, the grasshopper *Mestobregma plattei rubripenne* (Oedipodinae) was on two occasions seen to suddenly open its warningly coloured wings, partially in one instance and wholly in the other, stretch out its hind legs and freeze when attacked. Despite the fact that this is normal prey for *P. parkeri*, the wasp failed to drive home its attack and the grasshopper escaped. Since the warningly coloured wings would mean nothing to the wasp, it has to be assumed that it was the physical aspect of the grasshopper that deterred the wasp. Similar behaviour was observed in captivity, though its success seems something of a mystery.

In a survey of 25 species of Ghanaian mantids (Edmunds, 1972), the most common form of initial physical defence on being molested was to either run or fly or to exhibit a startle display. The leaf-mimicking species *Phyllocrania paradoxa* will run or fly when attacked or flatten itself against the substrate on which it is perched and remain perfectly still.

Molested stick insects often just drop off their perch and fall to the floor to escape but the Javanese species *Orxines macklotti* is a bit more spectacular in its avoidance of danger (Robinson, 1968). The insect is highly cryptic and remains pressed against lichen-covered substrates during daylight hours. When touched, both sexes jump backwards from their vertical resting position and fall to the ground, opening their wings and tegmina as they do so but folding them immediately on landing. In jumping from a vertical perch 120 cm high, males leaped backwards as far as about 60 cm horizontally and females as far as 40 cm. The wingless male of the Neotropical *Oncotophasma martini* also jumps backwards when escaping, but not the female (Robinson, 1968a). If grabbed by the head or thorax, the males exhibit a 'tail-bowing' display in which the tip of the abdomen is first lowered and then raised and curled forwards towards the head, in which position it is then moved slowly back and forth. It may well be that this is an intimidatory display, mimicking the tail-raising behaviour of a threatened scorpion; scorpions are sympatric with this particular stick insect. Again, in males, the ventral surface of each hind femur is armed with a row of strong spines and, when the insect is picked up, the hind limb is flexed at the femoro-tibial joint, driving these spines into the skin in much the same way as do some grasshoppers. That females lack spines and do not respond to being picked up by flexion of the hind limbs points to the male response as being defensive in character.

Although it has been known for a very long time that both termites and ants have defensive soldiers within their social groupings, the discovery of such a caste in the Hemiptera is as recent as 1977. In that year, it was reported from Japan that the woolly aphid *Colophina clematis* has a soldier caste consisting of first-instar individuals. Since then, soldiers have been reported from other aphid species, though it would appear that they have evolved independently at least four times within the Aphidoidea. *C. clematis* produces sterile first-instar soldiers on its secondary host, *Clematis* spp., as do two other species in the same genus, *C. monstrifica* and *C. arma*. *C. clematicola*, on the other hand, does not have special soldiers but, instead, first-instar larvae attack intruders with their stylets (Kurosu & Aoki, 1988). Not only do they attack the larvae of predacious insects but they also attack their eggs. On a *Clematis terniflora* plant infested with the aphid *Myzus varians* as well as with *C. clematicola*, the former were attacked by first instars of the latter species, presumably in a bid to reduce competition.

Pseudoregma alexanderi from Asia, which feeds upon bamboo, produces pseudoscorpion-like first-instar soldiers with a pair of sharp horns on the front of the head and enlarged grasping front legs. When predacious larvae of syrphid hover flies and hemerobiid brown lacewings were introduced to a *P. alexanderi* colony, they were attacked by the soldiers (Aoki *et al.*, 1981). A number of the latter attacked each time and the larva was either immobilized or fell to the ground with the aphids still attached. The soldiers clasp the larvae in their large front legs and insert their sharp horns into their body. When the two exist side by side on the same host plant, *P. alexanderi* soldiers also attack and kill individuals of the aphid *Astegopteryx bambucifoliae*. Horned first-instar larvae of the latter species have been seen to attack other insects but with their stylets not their horns (Aoki & Kurosu, 1989). A second Asian species of *Pseudoregma* also has pseudoscorpion-like first-instar soldiers similar to those of *P. alexanderi*. *P. bambucicola*, as its name implies, also lives on bamboo where its soldiers are produced to coincide with, and defend, winged individuals of the aphid's sexual phase (Sunose *et al.*, 1991). Their function is, however, seemingly pointless for, in Japan, this species has no known primary host to which the winged aphids can migrate. It must therefore be assumed that such a primary host did once exist, or that somewhere else in southern Asia, the primary host (a *Styrax* sp.) still grows and the life cycle is still carried out in full. Another southern Asian gall-forming aphid *Astegopteryx styracicola* was reported as long ago as 1930 as having the habit of biting human beings if it fell on to their bare skin. It is now known (Aoki *et al.*, 1977) that the second instar of this species is dimorphic, with both normal and peculiar hairy individuals. Microscopic examination of these hairy aphids indicates that they develop no further than the second instar and this, along with their biting habit, indicates that they are probably soldiers.

The aphid *Pemphigus spyrothecae*, which makes spiral galls on the leaf stalks of the black poplar, *Populus nigra*, in Europe, has also been found to have soldiers. The thick-legged first-instar soldiers of this species are able to protect the aphid colony from attack by a range of insect predators (Foster, 1990). In artificial galls, they were found to be capable of killing first-instar larvae of the coccinellid *Adalia bipunctata*, early third-instar larvae of the syrphid hover fly *Eupeodes corallae* and first to third instars of the cimicid flower bug *Anthocoris nemoralis*; all three are normally predators of aphids. In virtually all cases when the soldiers attack the predators, they themselves are also killed. Another natural predator of this aphid, the flower bug *Anthocoris minki*, was introduced into galls in the wild and the soldiers were effective not only in preventing the bug from entering the gall but also in being able to kill any that entered, though once again many soldiers died in the process. The loss of a few soldiers is of relatively little importance in terms of the survival of the other aphids within the gall, for a single *A. minki* is capable of consuming all of them. Unlike *Colophina clematicola* soldiers, those of *P. spyrothecae* were not seen to attack the cohabiting aphid *Chaitophorus leucomelas*.

Both male and female scorpionflies, when feeding kleptoparasitically in spider webs, have been observed to defend themselves against the rightful occupants (Thornhill, 1975). If a male scorpionfly is approached it will strike the spider with its enlarged genital bulb, which usually causes the latter to stop and then retreat. Similar behaviour is exhibited towards other scorpionflies trying to muscle in on its ill-gotten gains. Female scorpionflies lack the genital bulb and instead will strike a spider with their abdomen or else will butt it with their head.

Cerambycid beetles of the Neotropical genus *Hammaticherus* possess an unusual form of defence based on their elongate antennae. Both males and females of the Central American species *H. batus* have long, recurved spines on the third to sixth antennal segments (Silberglied & Aiello, 1976). When threatened, the beetle whips the antennae backwards and downwards to bring the spines into contact with the surface of an attacker. Rotation of the head then pulls the hooks deeply into position, in a human finger causing pain and sometimes drawing blood. As soon as the beetle is dropped, the hooked spines are quickly and easily withdrawn. The effectiveness of this form of attack on the soft snout of, e.g., a predatory mammal is easy to imagine.

In a manner better known in the Trichoptera and some Lepidoptera, a number of subfamilies within the Chrysomelidae have larvae which live in a case. *Exema canadensis*, for example, is a North American species of chlamisine beetle commonly found feeding on composites of the genera *Solidago* and *Aster*. In this case, the female beetle constructs the initial case by surrounding the newly laid egg with pieces of green faecal material (Root & Messina, 1983). When it is ready to emerge, the larva chews the cap off the egg case and then turns it over, presumably after it has cut the stalk by which the egg was attached to the substrate. The larva now curls itself within the case so that both its mouth and anus are at the single opening. As the larva feeds and grows, it increases the size of the case by adding its own faecal material around the rim of the original case formed by its mother. It has long legs, which can be extended laterally beyond the rim of the case as it moves round. If disturbed, it retracts its body and pulls the rim of the case hard down against the leaf on which it is feeding. The case is apparently a very effective deterrent to predation for no encased larvae were attacked during many hours of observation on them in the field.

Ants in three unrelated tribes have evolved a special 'trap-jaw' mechanism. The mandibles are extremely large and have pointed apical teeth. They are held open against a tendon spring mechanism, and when sensory hairs between them are touched, they spring shut on the prey. Majors of the Australian dacetine ant *Orectognathus versicolor* no longer appear to use their trap-jaw for prey capture because, although the mandibles are large, they have blunt apical teeth. Instead, they stand around with their mandibles in the cocked position and, if an alien ant touches the sensory hairs, they click shut on the intruder's shiny, sclerotized body with enough force to throw it through the air for several centimetres.

Colonies of the ant *Pheidole dentata* have developed a number of strategies to defend themselves against fire ants of the genus *Solenopsis* (Wilson, 1976). Fire ants, especially *S. germinata*, are one of the major enemies of *P. dentata* and it is of advantage if the latter can make a decisive counter-attack against the enemy as soon as they are detected. Minor workers of *P. dentata* lay down recruitment trails with secretions from the poison gland and the major workers are able to tell when trail-laying minors have come into contact with fire ants. The response of the major workers, who are the soldiers, is to immediately become more aggressive and to follow the worker trails back to the fire ants. If the *Solenopsis* are attacking in any numbers, then the tendency is for the defenders to withdraw towards the nest and fight along a shorter perimeter. If the fire-ant attack is massive, then the *Pheidole* just grab their brood and scatter in all directions. Another *Pheidole* species, *P. desertorum*, has a number of different nests but uses only one of these at a time. This appears to be a defence against the army ant *Neivamyrmex nigrescens* (Droual, 1984), for, when the latter attacks, the *P. desertorum* workers pick up their brood, evacuate the attacked nest and make their way to one of their alternative habitations.

A characteristic of the Neotropical leaf-cutting ants of the genus *Atta* is that leaf fragments being carried back to the nest by larger workers often bear a hitchhiker in the form of a minim worker. It may well be that one of the roles of these minims is to commence cleaning the leaf surface of fungal spores and other undesirable detritus, since it is important that only clean leaves end up in the fungus garden of these ants. Very recently (Feener Jr & Moss, 1990), another role has been discovered for these minim workers, that of defence. *Atta* and *Acromyrmex* ants are attacked by over 20 species of phorid fly, whose females deposit their eggs in the head-capsules of the leaf-carrying workers. The presence of the minim workers reduces both the probability of a parasitic fly landing on a leaf fragment and also the amount of time spent on it if it does land. This defensive role of the minim workers is supported by the fact that there are both daily and seasonal changes in the amount of hitchhiking which relates directly to the number of flies present.

Females of the aculeate Hymenoptera have evolved a very effective prey-killing and defensive structure in that the ovipositor forms a sting for the injection of venom. Why is it, however, that, when foraging, individual social wasps will ignore the presence of a potential predator such as a badger but will attack it *en masse* if the nest is disturbed? Some answers have been provided by an investigation into the attack response of the South American polybiine *Polybia occidentalis* in Brazil (Jeanne, 1981). When a nest of this species is disturbed, large numbers of adults make their way quickly to the outside surface of the nest. A small percentage of these adults may then take flight and attack and sting the intruder. The initial recruitment of individuals to the nest surface is due to venom, or at least one of its constituents, acting as an alarm pheromone. Wing-buzzing also takes place in a disturbed nest but this does not communicate alarm to other nest occupants. Once outside the nest, venom acts as a releaser for attack behaviour but does not elicit an attack response itself. This requires some visual stimulation, dark colour initiating a greater number of attacks than movement.

Within the larval phase of the encyrtid wasp *Copidosomopsis tanytmemus*, a parasite of the eggs and larvae of the Mediterranean flour moth, *Anagasta kuehniella*, there exists a morph which basically fulfils the role of the soldier caste in other insects. The female wasp lays a single egg in the host and a number of larvae arise by polyembryony, i.e. the initial egg cell divides a number of times to produce identical cells and each of these then gives rise to a larva; thus all of the larvae represent a single clone. Whereas the majority of the derived larvae are normal, a proportion of them are precocious: they have well-developed mouthparts and are very mobile. They never, however, complete their larval phase and always die in the host's haemolymph. Experiments have revealed (Cruz, 1986) that, when the host is also parasitized by the braconid wasp *Phanerotoma flavitestacea* or the ichneumonids *Venturia canescens* or *Trathala* sp., their larvae are either killed or injured by the precocious larvae of *C. tanytmemus*.

Whereas some polistine wasps go out of their way to deter ants there are others that apparently benefit from living in close association with ants, often those species which are associated with ant plants. Relatively defenceless wasps may also live in close association with the nests of more aggressive species. These cohabiting wasps apparently benefit by reduced predation from birds and from army ants. In a particular area of the Amazon rainforest in Peru, numerous species of vespid wasps were found to live in association with the ant *Allomerus octoarticulatus* on its melostomataceous ant-plant host, *Tococa guianensis*, and with *Pheidole* spp. ants on their melostomataceous ant-plant host, *Maieta poeppigii* (Heere *et al.*, 1986). Although workers of both ants were seen swarming all over the wasps' nests, in no instance was the wasps' brood seen to be attacked, despite the fact that a number of the wasps belonged to the genus *Mischocyttarus*. It was observed that building nests on these two particular ant plants eliminated attack by *Eciton* army ants, which actively avoided the two host trees, though they readily foraged on other trees in the same habitat. The army ants did not, however, avoid twigs from which formicaria of the symbiotic ants had been removed. They also readily attacked the brood in wasp nests on leaves of *T. guianensis* from which the formicaria had been removed, indicating that it was the presence of the other ant species that normally prevented their foraging on ant plants.

The Neotropical stingless bee *Trigona angustula* has developed a sophisticated form of defensive behaviour aimed at flying insects (Wittmann, 1985) and especially, it seems, against the kleptoparasitic lemon bee, *Lestrimelitta limao*. Groups of worker bees hover, facing one another, before the nest entrance, forming a sort of protective corridor down which all arrivals wishing to enter the nest have to fly. If another insect, such as a honey bee or a lemon bee, intrudes into this corridor, whether accidentally or on purpose, it is attacked by the hovering residents. They attack it and force it to the ground by hanging on to its wings with their jaws. When raiding a stingless-bee nest, the lemon bee releases citral, which overwhelms the nest's natural scents and allows it unlimited access. Citral appears to act as a recruitment pheromone to *T. angustula* for, if a lemon bee attempts to enter the nest, not only is it attacked but more workers emerge from the nest to assist.

STARTLE DISPLAYS AND FLASH COLORATION

A startle display involves the adoption of a posture which appears threatening and is usually accompanied by an apparent increase in size of the animal. It is generally accepted that these are anti-predator in nature and the fact that they often, though not always, coincide with flash coloration points to their being mainly directed towards vertebrate enemies. The colours involved in the flash display are much the same as those used aposematically to warn of distastefulness. Although a number of species which use flash colours may, indeed, be distasteful, just as many are not; the colour is used purely to frighten the predator. Startle or flash displays may be presented from the side, in which case they are referred to as lateral displays, or from the front, i.e. head-on, in which case they are called frontal displays.

▲ Whereas the adults of the African mantid *Polyspilota aeruginosa* use a lateral display of their warning colours, the nymphs, such as the one depicted here in Madagascar, use a frontal display revealing the bright colours on the undersides of their forelegs.

▼ Defensive associations between ants and other insects, especially homopteran bugs, are quite common. In the example illustrated, the green tree ants, *Oecophylla smaragdina*, are providing protection from potential predators for a butterfly larva of the lycaenid genus *Narathura*. In exchange, the ants receive a sugary secretion from the larva.

Within the Orthoptera, these types of display are to be found mainly in the tettigoniids and the acridids, many of whom are cryptically coloured. Startle displays in the katydids often involve raising of the legs and opening of the wings and both of these may then show flash colours. In Mexico, the katydid *Neobarrettia vannifera* is overall mainly green in colour and it walks around over shrubs in search of its prey. When disturbed, it opens its wings horizontally in what is effectively a frontal display to reveal the black and pale yellow polka-dot pattern on the hind wings; the wings are opened and closed repeatedly until the threat is withdrawn (pers. obs.). In other species of *Neobarrettia*, the wings may be raised towards the vertical, revealing a striped abdomen, and the front legs may be raised simultaneously in a threatening manner providing both lateral and frontal displays simultaneously. Another Central American species, *Balboa tibialis*, opens its brightly marked wings more to the vertical, at the same time elevating the rear end of the abdomen so that this display is again both frontal and lateral. Closure of the wings is very rapid in *N. vanifera* but is carried out slowly in *B. tibialis*. One of the most startling of all katydid displays is that of the Neotropical species *Tanusia brullaei*, which, with its wings folded, is a leaf mimic. When molested, it opens both pairs of wings, revealing on the hind pair two dark eyespots. *Sasima truncata* from the rainforests of New Guinea is a green leaf mimic with a heavily spined pronotal border forming a physical defence. When threatened, it suddenly raises its red and yellow hind legs vertically from beneath the wings (pers.-obs.). The Australian katydid *Acripeza reticulata* has reduced wings and, when threatened, 'stands on its head', opens its tegmina and reveals its red, black and blue striped abdomen (pers. obs.).

There is a variety of acridids from the world's deserts or their equivalent, sand dunes, which are cryptically coloured in shades of brown. If molested, they immediately take flight, at the same time revealing their bright red or bright blue hind wings, which are snapped shut again the moment they land. Most of these are highly edible but the distasteful acridids also have brightly coloured wings and, instead of flying, they raise them vertically in a lateral display as they remain seated on their food plant. The extremely repulsive African pyrgomorph *Dictyophorus spumans*, when disturbed, raises its short wing cases

▲ In Mexico the katydid *Neobarrettia vannifera*, when disturbed, opens its wings horizontally in what is effectively a frontal display to reveal the black and pale yellow polka-dot warning colour pattern on the hind wings; the wings are opened and closed repeatedly until the threat is withdrawn.

revealing the mainly black and yellow striped abdomen.

The stick insects, or walkingsticks, use startle display and flash colours very similar to those of the tettigoniids. The wings, which are sometimes brightly coloured in black and red with occasionally yellow, blue or orange, are raised vertically. In some species this opening of the wings occurs momentarily before they are snapped shut and the insect then falls to the ground. Alternatively, the wings may be held open in a lateral display, often for a considerable time. The Central American walkingstick *Metriotes diocles*, for example, has hind wings which are tessellated with dark-brown markings with a bright-blue spot at the base, ringed with black or dark brown, which extends on to the metathorax. In the most developed form of its startle display, *M. diocles* snaps the wings up into an erect position, suddenly revealing the eye-like blue spot, holds them there for up to 90 seconds and, just as suddenly, snaps them closed again. If molestation continues, the insect may fly off or walk away with the wings still in the raised position (Robinson, 1968). The large female of the Australian phasmid *Acrophylla titan* raises her dark-brown tessellated wings in a similar manner (pers. obs.), at the same time raising the head and thorax at an angle of nearly 45° from the abdomen.

The majority of mantids are cryptic when at rest but, when disturbed, many of these react with a startle display and/or flash coloration. Eliciting the whole display is not always easy, the simplest way to do so being to hold a monkey with its face just a short distance from the insect. In a survey of 25 Ghananian mantids (Edmunds, 1972), all of which were cryptically coloured and/or were grass- or twig-mimics, 15 species were found to give a startle display and/or use flash coloration when disturbed. The majority of these, i.e. nine species, give lateral startle displays, three give either type of display and only one gives just a frontal display. The remaining two species give no startle display but have parts of the legs and wings brightly coloured and these may therefore be flash colours. *Pseudocreobotra ocellata* is basically green and yellow but has a large black and yellow eyespot on each forewing. Under threat, the wings are raised to expose the eyespots and the white hind wings; the mantis may stridulate during its display. Whereas *P. ocellata* spends its time amongst foliage, the closely related African species *P. wahlbergi* sits on flowers and is able to alter its colour to match that of its background, the process taking several days. Like the previous species, on the wings it has large eyespots which are used in its startle display. *Polyspilota aeruginosa* is a species widespread in the rest of Africa and in Madagascar. It gives a lateral display during which the brightly coloured inner surfaces of the legs are turned towards the aggressor, the wings, which have some red coloration, are opened vertically to reveal red and black marks on the abdominal underside and the bright red of the jaws is also exposed. Nymphs employ a frontal display only, revealing the colour of the jaws and of the front legs (pers. obs.).

The lanternfly *Laternaria laternaria* (Fulgoridae) is a Neotropical species which spends its life on the trunks and branches of trees, feeding on the sap. At rest, it is highly cryptic but, when disturbed, it opens both pairs of wings laterally, revealing suddenly the large eyespots, one on the distal end of each hind wing (pers. obs.).

Flash and startle displays are fairly common amongst the moths and butterflies, with the latter usually involving a sudden presentation of large eyespots. Many

▼ The peanut bug, *Laternaria laternaria* (Fulgoridae), seen here in Costa Rica, is normally quite well camouflaged when at rest on the trunk of a tree. When disturbed, however, it opens its wings to reveal the large eyespots on the hind pair. It is believed that, to a predator, this appears as the face of a much larger animal staring back at it and is sufficient of a threat to cause it to retreat.

An alternative to flashing bright colours on the hind wings. Both pairs of wings are coloured cryptically or neutrally to expose a brightly coloured abdomen. The convolvulus hawk-moth, *Herse convolvuli* (Sphingidae), is very cryptic when at rest on the trunk of a tree but, when disturbed, it opens its wings to reveal the dorsal surface of the abdomen which is marked with alternate white, red and black stripes. The wings are kept open for a while and then are gradually closed (pers. obs.). Lymantriid moths of the genus *Euproctis*, such as the yellow-tail moth, *E. similis*, and the African bird-dropping mimic *E. conizona*, use a different ploy. If they are attacked, they open their wings a little and protrude the tip of the abdomen, which is covered in a thick layer of brightly coloured orange hairs, vertically from between them. During the startle display of the Neotropical saturniid *Dirphia avia*, the moth raises its wings upwards and forwards like a cloak over its head, at the same time arching its abdomen to reveal alternate lines of black cuticle and orange hairs.

▼ A number of insects, including some orthopterans and moths, raise their wings to expose their warningly coloured abdomen when they are threatened. Just such a behaviour is being exhibited by this saturniid moth *Dirphia avia* in Trinidad.

nymphaline butterflies have cryptically coloured underwing surfaces and very bright markings on the upper wing surfaces. When at rest, they tend to sit with the wings closed but, if touched, they open the wings, revealing bright flashes of colour. Since these butterflies are not known to be particularly unpalatable and they are not mimics, the flash display must be effective in its own right in deterring predators. The effect may be accentuated by the presence of eyespots on the upper wing surfaces; the peacock butterfly, *Nymphalis io*, for example, has an eyespot on all four wings. Not all eyespots on butterflies, however, perform a startle function. Many satyrines, for example, have eyespots on both their upper and lower wing surfaces so that one or other side is on permanent display. These serve to direct birds' beaks away from the delicate head to areas of the wing, which can be sacrificed while the butterfly makes good its escape.

In moths, the forewings are often cryptically coloured and are held folded over the hind wings, which may be brightly coloured and/or bear eyespots. The red underwing moth, *Catocala nupta* (Noctuidae), for example, is well camouflaged when at rest on a tree trunk with its wings folded. If touched, it flicks the forewings away from the hind wings, revealing the latter's startling red and black coloration momentarily. If further disturbed, it then flies off with a flash of red from its hind wings. A second noctuid, the large yellow underwing, *Noctua pronuba*, is less tolerant than its cousin and usually flies immediately it is touched, in this instance revealing its bright yellow and black hind wings. A number of arctiids, often referred to as tiger moths, have warningly coloured forewings

◀ When threatened by a nearby predator, flag-legged bugs (Coreidae) wave one or both of their brightly coloured rear legs in an arc. This probably serves a warning 'keep-off' function, but if a predator such as a bird does take a quick bite, it tends to be at the leg itself, which is expendable; these bugs are often seen with one or both rear legs missing. This is *Anisoscelis foliacea* in Peru.

▼ In their juvenile stages, many warningly coloured insects aggregate in dense clusters, thereby maximizing the 'keep-off' effect of the colours. If these *Pachylis* species squash-bug nymphs (Coreidae) from Mexico are molested, they squirt an unpleasant liquid from the anus into the face of their aggressor.

and, if these do not deter a predator, then the brightly coloured hind wings are brought into view in a flash display.

Moths from a number of different families have eyespots on the hind wings and these are normally kept covered when at rest by the often cryptically or neutrally coloured forewings. On being disturbed, the moth uncovers the eyespots, giving it, at least to our eyes and presumably to those of other vertebrate predators, the appearance of the face of a much larger creature. Such displays are to be found in saturniids, such as *Automeris* sp., and the sphingid eyed hawk-moth, *Smerinthus ocellata*. Saturniids of the genus *Saturnia*, the emperor moths, have gone one further in that they have eyespots on both pairs of wings so that one pair is in view all the time while the other pair are displayed when the moth is disturbed. Eye-spot displays are not, of course, restricted to adults because the caterpillars of a number of moths have eyespots on either side of the rear end of the abdomen. When threatened, they thrash the abdomen around in what has been interpreted as a snake-like display, which is seemingly good enough to deter at least some predators. The display of the large elephant hawk-moth, *Deilephila elpenor* (Sphingidae), was certainly good enough to convince a 5-year-old girl, for she rushed into the house saying that she had just found a snake in the garden (pers. obs.). Under what appear to be high selection pressures in the wild, the caterpillars of the Neotropical oxytenid moth *Oxytenis naemia* have evolved three separate defensive habits. Early instars are black and shaped to mimic a bird-dropping. The fourth instar is differently coloured and has an oily appearance and appears to mimic the droppings of a larger bird. The fifth instar resembles either a rolled-up green leaf or a rolled-up dead leaf and, as well as this, it has a pair of eyespots on the abdomen with which it performs a reasonable imitation of a snake when disturbed (Nentwig, 1985).

SOUND PRODUCTION

Bright colours as an aposematic cue to vertebrate predators that an animal is distasteful are well known; sound as an aposematic cue is, however, much less common but appears to be employed by arctiid moths. A number of these moths, when they detect the sounds made by hunting bats, produce a series of rapid clicks in response. These sounds deter the bats and, in an experiment, it was found that bats would even avoid a palatable food item such as a mealworm if it was presented to them accompanied by a recording of an arctiid moth. In a survey of Canadian arctiids (Fullard, 1977), of 24 species sampled from ultraviolet-light sources, 17 produced sounds under tactile stimulation. In general, there tends to be a relationship between time of year and sound production: those species which are on the wing in the early part of the summer, when bat activity is fairly low, tend to be silent while those in flight later in the summer, when the bats are more active, do produce sounds.

It is well known that a wide variety of insects which use stridulation during courtship will also stridulate if they are disturbed. The North American passalid beetle *Odontotaenius disjunctus* is a presocial species which lives in colonies in decaying logs. If disturbed, it produces a series of squeaks by rubbing part of the abdomen against the underside of the hind wing. Under experimental conditions, it was found (Buchler *et al.*, 1981) that the crow, *Corvus brachyrhynchos*, held back when attacking stridulating *O. disjunctus* and also took longer to eat them than silenced individuals. The grasshopper *Pareuprepocnemis syriaca*, on being handled, produces a series of chirps by rubbing the mandibles one against the other. In a laboratory investigation (Blondheim & Frankenberg, 1983), this grasshopper was fed to a number of different species of captive lizard. On being grasped by the head, the grasshopper began to chirp with the result that the lizard usually dropped it. In the wild situation, this would almost certainly give the grasshopper time to jump to freedom.

▲ It is not uncommon for the larvae of certain leaf beetles to cover themselves in frass and other detritus in order to hide their presence from searching predators. These larvae of a *Eugonycha* sp. chrysomelid from the Brazilian *campo cerrado* are using as camouflage material the trichomes which they have cut from the leaf surface of their *Solanum* host plant.

◀ The spider *Arachnura scorpionioides* from Madagascar incorporates a line of detritus down the centre of its web, often including one or two egg sacs as well. The spider then hangs head-downwards beneath this detritus, where it resembles yet another piece of dead leaf.

CAMOUFLAGING BEHAVIOUR

Although many invertebrates exhibit cryptic colouration, so that they blend in with their chosen backgrounds, there are some which are not marked in this way and, instead, camouflage themselves or, as in the case of some spiders, their egg sacs, with materials taken from their surroundings.

The larvae of many, though not all, chrysopids place detritus, such as pieces of plant material or prey remains, on the dorsal surface of the abdomen, where it forms an effective camouflaging layer. The larvae of *Ceraeochrysa cincta* from Florida live amongst colonies of the mealy bug *Plotococcus eugeniae* and prey upon it. Developing and adult females of the bug bear long, white lateral filaments of wax which gives them a rather stellate look. The chrysopid larva is not easy to distinguish amongst its prey for it covers its abdomen in a mass of material derived from them. A larva deprived of its mass of wax usually begins to re-build it immediately (Eisner & Silberglied, 1988). It usually approaches an adult female bug, grasps one or two filaments in its curved jaws and, by pulling on these, detaches a mass of wax powder and sometimes the detached filaments. The larva then bends its head backwards and deposits

▲ *Anaea* and *Prepona* butterfly caterpillars (Nymphalidae) from the Neotropical realm go to great lengths to become convincing mimics of withered leaf tips. After biting off the end of the leaf, the caterpillar extends its mid-rib outwards using silk glands in the mouthparts. Then it spends the day sitting head-downwards on its handiwork. Note how the caterpillar has incorporated some genuine leaf fragments into the false mid-rib at its rear, making a convincing lead-in to its body.

Cryptic coloration. This is of little use to those invertebrates which possess it if it is not coupled with the correct behaviour, in terms of stance or choice of a suitable background. For a cryptically coloured animal, the choice of a suitable background, with which it will blend rather than on which it will stand out, would seem to be of paramount importance. Sometimes, however, behaviour does not seem to fit the rules. For instance, a red underwing moth, *Catocala nupta*, was found during daylight at rest on the white wall of a house (pers. obs.). This moth has beautifully marked forewings which blend in perfectly with the background when it is on the bark of a tree but, against the white paint, this individual stuck out like a sore thumb. The reason for its being on the wall was quite straightforward: it had been attracted by an adjacent light which had been left on all night and presumably, when dawn broke, it was loath to move on. The observer, curious as to which species it was and needing to see the red flash of the hind wings, touched the moth, which duly displayed and then took flight. It circled several times past trees, a wooden fence and the wall of the house and then, much to the observer's surprise, landed on the white wall again. Further disturbance produced the same result, though, in a third instance, it flew off, risking an aerial attack from the local house-sparrow population. It may have returned to the wall because it was attracted by its own scent, which it had left behind on its previous visit, but this seems unlikely since both times it re-settled at some distance from its original position.

▲ Cryptic moths are often remarkably adept at selecting a background on which their camouflage will reap maximum benefit. This *Lophodes sinistraria* (Geometridae) from Australia has not only aligned itself perfectly with a twig, it has positioned its plump body on a blemish, which looks to be part of the moth itself.

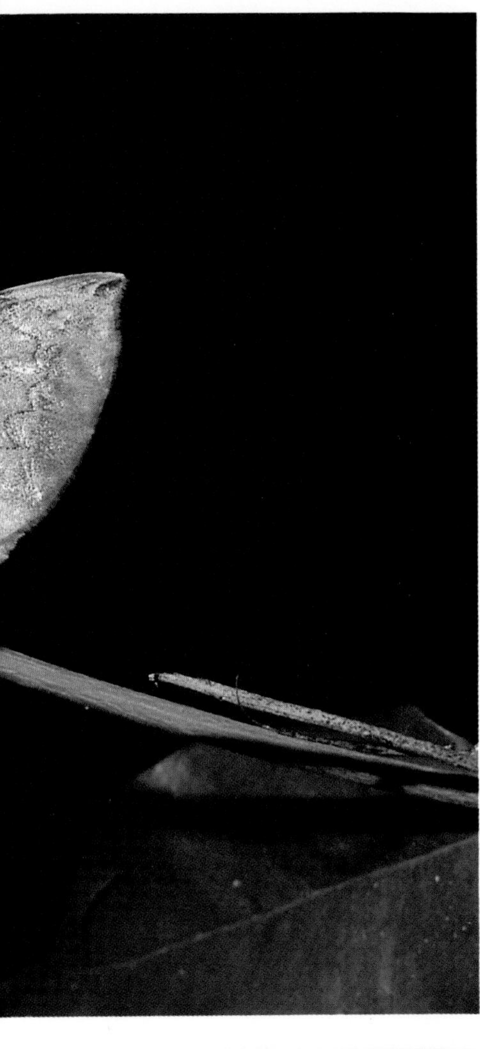

◀ One characteristic which betrays insects to their sharp-sighted enemies is their bilateral symmetry. Many moths, which are mimics of dead leaves, seek to destroy this by bending the body over to one side and adopting an asymmetric stance. This is the daytime resting pose of a *Rolepa* species (Lymantriidae) in a Brazilian forest.

▶ *Aethomerus cretatus* is a longhorn beetle which mimicks a bird-dropping, but of a different kind from that mimicked by *Ozodes* (below) from the same forest. *A. cretatus* adopts a stylized stance, with just one white front leg stuck out to resemble part of the dropping which has splatted outwards and the very thin antennae folded back neatly. This beetle totally refused to 'come alive', allowing itself to be rolled around on the leaf without doing more than waggle its antennae slightly.

▼ Many insects mimic bird-droppings, but this seldom requires much in the way of specialized behaviour – the insect just sits and looks splodgy. However, this *Ozodes* species longhorn beetle (Cerambycidae) from Brazil adopts an extreme stance in its efforts to mimic persuasively the mixture of plant seeds and chitinous insect remains which are frequently passed in a bird-dropping. Lying on its side, it curls its body slightly and exposes the enlarged femora on its legs; these resemble the seeds which are often passed through a bird unaltered. This beetle is very loath to forsake its act and 'admit' that it is alive; only after a great deal of gentle prodding will it stand up on its legs and walk across the leaf.

the mass of wax on the abdomen. The larva moves from one bug to another until the camouflaging mass of wax is complete. Similar behaviour is exhibited by the larvae of a number of leaf beetles, which cover their backs in frass and shed exuvia.

Ritualized behaviour on alighting is typical of a number of butterflies and the European grayling, *Hipparchia semele*, is a good example. The underwing markings of this species are highly cryptic but, displayed prominently near the outer edge of the underside of each forewing, is a black spot with a white centre, which resembles an eye. On being disturbed, the grayling takes off and flies for some distance before landing again. On landing, the underwing eyespots are shown for a moment and then immediately hidden beneath the hind wings. The butterfly now shuffles round to face the sun, tips over on to its side, thus eliminating its own shadow, and effectively disappears as it takes on the appearance of a dead leaf. Both authors can testify to the effectiveness of this display, in which choice of background is unimportant, for a dead leaf is a dead leaf whether it is lying in the middle of a mass of bright-pink heather flowers or amongst dead grass. A butterfly which shows even more complex behaviour (pers. obs.) is the Neotropical lycaenid *Arcas imperialis*. This species has brilliant emerald upper wing surfaces and iridescent green lower surfaces flecked with black. In flight, it resembles a blue spark flickering through the forest but it disappears abruptly as it shuts its wings on alighting on a leaf. Then, like the grayling, it slowly keels over on to its side and lies there like a piece of leaf which has been cut off from the canopy above and has lodged on a lower leaf. As an extra line of defence, as it is toppling over, it moves its hind wings against each other, thus activating a number of narrow black extensions at their rear end. These resemble the antennae on a false head.

FALSE-HEAD MIMICRY

Although most often to be seen in butterflies, for which it has been well documented, false-head mimicry is found in other groups. The basic idea seems to be to present a predator with a choice of two heads, one real and one false, with the animal often making the latter a more tempting target. Since predators usually aim for the head, this gives the victim a 50 per cent or greater chance of having its false head attacked, with little or no damage, rather than its real head, which would result in its death.

Although the few known mutillid-mimicking spiders are oriented the same way as their models, i.e. the head end of the spider is the head end of the model, at least one species does things back to front. The spider in question is *Orsima formica*, a remarkable 5 mm-long jumping spider from Borneo. The abdomen is separated into two regions by a narrow waist, so that the hindmost section represents the head of the wasp model, the foremost section the thorax, and the cephalothorax the model's abdomen. The species has unusually long

▲ 'False-head' mimicry requires both behavioural as well as morphological adaptations for it to be a success. In order to divert a predator's attention to the wing tips, which form the false head, *Zeltus amasa maximianus* (Lycaenidae) from Malaysia moves them up and down. This causes the streamers to move around as if they were antennae, while the black spots at their bases mimic eyes on the false head.

► (Top) 'False-head' mimicry is quite common in caterpillars, but is particularly sophisticated in this *Lirimiris* sp. moth (Notodontidae) from Costa Rica. When disturbed, the caterpillar curls its real head in close to its sides and inflates a head-like brown sac at its rear end. In view of its apparent warning coloration, it is likely that this is designed to ensure that a peck is aimed at the tough rear end. This could possibly release a distasteful chemical which would prevent any further molestation.

► (Bottom) The Asian salticid spider *Orsima formica* not only resembles a mutillid wasp but has also modified its behaviour to be similar to that of its model. It often stands with the abdomen raised so that the false 'head' is apparent. The head effect is enhanced by the movement of the posterior spinnerets mimicking the antennae. At intervals the abdomen is lowered to bring the jaw-mimicking anterior spinnerets into contact with the surface upon which the spider is standing, giving the effect of feeding. The mimicry is enhanced when these spinnerets are moved from side to side in the manner of insect jaws.

◀ A number of *Phrynarachne* species crab spiders (Thomisidae) are excellent mimics of bird-droppings. Typically, this Indian *P. tuberosa* is sitting in full view on top of a leaf. Sometimes the spider sits on a thin pad of white silk spun across the leaf. This species is especially interesting for its habit of increasing the perfection of its mimicry by destroying its natural bilateral symmetry. It does this by adopting an asymmetrical stance with its right-hand legs held out at an angle to its body, while the legs on the opposite side are held in close to the side.

It has been suggested that these *Phrynarachne* bird-dropping mimics may encourage prey to come within grabbing range of the front legs by disseminating a pheromone that mimics the odour of genuine dung. This indeed seems to be the case with this Indian species. The small black fly on the right is a *Sepsis* sp, a fly which normally breeds in rotting matter such as dung, and it is showing considerable interest in the spider.

▼ Many newly hatched stink-bug nymphs assemble on or near their old egg shells. However, the particularly neat arrangement in this Peruvian pentatomid is thought to be due to the likelihood that the nymphs are mimicking a limacodid moth caterpillar. These are renowned for their powerfully stinging spines, whose bristling movements are mimicked by the nymphs' habit of waving their antennae and legs in the air when disturbed.

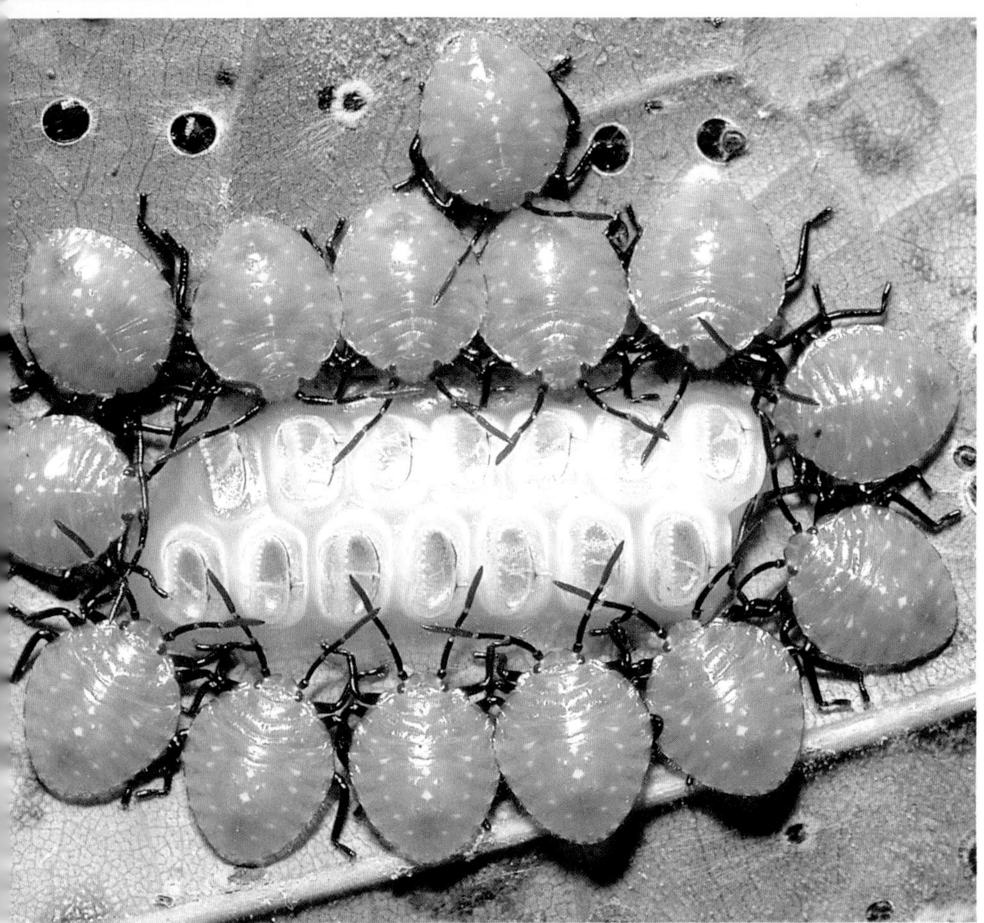

spinnerets for a salticid, one pair mimicking a pair of antennae and the other a pair of jaws. These are dark in colour, as are the hind legs, making the latter stand out as the forelegs of the model, since the other pairs of legs are pale in comparison. Apart from its overall insect-like appearance, the spider has also modified its behaviour to resemble that of a mutillid wasp (pers. obs.; Reiskind, 1976). It often stands with the abdomen raised so that the false 'head' is apparent. The head effect is enhanced by the movement of the posterior spinnerets mimicking the antennae. At intervals, the abdomen is lowered to bring the jaw-mimicking posterior spinnerets into contact with the surface upon which the spider is standing, giving the effect of feeding, especially so when these spinnerets are moved from side to side in the manner of insect jaws. It would appear that, if this spider indeed resembles a mutillid wasp to potential predators, then it derives its protection from Batesian mimicry, i.e. mimicking something distasteful or otherwise unpleasant or dangerous when you yourself are neither. Its behaviour, however, also provides it with a means of escape, for if a predator goes for the model 'head', this will retreat backwards in the opposite direction to that expected as the spider runs away forwards.

The Neotropical cerambycid *Oreodera glauca* from Central America has false-head markings on the rear end of the elytra (Silberglied & Aiello, 1976). Its hind legs are stretched out behind the end of the abdomen, giving the appearance of front legs stretched out in front of the head. The antennae are held tightly in against the side of the body, the narrow ends extending a good body length out beyond the tip of the abdomen, giving the impression that they are in fact protruding from the front of the normal head.

ACTIVE MIMICRY

Many invertebrates mimic objects, such as seeds, sticks, bird-droppings etc., without any particular accompanying behaviour. A number, however, mimic other invertebrates and also their behaviour and it is with these that we are concerned in this section.

It is unusual to find an animal that mimics two other different animals but this appears to be the case with the Neotropical cerambycid beetle *Acyphoderes sexualis* from Central America (Silberglied &

Aiello, 1976). This 'double mimic' has rather short antennae for a cerambycid and the abdomen is narrowed basally to form an apparent waist. In the way it waves its short antennae and moves around, it closely resembles a large ponerine ant. When picked up, the beetle initiates stinging movements with its abdomen and, like many ponerine ants, it also stridulates. When taking flight, its appearance changes suddenly to that of one of the aggressive polybiine wasps, typical of its habitat, both in shape and in the way it dangles its legs beneath it as it flies. On landing, it immediately resumes its ant guise.

A most unusual form of defensive mimicry has evolved in some species of tephritid flies. In North America, there are two genera of these flies which mimic the display behaviour of one of their main predators, salticid spiders. The snowberry fly, *Rhagoletis zephyria*, apparently mimics the common zebra spider, *Salticus scenicus* (Mather & Roitberg, 1987). The spider is basically black and white and, when displaying, it waves its anterior legs and palps in a characteristic fashion. The wings of the fly are also marked in black and white and, viewed from behind, bear a marked similarity to a front view of the salticid. When threatened, the fly waves its wings and dances from side to side in a manner akin to that of the displaying spider.

In a series of laboratory experiments, when *S. scenicus* individuals were presented with normal snowberry flies, snowberry flies with artificially blackened wings and house flies, the latter two were attacked significantly more often than the intact snowberry flies. Furthermore, when presented with displaying *R. zephyria*, the spider turned and fled a significantly greater number of times than when faced with non-displaying flies. A second species of tephritid does not apparently mimic a specific spider but is a general mimic of salticid display behaviour (Greene *et al.*, 1987). When disturbed, the fly raises its wings and waves them around in the manner of a salticid waving its front legs in display. *Zonosemata vittigera* is associated with *Solanum eleagnifolium* and, during the investigation of the fly's behaviour, 11 different species of salticid were found to live on the plant with it. In laboratory tests, individuals of these salticid spiders were presented with normal *Z. vittigera* individuals, *Z. vittigera* with their wings removed and then stuck back on again, *Z. vittigera* with their wings removed and replaced by house-fly wings,

▲ **Warningly coloured braconid wasps are a common sight in the rainforests of the Neotropical region. They have a characteristic mode of walking, flicking their partly opened wings upwards as they scour the forest for a possible host. They are mimicked by the remarkable assassin bug *Hiranetis braconiformis* (Reduviidae), which moves around with an identical gait. However, this individual from Peru reinforced its mimicry in a quite amazing way by raising just one rear leg in order to mimic an ovipositor. It only did this when actually threatened closely, replacing the leg on the leaf when the danger was no longer perceived to be close.**

► **The nest of this *Ropalidia* sp. wasp (close to *R. pomicolor* but possibly an undescribed species) from Madagascar has been carefully placed at the tip of a twig which the wasps have chewed off. With the wasps clustering closely on the nest, it comes to resemble a knobbly fruit of a type which is quite common in the island. When the author first saw the nest, he did indeed take it for a fruit. This is the dry season, and the nest is not active, the wasps just clustering quietly for several months; in some other Madagascan *Ropalidia*, the wasps leave the nest and crowd together beneath a nearby leaf for the duration of the dry season.**

◀ These *Phromnia rosea* flatid bugs from Madagascar are thought to be mimicking a spike of flowers. When molested by a bird, they 'explode' in all directions, thoroughly confusing their enemy.

house flies with their wings removed and replaced by *Z. vittigera* wings, and unadulterated house flies. When presented with intact tephritids and those with replaced wings (the latter were able to produce normal wing-waving responses despite the operation), the spiders, in most instances, stalked them initially but then waved their legs and backed off when the flies displayed. In the other three instances, the flies were attacked and killed. Since *Z. vittigera* with clear wings which they could wave and house flies with *Z. vittigera* wings which they could not wave are both attacked, it is clear from these experiments that a combination of both waving and markings on the wings is necessary to produce a convincing salticid display. This study also revealed that the tephritid's display had no effect on other predators tested, i.e. an oxyopid spider, a mantis, an assassin bug and a whiptail lizard, further indicating its specificity towards salticid spiders.

CHEMICAL DEFENCE

Strictly speaking, the term chemical defence implies that the animal concerned has synthesised or sequestered a substance which it then secretes when attacked. Within this section, however, are also included examples of animals 'sicking up' partially digested foods from their mouths or excreting waste from their rear ends in a form which is either repellent to or gums up the mouthparts and sense organs of an attacker. In some instances, the two are no doubt combined. As elsewhere in the book, only chemical-defence mechanisms where active behaviour is employed are described; invertebrates such as millipedes, which are simply unpleasant to the taste, are not included.

Although any spider which is able to will no doubt respond to attack by attempting to bite the attacker, parasitic wasps excepted, there is only one recorded instance of a spider deliberately using its venom as a repellent. Whilst censusing the Neotropical green lynx spider, *Peucetia viridans* (Oxyopidae), in the field, a researcher (Fink, 1984) noticed tiny drops of fluid on her face and hand. The liquid tastes bitter, irritates the eyes and feels cool upon the skin. A closer observation of the spider showed that females are capable of emitting a stream of venom from their fangs. The female spreads her chelicerae and spits the venom forwards, directing it by turning her head towards the target, which can be hit from as far as 20 cm. It is assumed that this spray serves a defensive role and its inaccuracy indicates that it may be aimed at larger predators, such as birds and lizards, rather than at other arthropods.

Most tettigonids use camouflage and mimicry as a defence, although a few species exhibit warning coloration and some of these may be distasteful. *Acanthoplus armativentris*, an armoured ground cricket from southern Africa, is not warningly coloured but does produce an offensive yellow liquid when threatened.

Chemical defence is much more common amongst members of the Acridoidea. Notable amongst these are the African pyrgomorphine grasshoppers (Acrididae), relatively cumbersome, poor-flying creatures, usually with a garishly patterned warning coloration. They spend most of their lives conspicuously feeding on their poisonous food plants, from which they both obtain nourishment and sequester toxins as a means of defence.

The southern African milkweed grasshopper, *Phymateus morbillosus*, as its common name implies, feeds upon poisonous Asclepiadaceae. When threatened, it spews forth a frothy liquid, containing sequestered cardiac glycosides, from openings situated near the bases of its hind legs. *Dictyophorus spumans*, from the same area of the world, also produces a foam which has such a nauseating stench that it surrounds the grasshopper in a protective chemical umbrella extending up to 1 m all around it (pers. obs). The nymphs of most of these chemically protected pyrgomorphs join together in large bands which roam around the countryside in search of food, protected by the ever-present disgusting aura which they trail around with them.

A North African pyrgomorph *Poecilocerus hieroglyphicus* feeds mainly on the asclepiad *Calotropis procera* in the wild. If attacked, this grasshopper emits a stream of whitish liquid, which is both irritant to human eyes and has a repugnant taste and smell (Abushama, 1972). The repugnatorial gland lies in the abdomen and its contents are expelled through an opening between the first and second abdominal tergites. The secretion is known to be effective against lizards, toads and mantids. Another acridoid family, the Eumastacidae, contains a

► If its warningly coloured threat display does not work, and a predator begins to molest this South African grasshopper, *Phymateus morbillosus* (Pyrgomorphinae), it produces copious quantities of foam, which has not only a foul taste but also a repulsive odour.

number of species which employ active chemical-defence behaviour.

The American lubber grasshopper, *Romalea microptera*, for example, initially hisses and foams at the mouth and thorax by mixing air with a solution of repellent chemicals. As the thousands of bubbles burst they disseminate a protective envelope of chemical mist around the grasshopper, in much the same way as in the pyrgomorphs. If this fails to deter an attacker, the *R. microptera* makes a last-ditch stand by disgorging a particularly repulsive droplet of chemical-rich fluid from its mouth.

The Southern walkingstick (stick insect), *Anisomorpha buprestoides*, from North America, is able to produce a defensive spray if it is threatened. In human beings, and presumably other vertebrates, this spray causes tear production and considerable discomfort if inhaled. The secretion is manufactured and stored in two elongate thoracic glands which open to the exterior just behind the head. Ants and beetles are instantly repelled by the spray, which *Anisomorpha* can aim very accurately, either from both glands or from a single gland, depending upon the direction of the stimulus. Mice sniff at the walkingstick and, upon being sprayed, retire immediately. Along with sprayed ants and beetles, they then spend some time in cleaning themselves but appear to suffer no long-term effects. In the laboratory situation, walkingsticks which emptied their glands but could not escape the attentions of the mice were eventually eaten. In the field, however, the withdrawal of an attacker gives the walkingstick time to make good its escape. In the laboratory, the spray does not deter opossums from holding on to walkingsticks until the glands are empty, whereupon they promptly eat them. *Anisomorpha* is apparently able to recognize blue jays, *Cyanocitta cristata*, for they are sprayed before they have even touched the insect.

The New Guinea stick insect *Eurycantha horrida* backs up its defensive display by releasing a brown liquid with an offensive odour. Secretions produced by the Neotropical phasmid *Isagoras* sp. are highly effective in deterring the attentions of the tamarin monkey, *Saguinus geoffroyi*, in Panama (Robinson, 1968).

A number of thrips raise the end of the abdomen when disturbed and exude strong-smelling droplets of liquid from the tip. *Bagnalliella yuccae* is a North American species which lives in aggregations in the leaf axils of its host plant, *Yucca filamentosa*. In the laboratory, aggregations of the thrips were submitted to attack by workers of the predatory ant *Monomorium minimum*, which also occurs on the yucca in the wild (Howard *et al.*, 1983). When attacked by the ant, the thrips raises the tip of the abdomen and, exuding the anal fluid, attempts to dab it on the ant. If it is successful, the ant withdraws and immediately begins to remove the exudate by wiping its body on the substrate and by grooming. Once the exudate, which contains the fruity-smelling t-decalactone, has been released, it also deters any other ants in the vicinity.

Larval chrysopids exude a drop of liquid from the anus if they are attacked by an insect such as an ant. They swing their flexible abdomens around and deposit the droplet on to the attacker. In a study in California (LaMunyon & Adams, 1987), the laboratory-reared larvae of five species of chrysopid, namely *Chrysoperla comanche*, *C. plorabunda*, *C. downesi*, *C. mojave* and *Eremochrysa punctinervis*, were placed in trails of the ant *Iridomyrmex humilis* in the field. All of them were attacked and all were able to defend themselves against the ant. In the laboratory, one of them, *C. comanche*, was then filmed in order to observe more closely its behaviour when confronted with an attacking ant. The film indicated that two methods of applying the defensive secretion are employed by the larva. If attacked from the front, it brings the tip of the abdomen over the head and then swings it from side to side until the blob of liquid contacts the ant. If the larva is attacked from the side, then it swings the abdomen laterally until the droplet touches the ant. The swing was found not to be under control of the brain, for an accurate response could be elicited from a decapitated larva if it was touched on one side. The effect of the blob of liquid is to coat the surface of the ant so that it has to take time off to clean itself up, giving the chrysopid larva a chance to escape. The number of consecutive times that a larva can use this form of defence decreases with increase in instar number. The secretion is replenished fairly quickly, about 80 per cent of it in 24 hours.

Scorpionflies of the genus *Panorpa* have been observed to regurgitate a brown liquid which helps them to free themselves from sticky silk when they invade spider webs to feed on the trapped prey which they contain. The resident spiders often do not take kindly to this invasion and will attack the scorpionfly. In response, the latter produces the same brown enteric fluid and attempts to dab it on to the spider (Thornhill, 1975). The fluid seems to be an effective deterrent when it is applied by the scorpionflies to their main hosts, the spiders *Micrathena gracilis*, *Mangora ornata* and *Leucage venusta*; under the same conditions, it is also effective against crab, jumping and wolf spiders as well as carpenter ants. If the scorpionflies are milked of their fluid then, when attacked, they are quickly killed by the spiders. When the fluid is applied to the spiders, they cease their attack and immediately begin to clean themselves.

Characteristic of larval papilionids are the osmateria, brightly coloured glands on the prothorax which are everted when they are threatened. It has always been assumed that the osmaterium is an effective defence against predators. In a carefully controlled field study carried out in Florida on the caterpillars of the zebra swallowtail butterfly, *Eurytides marcellus*, the limited usefulness of the osmaterium, at least in this species, has been revealed (Damman, 1986). It was found that natural enemies accounted for nearly all of the deaths of zebra-swallowtail caterpillars in the field. The main enemies were seen to be salticid spiders, ants, vespid wasps and the parasitic ichneumonid wasp *Trogus pennator*, though at no time were birds seen to take the caterpillars. The osmateria reduce predation on third-instar caterpillars during the spring but, in later broods, as the populations of predators increase in size, it has little effect upon predation levels. Ants and young salticid spiders abandon an attack when brushed with the osmateria but larger salticids and other larger spiders ignore them. The caterpillars do not even bother to evert the osmateria in the presence of the ichneumonid wasp; they just sit there and let it oviposit in them.

Caterpillars of a *Laetilia* sp. pyralid moth feed on scale insects of the genus *Toumeyella* on lignum vitae trees in Florida (Eisner *et al.*, 1972). The scale insects are tended by large numbers of the ant *Camponotus floridanus*, which vigorously protect them against potential predators. In order to repulse the ants, the caterpillars, when attacked, regurgitate fluid from their mouthparts which wets the ants and causes them to withdraw for a lengthy cleaning session. The thoracic glands of caterpillars of the North American variable oakleaf moth, *Heterocampa manteo* (Notodontidae), secrete formic acid strong enough to cause skin burns in man. In a serious outbreak of the moth in Missouri in 1971, these caterpillars defoliated 800 000 ha of oak trees but at no time were any birds seen feeding upon the pest (Kearby, 1975).

Within the family Carabidae are two subfamilies, some of whose members are referred to as bombardier beetles. The Brachininae contain the better-known examples but the genus *Goniotropis* of the subfamily Paussinae also deserves the common name of bombardier beetle. The name 'bombardier' comes from the beetles' long-known ability to produce an explosive discharge from the tip of its abdomen. An eighteenth-century observer noted that, when he picked up one of the beetles, it produced a small jet of blue smoke, accompanied by a slight report, and he was so startled that he dropped it. The same observer noted that the beetle was able to 'fire off' 20 times in succession when the abdomen was prodded with a pin. That this discharge was involved in a defensive role was always assumed to be the case and weight was added to this by an observation in the nineteenth century of a European *Brachinus crepitans* repelling a large carabid beetle.

The way in which these beetles employ their spray was not finally worked out in detail until nearly 100 years later (Eisner, 1958) with the North American species *Brachinus ballistarius* and this was followed by an elucidation of the way in which the mechanism works. The faint report that is heard when the beetle is stimulated is due to the rapid oxidation of hydrogen peroxide in a chitinous chamber which opens through a nozzle in the end of the abdomen. At the same time, hydroquinones in the chamber are oxidized to benzoquinones and toluquinones, with an accompanying rise in temperature to 100°C. The oxygen produced by oxidation of the hydrogen peroxide forces its way out of the chamber, its expansion aided by the rise in temperature, and this acts as a propellent for the quinones, which appear as a spray.

When the beetle is stimulated artificially at different points of its body, it is able to swing the tip of the abdomen round and discharge fairly accurately in the direction of the disturbance. In the laboratory, when individuals of the ant *Pogonomyrmex badius*

Chemical defence. In all but a small minority of adult Lepidoptera this is passive, i.e. they contain toxins sequestered by the larvae from the food plants or toxins manufactured within their own body. One genus of arctiid moth, *Rhodogastria* from Africa and Asia, initially uses a startle display to deter predators but if this fails it exudes an unpleasant yellow secretion in the form of a foam from its thorax; this is accompanied by a hissing noise. In some species, once the danger has passed, the moth manoeuvres its proboscis so that it can re-imbibe the foam. Butterflies of the African Acraeinae produce copious quantities of foam from their thoracic glands when handled. In *Acraea encedon*, this foam contains hydrogen cyanide.

▶ Moths from the African and Asian genus *Rhodogastria* (Arctiidae) ooze a distasteful froth from their thorax when they are molested by a potential predator. This individual is on Mount Kinabalu in Borneo.

attacked an anchored beetle, it responded by pointing its abdomen at the ant and discharging. The ant retreated and then underwent what were apparently a number of short seizures, each lasting a few seconds, during which its legs flailed the ground and prevented normal forward movement. The ant recovered within 2–3 minutes. Following the discharge, there was a period of about 40 seconds during which no other ant would go near the beetle. The ants are apparently repelled by the residual quinones on the beetle's surface and, in the wild, the time delay no doubt allows the beetle to make good its escape. A carabid beetle and a lycosid spider both reacted in much the same way to the discharge when they attacked the bombardier. A mantid caught the beetle when it was presented to it but the beetle discharged and was immediately released. All of the attackers spent some time in grooming following a discharge.

A study of the Central American paussine *Goniotropis nicaraguensis* (Eisner & Aneshansley, 1982) has shown it to have a defensive biology similar to that of the brachinines. *G. nicaraguensis* also secretes benzoquinones at 100°C by explosive oxidation of hydroquinones in a pair of specially adapted, two-chambered glands. *Goniotropis* fires a hot spray from these glands through a pair of orifices which open anteriorly to the tip of the abdomen. Associated with these orifices are flanges which direct the spray, depending upon the direction from which the beetle is stimulated; discharge always takes place on the side stimulated. If a hind leg is touched, the beetle lowers the tip of the abdomen slightly and the spray is directed obliquely downwards and slightly forwards of 90° to the side. If a foreleg is stimulated, the beetle appresses its abdominal tip against the wing cases and the spray is directed horizontally forwards. When aggressive ants (*Formica exsectoides*) were introduced to the beetles in the laboratory, the latter were immediately attacked but discharge of the defensive spray soon repelled the attackers. It was not necessary for the beetles to spray the ants repeatedly since the residual spray on and around the beetles was sufficient to deter further attacks by the ants.

The European staphylinid beetle *Deleaster dichrous* possesses not one but two pairs of defensive glands in the abdomen, which discharge at the same time when the beetle is attacked (Dettner *et al.*, 1985). One pair of glands, which are whitish in colour, secrete iridodial. On contact with the air, this substance forms an adhesive polymer. The other glands are reddish and secrete the toxic compound *p*-toluquinone as well as a variety of esters. In the laboratory, when the beetle comes into contact with a small insect, such as a fruit fly, it immediately bends the abdomen over dorsally and momentarily touches the intruder. The fly responds with an intensive bout of grooming but then dies within 1–2 hours. Ants are only attacked when they actually grab hold of a part of the beetle in their jaws. In several instances, the ant's antennae were seen to be gummed up and partially immobilized by the beetle's discharge.

Many species of longhorn beetles (Cerambycidae) emit strong-smelling defensive secretions. Early this century, for example, salicaldehyde was identified as being a major component of the odour produced by the European musk beetle, *Aromia moschata*. A study of two Australian cerambycids (Moore & Brown, 1971), *Stenocentrus ostricilla* and *Syllitus grammicus*, revealed that they produce almost identical secretions containing mainly toluene and *o*-cresol. The secretions are associated with paired mandibular glands and their reservoirs, the latter extending the length of the thorax. These glands open to the exterior through specially-shaped dispensers on the face. Since other members of the same genera produce almost identical odours it is assumed that they also have similar defence mechanisms.

Larvae of the North American myrmecophilous scarabaeid beetle *Cremastocheilus armatus* feed on ant brood and are able to defend themselves if attacked by host ants (Alpert & Ritcher, 1975). A threatened larva strikes at an ant with its mandibles, at the same time exuding a dark fluid. As it wriggles to escape, it also releases wet faecal pellets which have a deterrent effect on the ant. Both the pellets and the mouth-exudate have a strong, unpleasant odour. Adult beetles, when attacked, release a drop of fluid from the anus; this also has an unpleasant odour and is effective as a deterrent.

The gregarious larvae of the diprionid sawfly *Neodiprion sertifer* feed upon the leaves of conifers, such as Scots pine. If disturbed, they raise their front ends and emit a droplet of fluid from the mouth. This fluid has a strong smell of pine resin and an analysis of its constituents (Eisner *et al.*, 1974) has indicated that they are indeed sequestered from the pine needles upon which the larvae feed. The fluid is a deterrent to birds and, in the present study, was shown also to be effective against ants and spiders when tested in the laboratory situation.

Two species of Malayan ants, *Pachycondyla tridentata* and *P. insularis*, have enlarged venom glands from which they can secrete a clinging foam if attacked (Maschwitz *et al.*, 1981). If the ant is touched without restricting its movement the foam is produced in a long string 0.5 mm in diameter and up to 100 mm in length. If the ant's movements are restricted, by holding it in a pair of forceps, a pile of foam is released, the ant all the while trying to bring the gaster tip in contact with the 'attacker'. *Pheidole* sp. ant workers placed into the same area as *P. tridentata* workers attack them, eliciting the foaming response. The covering of foam causes the *Pheidole* attacker to withdraw to clean itself but it does not seem to suffer any long-term effect from the secretion. Studies in the laboratory revealed that the foam is exuded through the cloacal opening and that an average sized *P. tridentata* worker can produce 300 mm³ of foam from 1.5 mg dry weight of venom. A single stimulation causes the release of 3–8 mm³ of foam, although more can be produced from an artificially prolonged stimulus.

Observations that foraging ants, a constant threat to social-wasp broods in the tropics, did not appear to attack the nests of the Brazilian polistine wasp *Mischocyttarus drewseni* led to the discovery that the wasps use a form of chemical defence against intruders (Jeanne, 1970). Suspended on a narrow vertical stem, 2–3 cm long, the wasp constructs a simple comb made of paper. The female wasp then applies an ant-repellent secretion to the surface of the nest stem. At the anterior margin of the wasp's terminal gastral sternite, there is a small tuft of hairs which, in living wasps, often appears to be moist. The repellent is applied by rubbing this tuft of hairs on the nest stem. Though successful against intrusion from most types of ant it is of no use against attacks by *Eciton* spp. army ants.

Since 1970, a similar form of defence has been discovered in other polistines in the genera *Polistes* and *Ropalidia*, both of which bear hair tufts on the terminal abdominal sternite. Nests of the tropical and subtropical Asian species *R. fasciata* may be founded by either one or a number of foundresses. As with the previous species, the nest is suspended by a narrow petiole with a central core of fibrous material reinforced with layers of oral secretions. The latter give it strength and allow it to remain slimmer, as it grows in size, than if it were of fibrous material alone. Before embarking on a 'rubbing' (applying defensive chemical) session the female wasp first touches the petiole with her antennae before bringing her abdominal secretory organ into contact with it, a single stroke covering half of the petiole. Before moving round to rub the opposite side of the petiole, she may also rub part of the comb.

Observations on this species in Japan (Kojima, 1983) indicate that there are definite relationships between the rubbing-on of defensive chemicals, the stage of development of the nest and the number of foundresses, and also between chemical deposition and oral reinforcement of the petiole. In nests with a single foundress, she increases the number of times she rubs the petiole per hour as the number of developing young increase. It would seem that, the more the wasp has to lose in terms of investment in offspring, the more chemical she is prepared to lay down. In colonies with more than one foundress, rubbing is carried out less often and this may relate to the fact that there is more likely to be a wasp present physically to attack ants in this situation. In the field, it was observed that, when a foundress is present, she will attack any close-approaching ant and remove it with her mandibles. The presence of ants in the proximity of the nest does not seem, however, to precipitate rubbing behaviour in this species. Following strengthening of the petiole by licking, there is usually a follow-up rubbing session, presumably to replace defensive chemicals which were removed or covered over during the licking process. In mature single-foundress nests, about 50 per cent of rubbings occur in the 10 minutes before the female goes on a foraging bout. During this 10-minute interval, two or even three rubbing sessions are, on occasions, indulged in. Foundresses rarely depart more than an hour after the last rubbing has taken place.

Laboratory tests revealed that the substance produced by the wasp is able to deter one of their main enemies, the ant *Technomirmex albipes*. In these tests, chemicals produced by single foundresses were found to be fairly active for up to 1 hour from application but, after 2 hours, they had lost their effectiveness. An analysis of the chemicals produced by the wasp *Polistes fuscatus* indicates that one of the active ingredients is the ester, methyl palmitate (Post *et al.*, 1984). In field tests in the USA, using three species of ants sympatric with this wasp, synthetic methyl palmitate was found to be more effective against *Forelius pruinosus* and a *Pheidole* sp. than it was against *Solenopsis germinata*.

REFLEX BLEEDING

The adults of two North American Plecoptera, *Pteronarcys proteus* and *Peltoperla maria*, have been found to auto-haemorrhage when molested. In *Pteronarcys*, pinching of an appendage elicits bleeding from it alone, whereas pinching of the trunk causes bleeding from all legs. When a freshly emerged adult is seized in this way, the haemolymph is shot out forcibly to a distance of 25 cm, accompanied by a distinct and startling popping sound. *Peltoperla* responds to stimulation in a similar manner but without the apparent explosive auto-haemorrhage. Laboratory experiments reveal that, when the stonefly is attacked by ants, the haemolymph released soon clots, causing the attackers to retreat in order to clean themselves up; in the wild, this would give the stonefly time to escape.

Bibliography

General

Blum, M.S. & Blum, N.A. (eds) (1979) *Sexual selection and reproductive competition in insects.* Academic Press, London.

Vane-Wright, R.I. & Ackery, P.R. (eds) (1984) *The biology of butterflies.* Academic Press, London.

Wilson, E.O. (1971). *The insect societies.* Harvard University Press, Cambridge, Mass. 548 pp.

Sexual Behaviour

Adamo, S.A. & Chase, R. (1987) Courtship and copulation in the terrestrial snail *Helix aspersa. Can. J. Zool.* **66**, 1446–53.

Adler, P.H. & Adler, C.R. (1991) Mating behaviour and the evolutionary significance of mate guarding in three species of crane flies (Diptera: Tipulidae). *J. Insect Behav.***4**, 619–632.

Alcock, J. (1973). The mating behaviour of *Empis barbatoides* Melander and *Empis poplitea* Loew (Diptera: Empididae). *J. Nat. Hist.* **7**, 411–20.

Alcock, J. (1975) Male mating strategies of some philanthine wasps (Hymenoptera: Sphecidae). *J. Kans. Ent. Soc.* **48**, 532–45.

Alcock, J. (1975a) Territorial behaviour by males of *Philanthus multimaculatus* (Hymenoptera: Sphecidae) with a review of territoriality in male sphecids. *Anim. Behav.* **23**, 889–95.

Alcock, J. (1977) The courtship behaviour of *Heteropogon stonei* (Diptera: Asilidae). *J. Kans. Ent. Soc.* **50**, 238–43.

Alcock, J. (1979) Multiple mating in *Calopteryx maculata* (Odonata: Calopterygidae) and the advantage of non-contact guarding by males. *J. Nat. Hist.* **13**, 439–46.

Alcock, J. (1981) Notes on the reproductive behaviour of some Australian thynnine wasps (Hymenoptera: Tiphiidae). *J. Kans. Ent. Soc.* **54**, 681–93.

Alcock, J. (1981a) Lek territoriality in the tarantula hawk wasp *Hemipepsis ustulata* (Hymenoptera: Pompilidae). *Behav. Ecol. Sociobiol.* **8**, 309–17.

Alcock, J. (1982) Post-copulatory mate guarding by males of the damselfly *Hetaerina vulnerata* Selys (Odonata: Calopterygidae). *Anim. Behav.* **30**, 99–107.

Alcock, J. (1982a) Male reproductive behaviour in the anthomyiid fly *Hylemya alcathoe. Am. Midl. Nat.* **88**, 309–15.

Alcock, J. (1985) Hilltopping behaviour of the wasp *Pseudomasaris maculifrons* (Hymenoptera: Masaridae). *J. Kans. Ent. Soc.* **58**, 162–66.

Alcock, J. (1990) Oviposition resources, territoriality and male reproductive tactics in the dragonfly *Paltothemis lineatipes* (Odonata: Libellulidae). *Behaviour*, **113**, 251–65.

Alcock, J. (1990a) A large male competitive advantage in a lekking fly, *Hermetia comstocki* Williston (Diptera: Stratiomyidae). *Psyche*, **97**, 267–77.

Alcock, J. & Alcock, J.P. (1983) Male behaviour in two bumblebees, *Bombus nevadensis auricomus* and *B. griseicollis* (Hymenoptera: Apidae). *J. Zool. Lond.* **200**, 561–70.

Alcock, J., Eickwort, G.C. & Eickwort, K.R. (1977) The reproductive behaviour of *Anthidium maculosum* (Hymenoptera: Megachilidae) and the evolutionary significance of multiple copulations by females. *Behav. Ecol. Sociobiol.* **2**, 385–96.

Alcock, J. & Houston, T.F. (1987) Resource defense and alternative mating tactics in the banksia bee, *Hylaeus alcyoneus* (Erichson). *Ethology*, **76**, 177–88.

Alcock, J., Jones, C.E. & Buchmann, S.L. (1976) Location before emergence of the female bee *Centris pallida* by its male (Hymenoptera: Anthophoridae). *J. Zool. Lond.* **179**, 189–99.

Alcock, J., Jones, C.E. and Buchmann, S.L. (1977) Male mating strategies in the bee *Centris pallida* Fox (Anthophoridae: Hymenoptera). *Am. Nat.* **111**, 145–55.

Alcock, J. & Pyle, D.W. (1979) The complex courtship behaviour of *Physiphora demandata* (Diptera: Otitidae). *Z. Tierpsychol.* **49**, 352–62.

Alcock, J. & Schaefer, J.E. (1983) Hilltop territoriality in a Sonoran desert botfly (Diptera: Cuterebridae). *Anim. Behav.* **31**, 518–25.

Alcock, J. & Smith, A.P. (1987) Hilltopping, leks and female choice in the carpenter bee *Xylocopa (Neoxylocopa) varipuncta. J. Zool Lond.* **211**, 1–10.

Alexander, R.D. & Otte, D. (1967) Cannibalism during copulation in the brown bush cricket, *Hapithus agitator* (Gryllidae). *Fla Ent.* **50**, 80–87.

Andrews, R.H. & Bull, C.M. (1980) Mating behaviour in the Australian reptile tick *Aponomma hydrosauri. Anim. Behav.* **28**, 1280–86.

Andrews, R.H. & Bull, C.M. (1981) Inhibition of mating behaviour before feeding in the tick *Aponomma hydrosauri. Anim. Behav.* **29**, 518–22.

Anzenberger, G. (1977) Ethological study of African carpenter bees of the genus *Xylocopa* (Hymenoptera, Anthophoridae). *Z. Tierpsychol.* **44**, 337–74.

Austad, S.N. (1983) A game theoretical interpretation of male combat in the bowl and doily spider (*Frontinella pyramitela*). *Anim. Behav.* **31**, 59–73.

Austad, S.N. & Thornhill, R. (1986) Female reproductive variation in a nuptial-feeding spider, *Pisaura mirabilis. Bull. Br. Arachnol. Soc.* **7**, 48–52.

Austin, A.D. & Anderson, D.T. (1978) Reproduction and development of the spider *Nephila edulis* (Koch) (Araneidae: Araneae). *Aust. J. Zool.* **26**, 501–18.

Baker, R.R. (1972) Territorial behaviour of the nymphalid butterflies *Aglais urticae* and *Inachis io. J. Anim. Ecol.* **41**, 453–69.

Barrows, E.M. (1976) Mating behaviour in halictine bees (Hymenoptera: Halictidae). II. Microterritorial and patrolling behaviour in males of *Lasioglossum rohweri. Z. Tierpsychol.* **43**, 379–85.

Barrows, E.M. (1983) Male territoriality in the carpenter bee *Xylocopa virginica virginica. Anim. Behav.* **31**, 806–13.

Barrows, E.M., Chabot, M. R., Michener, C. D. & Snyder, T. P. (1976) Foraging and mating behaviour in *Perdita texana. J. Kans. Ent. Soc.* **49**, 275–79.

Barrows, E.M. & Gordh, G. (1978) Sexual behaviour in the Japanese beetle *Popillia japonica* and comparative notes on sexual behaviour in other scarabs. *Behav. Biol.* **23**, 341–54.

Barth, R.H. (1968) The mating behaviour of *Eurycotis floridiana* (Walker) (Blattaria. Blattoidea, Blattidae, Polyzosteriinae). *Psyche*, **75**, 274–84.

Bartlett, J. (1988) Male mating success and paternal care in *Nicrophorus vespilloides* (Coleoptera: Silphidae). *Behav. Ecol. Sociobiol.* **23**, 297–303.

Batra, S.W.T. (1972) Notes on the behavior and ecology of the mantispid, *Climaciella brunnea occidentalis. J. Kans. Ent. Soc.* **45**, 334–40.

Bell, P.D. (1979) Multimodal communication in the black-horned tree cricket *Oecanthus nigricornis* (Walker) (Orthoptera: Gryllidae). *Can. J. Zool.* **58**, 1861–68.

Bell, P.D. (1980) Opportunistic feeding by the female tree cricket, *Oecanthus nigricornis* (Orthoptera: Gryllidae). *Can. Ent.* **112**, 431–2.

Bennet-Clark, H.C., Leroy, Y. & Tsacas, L. (1980) Species and sex-specific songs and courtship behaviour in the genus *Zaprionus* (Diptera-Drosophilidae). *Anim. Behav.* **28**, 230–55.

Bennett, B. & Breed, M.D. (1985) The nesting biology, mating behaviour and foraging ecology of *Perdita opuntiae* (Hymenoptera: Andrenidae). *J. Kans. Ent. Soc.* **58**, 185–94.

Benson, W.W., Haddad, C.F.B. & Zikán, M. (1989) Territorial behaviour and dominance in some heliconiine butterflies (Nymphalidae). *J. Lepid. Soc.* **43**, 33–49.

Berg, C.O. & Valley, K. (1985) Nuptial feeding in *Sepedon* spp (Diptera: Sciomyzidae). *Proc. Ent. Soc. Wash.* **87**, 622–33.

Birch, M. (1970) Pre-courtship use of abdominal brushes by the nocturnal moth *Phlogophora meticulosa* (L.) (Lepidoptera: Noctuidae). *Anim. Behav.* **18**, 310–16.

Bland, R.G. (1987) Mating behaviour of the grasshopper *Melanoplus tequestae* (Orthoptera: Acrididae). *Fla Ent.* **70**, 483–7.

Blanke, V.R. (1975) Untersuchungen von Sexualverhalten von *Cyrtophora cicatrosa* (Stoliczka) (Araneae, Araneidae). *Z. Tierpsychol.* **37**, 62–74.

Bleckmann, H. & Bender, M. (1987) Watersurface waves generated by the male pisaurid spider *Dolomedes triton* (Walckenaer) during courtship behaviour. *J. Arachnol.* **15**, 363–9.

Blest, A.D. (1987) The copulation of a linyphiid spider *Baryphyma pratense*: does a female receive a blood-meal from her mate? *J. Zool. Lond.* **213**, 189–91.

Blickle, R.L. (1959) Observations on the hovering and mating of *Tabanus bishoppi* Stone (Diptera: Tabanidae). *Ann. Ent. Soc. Am.* **52**, 183–90.

Borgia, G. (1981) Mate selection in the fly *Scatophaga stercoraria*: female choice in a male-controlled system. *Anim. Behav.* **29**, 71–80.

Bornemissza, G.F. (1966) Observations on the hunting and mating behaviour of two species of scorpionflies (Bittacidae: Mecoptera). *Aust. J. Zool.* **14**, 371–82.

Boucher, L. & Huignard, J. (1987) Transfer of male secretions from the spermatophore to the female insect in *Caryedon serratus* (Ol.): analysis of the possible trophic role of these secretions. *J. Insect Physiol.* **33**, 949–57.

Brockmann, H.J. & Grafen, A. (1989) Mate conflict and male behaviour in a solitary wasp, *Trypoxylon (Trypargilum) politum* (Hymenoptera: Sphecidae). *Anim. Behav.* **37**, 232–55.

Brower, L.P., Brower, J.V.Z. & Cranston, F.P. (1965) Courtship behavior of the queen butterfly, *Danaus gilippus berenice* (Cramer). *Zoologica*, **50**, 1–39.

Brown, L. (1980) Aggression and mating success in males of the forked fungus beetle, *Bolitotherus cornutus* (Panzer) (Coleoptera: Tenebrioidae). *Proc. Ent. Soc. Wash.* **82**, 430–34.

Brown, L., Macdonell, J. & Fitzgerald, V.J. (1985) Courtship and female choice in the horned beetle *Bolitotherus cornutus* (Panzer) (Coleoptera:

Tenebrionidae). *Ann. Ent. Soc. Am.* **78**, 423–7.

Bruce, J.A. & Carico, J.E. (1988) Silk use during mating in *Pisaurina mira* (Walckenaer) (Araneae, Pisauridae). *J. Arachnol.* **16**, 1–4.

Buschinger, A. (1983) Sexual behaviour and slave raiding in the dulotic ant. *Harpagoxenus sublaevis* (Nyl.) under field conditions (Hym., Formicidae). *Insectes Soc.* **30**, 235–50.

Buschinger, A. & Alloway, T.M. (1979) Sexual behaviour in the slave-making ant *Harpagoxenus canadensis* M. R. Smith, and sexual pheromone experiments with *H. canadensis*, *H. americanus* (Emery), and *H. sublaevis* (Nylander) (Hymenoptera; Formicidae). *Z. Tierpsychol.* **49**, 113–19.

Cade, W.H. (1979) The evolution of alternative male reproductive strategies in field crickets (*Gryllus integer*). In: Blum, M. S. & Blum, N. A. (eds) *Sexual selection and reproductive competition in Insects*. Academic Press, London, pp. 343–79.

Campanella, P.J. & Wolf, L.L. (1974) Temporal leks as a mating system in a temperate zone dragonfly (Odonata: Anisoptera). I: *Plathemis lydia* (Drury). *Behaviour*, **51**, 49–87.

Carlberg, U. (1987) Reproduction behaviour of *Extatosoma tiaratum* (MacLeay) (Insecta: Phasmida). *Zool. Anz.* **219**, 331–6.

Carlson, G.D. & Copeland, J. (1988) Flash competition in male *Photinus macdermotti* fireflies. *Behav. Ecol. Sociobiol.* **22**, 271–6.

Case, J.F. (1980) Courting behaviour in a synchronously flashing, aggregative firefly, *Pteroptyx tener*. *Biol. Bull.* **159**, 613–25.

Cazier, M.A. & Linsley, E.G. (1963) Territorial behaviour among males of *Protoxaea gloriosa* (Fox) (Hymenoptera: Andrenidae). *Can. Ent.* **95**, 547–56.

Christenson, T.E. & Goist Jr, K.C. (1979) Costs and benefits of male–male competition in the orb weaving spider, *Nephila clavipes*. *Behav. Ecol. Sociobiol.* **5**, 87–92.

Chvála, M. (1980) Swarming rituals in two *Empis* and one *Bicellaria* species (Diptera, Empididae). *Acta Ent. Bohemoslovaca.* **77**, 1–15.

Clausen, I.H.S. (1987) On the biology and behavior of *Nephila senegalensis senegalensis* (Walckenaer, 1937). *Bull. Br. Arachnol. Soc.* **7**, 147–50.

Cohn, J. & Christenson, T.E. (1987) Utilization of resources by male golden orb-weaving spider *Nephila clavipes* (Araneae). *J. Arachnol.* **15**, 185–92.

Colyer, C.N. & Hammond, C.O. (1968) *Flies of the British Isles*. Frederick Warne, London, pp. 166–7.

Conner, J. (1989) Older males have higher insemination success in a beetle. *Anim. Behav.* **38**, 503–9.

Conner, W.E., Eisner, T., Van der Meer, R.K., Guerrero, A, Ghiringelli, D. & Meinwald, J. (1980) Sex attractant of an arctiid moth (*Utethesia ornatrix*): a pulsed chemical signal. *Behav. Ecol. Sociobiol.* **7**, 55–63.

Convey, P. (1989) Post-copulatory guarding strategies in the nonterritorial dragonfly *Sympetrum sanguineum* (Müller) (Odonata: Libellulidae). *Anim. Behav.* **37**, 56–63.

Courtney Smith, D. & Prokopy, R.J. (1982) Mating behaviour of *Rhagoletis mendax* (Diptera: Tephritidae) flies in nature. *Ann. Ent. Soc. Am.* **75**, 388–92.

Covalt Dunning, D, Byers, J.A. & Zanger, C.D. (1979) Courtship in two species of periodical cicadas, *Magicicada septendecim* and *Magicicada cassini*. *Anim. Behav.* **27**, 1073–90.

Coville, R.E. & Coville, P.L. (1980) Nesting biology and male behaviour of *Trypoxylon (Trypargilum) tenoctitlan* in Costa Rica (Hymenoptera: Sphecidae). *Ann. Ent. Soc. Am.* **73**, 110–19.

Cowan, D.P. (1979) Sibling matings in a hunting wasp: adaptive inbreeding? *Science, N.Y.* **205**, 1403–5.

Cowan, D.P. (1981) Parental investment in two solitary wasps *Ancistrocerus adabiatus* and *Euodynerus foraminatus* (Eumenidae: Hymenoptera). *Behav. Ecol. Sociobiol.* **9**, 95–102.

Cowan, D.P. (1986) Sexual behaviour of eumenid wasps (Hymenoptera: Eumenidae). *Proc. Ent. Soc. Wash.* **88**, 531–41.

Crankshaw, O.S. & Matthews, R.W. (1981) Sexual behaviour among parasitic *Megarhyssa* wasps (Hymenoptera: Ichneumonidae). *Behav. Ecol. Sociobiol.* **9**, 1–7.

Crespi, B.J. (1986) Size assessment and alternative fighting tactics in *Elaphrothrips tuberculatus* (Insecta: Thysanoptera). *Anim. Behav.* **34**, 1324–35.

Crespi, B.J. (1986a) Territoriality and fighting in a colonial thrips, *Hoplothrips pedicularius*, and sexual dimorphism in Thysanoptera. *Ecol. Ent.* **11**, 119–30.

Crespi, B.J. (1988) Risks and benefits of lethal male fighting in the colonial, polygynous thrips *Hoplothrips karnyi* (Insecta: Thysanoptera). *Behav. Ecol. Sociobiol.* **22**, 293–301.

Cress, D.C. (1966) Observations on the mating habits of *Pasimachus elongatus* (Coleoptera: Carabidae). *J. Kans. Ent. Soc.* **39**, 231–2.

Davidson, D.W. (1982) Sexual selection in harvester ants (Hymenoptera: Formicidae: *Pogonomyrmex*). *Behav. Ecol. Sociobiol.* **10**, 245–50.

Davies, N.B. (1978) Territorial defence in the speckled wood butterfly (*Pararge aegregia*): the resident always wins. *Anim. Behav.* **26**, 138–47.

Dennis, D.S. & Lavigne, R.J. (1976) Ethology of *Efferia varipes* with comments on species coexistence (Diptera: Asilidae). *J. Kans. Ent. Soc.* **49**, 48–62.

Dodson, G. (1986) Lek mating system and large male aggressive advantage in a gall-forming tephritid fly (Diptera: Tephritidae). *Ethology*, **72**, 99–108.

Donegan, J. & Ewing, A.W. (1980) Duetting in *Drosophila* and *Zaprionus* species. *Anim. Behav.* **28**, 1289.

Downes, J.A. (1970) The feeding and mating behaviour of the specialized Empidinae (Diptera); observations on four species of *Rhamphomyia* in the high Arctic and a general discussion. *Can. Ent.* **102**, 769–91.

Dunkle, S.W. (1991) Head damage from mating attempts in dragonflies (Odonata: Anisoptera). *Ent. News*, **102**, 37–41.

Eberhard, W.G. (1970) The natural history of the fungus gnats *Leptomorphus bifasciatus* (Say) and *L. subcaeruleus* (Coquillett) (Diptera: Mycetophilidae). *Psyche*, **77**, 361–83.

Eberhard, W.G. (1974) The natural history and behaviour of the wasp *Trigonopsis cameronii* Kohl (Sphecidae). *Trans. R. Ent. Soc. Lond.* **125**, 295–328.

Eberhard, W.G. (1975) The ecology and behaviour of a subsocial pentatomid bug and two scelionid wasps: strategy and counter-strategy in a host and its parasites. *Smithson. Contr. Zool.* **205**, 1–39.

Eberhard, W.G. (1977) Fighting behaviour of male *Golofa porteri* beetles (Scarabaeidae: Dynastinae). *Psyche*, **84**, 292–8.

Eberhard, W.G. (1979) The function of horns in *Podischnus agenor* (Dynastinae) and other beetles. In Blum, M. S. & Blum, N. A. (eds.) *Sexual selection and reproductive competition in insects*. Academic Press, London. p. 231.

Eberhard, W.G. (1981) The natural history of *Doryphora* sp. (Coleoptera, Chrysomelidae) and the function of its sternal horn. *Ann. Ent. Soc. Am.* **74**, 445–8.

Eberhard, W.G. (1988) Paradoxical post-coupling courtship in *Himantigera nigrifemorata* (Diptera: Stratiomyidae). *Psyche*, **95**, 115–22.

Eberhard, W.G. & Briceño L.R.D. (1983) Chivalry in pholcid spiders. *Behav. Ecol. Sociobiol.* **13**, 189–95.

Edgar, A.L. (1971) Studies in the biology and ecology of Michigan Phalangida (Opiliones). *Misc. Publs. Mus. Zool. Univ. Mich.* **144**, 1–64.

Edmunds, M. (1978) On the association between *Myrmarachne* spp. (Salticidae) and ants. *Bull. Br. Arachnol. Soc.* **4**, 149–60.

Edwards, G.B. (1981) Sound production by courting males of *Phidippus mystaceus* (Araneae: Salticidae). *Psyche*, **88**, 199–214.

Eickwort, G.C. (1977) Male territorial behaviour in the mason bee *Hoplitis anthocopoides* (Hymenoptera: Megachilidae). *Anim. Behav.* **25**, 542–54.

Elgar, M.A. & Nash, D.R. (1988) Sexual cannibalism in the garden spider *Araneus diadematus*. *Anim. Behav.* **36**, 1511–17.

Emlen, S.T. & Oring, L.W. (1977) Ecology, sexual selection, and the evolution of mating systems. *Science N.Y.* **197**, 215–22.

Evans, A.R. (1988) Mating systems and reproductive strategies in three Australian gryllid crickets: *Bobilla victoriae* Otte, *Balamara gidya* Otte and *Teleogryllus commodus* (Walker) (Orthoptera: Gryllidae: Nemobiinae; Trigonidiinae; Gryllinae). *Ethology*, **78**, 21–52.

Evans, D.A. & Matthews, R.W. (1976) Comparative courtship behaviour in two species of the parasitic chalcid wasp *Melittobia* (Hymenoptera: Eulophidae). *Anim. Behav.* **24**, 46–51.

Evans, H.E., O'Neill, K.M. & O'Neill, R.P. (1986) Nesting site changes and nocturnal clustering in the sand wasp *Bembecinus quinquespinosus* (Hymenoptera: Sphecidae). *J. Kans. Ent. Soc.* **59**, 280–86.

Feldman-Muhsam, B. & Borut, S. (1971) Copulation in ixodid ticks. *J. Parasit.* **57**, 630–34.

Fincke, O.M. (1985) Alternative mate-finding tactics in a nonterritorial damselfly (Odonata: Coenagrionidae). *Anim. Behav.* **33**, 1124–37.

Forsyth, A. & Alcock, J. (1990) Female mimicry and resource defense polygyny by males of a tropical rove beetle, *Leistrophus versicolor* (Coleoptera: Staphylinidae). *Behav. Ecol. Sociobiol.* **26**, 325–30.

Forsyth, A. & Montgomerie, R.D. (1987) Alternative reproductive tactics in the territorial damselfly *Calopteryx maculata*: sneaking by older males. *Behav. Ecol. Sociobiol.* **21**, 73–81.

Francke, O.F. (1979) Observations on the reproductive biology and life history of *Megacormus gertschii* Diaz (Scorpiones: Chactidae; Megacorminae). *J. Arachnol.* **7**, 223–30.

Freidberg, A. (1981) Mating behaviour of *Schistopterum moebiusi* Becker (Diptera: Tephritidae). *Israel J. Ent.* **15**, 89–95.

Freidberg, A. (1982) Courtship and post-mating behaviour of the fleabane gall fly *Spathulina tristis* (Diptera: Tephritidae). *Ent. Gener.* **7**, 273–85.

Freidberg, A. (1984) The mating behaviour of *Asteia elegantula* with biological notes on some other Asteiidae (Diptera). *Ent. Gener.* **9**, 217–24.

Freitag, R. (1974) Selection for a non-genitalic mating structure in female tiger beetles of the genus *Cicindela* (Coleoptera: Cicindelidae). *Can. Ent.* **106**, 561–8.

Frohlich, D.R. & Parker, F.D. (1985) Observations on the nest-building and reproductive behavior of a resin-gathering bee: *Dianthidium ulkei* (Hymenoptera: Megachilidae). *Ann. Ent. Soc. Am.* **78**, 804–10.

Fujisaki, K. (1981) Studies on the mating system of the winter cherry bug, *Acanthocoris sordidus* Thunberg (Heteroptera: Coreidae) II. Harem defence polygyny. *Researches Popul. Ecol. Kyoto Univ.* **23**, 262–79.

Gamboa, G. & Alcock, J. (1973) The mating behaviour of *Brochymena quadripustulata* (Fabricius). *Psyche*, **80**, 264–70.

Gilbert, L.E. (1976) Postmating female odour in *Heliconius* butterflies: a male-contributed antiaphrodisiac? *Science, N.Y.* **193**, 419–20.

Given, B.B. (1953) Evolutionary trends in the Thynninae (Hymenoptera: Tiphiidae) with special reference to feeding habits of Australian species. *Trans. R. Ent. Soc. Lond.* **105**, 1–10

Goodpasture, C. (1975) Comparative courtship behaviour and karyology in *Monodontomerus* (Hymenoptera: Torymidae). *Ann. Ent. Soc. Am.* **68**, 391–7.

Grant, G.G. & Brady, U.E. (1975) Courtship behaviour of phyctid moths. I. Comparison of *Plodia interpunctella* and *Cadra cautella* and role of male scent glands. *Can. Ent.* **53**, 813–26.

Greenfield, M.D. (1988) Interspecific acoustic interactions among katydids *Neoconocephalus*: inhibition-induced shifts in diel periodicity. *Anim. Behav.* **36**, 684–95.

Gwynne, D.T. (1977) Mating behaviour of *Neoconocephalus ensiger* (Orthoptera: Tettigoniidae) with notes on the calling song. *Can. Ent.* **109**, 237–42.

Gwynne, D.T. (1981) Sexual difference theory: mormon crickets show role reversal in mate choice. *Science,N.Y.* **231**, 779–80.

Gwynne, D.T. (1984) Sexual selection and sexual differences in mormon crickets (Orthoptera: Tettigoniidae), *Anabrus simplex*. *Evolution*, **38**, 1011–22.

Gwynne, D.T. (1987) Sex-biased predation and the risky mate-locating behaviour of male tick-tock cicadas (Homoptera: Cicadidae). *Anim. Behav.* **35**, 571–6.

Gwynne, D.T. (1988) Courtship feeding and the fitness of female katydids (Orthoptera: Tettigoni idae). *Evolution*, **42**, 545–55.

Gwynne, D.T. (1988a) Courtship feeding in katydids benefits the mating male's offspring. *Behav. Ecol. Sociobiol.* **23**, 373–7.

Gwynne, D.T. & Dadour, I.R. (1985) A new mechanism of sound production by courting male jumping spiders (Araneae: Salticidae, *Saitis michaelseni* Simon). *J. Zool. Lond.* **207**, 35–42.

Gwynne, D.T. and Rentz, C.F. (1983) Beetles on the bottle: male buprestids mistake stubbies for females (Coleoptera). *J. Aust. Ent. Soc.* **22**, 79–80.

Haacker, U. (1968) Das Sexualverhalten von *Sphaerotherium dorsale*. *Verh. Dt. Zool. Ges. Innsbruck* (= *Zool. Anz. Suppl.*), **32**, 454–63.

Haacker, U. (1968a) Sperma-transport beim Kugeltausendfüssler (*Sphaerotherium*). *Naturwiss.* **55**, 89.

Haacker, U. (1969) Spermaübertragung von *Glomeris*. *Naturwiss.* **56**, 467.

Haacker, U. (1974) Patterns of communication in courtship and mating behaviour of millipedes (Diplopoda). *Symp. Zool. Soc. Lond.* **32**, 317–28.

Hamilton, W.D. (1979) Wingless and fighting males in fig wasps and other insects. In Blum, M.S. & Blum, N.A. (eds.) *Sexual selection and reproductive competition in insects*. Academic Press, pp. 167–200.

Hamilton, W.J., Buskirk, R.E. & Buskirk, W.H. (1976) Social organization of the Namib desert tenebrionid beetle *Onymacris rugatipennis*. *Can. Ent.* **108**, 305–16.

Hamm, A.H. (1933) The epigamic behaviour and courtship of three species of Empididae. *Entomologists' Mon. Mag.* **69**, 113–7.

Hancock, R.G., Foster, W.A. & Yee, W.L. (1990) Courtship behaviour of the mosquito *Sabethes cyaneus* (Diptera Culicidae). *J. Insect Behav.* **3**, 401–12.

Harris, V.E., Todd, J.W., Webb, J.C. & Benner J.C. (1982) Acoustical and behavioural analysis of the songs of the southern green stink bug *Nezara viridula*. *Ann. Ent. Soc. Am.* **75**, 234–49.

Harvey, I.F. & Hubbard, S.F. (1987) Observations on the reproductive behaviour of *Orthemis ferruginea* (Fabricius) (Anisoptera: Libellulidae). *Odonatologica*, **16**, 1–8.

Headrick, D. & Goeden, R.D. (1989) Life history of *Pteromalus coloradensis* (Ashmead) (Hymenoptera: Pteromalidae) a parasite of *Paracantha gentilis* Hering (Diptera: Tephritidae) in *Cirsium* thistle capitula. *Proc. Ent. Soc. Wash.* **91**, 594–603.

Headrick, D. & Goeden, R.D. (1990) Life history of *Paracantha gentilis* (Diptera: Tephritidae). *Ann. Ent. Soc. Am.* **83**, 776–85.

Heady, S.E., Nault, L.R., Shambaugh, G.F. & Fairchild L. (1986) Acoustic and mating behaviour of *Dalbulus* leafhoppers (Homoptera: Cicadellidae). *Ann. Ent. Soc. Am.* **79**, 727–36.

Henry, C.S. (1979) Acoustical communication during courtship and mating in the green lacewing *Chrysopa carnea* (Neuroptera: Chrysopidae). *Ann. Ent. Soc. Am.* **72**, 68–79.

Henry, C.S. (1980) Acoustical communication in *Chrysopa rufilabris* (Neuroptera: Chrysopidae), a green lacewing with two distinct calls. *Proc. Ent. Soc. Wash.* **82**, 1–8.

Hespenheide, H.A. (1978) Prey, predatory and courtship behaviour of *Nannocyrtopogon neoculatus* Wilcox and Martin (Diptera: Asilidae). *J. Kans. Ent. Soc.* **51**, 449–56.

Hissmann, K. (1991) Phonotaxis of male crickets (*Gryllus campestris*) in a field population as an indication of territoriality (Orthoptera: Gryllidae). *J. Insect Behav.* **4**, 675–81.

Hobby, B.M. (1932) The epigamic behaviour of the male *Empis opaca* F., (Dipt., EMPIDIDAE). *Proc. R. Ent. Soc. Lond.* **6**, 67–8.

Hoikkala, A, Hoy, R.R. & Kaneshiro, K.Y. (1989) High-frequency clicks of Hawaiian picture-winged *Drosophila* species. *Anim. Behav.* **37**, 927–34.

Hölldobler, B. (1976) The behavioural ecology of mating in harvester ants (Hymenoptera: Formicidae: *Pogonomyrmex*). *Behav. Ecol. Sociobiol.* **1**, 405–23.

Hölldobler, B. & Haskins, C.P. (1977) Sexual calling behaviour in primitive ants. *Science, N.Y.* **189**, 793–4.

Hook, A.W. & Matthews, R.W. (1980) Nesting biology of *Oxybelus sericeus* with a discussion on nest guarding by male sphecid wasps (Hymenoptera). *Psyche*, **87**, 21–37.

Hughes, A.L. (1979) Reproductive behaviour and sexual dimorphism in the white-spotted sawyer *Monochamus scutellatus* (Say). *Coleopts Bull.* **33**, 45–7.

Hughes, A.L. (1981) Differential male mating success in the white spotted sawyer *Monochamus scutellatus* (Coleoptera: Cerambycidae). *Ann. Ent. Soc. Am.* **74**, 180–84.

Jackson, R.R. (1977) Courtship versatility in the jumping spider, *Phidippus johnsoni* (Araneae: Salticidae). *Anim. Behav.* **25**, 953–7.

Jackson, R.R. (1978) An analysis of alternative mating tactics of the jumping spider *Phidippus johnsoni* (Araneae: Salticidae). *J. Arachnol.* **5**, 185–230.

Jackson, R.R. (1982) The biology of *Portia fimbriata*, a web-building jumping spider (Araneae, Salticidae) from Queensland: intraspecific interactions. *J. Zool. Lond.* **196**, 295–305.

Jackson, R.R. (1982a) The biology of ant-like jumping spiders: intraspecific interactions of *Myrmarachne lupata* (Araneae, Salticidae). *Zool. J. Linn. Soc.* **76**, 293–319.

Jackson, R.R. (1985) The biology of *Simaetha paetula* and *S. thoracica*, web-building jumping spiders (Araneae, Salticidae) from Queensland: cohabitation with social spiders, utilization of silk, predatory behaviour and intraspecific interactions. *J. Zool. Lond. (B)*, **1**, 175–210.

Jayakar, S.D. (1966) Sexual behaviour in solitary Eumenid wasps. *J. Bombay Nat. Hist. Soc.* **63**, 760–63.

Jeanne, R.L. & Bermúdez, E.G.C. (1980) Reproductive behaviour of a male neotropical social wasp *Mischocyttarus drewsoni* (Hymenoptera: Vespidae). *J. Kans. Ent. Soc.* **53**, 271–6.

Jenkins, J. (1990) Mating behaviour of *Aciurina mexicana* (Aczél) (Diptera: Tephritidae). *Proc. Ent. Soc. Wash.* **92**, 66–75.

Jeppesen, L.L. (1976) The control of mating behaviour in *Helix pomatia* L. (Gastropoda: Pulmonata). *Anim. Behav.* **24**, 275–90.

Johnson, L.K. (1982) Sexual selection in a brentid weevil. *Evolution*, **36**, 251–62.

Johnson, L.K. (1983) Reproductive behaviour of *Claeoderes bivittata* (Coleoptera: Brentidae). *Psyche*, **90**, 135–49.

Kasuya, E. (1981) Male mating territory in a

Japanese paper wasp *Polistes jadwigae* DALLA TORRE (Hymenoptera, Vespidae). *Kontyû*, **49**, 607–14.

Kessler, A. (1971) Relation between egg production and food consumption in species of the genus *Pardosa* (Lycosidae: Araneae) under experimental conditions of food abundance and food shortage. *Oecologia*, **8**, 93–109.

Kimsey, L.S. (1978) Nesting and male behaviour in *Dynatus nigripes spinolae* (Lepeletier). *Pan-Pacif. Ent.* **54**, 65–7.

Kimsey, L.S. (1980) The behaviour of male orchid bees (Apidae, Hymenoptera, Insecta) and the question of leks. *Anim. Behav.* **28**, 996–1004.

Kirk, V.M. & Dupraz, B.J. (1972) Discharge by a female ground beetle, *Pterostichus lucublandus* (Coleoptera: Carabidae), used as a defense against males. *Ann. Ent. Soc. Am.* **65**, 513.

Klingel, H. (1960) Verhaltensbiologie der chilopoden *Scutigera coleopatra* L. ('Spinnenassel') und *Scolopendra cingulata* Latreille (Skolopender). *Z. Tierpsychol.* **17**, 11–30.

Klingel, H. (1962) Das Paarungsverhalten des malaischen Höhlentausendfüssers *Thereuopoda decipiens carvernicola* Verhoeff. *Zool. Anz.* **169**, 458–60.

Knapton, R.W. (1985) Lek structure and territoriality in the chryxus arctic butterfly, *Oeneis chryxus* (Satyridae). *Behav. Ecol. Sociobiol.* **17**, 389–95.

Krafft, B. (1978) The recording of vibratory signals performed by spiders during courtship. *Symp. Zool. Soc. Lond.* **42**, 59–67.

Kraus, W. & Lederhouse, R.C. (1983) Contact guarding during courtship in the tiger beetle *Cicindela marutha* Dow (Coleoptera: Cicindelidae). *Am. Midl. Nat.* **110**, 208–11.

Kukuk, P.F., Eickwort, G.C. & Wesley, R. (1985) Mate-seeking behaviour of *Dufourea novaeangliae* (Hymenoptera: Halictidae: Doufourinae): the effects of resource distribution. *J. Kans. Ent. Soc.* **58**, 142–50.

Kukuk, P.F. & Schwarz, M. (1988) Macrocephalic male bees as functional reproductives and probable guards. *Pan-Pacif. Ent.* **64**, 131–7.

Kyriacou, C.P. & Hall, J.C. (1982) The function of courtship song rhythms in *Drosophila*. *Anim. Behav.* **30**, 794–801.

Lanigan, P.J. & Barrows, E.M. (1977) Sexual behaviour of *Murgantia histrionica* (Hemiptera:Pentatomidae). *Psyche*, **84**, 191–7.

Lavallee, A.G. (1970) Courtship and mating habits of an asilid fly, *Cyrtopogon marginalis*. *Ann. Ent. Soc. Am.* **63**, 1199.

Lavigne, R. (1970) Courtship and predatory behaviour of *Cyrtopogon auratus* and *C. glarealis* (Diptera: Asilidae). *J. Kans. Ent. Soc.* **43**, 163–71.

Lavigne, R. (1970a) Courtship and predation behaviour of *Heteropogon maculinervis* (Diptera: Asilidae). *J. Kans. Ent. Soc.* **43**, 270–73.

Lavigne, R. (1972) Ethology of *Ablautus rufotibialis* on the Pawnee grasslands IBP site. *J. Kans. Ent. Soc.* **45**, 271–4.

Lavigne, R.J. & Dennis, D.S. (1975) Ethology of *Efferia frewingi* (Diptera: Asilidae). *Ann. Ent. Soc. Am.* **68**, 992–6.

Lederhouse, R.C. (1982) Territorial defense and lek behaviour of the black swallowtail butterfly *Papilio polyxenes*. *Behav. Ecol. Sociobiol.* **10**, 109–18.

Legg, G. (1977) Sperm transfer and mating in *Ricinoides hanseni* (Ricinulei: Arachnida). *J. Zool. Lond.* **182**, 51–61.

Leonard, S.H. & Ringo, J.M. (1978) Analysis of male mating patterns and courtship behaviour of *Brachymeria intermedia*. *Ann. Ent. Soc. Am.* **71**, 817–26.

Leuthold, R.H. & Bruinsma, O. (1977) Pairing behaviour in *Hodotermes mossambicus* (Isoptera). *Psyche*, **84**, 109–19.

Litte, M. (1979) *Mischocyttarus flavitarsis* in Arizona: social and nesting biology of a polistine wasp. *Z. Tierpsychol.* **50**, 282–312.

Lloyd, J.E. (1980) Male *Photuris* fireflies mimic sexual signals of their females' prey. *Science, N. Y.* **210**, 669–71.

Lloyd, J.E. (1980a.) Sexual selection: individuality, identification, and recognition in a bumblebee and other insects. *Fla Ent.* **64**, 89–94.

Lloyd, J.E. & Wing, S.R. (1983) Nocturnal aerial predation of fireflies by light-seeking fireflies. *Science, N.Y.* **222**, 634–5.

Loher, W. & Rence, B. (1978) The mating behaviour of *Teleogryllus commodus* (Walker) and its central and peripheral control. *Z. Tierpsychol.* **46**, 225–9.

Louw, S. (1987) Observations on the mating behaviour of *Platychelus brevis* Burmeister (Coleoptera: Scarabaeidae: Rutelinae). *J. Ent. Soc. Sth. Afr.* **50**, 258–9.

Lyal, C.H.C. (1986) Observations on zygopine weevil behaviour (Coleoptera: Curculionidae: Zygopinae). *J. Nat. Hist.* **20**, 789–98.

McAlpine, D.K. (1972) Observations on the sexual behaviour of some Australian Platystomatidae (Diptera, Schizophora). *Rec. Aust. Mus.* **29**, 1–9.

McAlpine, D.L. (1979) Agonistic behaviour in *Achias australis* (Diptera, Platystomatidae) and the significance of eyestalks. In Blum, M. S. & Blum W. A. (eds) *Sexual selection and reproductive competition in insects*. Academic Press, London.

McLain, D.K. (1981) Interspecific interference competition and mate choice in the soldier beetle *Chauliognathus pennsylvanicus*. *Behav. Ecol. Sociobiol.* **9**, 65–6.

McLain, D.S. (1989) Prolonged copulations as a post-insemination guarding tactic in a natural population of the ragwort seed bug. *Anim. Behav.* **38**, 659–64.

Magnus, D.B.E. (1950) Beobachten zur Balz und Eiablage des Kaisermantels *Argynnis paphia*. *Z. Tierpsychol.* **7**, 435–49.

Maier, C.T. & Waldbauer, G.P. (1979) Dual mate-seeking strategies in syrphid flies (Diptera: Syrphidae). *Ann. Ent. Soc. Am.* **72**, 54–61.

Mallet, J. (1984) Sex roles in the ghost moth *Hepialus humuli* (L.) and a review of mating in the Hepialidae (Lepidoptera). *Zool. J. Linn. Soc.* **79**, 67–82.

Mangan, R.L. (1979) Reproductive behaviour of the cactus fly, *Odontoloxozus longicornis*, male territoriality and female guarding as adaptive strategies. *Behav. Ecol. Sociobiol.* **4**, 265–78.

Marden J.H. (1987) In pursuit of females: following and contest behaviour by males of a Namib desert tenebrionid beetle, *Physadesmia globosa*. *Ethology*, **75**, 15–24.

Marden, J.H. & Waage, J.K. (1990) Escalated damselfly territorial contests are energetic wars of attrition. *Anim. Behav.* **39**, 954–9.

Marshall, L.D. & Alcock, J. (1981) The evolution of the mating system of the carpenter bee *Xylocopa varipuncta* (Hymenoptera: Anthophoridae). *J. Zool. Lond.* **193**, 315–24.

Martens, V.J. (1969) Die Sekretdarbietung während des Paarungsverhaltens von *Ischyropsalis* C. L. Koch (Opiliones). *Z. Tierpsychol.* **96**, 513–23.

Matthes, D. (1962) Excitationen und Paarungsverhalten mitteleuropäischen Malachiidae. *Z. Morph. Ökol. Tiere*, **51**, 375–546.

Mays, D.L. (1971) Mating behaviour of nemobiine crickets – *Hygronemobius, Nemobius*, and *Pteronemobius* (Orthoptera: Gryllidae). *Fla Ent.* **54**, 114–25.

Meads, M.J. (1976) Some observations on *Lasiorhynchus barbicornis* (Brentidae: Coleoptera). *N.Z. Ent.* **5**, 171–6.

Meijer, J. (1977) A glandular secretion in the ocular area of certain erigonine spiders (Araneae, Linyphiidae). *Bull. Br. Arachnol. Soc.* **3**, 251–3.

Merrett, P. (1988) Notes on the biology of the neotropical pisaurid, *Ancylometes bogotensis* (Keyserling) (Araneae: Pisauridae). *Bull. Br. Arachnol. Soc.* **7**, 197–201.

Michelsen, A. (1962) Observations on the sexual behaviour of some longicorn beetles, subfamily Lepturinae (Coleoptera, Cerambycidae). *Behaviour*, **22**, 152–66.

Miller, P.L. (1984) Alternative reproductive routines in a small fly, *Puliciphora borinquenensis* (Diptera: Phoridae). *Ecol. Ent.* **9**, 293–302.

Miller, P.L. (1991) Pre-tandem courtship in *Palpopleura sexmaculata* (Fabricius) (Anisoptera: Libellulidae). *Notul. Odonat.* **3**, 99–101.

Minch, E.W. (1979) Reproductive behaviour of the tarantula *Aphonopelma chalcodes* Chamberlin (Araneae: Theraphosidae). *Bull. Br. Arachnol. Soc.* **4**, 414–20.

Mitchell, P.L. (1980) Combat and territorial defense of *Acanthocephala femorata* (Hemiptera: Coreidae). *Ann. Ent. Soc. Am.* **73**, 404–8.

Moller, H. (1985) Tree wetas (*Hemideina crassicruris*) (Orthoptera: Stenopelmatidae) of Stephens' Island, Cook Strait. *N. Z. J. Zool.* **12**, 55–69.

Mook, J.H. & Bruggemann, C.G. (1968) Acoustical communication by *Lipara lucens* (Diptera, Chloropidae). *Entomologica Exp. Appl.* **11**, 397–402.

Morgan, F.D. (1977) Swarming behaviour of the Australian beetle *Heteronyx obesus*, with notes on related species. *Ecol. Ent.* **2**, 91–6.

Morris, G.K. (1971) Aggression in male conocephaline grasshoppers (Tettigoniidae). *Anim. Behav.* **19**, 132–7.

Morris, G.K. (1979) Mating systems, paternal investment and aggressive behaviour of acoustic Orthoptera. *Fla Ent.* **62**, 915.

Morris, G.K. (1980) Calling display and mating behaviour of *Copiphora rhinoceros* Pictet (Orthoptera: Tettigoniidae). *Anim. Behav.* **8**, 42–51.

Moser, J.C. (1967) Mating activities of *Atta texana* (Hymenoptera, Formicidae). *Insectes Soc.* **3**, 295–312.

Moulds, M.S. (1977) Field observations on behaviour of a north Queensland species of *Phytalmia* (Diptera: Tephritidae). *J. Aust. Ent. Soc.* **16**, 347–52.

Müller, J.K. & Eggert, A.K. (1989) Paternity assurance by 'helpful' males: adaptations to sperm competition in burying beetles. *Behav. Ecol. Sociobiol.* **24**, 245–9.

Muma, M.H. (1966) Mating behaviour in the

solpugid genus *Eremobates* Banks. *Anim. Behav.* **14**, 346–50.

Murray, M.G. (1987) The closed environment of the fig receptacle and its influence on conflict in the Old World fig wasp, *Philotrypesis pilosa*. *Anim. Behav.* **35**, 488–506.

Murray, M.G. (1989) Environmental constraints on fighting in flightless male fig wasps. *Anim. Behav.* **38**, 186–93.

Nelson, M.C. & Fraser, J. (1980) Sound production in the cockroach, *Gromphadorhina portentosa*: evidence for communication by hissing. *Behav. Ecol. Sociobiol.* **6**, 305–14.

Newkirk, M.R. (1970) Biology of the longtailed dance fly *Rhamphomyia longicauda* (Diptera: Empididae); a new look at swarming. *Ann. Ent. Soc. Am.* **63**, 1407–12.

Nickerson, J.C., Snyder, D.E. & Oliver, C.C. (1979) Acoustical burrows constructed by mole crickets. *Ann. Ent. Soc. Am.* **72**, 438–40.

Nilsson, L.A. & Rabakonandrianina, E. (1988) Chemical signalling and monopolization of nectar resources by territorial *Pachymelus limbatus* (Hymenoptera, Anthophoridae) male bees in Madagascar. *J. Zool. Lond.* **215**, 475–89.

Norval, R.A.I. (1974) Copulation and feeding in males of *Ixodes pilosus* Koch 1844 (Acarina: Ixodidae). *J. Ent. Soc. Sth. Afr.* **37**, 129–133.

Novak, J.A. & Foote, B.A. (1975) Biology and immature stages of fruit flies: the genus *Stenopa* (Diptera: Tephritidae). *J. Kans. Ent. Soc.* **48**, 42–52.

O'Neill, K.M. & Evans, H.E. (1981) Predation on conspecific males by females of the beewolf *Philanthus basilaris* Cresson (Hymenoptera: Sphecidae). *J. Kans. Ent. Soc.* **54**, 553–5.

O'Neill, K.M. & Evans, H.E. (1983) Alternative male mating tactics in *Bembecinus quinquespinosus* (Hymenoptera: Sphecidae): correlations with size and color variation. *Behav. Ecol. Sociobiol.* **14**, 39–46.

Ono, T., Siva-Jothy, M.T. & Kato, A. (1989) Removal and subsequent ingestion of rivals' semen during copulation in a tree cricket. *Ecol. Ent.* **14**, 195–202.

Orr, A.G. & Rutowski, R.L. (1991) The function of the sphragis in *Cressida cressida* (Fab.) (Lepidoptera, Papilionidae): a visual deterrent to copulation attempts. *J. Nat. Hist.* **25**, 703–10.

O'Toole, C. & Preston-Mafham, K.G. (1985) *Insects in camera.* Oxford University Press.

O'Toole, C. & Raw, A. (1991) *Bees of the world.* Blandford Press, London.

Otronen, M. (1984) The effect of differences in body size on the male territorial system of the fly *Dryomza analis*. *Anim. Behav.* **32**, 882–90.

Otte, D. (1972) An analysis of communication in the genus *Syrbula*. *Behaviour*, **72**, 291–322.

Otte, D. & Joern, A. (1975) Insect territoriality and its evolution: population studies of desert grasshoppers on creosote bushes. *J. Anim. Ecol.* **44**, 29–54.

Packer, L. (1986) The biology of a subtropical population of *Halictus ligatus* Say (Hymenoptera; Halictidae). II. Male behaviour. *Ethology*, **72**, 287–98.

Paetzel, M.M. (1973) Behaviour of the male *Trypoxylon rubrocinctum*. *Pan-Pacif. Ent.* **49**, 26–30.

Parker, G.A. (1970) Sperm competition and its evolutionary consequences in the insects. *Biol. Rev.* **45**, 525–67.

Parker, G.A. (1970a) The reproductive behaviour and nature of sexual selection in *Scatophaga stercoraria* L. (Diptera: Scatophagidae). IV. Epigamic recognition and competition between males for the possession of females. *Behaviour*, **37**, 113–39.

Parker, G.A. (1972) Reproductive behaviour of *Sepsis cynipsea* (L.) (Diptera: Sepsidae). I. A preliminary analysis of the reproductive strategy and its associated behaviour patterns. *Behaviour*, **41**, 172–206.

Parsons, M. (1983) Notes on the courtship of *Troides oblongomaculatus papuensis* (Papilionidae) in Papua New Guinea. *J. Lepid. Soc.* **37**, 83–5.

Pearson, D.L. (1988) Biology of tiger beetles. *Ann. Rev. Ent.* **33**, 123–47.

Peck, S.B. & Forsyth, A. (1982) Composition, structure and competitive behaviour in a guild of Ecuadorian rain forest dung beetles (Coleoptera: Scarabaeidae). *Can. J. Zool.* **60**, 1624–34.

Peckham, D.J. (1977) Reduction of miltogrammine cleptoparasitism by male *Oxybelus subulatus* (Hymenoptera: Sphecidae). *Ann. Ent. Soc. Am.* **70**, 823–8.

Pickford, R. & Gillott, C. (1972) Courtship behaviour of the migratory grasshopper, *Melanoplus sanguinipes* (Orthoptera: Acrididae). *Can. Ent.* **104**, 715–22.

Pinto, J.D. (1973) Sexual behaviour in the genus *Pleuropompha* LeConte: a new mating display in blister beetles (Coleoptera: Meloidae). *Can. Ent.* **105**, 957–69.

Pinto, J.D. (1975) Intra- and interspecific courtship behaviour in blister beetles of the genus *Tegrodera* (Meloidae). *Ann. Ent. Soc. Am.* **68**, 275–84.

Pinto, J.D. (1977) Comparative sexual behaviour of blister beetles in the subtribe Eupomphina (Coleoptera: Meloidae), and an evaluation of its taxonomic significance. *Ann. Ent. Soc. Am.* **70**, 937–51.

Pliske, T.E. (1975) Courtship behaviour of the monarch butterfly, *Danaus plexippus*. *Ann. Ent. Soc. Am.* **68**, 143–51.

Pliske, T.E. (1975a.) Courtship behaviour and use of chemical communication by males of certain species of ithomiine butterflies (Nymphalidae: Lepidoptera). *Ann. Ent. Soc. Am.* **68**, 935–42.

Polis, G.A. & Farley, R.D. (1979) Behaviour and ecology of mating in the cannibalistic scorpion, *Paruroctonus mesaensis* Stahnke (Scorpionida: Vaejovidae). *J. Arachnol.* **7**, 33–46.

Pont, A. (1987) 'The mysterious swarms of sepsid flies': an enigma solved? *J. Nat. Hist.* **21**, 305–17.

Post, D.C. & Jeanne, R.J. (1983) Male reproductive behaviour of the social wasp *Polistes fuscatus* (Hymenoptera: Vespidae). *Z. Tierpsychol.* **62**, 157–71.

Potter, D.A. (1981) Agonistic behaviour in male spider mites: factors affecting frequency and intensity of fighting. *Ann. Ent. Soc. Am.* **74**, 138–43.

Potter, D.A., Wrensch, D.L. & Johnston, D.E. (1976) Guarding, aggressive behaviour, and mating success in male twospotted spider mites. *Ann. Ent. Soc. Am.* **69**, 707–11.

Raw, A. (1975) Territoriality and scent marking by *Centris* males (Hymenoptera, Anthophoridae) in Jamaica. *Behaviour*, **54**, 311–21.

Raw, A. (1976) The behaviour of males of the solitary bee *Osmia rufa* (Megachilidae) searching for females. *Behaviour*, **56**, 279–85.

Richards, A.M. (1960) Observations on the New Zealand glowworm *Arachnocampa luminosa* (Skuse) 1890. *Trans. R. Soc. N. Z.* **88**, 559–74.

Richards, A.M. (1980) Sexual selection, guarding and sexual conflict in a species of Coccinellidae (Coleoptera). *J. Aust. Ent. Soc.* **19**, 26.

Ridsdill Smith, T.J. (1970) The behaviour of *Hemithynnus hyalinatus* (Hymenoptera: Tiphiidae), with notes on some other Thynninae. *J. Aust. Ent. Soc.* **9**, 196–208.

Ringo, J.M. (1974) Behavioural characters distinguishing two species of Hawaiian *Drosophila*, *Drosophila grimshawi* and *Drosophila pullipes* (Diptera: Drosophilidae). *Ann. Ent. Soc. Am.* **67**, 823.

Ringo, J.M. (1976) A communal display in Hawaiian *Drosophila* (Diptera: Drosophilidae). *Ann. Ent. Soc. Am.* **69**, 209–14.

Ritter Jr, H. (1964) Defense of mate and mating chamber in a wood roach. *Science, N.Y.* **143**, 1459–60.

Robacker, D.C. & Hart, W.G. (1985) Courtship and territoriality of laboratory-reared Mexican fruit flies, *Anastrepha ludens* (Diptera: Tephritidae), in cages containing host and nonhost trees. *Ann. Ent. Soc. Am.* **78**, 488–94.

Robertson, H.M. (1982) Mating behaviour and its relationship to territoriality in *Platycypha caligata* (Selys) (Odonata: Chlorocyphidae). *Behaviour*, **79**, 11–27.

Robinson, B. & Robinson, M.H. (1978) Developmental studies of *Argiope argentata* (Fabricius) and *Argiopa aemula* (Walckenaer). *Symp. Zool. Soc. Lond.* **42**, 31–40.

Robertson, H.M. (1985) Female dimorphism and mating behaviour in a damselfly, *Ischnura ramburi*: females mimicking males. *Anim. Behav.* **33**, 805–9.

Robinson, M.H. & Robinson B. (1973) Ecology and behaviour of the giant wood spider *Nephila maculata* (Fabricius) in New Guinea. *Smithson. Contr. Zool.* **149**, 1–176.

Robinson, M.H. & Robinson, B. (1976) The ecology and behaviour of *Nephila maculata*: a supplement. *Smithson. Contr. Zool.* **218**, 1–22.

Robinson, M.H. & Robinson, B. (1978a) The evolution of courtship systems in tropical areneid spiders. *Symp. Zool. Soc. Lond.* **42**, 17–29.

Robinson, M.H. & Robinson, B. (1979) By dawn's early light: matutinal mating and sex attractants in a Neotropical mantid. *Science, N.Y.* **205**, 825–7.

Robinson, M.H. & Robinson, B. (1980) Comparative studies of the courtship and mating behaviour of tropical araneid spiders. *Pacif. Insects Monogr.* **36**, 1–218.

Ross, K. & Smith, R.L. (1979) Aspects of the courtship behaviour of the black widow spider *Latrodectus hesperus* (Araneae: Theridiidae), with evidence for the existence of a contact sex pheromone. *J. Arachnol.* **7**, 69–77.

Rovner, J.S. (1967) Acoustic communication in a lycosid spider (*Lycosa rabida* Walckenaer). *Anim. Behav.* **15**, 273–81.

Rutowski, R.L. (1978) The courtship behaviour of the small sulphur butterfly *Eurema lisa* (Lepidoptera: Pieridae). *Anim. Behav.* **26**, 892–903.

Rutowski, R.L. (1981) Courtship behaviour of the

dainty sulfur butterfly, *Nathalis iole*, with a description of a new, facultative male display (Pieridae). *J. Res. Lepid.* **20**, 161–9.

Rutowski, R.L. (1983) The wing-waving display of *Eurema daira* males (Lepidoptera: Pieridae): its structure and role in successful courtship. *Anim. Behav.* **31**, 985–9.

Rutowski, R.L. & Alcock, J. (1980) Temporal variation in male copulatory behaviour in the solitary bee *Nomadopsis puellae* (Hymenoptera: Andrenidae). *Behaviour*, **73**, 175–88.

Rutowski, R.L. & Gilchrist, G.W. (1987) Courtship, copulation and oviposition in the chalcedon checkerspot, *Euphydryas chalcedona* (Lepidoptera: Nymphalidae). *J. Nat. Hist.* **21**, 1109–17.

Rutowski, R.L. & Gilchrist, G.W. (1988) Male mate-locating behaviour of the desert hackberry butterfly, *Asterocampa leila* (Nymphalidae). *J. Res. Lepid.* **26**, 1–12.

Rutowski, R.L. & Schaefer J. (1984) Courtship behaviour in the gulf fritillary, *Agraulis vanillae* (Nymphalidae). *J. Lepid. Soc.*, **38**, 23–31.

Sakaluk, S.K. (1985) Spermatophore size and its role in the reproductive behaviour of the cricket, *Gryllodes supplicans* (Orthoptera: Gryllidae). *Can. J. Zool.* **63**, 1652–6.

Sasakawa, M. & Sasakawa, T. (1981) Stridulatory organs of the minute pine bark beetle, *Cryphalus fulvus* Nijima (Coleoptera, Scolytidae), and the role of male sounds in aggregation behaviour. *Kontyû*, **49**, 461–9.

Sato, H. (1988) Further observations of the nesting behaviour of a subsocial ball-rolling scarab, *Kheper aegyptiorum*. *Kontyû*, **56**, 873–8.

Sato, H. & Imamori, M. (1986) Nidification of an African ball-rolling scarab, *Scarabaeus platynotus* Bates (Coleoptera, Scarabaeidae). *Kontyû*, **54**, 203–7.

Sato, H. & Imamori, M. (1987) Nesting behaviour of a subsocial African ball-roller *Kheper platynotus* (Coleoptera, Scarabaeidae). *Ecol. Ent.* **12**, 415–25.

Scarbrough, A.G. (1978) Ethology of *Cerotainia albipilosa* Curran (Diptera: Asilidae) in Maryland: courtship, mating and oviposition. *Proc. Ent. Soc. Wash.* **80**, 179–90.

Schal, C. & Bell, W.J. (1982) Ecological correlates of paternal investment of urates in a tropical cockroach. *Science, N.Y.* **218**, 171–3.

Schmitz, R.F. (1972) Behaviour of *Ips pini* during mating, oviposition, and larval development (Coleoptera: Scolytidae). *Can. Ent.* **104**, 1723–8.

Schömann, K. (1954) Das 'Paarungs' Verhalten von *Polyxenus lagurus* L. *Naturwiss.* **41**, 13.

Schömann, K. & Schaller, F. (1964) Das Paarungsverhalten von *Polyxenus lagurus* L. *Verh. Dt. Zool. Ges. Tübingen* (= *Zool. Anz. Suppl.*), **18**, 342–6.

Schöne H, & Tengö, J. (1981) Competition of males, courtship behaviour and chemical communication in the digger wasp *Bembix rostrata* (Hymenoptera, Sphecidae). *Behaviour*, **77**, 44–66.

Selander, R.B. (1984) On the bionomics, anatomy and systematics of *Wagneronota* (Coleoptera: Meloidae). *Proc. Ent. Soc. Wash.* **86**, 469–85.

Severinghaus, L.L., Kurtak, B.H. & Eickwort, G.C. (1981) The reproductive behaviour of *Anthidium manicatum* (Hymenoptera: Megachilidae) and the significance of size for territorial males. *Behav. Ecol. Sociobiol.* **9**, 51–8.

Shelly, T.E. & Weinberger, D.A. (1981) Courtship and diet of the Neotropical robber fly *Mallophora schwarzi* Curran (Diptera: Asilidae). *Pan-Pacif. Ent.* **57**, 380–84.

Silberglied, R.E. (1984) In: Vane-Wright, R.I. & Ackery, P.R. (eds) *The biology of butterflies*, Academic Press, London. pp. 216–17.

Sillén-Tullberg, B. (1981) Prolonged copulation: a male 'postcopulatory' strategy in a promiscuous species, *Lygaeus equestris* (Heteroptera: Lygaeidae). *Behav. Ecol. Sociobiol.* **9**, 283–9.

Simmons, L.W. (1986) Female choice in the field cricket *Gryllus bimaculatus* (De Geer). *Anim. Behav.* **34**, 1463–70.

Sivinski, J. (1978) Intrasexual aggression in the stick insects *Diapheromera veliei* and *D. covilleae* and sexual dimorphism in the Phasmatodea. *Psyche*, **85**, 395–403.

Sivinski, J. (1981) 'Love bites' in a lycid beetle. *Fla Ent.* **64**, 541.

Sivinski, J. (1984) Sexual conflict and choice in a phoretic fly, *Borborillus frigipennis* (Sphaeroceridae). *Ann. Ent. Soc. Am.* **77**, 232–5.

Sivinski, J., Burk, T. & Webb, J.C. (1984) Acoustic courtship signals in the Caribbean fruit fly, *Anastrepha suspensa* (Loew). *Anim. Behav.* **32**, 1011–16.

Slooten, E. & Lambert, D.M. (1983) Evolutionary studies of the New Zealand coastal mosquito *Opifex fuscus* (Hutton) I. Mating behaviour. *Behaviour*, **84**, 157–72.

Spangler, H.G. (1985) Sound production and communication by the greater wax moth (Lepidoptera: Pyralidae). *Ann. Ent. Soc. Am.*, **78**, 54–61.

Spangler, H.G. & Manley, D.G. (1978) Sounds associated with the mating behaviour of a mutillid wasp. *Ann. Ent. Soc. Am.* **71**, 389–92.

Spieth, H.T. (1966) Courtship behaviour of endemic Hawaiian *Drosophila*. *Univ. Tex. Publs*, **6615**, 245–313.

Spieth, H.T. (1981) *Drosophila heteroneura* and *Drosophila silvestris*: head shapes, behaviour and evolution. *Evolution*, **35**, 921–30

Spofford, M.G. & Kurczewski, F.E. (1985) Courtship and mating behaviour of *Phrosinella aurifacies* Downes (Diptera: Sarcophagidae: Miltogramminae). *Proc. Ent. Soc. Wash.* **87**, 273–82.

Steck, G.J. (1984) *Chaetostomella undosa* (Diptera: Tephritidae): biology, ecology, and larval description. *Ann. Ent. Soc. Am.* **77**, 669–78.

Steele, R.H. (1986) Courtship feeding in *Drosophila subobscura*. I. The nutritional significance of courtship feeding. *Anim. Behav.* **34**, 1087–98.

Steele, R.H. (1986a) Courtship feeding in *Drosophila subobscura*. II. Courtship feeding by males influences female mate choice. *Anim. Behav.* **34**, 1099–1108.

Steiner, A.L. (1978) Observations on spacing, lekking and aggressive behaviour of digger wasp males of *Eucerceris flavocincta* (Hymenoptera: Sphecidae: Cercerini). *J. Kans. Ent. Soc.* **51**, 492–8.

Stoltzfus, W.B. & Foote, B.A. (1965) The use of froth masses in courtship of *Eutreta*. *Proc. Ent. Soc. Wash.* **67**, 263–4.

Stratton, G.E. & Lowrie, D.C. (1984) Courtship behaviour and life cycle of the wolf spider *Schizocosa mccooki* (Araneae, Lycosidae). *J. Arachnol.* **12**, 223–8

Stratton, G.E. & Uetz, G.W. (1983) Communication via substratum-coupled stridulation and reproductive isolation in wolf spiders (Araneae: Lycosidae). *Anim. Behav.* **31**, 164–72.

Suter, R.B. (1990) Courtship and assessment of virginity by male bowl and doily spiders. *Anim. Behav.* **39**, 307–13.

Suter, R.B. & Renkes, G. (1982) Linyphiid spider courtship: releaser and attractant functions of a contact sex pheromone. *Anim. Behav.* **30**, 714–8.

Suzuki, Y. & Hiehata, K. (1985) Mating systems and sex ratios in the parasitoids, *Trichogramma dendrolimi* and *T. papilionis* (Hymenoptera: Trichogrammatidae). *Anim. Behav.* **33**, 1223–7.

Svensson, B.G. & Petersson, E. (1988) Non-random mating in the dance fly *Empis borealis*: the importance of male choice. *Ethology*, **79**, 307–16.

Szczytko, S.W. & Stewart, K.W. (1979) Drumming behaviour of four western Nearctic *Isoperla* (Plecoptera) species. *Ann. Ent. Soc. Am.* **72**, 781–6.

Tengö, J. (1979) Odour-released behaviour in *Andrena* male bees (Apoidea, Hymenoptera). *Zoon*, **7**, 15–48.

Tengö, J. & Bergström, G. (1977) Cleptoparasitism and odor mimetism in bees: do *Nomada* males imitate the odor of *Andrena* females? *Science, N.Y.* **196**, 1117–19.

Tengö, J., Eriksson, J., Borg-Karlson, A.K., Smith, B.H. & Dobson, H. (1988) Mate-locating strategies and multimodal communication in male mating behaviour of *Panurgus banksianus* and *P. calcaratus* (Apoidea, Andrenidae). *J. Kans. Ent. Soc.* **61**, 388–95.

Thornhill, R. (1976) Reproductive behaviour of the lovebug, *Plecia nearctica* (Diptera: Bibionidae). *Ann. Ent. Soc. Am.* **69**, 843–7.

Thornhill, R. (1977) The comparative predatory and sexual behaviour of hangingflies (Mecoptera: Bittacidae). *Occas. Pap. Mus. Zool. Univ. Mich.* **677**, 1–43.

Thornhill, R. (1978) Sexually selected predatory and mating behaviour of the hangingfly *Bittacus stigmaterus* (Mecoptera: Bittacidae). *Ann. Ent. Soc. Am.* **71**, 597–601.

Thornhill, R. (1979) Adaptive female-mimicking behaviour in a scorpionfly. *Science, N.Y.* **205**, 412–14.

Thornhill, R. (1980) Sexual selection within mating swarms of the lovebug, *Plecia neartica* (Diptera: Bibionidae). *Anim. Behav.* **28**, 405–12.

Thornhill, R. 1980a. Rape in *Panorpa* scorpionflies and a general rape hypothesis. *Anim. Behav.* **28**, 52–9.

Thornhill, R. & Alcock, J. (1983) *The evolution of insect mating systems*. Harvard University Press, Cambridge, Mass.

Thorpe, K.W. & Harrington, B.J. (1981) Sound production and courtship behaviour in the seed bug *Ligyrocoris diffusus*. *Ann. Ent. Soc. Am.* **74**, 369–73.

Tiemann, D.L. (1967) Observations on the mating behaviour of the western banded glowworm *Zaphipis integripennis* (LeConte). *Proc. Calif. Acad. Sci.* **35**, 235–64.

Tinbergen, N. (1941) Ethologische Beobachten am Samfalter, *Satyrus semele*. *J. Orn.* **89**, 132–44.

Tinkham, E.R. & Rentz, D.C. (1969) Notes on the bionomics and distribution of the genus *Stenopalmatus* in Central California with the description of a new species (Orthoptera: Gryllacrididae). *Pan-Pacif. Ent.* **45**, 4–14.

Toft, C.A. (1989) Population structure and mating system of a desert bee fly (*Lordotus pulchrissimus*; Diptera: Bombyliidae). 1. Male demography and interactions. *Oikos*, **54**, 345–58.

Toft, S. (1989) Mate guarding in two *Linyphia* species (Areneae: Linyphiidae). *Bull. Br. Arachnol. Soc.* **8**, 33–7.

Treusch, H.W. (1967) Bisher unbekanntes gezieltes Duftanbieten paarungsbereiter *Argynnis paphia* Weibchen. *Naturwiss.* **54**, 592.

Turillazzi, S. (1983) Extranidal behaviour of *Parischnogaster serrei* (Du Buysson) (Hymenoptera, Stenogastrinae). *Z. Tierpsychol.* **63**, 27–36.

Turillazzi, S. & Francescato, E. (1990) Patrolling behaviour and related secretory structures in the males of some stenogastrine wasps (Hymenoptera, Vespidae). *Insectes. Soc.* **37**, 146–57.

Turner, J.R.G. (1988) Sex, leks and fechts in swift moths *Hepialus* (Lepidoptera: Hepialidae): evidence for the hot shot moth. *Entomologist*, **107**, 90–95.

Van den Assam, J., Jachmann, F. & Simbolotti, P. (1980) Courtship behaviour of *Nasonia vitripennis* (Hym.: Pteromalidae): some qualitative experimental evidence for the role of pheromones. *Behaviour*, **75**, 300–307.

Waage, J.K. (1973) Reproductive behaviour and its relation to territoriality in *Calopteryx maculata* (Beauvois) (Odonata: Calopterygidae). *Behaviour*, **47**, 240–56.

Waage, J.K. (1979) Dual function of the damselfly penis: sperm removal and sperm transfer. *Science, N.Y.* **203**, 916–18.

Waage, J.K. (1979a) Adaptive significance of postcopulatory guarding of mates and nonmates by male *Calopteryx maculata* (Odonata). *Behav. Ecol. Sociobiol.* **6**, 147–54.

Walker, T.J. (1978) Post-copulatory behaviour of the two-spotted tree-cricket, *Neoxabea bipunctata*. *Fla Ent.* **61**, 39–40.

Watson, J. (1986) Transmission of a female sex pheromone thwarted by males in the spider *Linyphia litigiosa* (Linyphiidae). *Science, N.Y.* **233**, 219–21.

Wcislo, W.T. & Eberhard, W.G. (1989) Club fights in the weevil *Macromerus bicinctus* (Coleoptera: Curculionidae). *J. Kans. Ent. Soc.* **62**, 421–9.

Wedell, N. & Arak, A. (1989) The wartbiter spermatophore and its effect on female reproductive output (Orthoptera: Tettigoniidae, *Decticus verrucivorus*). *Behav. Ecol. Sociobiol.* **24**, 117–25.

Weeks, L. & Hespenheide, H.A. (1985) Predatory and mating behaviour of *Stichopogon* (Diptera: Asilidae) in Arizona. *Pan-Pac. Ent.* **61**, 95–104.

Wellso, S.G. (1966) Sexual attraction and biology of *Xenorhipis brendeli* (Coleoptera: Buprestidae). *J. Kans. Ent. Soc.* **39**, 242–5.

Wendelken, P.W. & Barth, R.H. (1985) On the significance of pseudofemale behaviour in the Neotropical cockroach genera *Blaberus*, *Archimandrita* and *Byrsotria*. *Psyche*, **92**, 493–501.

Weseloh, R.M. (1977) Mating behavior of the gypsy moth parasite *Apantales melanoscelus*. *Ann. Ent. Soc. Am.* **70**, 549–54.

Weygoldt, P. (1966) Spermatophore web formation in a pseudoscorpion. *Science N.Y.* **153**, 1647–9.

Weygoldt, P. (1970) Courtship behaviour and sperm transfer in the giant whip scorpion, *Mastigoproctus giganteus* (Lucas) (Uropygi, Thelyphonidae). *Behaviour*, **36**, 1–8.

Weygoldt, P. (1988) Sperm transfer and spermatophore morphology in the whip scorpion *Thelyphonus linganus* (Arachnida: Uropygi: Thelyphonidae). *J. Zool. Lond.* **215**, 189–96.

Wharton, R.A. (1986) Biology of the diurnal *Metasolpuga picta* (Kraepelin) (Solifugae, Solpugidae) compared with that of nocturnal species. *J. Arachnol.* **14**, 363–83.

Whitcomb, W.H., Hite, M. & Eason, R. (1966) Life history of the green lynx spider *Peucetia viridans* (Araneida: Oxyopidae). *J. Kans. Ent. Soc.* **39**, 259–67.

Whitehouse, M.E.A. (1986) The foraging behaviours of *Argyrodes antipodeana* (Theridiidae), a kleptoparasitic spider from New Zealand. *N. Z. J. Zool.* **13**, 151–68.

Wickler, W. & Seibt, U. (1985) Reproductive behaviour in *Zonocerus elegans* (Orthoptera: Pyrgomorphidae) with special reference to nuptial gift guarding. *Z. Tierpsychol.* **69**, 203–23.

Wickman, P.O. (1985) Territorial defence and mating success in males of the small heath butterfly, *Coenonympha pamphilus* L. (Lepidoptera: Satyridae). *Anim. Behav.* **33**, 1162–8.

Wickman, P.O. (1986) Courtship solicitation by females of the small heath butterfly, *Coenonympha pamphilus* (L.) (Lepidoptera: Satyridae) and their behaviour in relation to male territories before and after copulation. *Anim. Behav.* **34**, 153–7.

Wickman, P.O. & Wiklund, C. (1983) Territorial defence and its seasonal decline in the speckled wood butterfly (*Pararge aegeria*). *Anim. Behav.* **31**, 1206–16.

Wiklund, C. (1982) Behavioural shift from courtship solicitation to mate avoidance in female ringlet butterflies (*Aphantopus hyperanthus*) after copulation. *Anim. Behav.* **30**, 790–93.

Wilcox, R.S. (1972) Communication by surface waves. Mating behaviour of a water strider (Gerridae). *J. Comp. Physiol.* **80**, 255–66.

Willey, R.B. & Willey, R.L. (1970) The behavioural ecology of desert grasshoppers. I. Presumed sex-role reversal in flight displays of *Trimerotropis agrestis*. *Anim. Behav.* **18**, 473–7.

Willis, E.R. (1970) Mating behaviour of three cockroaches (*Latiblatella*) from Honduras. *Biotropica*, **2**, 120–28.

Willis, M.A. & Birch, M.C. (1982) Male lek formation and female calling in a population of the arctiid moth *Estigmene acrea*. *Science, N.Y.* **218**, 168–70.

Windsor, D.M. (1987) Natural history of a subsocial tortoise beetle, *Acromis sparsa* Boheman (Chrysomelidae, Cassidinae) in Panama. *Psyche*, **94**, 127–50.

Winter, U. & Buschinger, A. (1983) The reproductive biology of a slavemaker ant, *Epimyrma ravouxi*, and a degenerate slavemaker *E. kraussei* (Hymenoptera: Formicidae). *Ent. Gener.* **9**, 1–15.

Wittmann, D. & Scholz, E. (1989) Nectar dehydration by male carpenter bees as preparation for mating flights. *Behav. Ecol. Sociobiol.* **25**, 387–91.

Wood, T.K. (1974) Aggregating behaviour of *Umbonia crassicornis* (Homoptera: Membracidae). *Can. Ent.* **106**, 169–73.

Zdárek, J. & Kontev, C. (1975) Some ethological aspects of reproduction in *Eurygaster integriceps* (Heteroptera, Scutelleridae). *Acta Ent. Bohemoslovaca.* **72**, 239–48.

Zeh, D.W. (1987) Aggression, density, and sexual dimorphism in chernetid pseudoscorpions (Arachnida: Pseudoscorpionida). *Evolution.* **41**, 1072–87.

Egg-laying Behaviour

Aker, C.L. & Udovic, D. (1981) Oviposition and pollination behavior of the yucca moth, *Tegeticula maculata* (Lepidoptera: Prodoxidae), and its relation to the reproductive biology of *Yucca whipplei* (Agavaceae). *Oecologia*, **49**, 96–101.

Atsatt, P.R. (1981) Ant-dependent food plant selection by the mistletoe butterfly *Ogyris amaryllis* (Lycaenidae). *Oecologia*, **48**, 60–63.

Averill, A.L. & Prokopy, R.J. (1981) Oviposition deterring fruit marking pheromone in *Rhagoletis basiola*. *Fla Ent.* **64**, 222–6.

Braker, H.E. (1989) Oviposition in host plants by a tropical forest grasshopper (*Microtylopteryx hebardi*: Acrididae). *Ecol. Ent.* **14**, 141–8.

Brown, K.S., Damman, A.J. & Feeny, P. (1980) Troidine swallowtails (Lepidoptera: Papilionidae) in southeastern Brazil: natural history and foodplant relationships. *J. Res. Lepid.* **19**, 199–226.

Carlberg, U. (1989) Aspects of evolution and ecology in relation to defaecation and oviposition behaviour of *Extatosoma tiaratum* (MacLeay) (Insecta: Phasmida). *Biol. Zbl.* **107**, 541–51.

Chew, F.S. (1977) Evolution of pierid butterflies and their cruciferous foodplants. II. The distribution of eggs on potential foodplants. *Evolution*, **31**, 568–79.

Christenson, T.E. & Wenzl, P.A. (1980) Egg-laying of the golden silk spider *Nephila clavipes* L. (Araneae, Araneidae): Functional analysis of the egg sac. *Anim. Behav.* **28**, 1110–18.

Courtney, S.P. (1981) Coevolution of pierid butterflies and their cruciferous foodplants. III. *Anthocharis cardamines* (L.) survival, development and oviposition on different hostplants. *Oecologia*, **51**, 91–6.

Daanje, A. (1975) Some special features of the leaf-rolling technique of *Byctiscus populi* L. (Coleoptera, Rhynchitini). *Behaviour*, **53**, 285–316.

Damman, H. & Feeny, P. (1988) Mechanisms and consequences of selective oviposition by the zebra swallowtail butterfly. *Anim. Behav.* **36**, 563–73.

Dennis, D.S. & Lavigne, R.J. (1976) Ethology of *Efferia varipes* with comments on species coexistence (Diptera: Asilidae). *J. Kans. Ent. Soc.* **49**, 48–62.

Eason, R.R. (1969) Life history and behavior of *Pardosa lapidicina* Emerton. *J. Kans. Ent. Soc.* **42**, 339–60.

Eberhard, W.G. (1983) Behavior of adult bottle brush weevils (*Rhinostomus barbirostris*) (Coleoptera: Curculionidae). *Rev. Biol. Trop.* **31**, 233–44.

Eickwort, G.C. & Abrams, J. (1980) Parasitism of sweat bees in the genus *Agapostemon* by cuckoo bees in the genus *Nomada* (Hymenoptera: Halictidae, Anthophoridae). *Pan-Pacif. Ent.* **56**, 144–52.

Fincke, O.M. (1986) Underwater oviposition in a damselfly (Odonata: Coenagrionidae) favors male vigilance, and multiple mating by females. *Behav. Ecol. Sociobiol.* **18**, 405–12.

Fowler, H.G. (1987) Field behavior of *Euphasiopteryx depleta* (Diptera: Tachinidae): phonotactically orienting parasitoids of mole crickets (Orthoptera: Gryllotalpidae: *Scapteriscus*). *J. N. Y. Ent. Soc.* **95**, 474–80.

Frank, S.A. (1984) The behavior and morphology of the fig wasps *Pegoscapus assuetus* and *P. jimenezi*: descriptions and suggested behavioral characters for phylogenetic studies. *Psyche*, **91**, 289–304.

Gonzalez Soriano, E. (1986) *Dythemis cannacrioides* Calvert, a libellulid with unusual ovipositing behaviour (Anisoptera). *Odonatologica*, **15**, 147–50.

Grossmueller, D.W. & Lederhouse, R.C. (1985) Oviposition site selection: an aid to rapid growth and development in the tiger swallowtail butterfly, *Papilio glaucus*. *Oecologia*, **66**, 68–73.

Grout, T.G. & Brothers, D.J. (1982) Behaviour of a parasitic pompilid wasp (Hymenoptera). *J. Ent. Soc. Sth. Afr.* **45**, 217–20.

Hackwell, G.A. (1973) Biology of *Lasconotus subcostulatus* (Coleoptera: Colydiidae) with special reference to feeding behavior. *Ann. Ent. Soc. Am.* **66**, 62–5.

Henning, S.F. (1983) Chemical communication between lycaenid larvae (Lepidoptera: Lycaenidae) and ants (Hymenoptera: Formicidae). *J. Ent. Soc. Sth. Afr.* **46**, 341–66.

Henning, S.F. (1984) Life history and behaviour of the rare myrmecophilous lycaenid *Erikssonia acraeina* Trimen (Lepidoptera: Lycaenidae). *J. Ent. Soc. Sth. Afr.* **47**, 337–42.

Janzen, D.H. (1984) Natural history of *Hylesia lineata* (Saturniidae: Hemileucinae) in Santa Rosa National Park, Costa Rica. *J. Kans. Ent. Soc.* **57**, 490–514.

Kitching, R.L. (1987) Aspects of the natural history of the lycaenid butterfly *Allotinus major* in Sulawesi. *J. Nat. Hist.* **21**, 535–44.

Knisley, C.B., Reeves, D.L. & Stephens, G.T. (1989) Behavior and development of the wasp *Pterombrus rufiventris hyalinatus* Krombein (Hymenoptera: Tiphiidae), a parasite of larval tiger beetles (Coleoptera: Cicindelidae). *Proc. Ent. Soc. Wash.* **91**, 179–84.

Larsen, T.B. (1988) Differing oviposition and larval feeding strategies in two *Colotis* butterflies sharing the same food plant. *J. Lepid. Soc.* **42**, 57–8.

Lavigne, R.J. & Bullington, S.W. (1984) Ethology of *Laphria fernaldi* (Back) (Diptera: Asilidae) in southeast Wyoming. *Proc. Ent. Soc. Wash.* **86**, 326–36.

Lavigne, R.J. & Dennis, D.S. (1975) Ethology of *Efferia frewingi* (Diptera: Asilidae). *Ann. Ent. Soc. Am.* **68**, 992–9.

Lavigne, R.J. & Dennis, D.S. (1980) Ethology of *Proctacanthella leucopogon* in Mexico (Diptera: Asilidae). *Proc. Ent. Soc. Wash.* **82**, 260–68.

Londt, J.G.H. & Harris, M.S. (1987) Notes on the biology and immature stages of *Neolophonotus dichaetus* Hull in South Africa (Diptera: Asilidae). *J. Ent. Soc. Sth. Afr.* **50**, 427–34.

McCaffery, A.R. & Page, W.W. (1982) Oviposition behaviour of the grasshopper *Zonocerus variegatus*. *Ecol. Ent.* **7**, 85–90.

Martens, A. (1991) Plasticity of mate-guarding and oviposition behaviour in *Zygonyx natalensis* (Martin) (Anisoptera: Libellulidae). *Odonatologica*, **20**, 293–302.

Mason, M.L. & Lawson, F.A. (1982) Biology of the American aspen beetle (Coleoptera: Chrysomelidae: *Gonioctena americana* (Schaeffer)) in the Medicine Bow National Forest, Wyoming. *J. Kans. Ent. Soc.* **55**, 779–88.

Miller, P.L. & Miller, A.K. (1985) Rates of oviposition and some other aspects of reproductive behaviour in *Tholymis tillarga* (Fabricius) in Kenya (Anisoptera: Libellulidae). *Odonatologica*, **14**, 287–99.

Miller, P.L. & Miller, A.K. (1988) Reproductive behaviour and two modes of oviposition in *Phaon iridipennis* (Burmeister) (Zygoptera: Calopterygidae). *Odonatologica*, **17**, 187–94.

Monteith G.B. & Storey, R.I. (1981) The biology of *Cephalodesmius*, a genus of dung beetles which synthesizes 'dung' from plant material (Coleoptera: Scarabaeidae: Scarabaeinae). *Mem. Qd Mus.* **20**, 253–77.

Mourier, H. & Banegas, A.D. (1970) Observations on the oviposition and the ecology of the eggs of *Dermatobia hominis* (Diptera: Cuterebridae). *Vidensk. Meddr. Dansk Naturh. Foren.* **33**, 59–68.

Palmer, M.K. (1982) Biology and behavior of two species of *Anthrax* (Diptera; Bombyliidae), parasitoids of the larvae of tiger beetles (Coleoptera: Cicindelidae). *Ann. Ent. Soc. Am.* **75**, 61–70.

Papaj, D.R. & Rausher, M.D. (1987) Components of conspecific host discrimination behavior in the butterfly *Battus philenor*. *Ecology*, **68**, 245–53.

Pappas, C.D. & Pappas, L.G. (1982) Observations on the egg raft formation behavior of *Culiseta inornata*. *Ann. Ent. Soc. Am.* **75**, 393–4.

Peckham, D.J. (1977) Reduction of miltogrammine cleptoparasitism by male *Oxybelus subulatus* (Hymenoptera; Sphecidae). *Ann. Ent. Soc. Am.* **70**, 823–8.

Pierce, N.E. & Elgar, M.A. (1985) The influence of ants on host plant selection by *Jalmenus evagoras*, a myrmecophilous lycaenid butterfly. *Behav. Ecol. Sociobiol.* **16**, 209–22.

Rausher, M.D. (1980) Host abundance, juvenile survival, and oviposition preference in *Blattus philenor*. *Evolution*, **34**, 342–55.

Rice, M.E. (1989) Branch girdling and oviposition biology of *Oncideres pustulatus* (Coleoptera: Cerambycidae) on *Acacia farnesiana*. *Ann. Ent. Soc. Am.* **82**, 181–6.

Ridsdill Smith, T.J. (1970) The behaviour of *Hemithynnus hyalinatus* (Hymenoptera: Tiphiidae), with notes on some other Thynninae. *J. Aust. Ent. Soc.* **9**, 196–208.

Rotheray, G.E. (1979) The biology and host searching behaviour of a Cynipoid parasite of aphidophagous syrphid larvae. *Ecol. Ent.* **4**, 75–82.

Rotheray, G.E. (1981) Host searching and oviposition behaviour of some parasitoids of aphidophagous Syrphidae. *Ecol. Ent.* **6**, 79–87.

Rothschild, M. & Schoonhoven, L.M. (1977) Assessment of egg load by *Pieris brassicae* (Lepidoptera: Pieridae). *Nature, Lond.* **266**, 352–5.

Sakurai, K. (1986) Multiple oviposition by *Apoderus balteatus* (Coleoptera, Attelabidae) is associated with larger leaves. *Ecol. Ent.* **11**, 319–23.

Schmitz, R.F. (1972) Behavior of *Ips pini* during mating, oviposition, and larval development (Coleoptera: Scolytidae). *Can. Ent.* **104**, 1723–8.

Schreiber, E.T. & Lavigne, R.J. (1986) Ethology of *Asilus gilvipes* (Hine) (Diptera: Asilidae) associated with small mammal burrows in southeastern Wyoming. *Proc. Ent. Soc. Wash.* **88**, 711–19.

Spofford, M.G., Kurczewski, F.E. & Peckam, D.J. (1986) Cleptoparasitism of *Tachysphex terminatus* (Hymenoptera: Sphecidae) by three species of Miltogrammini (Diptera: Sarcophagidae). *Ann. Ent. Soc. Am.* **79**, 350–58.

Spradbery, J.P. (1970) Host finding by *Rhyssa persuasoria* (L.), an ichneumonid parasite of siricid woodwasps. *Anim. Behav.* **18**, 103–14.

Spradbery, J.P. (1977) The oviposition biology of siricid woodwasps in Europe. *Ecol. Ent.* **2**, 225–30.

Stanton, M.L. (1982) Searching in a patchy environment: foodplant selection by *Colias p. eriphyle* butterflies. *Ecology*, **63**, 839–53.

Tengö, J. & Bergström, G. (1977) Cleptoparasitism and odor mimetism in bees: do *Nomada* males imitate the odor of *Andrena* females? *Science, N.Y.* **196**, 1117–19.

Torchio, P.F. (1989) Biology, immature development, and adaptive behavior of *Stelis montana*, a cleptoparasite of *Osmia* (Hymenoptera: Megachilidae). *Ann. Ent. Soc. Am.* **82**, 616–32.

Torchio, P.F. & Burdick, D.J. (1988) Comparative notes on the biology and development of *Epeolus compactus* Cresson, a cleptoparasite of *Colletes kincaidii* Cockerell (Hymenoptera: Anthophoridae, Colletidae). *Ann. Ent. Soc. Am.* **81**, 626–36.

Turillazzi, S. & Pardi, L. (1982) Social behaviour of *Parischnogaster nigricans serrei* (Hymenoptera: Vespoidea) in Java. *Ann. Ent. Soc. Am.* **75**, 657–64.

Waage, J.K. & Montgomery, G.G. (1976) *Cryptoses choloepi*: a coprophagous moth that lives on a sloth. *Science, N.Y.* **193**, 157–8.

Waloff, N. (1974) Biology and behaviour of some species of Drynidae (Hymenoptera). *J. Ent.* **49**, 97–109.

Webster, R.P. & Nielsen, M.C. (1984) Myrmecophily in the Edward's hairstreak butterfly *Satyrium edwardsi* (Lycaenidae). *J. Lepid. Soc.* **38**, 124–33.

Wood, J.R., Resh, V.H. & McEwan, E.M. (1982) Egg masses of Nearctic sericostomatid caddisfly genera (Trichoptera). *Ann. Ent. Soc. Am.* **75**, 430–34.

Yeargan, K.V. & Braman, S.K. (1986) Life history of the parasite *Diolcogaster facetosa* (Weed) (Hymenoptera: Braconidae) and its behavioral adaptation to the defensive response of a lepidopteran host. *Ann. Ent. Soc. Am.* **79**, 1029–33.

Yuma, M. & Hori, M. (1981) Gregarious oviposition of *Luciola cruciata* Motschulsky (Coleoptera: Lampyridae). *Physiol. Ecol. Jap.* **18**, 93–112.

Parental Care

Akre, R.D., Garnett, W.B., Macdonald, J.F., Greene, A. & Landolt, P. (1976) Behaviour and colony development of *Vespula pensylvanica* and *V. atropilosa* (Hymenoptera: Vespidae). *J. Kans. Ent. Soc.* **49**, 63–84.

Albans, K.R., Aplin, R.T., Brehcist, J., Moore, J.F. & O'Toole, C. (1980) Dufour's gland and its role in the secretion of nest cell lining in bees of the genus *Colletes* (Hymenoptera: Colletidae). *J. Chem. Ecol.* **6**, 549–64.

Alloway, T.M. (1979) Raiding behaviour of two species of slavemaking ants, *Harpagoxenus americanus* (Emery) and *Leptothorax duloticus* (Wesson) (Hymenoptera: Formicidae). *Anim. Behav.* **27**, 202–10.

Alpert, G.D. & Akre, R.D. (1973) Distribution, abundance and behavior of the inquiline ant *Leptothorax diversipilosus*. *Ann. Ent. Soc. Am.* **66**, 753–60.

Anastos, G., Kaufman, T.S. & Kadarsan, S. (1973) An unusual reproductive process in *Ixodes kopsteini* (Acarina: Ixodidae). *Ann. Ent. Soc. Am.* **66**, 483–4.

Austad, S.N. & Thornhill, R. (1986) Female reproductive variation in a nuptial-feeding spider, *Pisaura mirabilis*. *Bull. Br. Arachnol. Soc.* **7**, 48–52.

Ballard, E. & Holdaway, F.G. (1926) The life-history of *Tectacoris lineola*, F., and its connection with internal boll rots in Queensland. *Bull. Ent. Res.* **16**, 329–46.

Bartlett, J. (1987) Filial cannibalism in burying beetles. *Behav. Ecol. Sociobiol.* **21**, 179–83.

Bartlett, J. (1988) Male mating success and paternal care in *Nicrophorus vespilloides* (Coleoptera: Silphidae). *Behav. Ecol. Sociobiol.* **23**, 297–303.

Batra, S.W.T. (1966) Social behavior and nests of some nomiine bees in India (Hymenoptera, Halictidae). *Insectes Soc.* **13**, 145–54.

Batra, S.W.T., Fales, H.M., Hefetz, A. & Shaw, G.J. (1980) *Anthophora* bees: unusual glycerides from maternal Dufour's glands serve as larval food and cell lining. *Science, N.Y.* **207**, 1095–7.

Bellés, X. & Favila, M.E. (1983) Protection chimique du nid chez *Canthon cyanellus cyanellus* LeConte (Col. Scarabaeidae). *Bull. Soc. Ent. Fr.* **88**, 602–7.

Bourke, A.F.G. (1988) Dominance orders, worker reproduction, and queen-worker conflict in the slave-making ant *Harpagoxenus sublaevis*. *Behav. Ecol. Sociobiol.* **23**, 323–33.

Bradoo, B.L. (1967) Observations on the life history and bionomics of *Oligotoma ceylonica ceylonica* Enderlein (Oligotomidae, Embioptera), commensal in the nest of the social spider *Stegodyphus sarasinorum*. *J. Bomb. Nat. Hist. Soc.* **64**, 447–54.

Breed, M.D. & Gamboa, G.J. (1977) Behavioral control of workers by queens in primitively eusocial bees. *Science, N.Y.* **195**, 694–6.

Bristow, C.M. (1983) Treehoppers transfer parental care to ants: a new benefit of mutualism. *Science, N.Y.* **220**, 552–3.

Brown, R.L. (1976) Behavioural observation on *Aethalion reticulatum* (Hem. Aethalionidae) and associated ants. *Insectes Soc.* **23**, 99–108.

Buckle, G.R. (1985) Increased queen-like behaviour of workers in large colonies of the sweat bee *Lasioglossum zephyrum*. *Anim. Behav.* **33**, 1275–80.

Burgess, J.W. (1978) Social behavior in group-living spider species. *Symp. Zool. Soc. Lond.* **42**, 69–78.

Cameron, S.A. (1986) Brood care by males of *Polistes major* (Hymenoptera: Vespidae). *J. Kans. Ent. Soc.* **59**, 183–5.

Cane, J.H. (1981) Dufour's gland secretion in the cell linings of bees (Hymenoptera: Apoidea). *J. Chem. Ecol.* **7**, 403–10.

Carroll, C.R. (1978) Beetles, parasitoids and tropical morning glories: a study in host discrimination. *Ecol. Ent.* **3**, 79–85.

Cervo, R., Lorenzi, M.C. & Turillazzi, S. (1990) *Sulcopolistes atrimandibularis*, social parasite and predator of an alpine *Polistes* (Hymenoptera, Vespidae). *Ethology*, **86**, 71–8.

Choe, J.C. (1989) Maternal care in *Labidomera suturella* Chevrolat (Coleoptera: Chrysomelidae: Chrysomelinae) from Costa Rica. *Psyche*, **96**, 63–7.

Christenson, T.E. (1984) Behaviour of colonial and solitary spiders of the theridiid species *Anelosimus eximius*. *Anim. Behav.* **32**, 725–34.

Cowan, D.P. (1981) Parental investment in two solitary wasps *Ancistrocerus adiabatus* and *Euodynerus foraminatus* (Eumenidae: Hymenoptera). *Behav. Ecol. Sociobiol.* **9**, 95–102.

Crespi, B.J. (1990) Subsociality and female reproductive success in a mycophagous thrips: an observational and experimental analysis. *J. Insect Behav.* **3**, 61–72.

Deyrup, M., Cronin, J.T. & Kurczewski, F.E. (1988) *Allochares azureus*: an unusual wasp exploits unusual prey (Hymenoptera: Pompilidae; Arachnida: Filistatidae). *Psyche*, **95**, 265–81.

Dias, B.F. de Souza. (1975) Comportamento pré-social de sinfitas do Brasil central. I. *Themos olfersii* (Klug) (Hymenoptera, Argidae). *Studia Ent.* **18**, 401–32.

Dias, B.F. de Souza. (1976) Comportamento pré-social de sinfitas do Brasil central. II. *Dielocerus diasi* Smith, 1975 (Hymenoptera, Argidae). *Studia Ent.* **19**, 461–501.

Diesel, R. (1989) Parental care in an unusual environment: *Metapaulias depressus* (Decapoda: Grapsidae), a crab that lives in epiphytic bromeliads. *Anim. Behav.* **38**, 561–75.

Doutt, R.L. (1973) Maternal care of immature progeny by parasitoids. *Ann. Ent. Soc. Am.* **66**, 486–7.

Eberhard, W.G. (1974) Maternal behaviour in a South American *Lyssomanes*. *Bull. Br. Arachnol. Soc.* **3**, 51.

Eberhard, W.G. (1974a) The natural history and behaviour of the wasp *Trigonopsis cameronii* Kohl. (Sphecidae). *Trans. R. Ent. Soc. Lond.* **125**, 295–328.

Eberhard, W.G. (1975) The ecology and behaviour of a subsocial pentatomid bug and two scelionid wasps: strategy and counter-strategy in a host and its parasites. *Smithson. Contr. Zool.* **205**, 1–39.

Eberhard, W.G. (1986) Possible mutualism between females of the subsocial membracid *Polyglypta dispar* (Homoptera). *Behav. Ecol. Sociobiol.* **19**, 447–53.

Edgerly, J.S. (1987) Maternal behaviour of a web-spinner (Order Embiidina). *Ecol. Ent.* **12**, 1–11.

Edgerly, J.S. (1988) Maternal behaviour of a web-spinner (Order Embiidina): mother–nymph associations. *Ecol. Ent.* **13**, 263–72.

Edwards, J.S. (1962) Observations on the development and predatory habit of two Reduviid Heteroptera,, *Rhinocoris carmelite* Stäl and *Platymeris rhadamanthus* Gerst. *Proc. R. Ent. Soc. Lond.* **37**, 89–98.

Edwards, P.B. & Aschenborn, H.H. (1988) Male reproductive behaviour of the African ball-rolling dung beetle *Kheper nigroaeneus* (Coleoptera: Scarabaeidae). *Coleopts Bull.* **42**, 12–27.

Edwards, P.B. & Aschenborn, H.H. (1989) Maternal care of a single offspring in the dung beetle *Kheper nigroaeneus*: the consequences of extreme parental investment. *J. Nat. Hist.* **23**, 17–27.

Eickwort, G.C. (1975) Nest-building behaviour of the mason bee *Hoplitis anthocopoides* (Hymenoptera: Megachilidae). *Z. Tierpsychol.* **37**, 237–54.

Eickwort, G.C. (1986) First steps into eusociality: the sweat bee *Dialictus lineatulus*. *Fla Ent.* **69**, 742–54.

Eickwort, G.C., Kukuk, P.L. & Wesley, F.R. (1986) The nesting biology of *Dufourea novaeangliae* (Hymenoptera: Halictidae) and the systematic position of the Dufoureinae based on behaviour and development. *J. Kans. Ent. Soc.* **59**, 103–20.

Eisner, T., Aneshansley, D., Eisner, M., Rutowski, R., Chong, B. & Meinwald, J. (1974) The soldier of the ant *Camponotus (Colobopsis) fraxinicola* as a trophic caste. *Psyche*, **81**, 182–8.

Evans, H.E. (1965) Simultaneous care of more than one nest by *Ammophila azteca* Cameron (Hymenoptera, Sphecidae). *Psyche*, **72**, 8–23.

Evans, H.E. (1973) Burrow sharing and nest transfer in the digger wasp *Philanthus gibbosus* (Fabricius). *Anim. Behav.* **21**, 302–8.

Faeth, F.H. (1989) Maternal care in a lace bug *Corythucha hewitti* (Hemiptera: Tingidae). *Psyche*, **20**, 101–10.

Field, J. (1989) Intraspecific parasitism and nesting success in the solitary wasp *Ammophila sabulosa*. *Behaviour*, **110**, 23–46.

Fink, L.S. (1986) Costs and benefits of maternal behaviour in the green lynx spider (Oxyopidae, *Peucetia viridans*). *Anim. Behav.* **34**, 1051–60.

Fink, L.S. (1987) Green lynx spider egg sacs: sources of mortality and the function of female guarding (Araneae, Oxyopidae). *J. Arachnol.* **15**, 231–9.

Forsyth, A. (1981) Swarming activity of polybiinae social wasps (Hymenoptera: Vespidae: Polybiini). *Biotropica*, **13**, 93–9.

Frohlich, D.R. (1983) On the nesting biology of *Osmia (Chenosmia) bruneri* (Hymenoptera: Megachilidae). *J. Kans. Ent. Soc.* **56**, 123–30.

Frohlich, D.R. & Parker, F.D. (1983) Nest building behaviour and development of the sunflower leafcutter bee: *Eumegachile (Sayapis) pugnata* (Say) (Hymenoptera: Megachilidae). *Psyche*, **90**, 193–207.

Frohlich, D.R. & Parker, F.D. (1985) Observations on the nest-building and reproductive behaviour of a resin-gathering bee: *Dianthidium ulkei* (Hymenoptera: Megachilidae). *Ann. Ent. Soc. Am.* **78**, 804–10.

Halffter, G., Halffter, V. & Huerta, C. (1983) Comportement sexual et nidification chez *Canthon cyanellus cyanellus* Le Conte. *Bull. Soc. Ent. Fr.* **88**, 585–93.

Hamilton, W.D. (1964) The genetical evolution of social behaviour I, II. *J. Theor. Biol.* **7**, 1–52.

Hansell, M.H. (1982) Brood development in the subsocial wasp *Parischnogaster mellyi* (Saussure) (Stenogastrinae, Hymenoptera). *Insectes Soc.* **29**, 3–14.

Hansell, M.H. (1987) Elements of eusociality in colonies of *Eustenogaster calyptodoma* (Sakagami

& Yoshikawa) (Stenogastrinae, Vespidae). *Anim. Behav.* **35**, 131–41.

Hemmingsen, A.M. (1947) Plant bug guarding eggs and offspring and shooting anal jets (*Physomerus grossipes* F., Coreidae). *Ent. Medd.* **25**, 200.

Hinton, H.E. (1944) Some general remarks on subsocial beetles, with notes on the biology of the staphylinid, *Platystethus arenarius* (Fourcroy). *Proc. R. Ent. Soc. Lond. (A)*. **19**, 115–28.

Hölldobler, B. (1982) Communication, raiding behaviour and prey storage in *Cerapachys* (Hymenoptera: Formicidae). *Psyche*, **89**, 3-.

Holloway, B.A. (1976) A new bat-fly family from New Zealand (Diptera: Mystacinobiidae). *N. Z. J. Zool.* **3**, 279–301.

Ishay, J.S., Dotan, Z.A. & Pinchasov, A. (1983) Combativeness among oriental hornet queens. *Insectes Soc.* **30**, 57–69.

Jackson, R.R. (1982) The biology of *Portia fimbriata*, a web-building jumping spider (Araneae, Salticidae) from Queensland: intraspecific interactions. *J. Zool. Lond.* **196**, 295–305.

Jackson, R.R. (1985) The biology of *Euryattus* sp. indet., a web-building jumping spider (Araneae, Salticidae) from Queensland: utilization of silk, predatory behaviour and intraspecific interactions. *J. Zool. Lond. (B)* **1**, 145–73.

Jacson, C.C. & Joseph, K.J. (1973) Life, history, bionomics and behaviour of *S. sarasinorum* Karsch. *Insectes Soc.* **30**, 189–204.

Jeanne, R.L. (1970) Descriptions of the nests of *Pseudochartergus fuscatus* and *Stelopolybia testacea*, with a note of a parasite of *S. testacea* (Hymenoptera, Vespidae). *Psyche*, **77**, 54–69.

Jeanne, R.L. (1977) Behaviour of the obligate social parasite *Vespula arctica* (Hymenoptera: Vespidae). *J. Kans. Ent. Soc.* **50**, 541–57.

Jeanne, R.L. (1980) Evolution of social behaviour in the Vespidae. *Ann. Rev. Ent.* **25**, 371–96.

Jeanne, R.L. (1981) Chemical communication during swarm emigration in the social wasp *Polybia sericea* (Olivier). *Anim. Behav.* **29**, 102–13.

Jeanne, R.L. (1986) The organization of work in *Polybia occidentalis*: costs and benefits of specialization in a social wasp. *Behav. Ecol. Sociobiol.* **19**, 333–41.

Kearns, R.S. & Yamamoto, R.T. (1981) Maternal behaviour and alarm response in the eggplant lace bug *Gargaphia solani* Heidemann (Tingidae: Heteroptera). *Psyche*, **88**, 215–30.

Kiester, A.R. & Strates, E. (1984) Social behaviour in a thrips from Panama. *J. Nat. Hist.* **18**, 303–14.

Klahn, J. (1988) Intraspecific comb usurpation in the social wasp *Polistes fuscatus*. *Behav. Ecol. Sociobiol.* **23**, 1–8.

Klemperer, H.G. (1981) Nest construction and larval behaviour of *Bubas bison* (L.) and *Bubas bubalus* (Ol.) (Coleoptera, Scarabaeidae). *Ecol. Ent.* **6**, 23–33.

Klemperer, H.G. (1982) Nest construction and larval behaviour of *Onitis belial* and *Onitis ion* (Coleoptera, Scarabaeidae). *Ecol. Ent.* **7**,291–7.

Klemperer, H.G. (1982a). Normal and atypical nesting behaviour of *Copris lunaris* (L.): comparison with related species (Coleoptera, Scarabaeidae). *Ecol. Ent.* **7**, 69–83.

Klemperer, H.G. (1983) Subsocial behaviour in *Oniticellus cinctus* (Coleoptera, Scarabaeidae): effect of the brood on parental care and oviposition. *Physiol. Ent.* **8**, 393–402.

Klemperer, H.G. (1984) Nest construction, fighting and larval behaviour in a geotrupine dung beetle, *Ceratophys hoffmannseggi (Coleoptera: Scarabaeidae)*. *J. Zool. Lond.* **204**, 119–27.

Klemperer, H.G. (1986) Life history and parental behaviour of a dung beetle from neotropical rainforest, *Copris laeviceps* (Coleoptera, Scarabaeidae). *J. Zool. Lond.* **209**, 319–26.

Krombein, K.V. (1978) Biosystematic studies of Ceylonese wasps III. Life history, nest and associates of *Paraleptomenes mephitis* (Cameron) (Hymenoptera: Eumenidae). *J. Kans. Ent. Soc.* **51**, 721–34.

Kurczewski, F.E. & Kurczewski, E.J. (1984) Mating and nesting behaviour of *Tachytes intermedius* (Viereck) (Hymenoptera: Sphecidae). *Proc. Ent. Soc. Wash.* **86**, 176–84.

Lamb, R.J. (1976) Parental behaviour in the Dermaptera with special reference to *Forficula auricularia* (Dermaptera: Forficulidae). *Can. Ent.* **108**, 609–19.

Linsenmair, K.E., (1972) Die Bedeutung familienspezifischer 'Abzeichen' für den Familienzusammenhalt bei der sozialen Wüstenassel *Hemilepistus reaumeri* Audouin u. Savigny (Crustacea, Isopoda, Oniscoidea). *Z. Tierpsychol.* **31**, 131–62.

Linsenmair, K.E. (1984) Comparative studies on the social behaviour of the desert isopod *Hemilepistus reaumeri* and of a *Porcellia* species. *Symp. Zool. Soc. Lond.* **53**, 423–53.

Litte, M. (1979) *Mischocyttarus flavitarsis* in Arizona: social and nesting biology of a polistine wasp. *Z. Tierpsychol.* **50**, 282–312.

Lounibos, L.P. (1983) Behavioral convergences among fruit-husk mosquitoes. *Fla Ent.* **66**, 32–41.

Lounibos, L.P. & Machado-Allison, C.E. (1987) Female brooding protects mosquito eggs from rainfall. *Biotropica*, **19**, 83–5.

Lubin, Y.D. (1982) Does the social spider *Achaearanea wau* (Theridiidae), feed its young? *Z. Tierpsychol.* **60**, 127–34.

McCaffery, A.R. & Page, W.W. (1982) Oviposition behaviour of the grasshopper *Zonocerus variegatus*. *Ecol. Ent.* **7**, 85–90.

McLay, C.L. & Hayward, T.L. (1987) Reproductive biology of the intertidal spider *Desis marina* (Araneae: Desidae) on a New Zealand rocky shore. *J. Zool. Lond.* **211**, 357–72.

Marino Piccioli, M.T. (1968) The extraction of the larval peritrophic sac by the adults in *Belonogaster*. *Monitore Zool. Ital. (NS). Suppl.* **2**, 203–6.

Maschwitz, U., Dorow, W.H.O. & Botz, T. (1990) Chemical composition of the nest walls, and nesting behaviour, of *Ropalidia (Icarielia) opifex* van der Vecht, 1962 (Hymenoptera: Vespidae), a Southeast Asian social wasp with translucent nests. *J. Nat. Hist.* **24**, 1311–19.

Maschwitz, U. & Gutmann, Ch. (1979) Trail and alarm pheromones in *Elasmucha grisea* (Heteroptera : Acanthosomidae). *Insectes Soc.* **26**, 101–11.

Maschwitz, U. & Hänel, H. (1985) The migrating herdsman *Dolichoderus (Diabolus) cuspidatus*: an ant with a novel mode of life. *Behav. Ecol. Sociobiol.* **17**, 171–84.

Maschwitz, U., Steghaus-Kovac, S., Gaube, R. & Hänel, H. (1989) A South East Asian ponerine ant of the genus *Leptogenys* (Hym., Form.) with army ant life habits. *Behav. Ecol. Sociobiol.* **24**, 305–16.

Matthews, E.G. (1963) Observations on the ball-rolling behaviour of *Canthon pilularius* (L.) (Coleoptera, Scarabaeidae). *Psyche*, **70**, 75–93.

Matthews, R.W. (1983) Biology of a new *Trypoxylon* that utilizes nests of *Microstigmus* in Costa Rica (Hymenoptera: Sphecidae). *Pan-Pacif. Ent.* **59**, 152–62.

Matthews, R.W., Saunders, R.A. & Matthews, J.R. (1981) Nesting behavior of the sand wasp *Stictia maculata* (Hymenoptera: Sphecidae) in Costa Rica. *J. Kans. Ent. Soc.* **54**, 249–54.

Matthews, R.W. & Starr, C.K. (1984) *Microstigmus comes* wasps have a method of nest construction unique among social insects. *Biotropica*, **16**, 55–8.

Messer, A.C. (1984) *Chalicodoma pluto*: the world's largest bee rediscovered living communally in termite nests (Hymenoptera: Megachilidae). *J. Kans. Ent. Soc.* **57**, 165–8.

Messer, A.C. (1985) Fresh dipterocarp resins gathered by megachilid bees inhibit growth of pollen-associated fungi. *Biotropica*, **17**, 175–6.

Michener, C.D. & Lange, R.B. (1958) Distinctive type of primitive social behavior among bees. *Science, N.Y.* **127**, 1046–7.

Mitchell, R.W. (1971) Egg and young guarding by a Mexican cave-dwelling harvestman, *Hoplobunus boneti* (Arachnida). *Swest. Nat.* **15**, 392–5.

Moffett, M.W. (1984) Swarm raiding in a myrmicine ant. *Naturwiss.* **71**, 588–90.

Moffett, M.W. (1988) Foraging behavior in the Malayan swarm-raiding ant *Pheidologeton silenus* (Hymenoptera: Formicidae: Myrmicinae). *Ann. Ent. Soc. Am.* **81**, 356–61.

Monteith, G.B. & Storey, R.I. (1981) The biology of *Cephalodesmius*, a genus of dung beetles which synthesizes 'dung' from plant material (Coleoptera: Scarabaeidae: Scarabaeinae). *Mem. Qd Mus.* **20**, 253–77.

Mora, G. (1990) Paternal care in a neotropical harvestman, *Zygopachylus albomarginis* (Arachnida, Opiliones: Gonyleptidae). *Anim. Behav.* **39**, 582–93.

Müller, J.K., Eggert, A.K. & Dressel, J. (1990) Intraspecific brood parasitism in the burying beetle, *Necrophorus vespilloides (*Coleoptera: Silphidae*)*. *Anim. Behav.* **40**, 491–9.

Nafus, D.M. & Schreiner, I.H. (1988) Parental care in a tropical nymphalid butterfly *Hypolimnas anomala*. *Anim. Behav.* **36**, 1425–31.

Nalepa, C.A. (1984) Colony composition, protozoan transfer and some life history characteristics of the woodroach *Cryptocercus punctulatus* Scudder (Dictyoptera: Cryptocercidae). *Behav. Ecol. Sociobiol.* **14**, 273–9.

Nault, L.R., Montgomery, M.E. & Bowers, W.S. (1976) Ant–aphid association: role of aphid alarm pheromone. *Science, N.Y.* **194**, 1349–51.

Nault, L.R., Wood, T.K. & Goff, A.M. (1974) Treehopper alarm pheromones. *Nature, Lond.* **249**, 387–8.

Naumann, N.G. (1975) Swarming behaviour: evidence for communication in social wasps. *Science, N.Y.*. **189**, 642–4.

Norden, B, Batra, S.W.T., Fales, H.M., Hefetz, A. & Shaw, G.J. (1980) *Anthophora* bees: unusual glycerides from maternal Dufour's glands serve as larval food and cell lining. *Science, N.Y.* **207**, 1095–7.

Odhiambo, T.R. (1959) An account of parental care in *Rhinocoris albopilosus* Signoret (Hemiptera–Heteroptera: Reduviidae), with notes on its life history. *Proc. R. Ent. Soc. Lond. (A)*, **34**, 175–85.

Oniki, Y. (1970) Brazilian sphecid wasps in occupied hummingbird nests. *J. Kans. Ent. Soc.* **43**, 354–6.

O'Toole, C. & Raw, A. (1991) *Bees of the world*. Blandford Press, London.

Page Jr, R.E. & Erickson Jr. E.H. (1988) Reproduction by worker honey bees (*Apis mellifera* L.). *Behav. Ecol. Sociobiol.* **23**, 117–26.

Pages, J. (1967) Données sur la biologie de *Dipljapyx humberti* (Grassi). *Rev. Ecol. Biol. Soc.* **4**, 187–281.

Palmer, T.J. (1978) A horned beetle which fights. *Nature, Lond.* **274**, 583–4.

Parker, A.H. (1965) The maternal behaviour of *Pisilus tipuliformis* Fabricius (Hemiptera: Reduviidae). *Entomologia Exp. Appl.* **8**, 13–19.

Pratte, M. & Jeanne, R.L. (1984) Antennal drumming behaviour of *Polistes* wasps (Hymenoptera: Vespidae). *Z. Tierpsychol.* **66**, 177–88.

Pukowski, E. (1933) Ökologische Untersuchungen an *Necrophorus F. Z. Morph. Okol. Tiere*, **27**, 418–86.

Ralston, J.S. (1977) Egg guarding by male assassin bugs of the genus *Zelus* (Hemiptera: Reduviidae). *Psyche*, **84**, 103–7.

Randall, J.B. (1977) New observations of maternal care exhibited by the green lynx spider, *Peucetia viridans* Hentz (Araneida: Oxyopidae). *Psyche*, **84**, 286–91.

Reed, H.C. & Akre, R.D. (1983) Colony behaviour of the obligate social parasite *Vespula austriaca* (Panzer) (Hymenoptera: Vespidae). *Insectes Soc.* **30**, 259–73.

Rettenmeyer, C.W., Topoff, H. & Mirenda, J. (1978) Queen retinues of army ants. *Ann. Ent. Soc. Am.* **71**, 519–28.

Roberts, R.B. (1971) Biology of the crepuscular bee *Ptiloglossa guinnae* n. sp. with notes on associated bees, mites and yeasts. **4**, *J. Kans. Ent. Soc.* **44**, 283–94.

Rodriguez, C.A. & Stella Guerrero, B. (1976) La historia natural y el comportamiento de *Zygopachylus albomarginis* (Chamberlain) (Arachnida, Opiliones: Gonyleptidae). *Biotropica*, **8**, 242–7.

Roeloffs, R. & Riechert, S.E. (1988) Dispersal and population-genetic structure of the cooperative spider, *Agelena consociata*, in West African rainforest. *Evolution*, **42**, 173–83.

Rosenheim, J.A. (1987) Nesting behaviour and bionomics of a solitary ground-nesting wasp, *Ammophila dysmica* (Hymenoptera: Sphecidae): influence of parasite pressure. *Ann. Ent. Soc. Am.* **80**, 739–49.

Ross, K.G. & Matthews, R.W. (1989) New evidence for eusociality in the sphecid wasp *Microstigmus comes*. *Anim. Behav.* **38**, 613–19.

Roth, L.M. (1981) The mother–offspring relationship of some blaberid cockroaches (Dictyoptera: Blattaria: Blaberidae). *Proc. Ent. Soc. Wash.* **83**, 390–98.

Rubink, W.L. & Evans, H.E. (1979) Notes on the nesting behaviour of the bethylid wasp, *Epyris eriogoni* Kiefer, in southern Texas. *Psyche*, **86**, 313–19.

Saito, Y. (1986) Biparental defence in a spider mite (Acari: Tetranychidae) infesting *Sasa* bamboo. *Behav. Ecol. Sociobiol.* **18**, 377–86.

Sakagami, S.F. & Maeta, Y. (1977) Some presumably presocial habits of Japanese *Ceratina* bees, with notes on various social types in Hymenoptera. *Insectes Soc.* **24**, 319–43.

Sato, H. & Imamori, M. (1986) Nidification of an African ball-rolling scarab, *Scarabaeus platynotus* Bates (Coleoptera, Scarabaeidae). *Kontyû*, **54**, 203–7.

Sato, H. & Imamori, M. (1987) Nesting behaviour of a subsocial African ball-roller *Kheper platynotus* (Coleoptera: Scarabaeidae). *Ecol. Ent.* **12**, 415–25.

Schorr, H. (1957) Zur Verhaltensbiologie und Symbiose von *Brachypelta aterrima* Först. (Cydnidae, Heteroptera). *Z. Morph. Ökol. Tiere*, **45**, 561–602.

Schuster, J.L. & Schuster, L.B. (1985) Social behaviour in passalid beetles (Coleoptera: Passalidae): cooperative brood care. *Fla Ent.* **68**, 266–72.

Seeley, T.D. (1979) Queen substance dispersal by messenger workers in honeybee colonies. *Behav. Ecol. Sociobiol.* **5**, 391–415.

Seelinger, G. & Seelinger, U. (1983) On the social organisation, alarm and fighting in the primitive cockroach *Cryptocercus punctulatus* Scudder. *Z. Tierpsychol.* **62**, 315–33.

Seibt, U. & Wickler, W. (1987) Gerontophagy versus cannibalism in the social spiders *Stegodyphus mimosarum* Pavesi and *Stegodyphus dumicola* Pocock. *Anim. Behav.* **35**, 1903–4.

Sites, R.W. & McPherson, J.E. (1982) Life history and laboratory rearing of *Sehirus cinctus cinctus* (Hemiptera: Cydnidae), with descriptions of immature stages. *Ann. Ent. Soc. Am.* **75**, 210–15.

Skaife, S.H. (1979) *African insect life*. Country Life Books, London.

Smith, R.L. (1976) Male brooding behaviour of the water bug *Abedus herberti* (Hemiptera: Belostomatidae). *Ann. Ent. Soc. Am.* **69**, 740–7.

Smith, R.L. (1979) Paternity assurance and altered roles in the mating behaviour of a giant water bug, *Abedus herberti* (Heteroptera: Belostomatidae). *Anim. Behav.* **27**, 716–25.

Strassmann, J.E. (1981) Wasp reproduction and kin selection: reproductive competition and dominance hierarchies among *Polistes annularis* foundresses. *Fla Ent.* **64**, 74–88.

Strassmann, J.E. & Meyer, D.C. (1983) Gerontocracy in the social wasp, *Polistes exclamans*. *Anim. Behav.* **31**, 431–8.

Stuart, R.J. & Alloway, T.M. (1985) Behavioural evolution and domestic degeneration in obligatory slave-making ants (Hymenoptera: Formicidae: Leptothoracini). *Anim. Behav.* **33**, 1080–88.

Tachikawa, S. & Schaefer, C.W. (1985) Biology of *Parastrachia japonensis* (Hemiptera: Pentatomoidea: ?idae). *Ann. Ent. Soc. Am.* **78**, 387–97.

Tallamy, D.W. (1985) 'Egg-dumping' in lace bugs (*Gargaphia solani*, Hemiptera: Tingidae). *Behav. Ecol. Sociobiol.* **17**, 357–62.

Tallamy, D.W. (1986) Age specificity of 'egg dumping' in *Gargaphia solani* (Hemiptera: Tingidae). *Anim. Behav.* **34**, 599–603.

Tallamy, D.W. & Denno, R.F. 1981 Maternal care in *Gargaphia solani* (Hemiptera: Tingidae). *Anim. Behav.* **29**, 771–8.

Tallamy, D.W. & Wood, T.K. (1986) Convergence patterns in subsocial insects. *Ann. Rev. Ent.* **31**, 369–90.

Tanaka, Y. (1985) Alternative manners of prey-carrying in the fossorial wasp, *Oxybelus strandi* YASUMATSU (Hymenoptera, Sphecoidea). *Kontyû*, **53**, 277–83.

Thiele, H.V. (1977) *Carabid beetles in their environment*. Springer Verlag, pp. 76–81.

Topoff, H. & Mirenda, J. (1978) Precocial behaviour of callow workers of the army ant *Neivamyrmex nigrescens*: importance of stimulation by adults during mass recruitment. *Anim. Behav.* **26**, 698–706.

Torchio, P.F. (1989) In-nest biologies and development of immature stages of three *Osmia* species (Hymenoptera: Megachilidae). *Ann. Ent. Soc. Am.* **82**, 599–615.

Torchio, P.F., Trostle, G.E. & Burdick, D.J. (1988) The nesting biology of *Colletes kincaidii* Cockerell (Hymenoptera: Colletidae) and development of its immature forms. *Ann. Ent. Soc. Am.* **81**, 605–25.

Treat, A.E. (1956) Social organization in the moth ear mite (*Myrmonyssus phaloenodectes*). *Proc. 10th. Int. Congr. Ent.* **2**, 475–80.

Trumbo, S.T. (1990) Reproductive benefits of infanticide in a biparental burying beetle *Nicrophorus orbicollis*. *Behav. Ecol. Sociobiol.* **27**, 269–73.

Tsuji, K. (1988) Obligate parthenogenesis and reproductive division of labor in the Japanese queenless ant *Pristomyrmex pungens*. Comparison of intranidal and extranidal workers. *Behav. Ecol. Sociobiol.* **23**, 247–55.

Turillazzi, S., Cervo. R. & Cavallari, I. (1990) Invasion of the nest of *Polistes dominulus* by the social parasite *Sulcopolistes sulcifer* (Hymenoptera, Vespidae). *Ethology*, **84**, 47–59.

Turillazzi, S. & Pardi, L. (1982) Social behaviour of *Parischnogaster nigricans serrei* (Hymenoptera: Vespoidea) in Java. *Ann. Ent. Soc. Am.* **75**, 657–64.

Visscher, P.K. (1983) The honey bee way of death: necrophoric behaviour in *Apis mellifera* colonies. *Anim. Behav.* **31**, 1070–76.

Weaving, A.J.S. (1984) Nesting behaviour of *Ammophila dolichodera* Kohl (Hymenoptera: Sphecidae). *J. Ent. Soc. Sth. Afr.* **47**, 303–8.

West, M.J. & Alexander, R.D. (1963) Sub-social behaviour in a burrowing cricket *Anurogryllus muticus* (De Geer). *Ohio J. Sci.* **63**, 19–25.

West Eberhard, M.J. (1977) The establishment of reproductive dominance in social wasp colonies. *Proc. Int. Congr. Int. Union Study Soc. Insects. 8th, Wagenringen*, pp. 223–7.

Wille, A. (1983) Biology of the stingless bees. *Ann. Rev. Ent.* **28**, 41–64.

Wilson, E.O. (1971) *The insect societies*. Harvard University Press, Cambridge, Mass. 548 pp.

Wilson, E.O. (1974) The soldier of the ant *Camponotus (Colobopsis) fraxinicola* as a trophic caste. *Psyche*, **81**, 182–8.

Windsor, D.M. (1987) Natural history of a subsocial tortoise beetle, *Acromis sparsa* Boheman (Chrysomelidae, Cassidinae) in Panama. *Psyche*, **94**, 127–50.

Wood, T.K. (1974) Aggregating behaviour of *Umbonia crassicornis* (Homoptera: Membracidae). *Can. Ent.* **106**, 169–73.

Wood, T.K. (1976) Biology and presocial behaviour of *Platycotis vittata* (Homoptera: Membracidae). *Ann. Ent. Soc. Am.* **69**, 807–11.

Wood, T.K. (1976a) Alarm behaviour of brooding female *Umbonia crassicornis* (Homoptera: Membracidae). *Ann. Ent. Soc. Am.* **69**, 340–44.

Wood, T.K. (1977) Role of parent females and attendant ants in the maturation of the treehopper, *Entylia bactriana* (Homoptera: Membracidae). *Sociobiology*, **2**, 257–72.

Wood, T.K. (1978) Parental care in *Guayaquila compressa* Walker (Homoptera: Membracidae). *Psyche*, **85**, 135–45.

Wyatt, T.D. (1986) How a subsocial intertidal beetle, *Bledius spectabilis*, prevents flooding and anoxia in its burrow. *Behav. Ecol. Sociobiol.* **19**, 323–31.

Feeding Behaviour

Adis, J. (1984) Eco-entomological observations from the Amazon. V. Feeding habits of Neotropical 'bee killers' and 'resin bugs' (Apiomerinae: Reduviidae: Hemiptera). *Rev. Biol. Trop.* **32**, 151–3.

Akimoto, S. (1981) Gall formation by *Eriosoma* fundatrices and gall parasitism in *Eriosoma yangi* (Homoptera, Pemphigidae). *Kontyû*, **49**, 426–36.

Akre, R.D. & Bruce Hill, W. (1973) Behavior of *Adranes taylori*, a myrmecophilous beetle associated with *Lasius sitkaensis* in the Pacific Northwest. *J. Kans. Ent. Soc.* **46**, 526–36.

Akre, R.D., Garnett, W. B. & Zack, R.S. (1988) Biology and behavior of *Microdon piperi* in the Pacific Northwest (Diptera: Syrphidae). *J. Kans. Ent. Soc.* **61**, 441–52.

Alcock, J. (1974) The behaviour of *Philanthus crabroniformis* (Hymenoptera: Sphecidae). *J. Zool. Lond.* **173**, 233–46.

Alpert, G.D. & Akre, R.D. (1973) Distribution, abundance and behavior of the inquiline ant *Leptothorax diversipilosus*. *Ann. Ent. Soc. Am.* **66**, 753–60.

Alpert, G.D. & Ritcher, P.O. (1975) Notes on the life cycle and myrmecophilous adaptations of *Chremastocheilus armatus* (Coleoptera: Scarabeidae). *Psyche*, **82**, 283–91.

Ambrose, D.P. & Livingstone, D. (1979) On the bioecology of *Lophocephala guerini* Lap. (Reduviidae: Harpactorinae) a coprophagous reduviid from the Palghat Gap, India. *J. Nat. Hist.* **13**, 581–8.

Austin, A.D. & Blest, A.D. (1979) The biology of two Australian species of dinopid spider. *J. Zool. Lond.* **189**, 145–56.

Bänziger, H. & Fletcher, D.S. (1985) Three new zoophilous moths of the genus *Scopula* (Lepidoptera: Geometridae) from South-east Asia. *J. Nat. Hist.* **19**, 851–60.

Barber, J.T, Ellgaard, E.G., Thien, L.B. & Stack, A.E. (1989) The use of tools for food transportation by the imported fire ant *Solenopsis invicta*. *Anim. Behav.* **38**, 550–52.

Batra, L.R. & Batra, S.W.T. (1966) Fungus-growing termites of tropical India and associated fungi. *J. Kans. Ent. Soc.* **39**, 725–38.

Batra, S.W.T. (1972) Notes on the behavior and ecology of the mantispid, *Climaciella brunnea occidentalis*. *J. Kans. Ent. Soc.* **45**, 334–40.

Benson, W.W., Brown, K.S, & Gilbert, L.E. (1976) Coevolution of plants and herbivores: passion flower butterflies. *Evolution*, **29**, 659–80.

Bleckmann, H. & Lotz, T. (1987) The vertebrate-catching behaviour of the fishing spider *Dolomedes triton* (Araneae, Pisauridae). *Anim. Behav.* **35**, 641–51.

Boggs, C.L., Smiley, J.T. & Gilbert, L.E. (1981) Patterns of pollen exploitation by *Heliconius* butterflies. *Oecologia*, **48**, 284–9.

Bornemissza, G.F. (1966) Observations on the hunting and mating behaviour of two species of scorpion flies (Bittacidae: Mecoptera). *Aust. J. Zool.* **14**, 371–82.

Bradoo, B.L. (1980) Feeding behaviour and recruitment display in the social spider *Stegodyphus sarasinorum* Karsch (Araneae, Eresidae). *Tjidschr. v. Ent.* **123**, 89–104.

Brownell, P. & Farley, R.D. (1979) Prey-localizing behaviour of the nocturnal desert scorpion, *Paruroctonus mesaensis*: orientation to substrate vibrations. *Anim. Behav.* **27**, 185–93.

Bub, K. & Bowerman, R.F. (1979) Prey capture by the scorpion *Hadrurus arizonensis* Ewing (Scorpiones: Vaejovidae). *J. Arachnol.* **7**, 243–53.

Buchmann, S.L. (1980) Preliminary anthecological observations on *Xiphidium caeruleum* Aubl. (Monocotyledoneae: Haemodoraceae) in Panama. *J. Kans. Ent. Soc.* **53**, 685–99.

Buchmann, S.L. & Buchmann, M.D. (1981) Anthecology of *Mouriri myrtilloides* (Melastomataceae: Memecyleae), an oil flower in Panama. *Reprod. Bot.* **7**, 7–24.

Burgess, J. W. (1979) Web-signal processing for tolerance and group predation in the social spider *Mallos gregalis* Simon. *Anim. Behav.* **27**, 157–64.

Büttiker, W. & Bezuidenhout, J.D. (1974) First records of eye-frequenting Lepidoptera from South West Africa. *J. Ent. Soc. Sth. Afr.* **37**, 73–8.

Cammaerts, R., Detrain, C. & Cammaerts, M.-C. (1990) Host trail following by the myrmecophilous beetle *Edaphopaussus favieri* (Fairmaire) (Carabidae Paussinae). *Insectes Soc.* **37**, 200–211.

Carico, J.E. (1978) Predatory behaviour in *Euryopis funebris* (Hentz) (Araneae: Theridiidae) and the evolutionary significance of web reduction. *Symp. Zool. Soc. Lond.* **42**, 51–8.

Casper, G.S. (1985) Prey capture and stinging behavior in the emperor scorpion, *Pandinus imperator* (Koch) (Scorpiones, Scorpionidae). *J. Arachnol.* **13**, 277–83.

Chadab, R. & Rettenmeyer, C.W. (1975) Mass recruitment by army ants. *Science, N.Y.* **188**, 1124–5.

Christenson, T.E. (1984) Behaviour of colonial and solitary spiders of the theridiid species *Anelosimus eximius*. *Anim. Behav.* **32**, 725–34.

Claassens, A.J.M. (1976) Observations on the myrmecophilous relationships and the parasites of *Lepidochrysops methymna methymna* (Trimen) and *L. trimeni* (Bethune-Baker) (Lepidoptera: Lycaenidae). *J. Ent. Soc. Sth. Afr.* **39**, 279–89.

Clyne, D. (1973) Notes on the web of *Poecilopachys australasia* (Griffin & Pidgeon, 1833) (Araneida: Argiopidae). *Aust. Ent. Mag.* **1**, 23–9.

Coddington, J. & Valerio, C.E. (1980) Observations on the web and behavior of *Wendilgarda* spiders (Araneae: Theridiosomatidae). *Psyche*, **87**, 93–105.

Collins, N.M. (1979) Observations on the foraging activity of *Hospitalitermes umbrinus* (Haviland), (Isoptera: Termitidae) in the Gunong Mulu National Park, Sarawak. *Ecol. Ent.* **4**, 231–8.

Corbet, P.S. & Miller, P.L. (1991) 'Accompanying' behaviour as a means of prey acquisition by *Brachythemis leucosticta* (Burmeister) and other Anisoptera. *Odonatologica*, **20**, 29–36.

Coyle, F.A. (1981) Notes on the behaviour of *Ummidia* trapdoor spiders (Araneae, Ctenizidae): burrow construction, prey capture and the functional morphology of the peculiar hind tibia. *Bull. Br. Arachnol. Soc.* **5**, 159–65.

Craig, C.L. & Bernard, G.D. (1990) Insect attraction to ultraviolet-reflecting spider webs and web decorations. *Ecology*, **71**, 616–23.

Curtis, B.A. (1985) The dietary spectrum of the Namib Desert dune ant *Camponotus detritus*. *Insectes Soc.* **32**, 78–85.

Davis, R.M. & Russell, M.P. (1969) Commensalism between *Ranzovius moerens* (Reuter) (Hemiptera: Miridae) and *Hololena curta* (McCook) (Araneida: Agelenidae). *Psyche*, **76**, 262–9.

DeVries, P.J. (1979) Pollen-feeding rainforest *Parides* and *Battus* butterflies in Costa Rica. *Biotropica*, **11**, 237–8.

DeVries, P.J. (1988) The use of epiphylls as larval hostplants by the neotropical riodinid butterfly, *Sarota gyas*. *J. Nat. Hist.* **22**, 1447–50.

Deyrup, M.A. (1988) Pollen-feeding in *Poecilognathus punctipennis* (Diptera: Bombyliidae). *Fla Ent.* **71**, 597–605.

Dillon, P.M, Lowrie, S. & McKey, D. (1983) Disarming the 'evil woman': petiole constriction by a sphingid larva circumvents mechanical defenses of its host plant *Cnidoscolus urens* (Euphorbiaceae). *Biotropica*, **15**, 112–16.

Dorow, W.H.O. & Maschwitz, U. (1990) The *arachne*-group of *Polyrhachis* (Formicidae, Formicinae): Weaver ants cultivating Homoptera on bamboo. *Insectes Soc.* **37**, 73–89.

Eberhard, W.G. (1970) The natural history of the fungus gnats *Leptomorphus bifasciatus* (Say) and *L. subcaeruleus* (Coquillett) (Diptera: Mycetophilidae). *Psyche*, **77**, 361–83.

Eberhard, W.G. (1977) Aggressive chemical mimicry by a bolas spider. *Science, N.Y.* **198**, 1173–5.

Eberhard, W.G. (1979) *Argyrodes attenuatus* (Theridiidae): a web that is not a snare. *Psyche*, **86**, 407–13.

Eberhard, W.G. (1980) The natural history and behavior of the bolas spider *Mastophora dizzydeani* sp. n. (Araneidae). *Psyche*, **87**, 143–69.

Eberhard, W.G. (1981) Notes on the natural history of *Taczanowskia* sp. (Araneae: Araneidae). *Bull. Br. Arachnol. Soc.* **5**,175–6.

Eberhard, W.G. (1981a) The single line web of *Phoroncidia studo* Levi (Araneae: Theridiidae): a prey attractant? *J. Arachnol.* **9**, 229–32.

Eisner, T., Kafatos, F.C. & Linsley, E.G. (1962) Lycid predation by mimetic adult cerambycidae (Coleoptera). *Evolution*, **16**, 316–24.

Feener Jr, D.H. (1988) Effects of parasites on foraging and defense behavior of a termitophagous ant, *Pheidole titanis* Wheeler (Hymenoptera:

Formicidae). *Behav. Ecol. Sociobiol.* **22**, 421–7.

Fiedler, K. & Maschwitz, U. (1989) The symbiosis between the weaver ant, *Oecophylla smaragdina*, and *Anthene emolus*, an obligate myrmecophilous lycaenid butterfly. *J. Nat. Hist.* **23**, 833–46.

Filia, B. & Maschwitz, U. (1990) Studies on the South East Asian ant-plant association *Crematogaster borneensis/Macaranga*: adaptations of the ant partner. *Insectes Soc.* **37**, 212–31.

Fitzgerald, T.D. & Peterson, S.C. (1983) Elective recruitment by the eastern tent caterpillar (*Malacosoma americanum*). *Anim. Behav.* **31**, 417–23.

Ford, M.J. (1978) Locomotory activity and the predation strategy of the wolf spider *Pardosa amentata* (Clerck) (Lycosidae). *Anim. Behav.* **26**, 31–5.

Forsyth, A. & Alcock, J. (1990) Ambushing and prey-luring as alternative foraging tactics of the fly-catching rove beetle *Leistotrophus versicolor* (Coleoptera: Staphylinidae). *J. Insect Behav.* **3**, 703–18.

Fowler, H.G. (1981) Behavior of two myrmecophiles of Paraguayan leaf-cutting ants. *Revta Chil. Ent.* **11**, 69–72.

Gilbert, L.E. (1972) Pollen feeding and reproductive biology of *Heliconius* butterflies. *Proc. Natn. Acad. Sci. U.S.A.* **69**, 1403–7.

Gilbert, L.E. (1976) Adult resources in butterflies: African lycaenid *Megalopalpus* feeds on larval nectary. *Biotropica*, **8**, 282–3.

Gotwald Jr, W.H. & Levieux, J. (1972) Taxonomy and biology of a new West African ant belong to the genus *Amblyopone* (Hymenoptera: Formicidae). *Ann. Ent. Soc. Am.* **65**, 383–96.

Greenfield, M.D., Shelly, T.E. & Gonzalez-Coloma, A. (1989) Territory selection in a desert grasshopper: the maximization of conversion efficiency on a chemically defended shrub. *J. Anim. Ecol.* **58**, 761–71.

Griswold, C.E. (1983) *Tapinillus longipes* (Taczanowski), a web-building lynx spider from the American tropics (Araneae: Oxyopidae). *J. Nat. Hist.* **17**, 979–85.

Gwynne, D.T. (1979) Nesting biology of the spider wasps (Hymenoptera: Pompilidae) which prey on burrowing wolf spiders (Araneae: Lycosidae, *Geolycosa*). *J. Nat. Hist.* **13**, 681–92.

Hackwell, G.A. (1973) Biology of *Lasconotus subcostulatus* (Coleoptera: Colydiidae) with special reference to feeding behavior. *Ann. Ent. Soc. Am.* **66**, 62–5.

Hawkins, B.A. & Goeden, R.D. (1982) Biology of a gall-forming *Tetrastichus* (Hymenoptera: Eulophidae) associated with gall midges on saltbush in southern California. *Ann. Ent. Soc. Am.* **75**, 444–7.

Henderson, G. & Akre R.D. (1986) Biology of the myrmecophilous cricket, *Myrmecophila manni* (Orthoptera: Gryllidae). *J. Kans. Ent. Soc.* **53**, 454–67.

Henry, C.S. (1977) The behavior and life histories of two North American ascalaphids. *Ann. Ent. Soc. Am.* **70**, 179–95.

Hölldobler, B. (1970) *Steatoda fulva* (Theridiidae), a spider that feeds on harvester ants. *Psyche*, **77**, 202–8.

Hölldobler, B. (1981) Foraging and spatiotemporal territories in the honey ant *Myrmecocystus mimicus* Wheeler (Hymenoptera: Formicidae). *Behav. Ecol. Sociobiol.* **9**, 301–14.

Hölldobler, B. (1982) Communication, raiding behavior and prey storage in *Cerapachys* (Hymenoptera: Formicidae). *Psyche*, **89**, 3–23.

Holloway, B.A. (1976) Pollen-feeding in hoverflies (Diptera: Syrphidae). *N. Z. J. Zool.* **3**, 339–50.

Howard, R.W. (1978) Proctodeal feeding by termitophilous Staphylinidae associated with *Reticulotermes virginicus* (Banks). *Science, N.Y.* **201**, 541–3.

Hulley, P.E. (1988) Caterpillar attacks plant mechanical defence by mowing trichomes before feeding. *Ecol. Ent.* **13**, 239–41.

Itô, Y. & Hattori, I. (1982) A kleptoparasitic moth, *Nola innocua*, attacking aphid galls. *Ecol. Ent.* **7**, 475–8.

Jackson, J.F. (1974) Goldschmidt's dilemma resolved: notes on the larval behavior of a new Neotropical web-spinning mycetophilid (Diptera). *Am. Midl. Nat.* **92**, 240–45.

Jackson, R.R. (1985) The biology of *Simaetha paetula* and *S. thoracica*, web-building jumping spiders (Araneae, Salticidae) from Queensland: cohabitation with social spiders, utilization of silk, predatory behaviour and intraspecific interactions. *J. Zool. Lond. (B)*, **1**, 175–210.

Jackson, R.R. (1987) The biology of *Olios* spp., huntsman spiders (Araneae, Sparassidae) from Queensland and Sri Lanka: predatory behaviour and cohabitation with social spiders. *Bull. Br. Arachnol. Soc.* **7**, 133–6.

Jackson, R.R. & Blest, A.D. (1982) The biology of *Portia fimbriata*, a web-building jumping spider (Araneae, Salticidae) from Queensland: utilization of webs and predatory versatility. *J. Zool. Lond.* **196**, 255–93.

Jackson, R.R. & Brassington, R.J. (1987) The biology of *Pholcus phalangioides* (Araneae, Pholcidae): predatory versatility, araneophagy and aggressive mimicry. *J. Zool. Lond.* **211**, 227–38.

Jackson, R.R. & Pollard, S.D. (1990) Web-building and predatory behaviour of *Spartaeus spinimanus* and *Spartaeus thailandicus*, primitive jumping spiders (Araneae, Salticidae) from South-east Asia. *J. Zool. Lond.* **220**, 561–7.

Jackson, R.R. & Van Olphen, A. (1991) Prey-capture techniques and prey preferences of *Corythalia canosa* and *Pystira orbiculata*, ant-eating jumping spiders (Araneae, Salticidae). *J. Zool. Lond.* **223**, 577–91.

Jackson, R.R. & Whitehouse, M.E.A. (1986) The biology of New Zealand and Queensland pirate spiders (Araneae, Mimetidae): aggressive mimicry, araneophagy and prey specialization. *J. Zool. Lond.* **210**, 279–303.

Jackson, R.R. & Wilcox, R.S. (1990) Aggressive mimicry, prey-specific predatory behaviour and predator-recognition in the predator–prey interactions of *Portia fimbriata* and *Euryattus* sp., jumping spiders from Queensland. *Behav. Ecol. Sociobiol.* **26**, 111–19.

Jaffe, K. & Howse, P.E. (1979) The mass recruitment system of the leaf cutting ant, *Atta cephalotes* (L.). *Anim. Behav.* **27**, 930–39.

Jander, R. & Daumer, K. (1974) Guide-line and gravity orientation of blind termites foraging in the open (Termitidae: Macrotermes, Hospitalitermes). *Insectes Soc.* **21**, 45–69.

Janzen, D.H. (1966) Coevolution of mutualism between ants and acacias in Central America. *Evolution*, **20**, 249–75.

Janzen, D.H. (1972) Protection of *Barteria* (Passifloraceae) by *Pachysmia* ants (Pseudomyrmecinae) in a Nigerian rain forest. *Ecology*, **53**, 885–92.

Janzen, D.H. (1975) *Pseudomyrmex nigripilosa*: a parasite of mutualism. *Science, N.Y.* **188**, 936–37.

Jarman, E.A.R. & Jackson, R.R. (1986) The biology of *Taieria erebus* (Araneae, Gnaphosidae), an araneophagic spider from New Zealand; silk utilisation and predatory versatility. *N. Z. J. Zool.* **13**, 521–41.

Jones, H.D., Darlington, J.P.E.C. & Newson, R.M. (1990) A new species of land planarian preying on termites in Kenya (Platyhelminthes: Turbellaria: Tricladida: Terricola). *J. Zool. Lond.* **220**, 249–56.

Kirk, W.D.J. (1984) Pollen-feeding in thrips (Insecta: Thysanoptera). *J. Zool. Lond.* **204**, 107–17.

Kistner, D.H. (1973) The termitophilous Staphylinidae associated with *Grallatotermes* in Africa; their taxonomy, behavior, and a survey of their glands of external secretion. *Ann. Ent. Soc. Am.* **66**, 197–222.

Kitching, R.L. (1987) Apsects of the natural history of the lycaenid butterfly *Allotinus major* in Sulawesi. *J. Nat. Hist.* **21**, 535–44.

Klemperer, H.G. (1980) Kleptoparasitic behaviour of *Aphodius rufipes* (L.) larvae in nests of *Geotrupes spiniger* Marsh. (Coleoptera, Scarabeidae). *Ecol. Ent.* **5**, 143–51.

Kurczewski, F.E. (1966) Behavioral notes on two species of *Tachytes* that hunt pygmy mole-crickets. *J. Kans. Ent. Soc.* **39**, 147–55.

Kurczewski, F.E. & Kurczewski, E.J. (1984) Mating and nesting behavior of *Tachytes intermedius* (Viereck) (Hymenoptera: Sphecidae). *Proc. Ent. Soc. Wash.* **86**, 176–84.

Kurczewski, F.E. & Spofford, M.G. (1986) Observations on the nesting behaviors of *Tachytes parvus* Fox and *T. obductus* Fox (Hymenoptera: Sphecidae). *Proc. Ent. Soc. Wash.* **88**, 13–24.

Lawrence, E.G. (1976) Adult resources in butterflies: African lycaenid *Megalopalpus* feeds on larval nectary. *Biotropica*, **8**, 282–3.

Leuthold, R.H., Badertscher, S. & Imboden, H. (1989) The inoculation of newly formed fungus comb with *Termitomyces* in *Macrotermes colonies* (Isoptera, Macrotermitinae). *Insectes Soc.* **36**, 328–38.

Lin, N. (1978) Defended hunting territories and hunting behavior of females of *Philanthus gibbosus* (Hymenoptera: Sphecidae). *Proc. Ent. Soc. Wash.* **80**, 234–9.

Lloyd, J.E. (1975) Aggressive mimicry in *Photuris* fireflies: signal repertoires by femmes fatales. *Science, N.Y.* **187**, 452–3.

Lloyd, J.E. & Wing, S.R. (1983) Nocturnal aerial predation of fireflies by light-seeking fireflies. *Science, N.Y.* **222**, 634–5.

Longhurst, C., Baker, R. & Howse, P.E. (1979) Termite predation by *Megaponera foetens* (Fab.) (Hymenoptera: Formicidae). *J. Chem. Ecol.* **5**, 703–19.

McMahan, E.A. (1983) Adaptations, feeding preferences, and biometrics of a termite-baiting assassin bug (Hemiptera: Reduviidae). *Ann. Ent. Soc. Am.* **76**, 483–6.

Main, B.Y. (1982) Notes on the reduced web, behaviour and prey of *Arcys nitidiceps* Simon (Araneidae) in south western Australia. *Bull. Br. Arachnol. Soc.* **5**, 425–32.

Maschwitz, U., Dumpert, K. & Tuck, K.R. (1986) Ants feeding on anal exudate from tortricid larvae: a new type of trophobiosis. *J. Nat. Hist.* **20**, 1041–50.

Maschwitz, U., Fiala, B. & Dolling, W.R. (1987) New trophobiotic symbioses of ants with South East Asian bugs. *J. Nat. Hist.* **21**, 1097–1107.

Maschwitz, U. & Hänel, H. (1985) The migrating herdsman *Dolichoderus (Diabolus) cuspidatus*: an ant with a novel mode of life. *Behav. Ecol. Sociobiol.* **17**, 171–84.

Masuko, K. (1986) Larval haemolymph feeding: a nondestructive parental cannibalism in the primitive ant *Amblyopone silvestrii* Wheeler (Hymenoptera: Formicidae). *Behav. Ecol. Sociobiol.* **19**, 249–56.

Masuko, K. (1989) Larval haemolymph feeding in the ant *Leptanilla japonica* by use of a specialised duct organ the 'larval haemolymph tap' (Hymenoptera: Formicidae). *Behav. Ecol. Sociobiol.* **24**, 127–32.

Mill, A.E. (1984) Predation by the ponerine ant *Pachycondyla commutata* on termites of the genus *Syntermes* in Amazonian rain forest. *J. Nat. Hist.* **18**, 405–10.

Moffet, M.W. (1984) Swarm raiding in a myrmicine ant. *Naturwiss.* **71**, 588–90.

Nentwig, W. (1985) Social spiders catch large prey: a study of *Anelosimus eximius* (Araneae: Theridiidae). *Behav. Ecol. Sociobiol.* **17**, 79–85.

Oliveira, P.S. & Sazima, I. (1984) The adaptive bases of ant-mimicry in a neotropical aphantochilid spider (Araneae: Aphantochilidae). *Biol. J. Linn. Soc.* **22**, 145–55.

Opell, B.D. (1990) Material investment and prey capture potential of reduced spider webs. *Behav. Ecol. Sociobiol.* **26**, 375–81.

Petrunkevitch, A. (1952) The spider and the wasp. *Sci. Am.* **187**, 20–23.

Pierce, N.E. & Elgar, M.A. (1985) The influence of ants on host plant selection by *Jalmenus evagoras*, a myrmecophilous lycaenid butterfly. *Behav. Ecol. Sociobiol.* **16**, 209–22.

Polis, G.A. (1979) Prey and feeding phenology of the desert sand scorpion *Paruroctonus mesaensis* (Scorpionidae: Vaejovidae). *J. Zool. Lond.* **188**, 333–46.

Porter, S.D. & Eastmond, D.A. (1982) *Euryopis coki* (Theridiidae), a spider that preys on *Pogonomyrmex* ants. *J. Arachnol.* **10**, 275–7.

Rathcke, B.J. & Poole, R.W. (1975) Coevolutionary race continues: butterfly larval adaptation to plant trichomes. *Science, N.Y.* **187**, 175–6.

Ray, T.S. & Andrews, C.C. (1980) Antbutterflies: butterflies that follow army ants to feed on antbird droppings. *Science, N.Y.* **210**, 1147–8.

Redborg, K.E. (1982) Interference by the mantispid *Mantispa uhleri* with the development of the spider *Lycosa rabida*. *Ecol. Ent.* **7**, 187–96.

Redborg, K.E. & Macleod, E.G. (1983) *Climaciella brunnea* (Neuroptera: Mantispidae): a mantispid that obligately boards spiders. *J. Nat. Hist.* **17**, 63–73.

Renner, S. (1983) The widespread occurrence of anther destruction by *Trigona* bees in Melastomataceae. *Biotropica*, **15**, 251–6.

Rentz, D.C. (1970) An observation of the feeding behavior of a sand-treader cricket. *Ent. News* **81**, 289–91.

Richards, A.M. (1962) Feeding behaviour and enemies of Rhaphidophoridae (Orthoptera) from Waitomo Caves, New Zealand. *Trans. R.. Soc. N.Z.* **2**, 121–9.

Robinson, M.H. (1977) Symbiosis between insects and spiders: an association between lepidopteran larvae and the social spider *Anelosimus eximius* (Araneae: Theridiidae). *Psyche*, **84**, 225–32.

Robinson, M.H. & Robinson, B. (1971) The predatory behaviour of the ogre-faced spider *Dinopis longipes* F. Cambridge (Araneae: Dinopidae). *Am. Midl. Nat.* **85**, 85–96.

Robinson, M.H. & Robinson, B. (1975) Evolution beyond the orb web: the web of the araneid spider *Pasilobus* sp., its structure, operation and construction. *Zool. J. Linn. Soc.* **56**, 302–14.

Robinson, M.H. & Robinson, B. (1977) Associations between flies and spiders: bibiocommensalism and dipsoparasitism. *Psyche*, **84**, 150–57.

Robinson, M.H. & Valerio, C.E. (1977) Attacks on large or heavily defended prey by tropical salticid spiders. *Psyche*, **84**, 1–10.

Roble, S.M. (1985) Submergent capture of *Dolomedes triton* (Araneae, Pisauridae) by *Anoplius depressipes* (Hymenoptera, Pompilidae). *J. Arachnol.* **13**, 391–2.

Roubik, D.W. (1982) Obligate necrophagy in a social bee. *Science, N.Y.* **217**, 1059–60.

St J. Read, V.M. & Hughes, R.N. (1987) Feeding behaviour and prey choice in *Macroperipatus torquatus* (Onychophora). *Proc. R. Soc. Lond.* (*B*) **230**, 483–506.

Schlising, R.A. (1972) Foraging and nest provisioning behavior of the oligolectic bee, *Diadasia bituberculata*. *Pan-Pacif. Ent.* **48**, 175–88.

Schuster, J.C. (1981) Stingless bees attending honeydew-producing tree-hoppers in Guatemala. *Fla Ent.* **64**, 192.

Seeley, T.D. (1989) Social foraging in honey bees: how nectar foragers assess their colony's nutritional status. *Behav. Ecol. Sociobiol.* **24**, 181–99.

Sivinski, J. (1982) Prey attraction by luminous larvae of the fungus gnat *Orfelia fultoni*. *Ecol. Ent.* **7**, 443–6.

Sivinski, J. & Stowe, M. (1980) A kleptoparasitic cecidomyiid and other flies associated with spiders. *Psyche*, **87**, 337–48.

Smith, D.R. (1982) Reproductive success of solitary and communal *Philoponella oweni* (Araneae: Uloboridae). *Behav. Ecol. Sociobiol.* **11**, 149–54.

Snoddy, E.L., Humphreys, W.J. & Blum, M.S. (1976) Observations on the behavior and morphology of the spider predator *Stenolemus lanipes* (Hemiptera: Reduviidae). *J. Georgia Ent. Soc.* **11**, 55–8.

Stout, J. (1979) An association of an ant, a mealy bug, and an understorey tree from a Costa Rican rain forest. *Biotropica*, **11**, 309–11.

Stowe, M.K. (1978) Observations of two nocturnal orbweavers that build specialized webs: *Scoloderus cordatus* and *Wixia ectypa* (*Araneae: Araneidae*). *J. Arachnol.* **6**, 141–6.

Tanaka, K. (1984) Rate of predation by a kleptoparasitic spider, *Argyrodes fissifrons*, upon a large host spider, *Agelena limbata*. *J. Arachnol.* **12**, 363–7.

Thornhill, R. (1975) Scorpionflies as kleptoparasites of web-building spiders. *Nature, London*, **258**, 709–11.

Thornhill, R. (1977) The comparative predatory and sexual behaviour of hangingflies (Mecoptera: Bittacidae). *Occ. Pap, Mus. Zool. Univ. Mich.* No. 677, 1–49.

Thornhill, R. (1978) Sexually selected predatory and mating behaviour of the hangingfly, *Bittacus stigmaterus* (Mecoptera: Bittacidae). *Ann. Ent. Soc. Am.* **71**, 597–601.

Thorp, R.W. & Briggs, D.L. (1980) Bees collecting pollen from other bees (Hymenoptera: Apoidea). *J. Kans. Ent. Soc.* **53**, 166–70.

Thorp, R.W. & Estes, J.R. (1975) Intrafloral behaior of bees on flowers of *Cassia fasciculata*. *J. Kans. Ent Soc.* **48**, 166–70.

Topoff, H.R. (1969) A unique predatory association between carabid beetles of the genus *Helluomorphoides* and colonies of the army ant *Neivamyrmex nigrescens*. *Psyche*, **76**, 375–81.

Uetz, G.W., Kane, T.C. & Stratton, G.E. (1982) Variation in the social grouping tendency of a communal web-building spider. *Science, N.Y.* **217**, 547–9.

Van Berkum, F.H. (1982) Natural history of a tropical, shrimp-eating spider (Pisauridae). *J. Arachnol.* **10**, 117–121.

Vasconcellos-Neto J. & Lewinsohn, T.M. (1984) Discrimination and release of unpalatable butterflies by *Nephila clavipes*, a neotropical orbweaving spider. *Ecol. Ent.* **9**, 337–44.

Vogel, S. (1981) Abdominal oil-mopping – a new type of foraging in bees. *Naturwiss.* **68**, 627–8.

Vogel, S. & Michener, C.D. (1985) Long bee legs and oil-producing floral spurs, and a new *Rediviva* (Hymenoptera, Melittidae; Scrophulariaceae). *J. Kans. Ent. Soc.* **58**, 359–64.

Vollrath, F. (1978) A close relationship between two spiders (Arachnida, Araneidae): *Curimagua bayano* synecious on a *Diplura species*. *Psyche*, **85**, 347–53.

Vollrath, F. (1979) Behaviour of the kleptoparasitic spider *Argyrodes elevatus* (Araneae, Theridiidae). *Anim. Behav.* **27**, 515–21.

Waller, D.A. (1980) Leaf-cutting ants and leaf-riding flies. *Ecol. Ent.* **5**, 305–6.

Ward, P.I. & Enders, M.M. (1985) Conflict and cooperation in the group feeding of the social spider *Stegodyphus mimosarum*. *Behaviour*, **94**, 167–82.

Weber, N. (1972) The fungus culturing behavior of ants. *Am. Zoo.* **12**, 577–87.

Webster, R.P. & Nielsen, M.C. (1984) Myrmecophily in the Edward's hairstreak butterfly *Satyrium edwardsii* (Lycaenidae). *J. Lepid. Soc.* **38**, 124–33.

Weeks, L. & Hespenheide, H.A. (1985) Predatory and mating behaviour of *Stichopogon* (Diptera: Asilidae) in Arizona. *Pan-Pacific Ent.* **61**, 95–104.

Wharton, R.A. (1986) Biology of the diurnal *Metasolpuga picta* (Kraepelin) (Solifugae, Solpugidae) compared with that of nocturnal species. *J. Arachnol.* **14**, 363–84.

Wheeler Jr., A.G. & McCaffrey, J.P. (1984) *Ranzovius contubernalis*: seasonal history, habits, and description of fifth instar, with speculation on the origin of spider commensalism in the genus *Ranzovius* (Hemiptera: Miridae). *Proc. Ent. Soc. Wash.* **86**, 68–80.

Whitehouse, M.E.A. (1986) The foraging behaviours of *Argyrodes antipodiana* (Theridiidae), a kleptoparasitic spider from New Zealand. *N. Z. J. Zool.* **13**, 151–68.

Whitehouse, M.E.A. (1987) 'Spider eat spider': the predatory behavior of *Rhomphaea* sp. from New Zealand. *J. Arachnol* **15**, 355–62.

Wille, A. (1983) Biology of the stingless bees. *Ann. Rev. Ent.* **28**, 41–64.

Williams, T. & Franks, N.R. (1988) Population size and growth rate, sex ratio and behaviour in the ant isopod, *Platyarthrus hoffmannseggi*. *J. Zool., Lond.* **215**, 703–17.

Willmer, P.G. (1986) Foraging patterns and water balance problems of optimization of a xerophilic bee, *Chalcidoma sicula*. *J. Anim. Ecol.* **55**, 941–62.

Wing, K. (1983) *Tutelina similis* (Araneae: Salticidae): an ant mimic that feeds on ants. *J. Kans. Ent. Soc.* **56**,55–8.

Young, A.M. (1980) Feeding and oviposition in the giant tropical damselfly *Megaloprepus coerulatus* (Drury) in Costa Rica. *Biotropica*, **12**, 237–9.

Defensive Behaviour

Abushama, F.T.E. (1972) The repugnatorial gland of the grasshopper *Poecilocerus hieroglyphicus* (Klug). *J. Ent.* **47**, 95–100.

Alpert, G.D. & Ritcher, P.O. (1975) Notes on the life cycle and myrmecophilous adaptations of *Cremastocheilus armatus* (Coleoptera: Scarabaeidae). *Psyche*, **82**, 283–91.

Aoki, S., Akimoto, S. & Yamane, S. (1981) Observation on *Pseudoregma alexanderi* (Homoptera: Pemphigidae), an aphid species producing pseudoscorpion-like soldiers on bamboos. *Kontyû*, **49**, 355–66.

Aoki, S. & Kurosu, U. (1989) A bamboo horned aphid attacking other insects with its stylets. *Jap. J. Ent.* **57**, 663–5.

Aoki, S., Yamane, S. & Kiuchi, M. (1977) On the biters of *Astegopteryx styracicola* (Homoptera, Aphidoidea). *Kontyû*, **45**, 563–70.

Benfield, E.F. (1974) Autohaemorrhage in two stoneflies (Plecoptera) and its effectiveness as a defense mechanism. *Ann. Ent. Soc. Am.* **67**, 739–42.

Blondheim, S.A. & Frankenberg, E. (1983) 'Protest' sounds of a grasshopper: predator-deterrent signal? *Psyche*, **90**, 387–92.

Buchler, E.R., Wright, T.B. & Brown, E.D. (1981) On the functions of stridulation by the passalid beetle *Odontotaenius disjunctus* (Coleoptera: Passalidae). *Anim. Behav.* **29**, 483–6.

Cruz, Y.P. (1986) The defender role of the precocious larvae of *Copidosomopsis tanytmemus* Caltagirone (Encyrtidae, Hymenoptera). *J. Exp. Zool.* **237**, 309–18.

Damman, H. (1986) The osmaterial glands of the swallowtail butterfly *Eurytides marcellus* as a defence against natural enemies. *Ecol. Ent.* **11**, 261–5.

Deshefy, G.S. (1981) 'Sailing' behaviour in the fishing spider *Dolomedes triton* (Walckenaer). *Anim. Behav.* **29**, 965–6.

Dettner, K., Schwingler, G. & Wunderle, P. (1985) Sticky secretion from two pairs of defensive glands of rove beetle *Deleaster dichrous* (Grav.) (Coleoptera: Staphylinidae). *J. Chem. Ecol.* **11**, 859–83.

Droual, R. (1984) Anti-predator behaviour in the ant *Pheidole desertorum*: the importance of multiple nests. *Anim. Behav.* **32**, 1054–58.

Edmunds, M. (1972) Defensive behaviour in Ghanaian praying mantids. *Zool. J. Linn. Soc.* **51**, 1–32.

Eisner, T. (1958) The protective role of the spray mechanism of the bombardier beetle *Brachynus ballistarius* Lec. *J. Insect Physiol.* **2**, 215–20.

Eisner, T. & Aneshansley, D.J. (1982) Spray aiming in bombadier beetles: jet deflection by the Coanda effect. *Science, N. Y.* **215**, 83–5.

Eisner, T., Johnessee, J.S., Carrel, J., Hendry, L.B. & Meinwald, J. (1974) Defensive use by an insect of a plant resin. *Science, N.Y.* **184**, 996–9.

Eisner, T., Jutro, P., Aneshansley, D.J. & Niedhauk, R. (1972) Defense against ants in a caterpillar that feeds on ant-guarded scale insects. *Ann. Ent. Soc. Am.* **65**, 987–8.

Eisner, T. & Silberglied, R.E. (1988) A chrysopid larva that cloaks itself in mealybug wax. *Psyche*, **95**, 15–19.

Feener, Jr., D.H & Moss, K.A.G. (1990) Defense against parasites by hitchhikers in leaf-cutting ants: a quantitative assessment. *Behav. Ecol. Sociobiol.* **26**, 17–29.

Fink, L.S. (1984) Venom spitting by the green lynx spider *Peucetia viridans* (Araneae, Oxyopidae). *J. Arachnol.* **12**, 372–3.

Foster, W.A. (1990) Experimental evidence for effective and altruistic colony defence against natural predators by soldiers of the gall-forming aphid *Pemphigus spyrothecae* (Homoptera: Pemphigidae). *Behav. Ecol. Sociobiol.* **27**, 421–30.

Fullard, J.H. (1977) Phenology of sound-producing arctiid moths and the activity of insectivorous bats. *Nature, Lond.* **267**, 42–3.

Greene, E., Orsak, L.J. & Whitman, D.W. (1987) A tephritid fly mimics the territorial displays of its jumping spider predators. *Science, N.Y.* **236**, 310–12.

Heere, E.A., Windsor, D.M. & Foster, R.B. (1986) Nesting associations of wasps and ants on lowland Peruvian ant-plants. *Psyche*, **93**, 321–9.

Howard, D.F., Blum, M.S. & Fales, H.M. (1983) Defense in thrips: forbidding fruitiness of a lactone. *Science, N.Y.* **220**, 335–6.

Jeanne, R.L. (1970) Chemical defense of brood by a social wasp. *Science, N.Y.* **168**, 1465–66.

Jeanne, R.L. (1981) Alarm recruitment, attack behavior, and the role of the alarm pheromone in *Polybia occidentalis* (Hymenoptera: Vespidae). *Behav. Ecol. Sociobiol.* **9**, 143–8.

Kearby W.H. (1975) Variable oakleaf caterpillar larvae secrete formic acid that causes skin lesions (Lepidoptera: Notodontidae). *J. Kans. Ent. Soc.* **48**, 280–82.

Kojima, J. (1983) Defense of the pre-emergence colony against ants by means of a chemical barrier in *Ropalidia fasciata* (Hymenoptera, Vespidae). *Jap. J. Ecol.* **33**, 213–23.

Kurosu, U. & Aoki, S. (1988) Monomorphic first instar larvae of *Colophina clematicola* (Homoptera, Aphidoidea) attack predators. *Kontyû*, **56**, 867–71.

LaMunyon, C.W. & Adams, P.A. (1987) Use and effect of an anal defensive secretion in larval Chrysopidae (Neuroptera). *Ann. Ent. Soc. Am.* **80**, 804–8.

Maschwitz, U., Jessen, K. & Maschwitz, E. (1981) Foaming in *Pachycondyla*: a new defense mechanism in ants. *Behav. Ecol. Sociobiol.* **9**, 79–81.

Mather, M.H. & Roitberg, B.D. (1987) A sheep in wolf's clothing: tephritid flies mimic spider predators. *Science, N.Y.* **236**, 308–10.

Moore, B.P. & Brown, W.V. (1971) Chemical defence in longhorn beetles of the genera *Stenocentrus* and *Syllitus* (Coleoptera: Cerambycidae). *J. Aust. Ent. Soc.* **10**, 230–2.

Nentwig, W. (1985) A tropical caterpillar mimics faeces, leaves and a snake (Lepidoptera: Oxytenidae: *Oxytenis naemia*). *J. Res. Lepid.* **24**, 136–41.

Petrunkevitch, A. (1952) The spider and the wasp. *Scient. Am.* **187**, 20–23.

Post, D.C., Mohamed, M.A, Coppel, H.C. & Jeanne, R.L. (1984) Identification of ant repellent allomone produced by social wasp *Polistes fuscatus* (Hymenoptera: Vespidae). *J. Chem. Ecol.* **10**, 1799–1807.

Reiskind, J. (1976) *Orsima formica*: a Bornean salticid mimicking an insect in reverse. *Brit. Arachnol. Soc. Bull.* **3**, 235–6.

Robinson, M.H. (1968) The defensive behaviour of the Javanese stick insect, *Orxines macklotti* De Haan, with a note on the startle display of *Metriotes diocles* Westw. (Phasmatodea, Phasmidae). *Entomologists' Mon. Mag.* **104**, 46–54.

Robinson, M.H. (1968a) The defensive behaviour of the stick insect *Oncotophasma martini* (Griffini) (Orthoptera: Phasmitidae). *Proc. R. Ent. Soc. Lond. (A)*, **43**, 183–7.

Root, R.B. & Messina, F.J. (1983) Defensive adaptations and natural enemies of a case-bearing beetle, *Exema canadensis* (Coleoptera: Chrysomelidae). *Psyche*, **90**, 67–80.

Silberglied, R.E. & Aiello, A. (1976) Defensive adaptations of some neotropical long-horned beetles (Coleoptera, Cerambydicae): antennal spines, tergiversation, and double mimicry. *Psyche*, **83**, 256–62.

Steiner, A.L. (1981) Anti-predator strategies. 2. Grasshoppers (Orthoptera, Acrididae) attacked by *Prionyx parkeri* and some *Tachysphex* wasps (Hymenoptera, Sphecinae and Larrinae): a descriptive study. *Psyche*, **88**, 1–24.

Sunose, T., Yamane, S., Tsuda, K. & Takasu, K. (1991) What do soldiers of *Pseudoregma bambucicola* (Homoptera, Aphidodea) defend? *Jap. J. Ent.* **59**, 141–8.

Thornhill, R. (1975) Scorpionflies as kleptoparasites of web-building spiders. *Nature, Lond.* **258**, 709–11.

Vetter, R.S. (1980) Defensive behavior of the black widow spider *Latrodectus hesperus* (Araneae: Theridiidae). *Behav. Ecol. Sociobiol.* **7**, 187–93.

Wilson, E.O. (1976) The organization of colony defense in the ant *Pheidole dentata* Mayr (Hymenoptera: Formicidae). *Behav. Ecol. Sociobiol.* **1**, 63–81.

Wittmann, D. (1985) Aerial defense of the nest by workers of the stingless bee *Trigona (Tetragonisca) angustula* (Latreille) (Hymenoptera: Apidae). *Behav. Ecol. Sociobiol.* **16**, 111–114.

INDEX OF SCIENTIFIC NAMES

Page numbers in *italic* refer to line illustrations; page numbers in bold refer to colour photographs.

INDEX OF COMMON NAMES

Page numbers in *italic* refer to line illustrations; page numbers in bold refer to colour photographs.

GENERAL INDEX

Page numbers in *italic* refer to line illustrations; page numbers in bold refer to colour photographs.